The Fundamentals of
Stellar Astrophysics

The Fundamentals of Stellar Astrophysics

George W. Collins II

W. H. FREEMAN AND COMPANY • NEW YORK

Library of Congress Cataloging-in-Publication Data

Collins, George W. (George William), 1937–
  The fundamentals of stellar astrophysics.

  Includes bibliographies and index.
  1. Stars. 2. Astrophysics. I. Title.
QB801.C615 1989 523.8 88-30949
ISBN 0-7167-1993-2

Printed in the United States of America

1 2 3 4 5 6 7 8 9 0 VB 7 6 5 4 3 2 1 0 8 9

*To TR, who understands that the study
of theoretical astrophysics is an
affliction, not an occupation*

Contents

. . .

CHAPTER 10
Solution of the Equation of Radiative Transfer   253

Chapter 11
Environment of the Radiation Field
291

11.1 Statistics of the Gas and the Equation of State
292

Chapter 13
Formation of Spectral Lines

Chapter 14
Shape of Spectral Lines

Chapter 15
Breakdown of Local Thermodynamic Equilibrium

Preface

. . .

Since I began studying this subject some 30 years ago, its development has continued at a slow and measured pace. There have been few of the breakthroughs or leaps forward that characterize the early development of a discipline. Perhaps that is because the foundations of the understanding of stars was provided by the generation that preceded mine. Names like Eddington, Milne, Schuster, Schwarzschild, Cowling, Chandrasekhar, and many others echo down through the history of this subject as the definers and elucidators of stellar structure. The outline of the theory of the structure and evolution of the stars clearly has belonged to the first half of the twentieth century. In the second half of this century, we have seen that outline filled in so that there are very few aspects of either a star's structure or life history for which our understanding is incomplete. Certainly the advent of pulsars, black holes, and the other unusual objects that are often called stars has necessitated broadening the scope of the theory of stellar astrophysics. Then there are areas concerning both the birth and death of stars that largely elude our understanding. But the overall picture of the structure and evolution of most stars now seems, in the main, to be well understood.

When I say that "the overall picture of the structure and evolution of stars is, in the main, understood," I do not imply that there is not much to be learned. Nothing should humble a theorist more than supernova 1987A, whose progenitor was a blue supergiant, when the conventional wisdom said it should be a red supergiant. Theorists instantly explained such a result with the benefit of perfect hindsight, but the event should give us pause for thought. It was indeed a massive star that exploded, and contemporary models firmly rooted in the physics described in this text and elsewhere describe the event qualitatively quite well. Even with such an unexpected event the basic picture has been confirmed, but as time passes, the picture will become clearer. It is even likely that the outlines of the picture defined in the twentieth century may be resolved in an unanticipated manner in the twenty-first century. But the fundamentals of that picture are unlikely to change.

I suspect that there are few astronomers alive who would not be astounded if we found that stars do *not* form from the interstellar medium, burn hydrogen as main sequence stars for 90 percent of their life, and undergo complex but understandable changes during the last moments of their life. It is in this sense that the foundations of stellar astrophysics are understood. I am convinced that there will continually be surprises as we probe more carefully into the role of rotation, magnetic fields, and companion-induced distortion in the structure and evolution of stars. But the understanding of these issues will be built on the foundation of spherical stars that I have attempted to present in this book, and it is this foundation that must be understood before one can move on to more complicated problems.

The general speculation and excitement that encompassed the growing theory of stellar structure 50 years ago has moved on to the poorly understood realm of the galaxies and cosmology. The theoretical foundations of galactic structure seem to be in a state akin to that of stellar structure in the early part of this century, while recent developments in cosmology may actually have elevated that discipline to the status of a science. The pressure exerted by the burgeoning information from these areas on graduate curricula has provided a substantial squeeze on the more traditional aspects of an astronomer's education.

This is as it should be. If a discipline does not develop and expand, it will stagnate. Change is the hallmark of any vital intellectual enterprise. Few graduate programs in the United States now offer courses in celestial mechanics. Yet half a century ago, no one would have been called an astronomer who could not determine planetary positions from the orbital elements or determine those elements from several independent observations. However, today we all know where to look for that information if we ever actually have to perform such a task. Such is the evolution of that subject matter we call astronomy. It is a time-dependent thing, for one individual can only hold so much information in mind at one time. Thus a course of study in stellar astrophysics that used to cover 2 years is now condensed into 1 year or less, and this pressure can only increase. I have always felt that in addition to discovering "new" things about the universe, it is important to "sift and winnow" the old to save that which will be important for the understanding of the new. This is a responsibility that all academicians have, and it must be assumed if the next generation is to have the limited room of their minds filled with the essentials of the old so that they may continue the development of the new. Such is the basic motivation for this book.

Over the years, a number of books have been written about various aspects of stellar astrophysics, and many have deservedly become classics. It is not my intention to compete with these classics; indeed, the reader will find them referenced often, and it is my sincere hope that the reader will take the time to read and learn from them. A major purpose of this effort is to make, in some cases, that reading a little easier. Thus the aim of this book differs from others used as graduate texts in astronomy. Traditionally, they have taken a discipline as far as it could be developed and in some cases beyond. That is not my intent. Instead, I present the material required to understand contemporary research in a variety of areas related to the study of stars. For example, it would be fruitless to attempt to grapple with contemporary work in the theory of nonradial oscillations without understanding the basis for pulsation theory such as is given in Chapter 8.

As stellar astrophysics has developed, attention has focused on the details and refinements that make current models of stars so quantitatively accurate. While this accuracy is important for the advancement of the subject, it can form a barrier to the understanding of its foundations. Thus, I have left many of these details to others in the hopes that the student interested in advancing the understanding of stellar astrophysics will search them out. Some will observe that I have not sifted and winnowed enough and that too many of the blind alleys and unproductive directions of development have been included. This may be so, for it is difficult to shrug off those formalisms which one has struggled with and found rewarding in youth. I leave further sifting to the next generation. Suffice it to say that I have included in this book what I feel is either necessary or at least enjoyable for the understanding of stars.

This book is aimed at first-year graduate students or very advanced undergraduates. I assume throughout that they have considerable factual knowledge of stars and astronomy. Readers should be acquainted with the Hertzsprung-Russell diagram and know something of the ranges of the parameters that define stars. The student who wants to make a contribution to astronomy must understand how this knowledge about stars was gained from observation. Only then can the accuracy of that knowledge be assessed, and without such an assessment, deception of self and perhaps others is guaranteed.

Given such a background, I shall attempt to describe the development of a nearly axiomatic theory of stellar structure that is consistent with what we know about stars. This theory is incomplete for there is much that we still do not know. The terminal phases of stellar evolution are treated schematically. The structure of distorted stars is barely touched, and the theory of evolution of close binaries is ignored entirely. The decision to downplay or ignore this material does not arise from a disdain of these subjects but is simply a question of time and space. It is my sincere hope that the student, upon finishing this book, will seize some of these areas for future research and, being interested and prepared, pick up the gauntlet and advance the subject. At the end of the sections on stellar interiors and atmospheres, I have included several topics that represent logical extensions of the traditional theory of stellar structure. The treatment of these topics should not be considered either complete or exhaustive but merely illustrative, for the selection of the subjects was dictated by personal interest as opposed to fundamental importance. In a curriculum pressed for time, some topics can be safely ignored.

The relatively complete foundation of the theory of stellar structure has one minor psychological drawback that results from a contemporary penchant in some of the physical sciences. The rapid development of astronomy into new areas of research during the past two decades has tended to produce research papers that emphasize only the most contemporary work. Thus papers and books that reference older works are likely to be regarded as out of date. My tendency, nevertheless, is to give original references wherever possible. Thus the reader will find that many of the references in this book date to the middle part of this century or earlier. Hopefully the reader will forgive me and remember that this book is about the fundamentals of stellar astrophysics, and is not intended to bring the reader to the current state of research effort in stellar astronomy. To answer the need of a student who wishes to go beyond an introduction, I have included some additional references at the end of some chapters that represent reviews of a few of the more contemporary concerns of stellar structure.

Inevitably, some people will feel that more problems in stellar structure and atmospheres should be discussed. I can only counter by saying that it is easy to add but difficult to take away. For any topic that you might add, find one that you would remove without endangering the basic understanding of the student. Regrettably, only a finite amount of time and space can be devoted to the teaching of this subject, and the hard choices are not what to include but what to leave out. With the exception of a few topics that I included purely for my own enjoyment, I regard the vast majority of this book as fundamental to the understanding of stars.

To those who would say, "Yes, models are well understood, but models are not stars," I would shout "Amen!" I have spent most of my professional career modeling the outer layers of distorted stars, and I am acutely aware of the limitations of such models. Nevertheless, modeling as a mode for understanding nature is becoming a completely acceptable method. For stellar astrophysics, it has been an extremely productive approach. When combined critically with observation, modeling can provide an excellent avenue toward the understanding of how things work. Indeed, if pressed in a thoughtful way, most would find that virtually any comparison of theory with observation or experiment involves the modeling of some aspect of the physical world. Thus while one must be ever mindful of the distinction between models of the real world and the world itself, one cannot use that distinction as an excuse for failing to try to describe the world.

For the student who feels that it is unnecessary to understand all this theory simply to observe the stars, ask yourself how you will decide what you will observe. If that does not appear to be a significant question, then consider another line of endeavor. For those who suffer through this material on their way to what they perceive as the more challenging and interesting subjects of galaxies and cosmology, consider the following argument.

While the fascinating areas of galactic and extragalactic astronomy deservedly fill larger and larger parts of the graduate curricula, let us not forget that galaxies are made of stars and ultimately our conception of the whole can be no better than our understanding of the parts. In addition, the physical principles that govern stars are at work throughout the universe. Stars are the basic building blocks from which our larger world is made and remain the fundamental probes with which we test our theories of that world. The understanding of stars and the physical principles that rule their existence is, and I believe will remain, central to our understanding of the universe.

Do not take this argument as an apology for the study of stars. The opposite is true, for I feel some of the most difficult problems in astronomy involve the detailed understanding of stars. Consider the following example. Nearly 25 years ago, Thomas Gold described the basic picture and arguments for believing that pulsars are spinning magnetic neutron stars. In the main he was correct, although many details of his picture have been changed. However, we do not yet have a fully self-consistent picture of pulsars in spite of the efforts of a substantial number of very bright and dedicated astronomers. Such a picture is very difficult to formulate. Without it, our understanding of pulsars will not be complete, but that is not to say that the basic picture of a pulsar as a spinning magnetic neutron star is wrong. Rather, it is still incomplete. The solutions to many problems of stellar astrophysics remain incomplete. Such problems are not to be undertaken by the timid for they are demanding in the extreme. Nor should addressing them be regarded as merely filling in details. This is the excuse of the dilettante who would be well advised to follow Isaac Newton's admonition: "To explain all nature is too difficult a task for any man or even for any one age. 'Tis much better to do a little with certainty, and leave the rest for others that come after you than to explain all things."

I believe that many astronomers will choose, as I have chosen, to spend the majority of their professional careers involved in the study of stars themselves. It is my hope that they will recognize the fundamental nature of the material in this book and use it to attack the harder problems of today and the future.

I cannot conclude this preface without some acknowledgment of those who made this effort possible. Anyone who sets out to codify that which he or she has spent the greater part of life acquiring cannot expect to achieve any measure of success unless he or she is surrounded by an understanding family and colleagues. Particular thanks are extended to the students of stellar interiors and atmospheres at The Ohio State University who used this book in its earliest form and found and eliminated numerous errors. Many more were revealed by the core of reviewers who scrutinized the text. My thanks to Richard Boyd, Joe Cassinelli, George Field, Arne Henden, John Mathis, Peter Mészáros, Dimitri Mihalas, Donald Osterbrock, Michael Sitko, and Sydney Wolf for being members of that core. Their comments and constructive criticism were most helpful in shaping this book. The remaining shortcomings, mistakes, and blunders are mine and mine alone.

Certainly little of the knowledge contained here originated from my own efforts. I have merely chewed and digested material fed to me by mentors dedicated to the search and preservation of that body of knowledge known as astronomy. To name them all would require considerable space, possibly be construed as self-serving, and perhaps be embarrassing to some of them. Nevertheless, they have my undying admiration and gratitude for passing on some of what they know and sharing with me those most precious of commodities—their knowledge and wisdom, and most of all, their time.

George W. Collins II

The Ohio State University

Part I

Stellar Interiors

1

Introduction and Fundamental Principles

. . .

The development of a relatively complete picture of the structure and evolution of the stars has been one of the great conceptual accomplishments of the twentieth century. While questions still exist concerning the details of the birth and death of stars, scientists now understand over 90 percent of a star's life. Furthermore, our understanding of stellar structure has progressed to the point where it can be studied within an axiomatic framework comparable to those of other branches of physics. It is within this axiomatic framework that we will study stellar structure and stellar spectra—the traditional source of virtually all information about stars.

This book is divided into two parts: stellar interiors and stellar atmospheres. While the division between the two is fairly arbitrary, it is a traditional division separating regimes where different axioms apply. A similar distintion exists between the continuum and lines of a stellar spectrum. These distinctions represent transition zones where one physical process dominates over another. The transition in nature is never abrupt and represents a difference in degree rather than in kind.

We assume that the readers know what stars are, that is, have a working knowledge of the Hertzsprung-Russell diagram and of how the vast wealth of knowledge contained in it has been acquired. Readers should understand that most stars are basically spherical and should know something about the ranges of masses, radii, and luminosities appropriate for the majority of stars. The relative size and accuracy of the stellar sample upon which this information is based must be understood before a theoretical description of stars can be believed. However, the more we learn about stars, the more the fundamentals of our theoretical description are confirmed. The history of stellar astrophysics in the twentieth century can be likened to that of a photographer steadily sharpening the focus of the camera to capture the basic nature of stars.

In this book, the basic problem of stellar structure under consideration is the determination of the run of physical variables that describe the local properties of stellar material with position in the star. In general, the position in the star is the independent variable(s) in the problem, and other parameters such as the pressure P, temperature T, and density ρ are the dependent variables. Since these parameters describe the state of the material, they are often referred to as *state variables*. Part I of this book discusses these parameters alone. In Part II, when we arrive near the surface of the star, we shall also be interested in the detailed distribution of the photons, particularly as they leave the star.

Although there are some excursions into the study of nonspherical stars, the main thrust of this book is to provide a basis for understanding the structure of spherical stars. Although the proof is not a simple one, it would be interesting to show that the equilibrium configuration of a gas cloud confined solely by gravity is that of a sphere. However, instead of beginning this book with a lengthy proof, we simply take the result as an axiom that all stars dominated by gravity alone are spherical.

We describe these remarkably stable structures in terms of microphysics, involving particles and photons which are largely in equilibrium. Statistical mechanics is the general area of physics that deals with this subject and contains the axioms that form the basis for stellar astrophysics. Our discussion of stellar structure centers on the interaction of light with matter. We must first describe the properties of the space in which the interaction will take place. It is not the normal euclidean three-dimensional space of intuition, but a higher-dimension space. This higher-dimension space, called *phase space*, includes the momentum distribution of the particles which make up the star as well as their physical location.

1.1 Stationary or "Steady" Properties of Matter

a *Phase Space and Phase Density*

Consider a volume of physical space that is small compared to the physical system in question, but still large enough to contain a statistically significant number of particles. The range of physical space in which this small volume is embedded may be infinite or finite as long as it is significantly larger than the small volume. First let a set of three cartesian coordinates x_1, x_2, and x_3 represent the spatial part of the volume. Then allow the additional three cartesian coordinates v_1, v_2, and v_3 represent the components of the velocity of the particles. Coordinates v_1, v_2, and v_3 are orthogonal to the spatial coordinates. This simply indicates that the velocity and position are assumed to be uncorrelated. It also provides for a six-dimensional space which we call *phase space*. The volume of the space is

$$dV = dx_1\, dx_2\, dx_3\, dv_1\, dv_2\, dv_3 \qquad (1.1.1)$$

If the number of particles in the small volume dV is N, then we can define a parameter f, known as the *phase density*, by

$$f(x_1, x_2, x_3, v_1, v_2, v_3)\, dV = N \qquad (1.1.2)$$

The manner in which a number of particles can be arranged in an ensemble of phase space volumes is described in Figure 1.1.

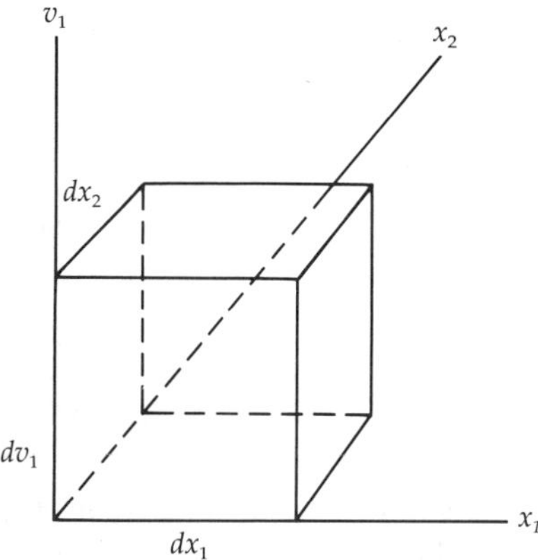

Figure 1.1 shows part of a small differential volume of phase space. Remember that the position and velocity coordinates are orthogonal to each other.

b Macrostates and Microstates

A *macrostate* of a system is said to be specified when the number of particles in each phase space volume dV is specified. That is, if the phase density is specified everywhere, then the macrostate of the system has been specified. Later we shall see how the phase density can be used to specify all the physical properties of the system.

To discuss the notion of a *microstate*, it must be assumed that there is a perceptible difference between particles, because in a microstate, in addition to the number of particles in each volume, it makes a difference *which* particles are in which volumes. If the specification of individual particles can be accomplished, then it can be said that a microstate has been specified. Clearly one macrostate could consist of many microstates. For example, the number of balls on a pool table might be said to be a macrostate, whereas the specification of which balls they are would denote a specific microstate. In a similar manner, the distribution of suits of playing cards in a bridge hand might be said to represent a macrostate, but the specific cards in each suit would specify the microstate.

c Probability and Statistical Equilibrium

If we were to create macrostates by assembling particles by randomly throwing them into various microstates, then the macrostate most likely to occur is the one with the greatest number of microstates. That is why a bridge hand consisting of 13 spades occurs so rarely compared to a hand with four spades and three hearts, three diamonds, or three clubs. If we consider a system where the particles are continually moving from one phase space volume to another, say, by collisions, then the most likely macrostate is the one with the largest number of associated microstates. There is an implicit assumption here that *all microstates are equally probable*. Is this reasonable?

Imagine a case where all the molecules in a room are gathered in one corner. This represents a particular microstate—a particularly unlikely one, we would think. Through random motions, it would take an extremely long time for the particles to return to that microstate. However, given the position and velocity of each particle in an ordinary room of gas, is this any more unlikely than each particle returning to that specific position with the same velocity? The answer is no. Thus, if each microstate is equally probable, then the associated macrostates are not equally probable, and it makes sense to search for the most probable macrostate of a system. In a system which is continually rearranging itself by collisions, the most probable macrostate becomes the most likely state in which to find the system. A system which is in its most probable macrostate is said to be in *statistical equilibrium*.

Many things can determine the most probable macrostate. Certainly the total number of particles allowed in each microstate and the total number of particles

4	3	2	1	0	Volume 1
0	1	2	3	4	Volume 2

Figure 1.2 Phase space composed of only two cells in which four particles reside. All possible macrostates are illustrated.

available to distribute will be important in determining the total number of microstates in a given macrostate. In addition, quantum mechanics places some conditions on our ability to distinguish particles and even limits how many of certain kinds of particles can be placed in a given volume of phase space. But, for the moment, let us put aside these considerations and concentrate on calculating the number of microstates in a particular macrostate.

Consider a simple system consisting of only two phase space volumes and four particles (see Figure 1.2). There are precisely five different ways that the four particles can be arranged in the two volumes. Thus there are five macrostates of the system. But which is the most probable? Consider the second macrostate in Figure 1.2 (that is, $N_1 = 3$, $N_2 = 1$). Here we have three particles in one volume and one particle in the other volume. If we regard the four particles as individuals, then there are four different ways in which we can place those four particles in the two volumes so that one volume has three and the other volume has only one (see Figure 1.3). Since the order in which the particles are placed in the volume does not matter, all permutations of the particles in any volume must be viewed as constituting the same microstate.

Now if we consider the total number of particles N to be arranged sequentially among m volumes, then the total number of sequences is simply $N!$. However, within each volume (say, the ith volume), N_i particles yield $N_i!$ indistinguishable sequences which must be removed when the allowed number of microstates is counted. Thus the total number of allowed microstates in a given macrostate is

$$W = \frac{N!}{\displaystyle\prod_{i=1}^{m} N_i!} \tag{1.1.3}$$

abc	abd	acd	bcd	Volume 1
d	c	b	a	Volume 2

Figure 1.3 One macrostate in Figure 1.2, specifically the state where $N_1 = 3$, and $N_2 = 1$. All the allowed microstates for distinguishable particles are shown.

For the five macrostates shown in Figure 1.2, the number of possible microstates is

$$W_{4,0} = \frac{4!}{4!0!} = 1$$

$$W_{3,1} = \frac{4!}{3!1!} = 4$$

$$W_{2,2} = \frac{4!}{2!2!} = 6 \tag{1.1.4}$$

$$W_{1,3} = \frac{4!}{1!3!} = 4$$

$$W_{0,4} = \frac{4!}{0!4!} = 1$$

Clearly $W_{2,2}$ is the most probable macrostate of the five. The particle distribution of the most probable macrostate is unique and is known as the *equilibrium macrostate*.

In a physical system where particle interactions are restricted to those between particles which make up the system, the number of microstates within the system changes after each interaction and, in general, increases, so that the macrostate of the system tends toward that with the largest number of microstates—the equilibrium macrostate. In this argument we assume that the interactions are uncorrelated and random. Under these conditions, a system which has reached its equilibrium macrostate is said to be in *strict thermodynamic equilibrium* (STE). Note that interactions among particles which are not in strict thermodynamic equilibrium will tend to drive the system away from strict thermodynamic equilibrium and toward a different statistical equilibrium distribution. This is the case for stars near their surfaces.

The statistical distribution of microstates versus macrostates given by equation (1.1.3) is known as the *Maxwell-Boltzmann statistics,* and it gives excellent results for a classical gas in which the particles can be regarded as distinguishable. In a classical world, the position and momentum of a particle are sufficient to make it distinguishable from all other particles. However, the quantum mechanical picture of the physical world is quite different. So far, we have neglected both the Heisenberg uncertainty principle and the Pauli exclusion principle.

d Quantum Statistics

Within the realm of classical physics, a particle occupies a point in phase space, and in some sense all particles are distinguishable by their positions and velocities. The phase space volumes are indeed differential and arbitrarily small. However, in the quantum mechanical view of the physical world, there is a limit to how well the position and momentum (velocity, if the mass is known) of any particle can be determined. Within that phase space volume, particles are indistinguishable. This limit is known as the *Heisenberg uncertainty principle*, and it is stated as follows:

$$\Delta p \, \Delta x \geq \frac{h}{2\pi} \equiv \hbar \tag{1.1.5}$$

Thus the minimum phase space volume which quantum mechanics allows is of the order of h^3. To return to our analogy with Maxwell-Boltzmann statistics, let us subdivide the differential cell volumes into compartments of size h^3 so that the total number of compartments is

$$n = \frac{dx_1 dx_2 dx_3 dp_1 dp_2 dp_3}{h^3} \tag{1.1.6}$$

Let us redraw the example in Figure 1.3 so that each cell in phase space is subdivided into four compartments within which the particles are indistinguishable. Figure 1.4 shows the arrangement of the four particles for the $W_{3,1}$ macrostate for which there were only four allowed microstates under Maxwell-Boltzmann

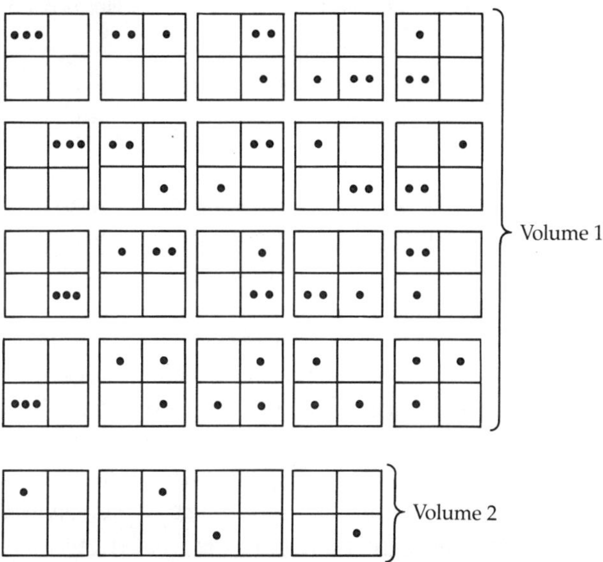

Figure 1.4 The same macrostate as Figure 1.3 only now the cells of phase space are subdivided into four compartments within which particles are indistinguishable. All of the possible microstates are shown for the four particles.

statistics. Since the particles are now distinguishable within a cell, there are 20 separate ways to arrange the three particles in volume 1 and 4 ways to arrange the single particle in volume 2. The total number of allowed microstates for $W_{3,1}$ is 20×4, or 80. Under these conditions the total number of microstates per macrostate is

$$W = \prod_i W_i \qquad (1.1.7)$$

where W_i is the number of microstates per cell of phase space, which can be expressed in terms of the number of particles N_i in that cell.

Let us assume that there are n compartments in the ith cell which contains N_i particles. Now we have to arrange a sequence of $n + N_i$ objects, since we have to consider both the particles and the compartments into which they can be placed. However, not all sequences are possible since we must always start a sequence with a compartment. After all, we have to put the particle somewhere! Thus there are n sequences with $N_i + n - 1$ items to be further arranged. So there are $n(N_i + n - 1)!$ different ways to arrange the particles and compartments. We must eliminate all the permutations of the compartments because they reside within a cell and therefore represent the same microstate. But there are just $n!$ of these. Similarly, the order in which the particles are added to the cell volume is just as irrelevant to the final microstate as it was under Maxwell-Boltzmann statistics. And so we must eliminate all the permutations of the N_i particles, which is just $N_i!$. Thus the number of microstates allowed for a given macrostate becomes

$$W_{\text{B-E}} = \prod_i \frac{n(N_i + n - 1)!}{N_i! n!} = \prod_i \frac{(N_i + n - 1)!}{N_i!(n - 1)!} \qquad (1.1.8)$$

The subscript "B-E" on W indicates that these statistics are known as *Bose-Einstein statistics*, which allow for the Heisenberg uncertainty principle and the associated limit on the distinguishability of phase space volumes. We have assumed that an unlimited number of particles can be placed within a volume h^3 of phase space, and those particles for which this is true are called *bosons*. Perhaps the most important representatives of the class of particles for stellar astrophysics are the photons. Thus, we may expect the statistical equilibrium distribution for photons to be different from that of classical particles described by Maxwell-Boltzmann statistics.

Within the domain of quantum mechanics, there are further constraints to consider. Most particles such as electrons and protons obey the Pauli exclusion principle, which basically says that there is a limit to the number of these particles that can be placed within a compartment of size h^3. Specifically, only one particle with a given set of quantum numbers may be placed in such a volume. However, two electrons which have their spins arranged in opposite directions but are otherwise identical can fit within a volume h^3 of phase space. Since we can put no more than two of these particles in a compartment, let us consider phase space to be made up of $2n$ half-compartments which are either full or empty. We could say that there are no more than $2n$ things to be arranged in sequence and therefore no more than $2n!$ allowed microstates. But, since each particle has to go somewhere, the number of filled compartments which have $N_i!$ indistinguishable permutations is just N_i. Similarly, the number of indistinguishable permutations of the empty compartments is $(2n - N_i)!$. Taking the product of all the allowed microstates for a given macrostate, we find that the total number of allowed microstates is

$$W_{\text{F-D}} = \prod_i \frac{2n!}{N_i!(2n - N_i)!} \tag{1.1.9}$$

The subscript "F-D" refers to Enrico Fermi and P. A. M. Dirac, who were responsible for the development of these statistics. These are the statistics we can expect to be followed by an electron gas and all other particles that obey the Pauli exclusion principle. Such particles are normally called *fermions*.

e Statistical Equilibrium for a Gas

To find the macrostate which represents a steady equilibrium for a gas, we follow basically the same procedures regardless of the statistics of the gas. In general, we wish to find that macrostate for which the number of microstates is a maximum. So by varying the number of particles in a cell volume we will search for $\delta W = 0$. Since $\ln W$ is a monotonic function of W, any maximum of $\ln W$ is a maximum of W. Thus we use the logarithm of equations (1.1.7) through (1.1.9) to search for the most probable macrostate of the distribution functions. These are

$$\ln W_{\text{M-B}} = \ln N! - \sum_i \ln N_i!$$

$$\ln W_{\text{B-E}} = \sum_i \ln (n + N_i - 1)! - \ln N_i! - \ln (n - 1)! \tag{1.1.10}$$

$$\ln W_{\text{F-D}} = \sum_i \ln (2n)! - \ln (2n - N_i)! - \ln N_i!$$

The use of logarithms also makes it easier to deal with the factorials through the use of Stirling's formula for the logarithm of a factorial of a large number:

$$\ln N! \approx N \ln N - N \tag{1.1.11}$$

For a given volume of gas, $\delta N = \delta n = 0$. The variations of equations (1.1.10) become

$$\delta \ln W_{\text{M-B}} = \sum_i \ln N_i \, \delta N_i = 0$$

$$\delta \ln W_{\text{B-E}} = \sum_i \ln \frac{n + N_i}{N_i} \, \delta N_i = 0 \tag{1.1.12}$$

$$\delta \ln W_{\text{F-D}} = \sum_i \ln \frac{2n - N_i}{N_i} \, \delta N_i = 0$$

These are the equations of condition for the most probable macrostate for the three statistics which must be solved for the particle distribution N_i. We have additional constraints, which arise from the conservation of the particle number and energy, on the system which have not been directly incorporated into the equations of condition. These can be stated as follows:

$$\delta \left(\sum_i N_i \right) = \delta N = 0 \qquad \delta \left(\sum_i w_i N_i \right) = \sum_i w_i \, \delta N_i = 0 \tag{1.1.13}$$

where w_i is the energy of an individual particle. Since these additional constraints represent new information about the system, we must find a way to incorporate them into the equation of condition. A standard method for doing this is known as the *method of Lagrange multipliers*. Since equations (1.1.13) represent quantities which are zero, we can multiply them by arbitrary constants and add them to equations (1.1.12) to get

M-B:
$$\sum_i (\ln N_i - \ln \alpha_1 + \beta_1 w_i) \, \delta N_i = 0$$

B-E:
$$\sum_i \left(\ln \frac{n + N_i}{N_i} - \ln \alpha_2 - \beta_2 w_i \right) \delta N_i = 0 \tag{1.1.14}$$

F-D:
$$\sum_i \left(\ln \frac{2n - N_i}{N_i} - \ln \alpha_3 - \beta_3 w_i \right) \delta N_i = 0$$

Each term in equations (1.1.14) must be zero since the variations in N_i are arbitrary and any such variation must lead to the stationary, most probable, macrostate.

Thus,

M-B:
$$\frac{N_i}{\alpha_1} = \exp\left(-w_i\beta_1\right)$$

B-E:
$$\frac{n}{N_i} = \alpha_2 \exp w_i\beta_2 - 1 \tag{1.1.15}$$

F-D:
$$\frac{2n}{N_i} = \alpha_3 \exp w_i\beta_3 + 1$$

All that remains is to develop a physical interpretation of the undetermined parameters α_j and β_j. Let us look at Maxwell-Boltzmann statistics for an example of how this is done. Since we have not said what β_1 is, let us call it $1/(kT)$. Then

$$N_i = \alpha_1 e^{-w_i/(kT)} \tag{1.1.16}$$

If the cell volumes of phase space are not all the same size, it may be necessary to weight the number of particles to adjust for the different cell volumes. We call these weight functions g_i. Then

$$N = \sum_i g_iN_i = \alpha_1 \sum_i g_i e^{-w_i/(kT)} \equiv \alpha_1 U(T) \tag{1.1.17}$$

The parameter $U(T)$ is called the *partition function*, and it depends on the composition of the gas and parameter T alone. Now if the total energy of the gas is E, then

$$E = \sum_i g_iw_iN_i = \sum_i w_ig_i\alpha_1 e^{-w_1/(kT)} = \frac{\sum\limits_i w_ig_iNe^{-w_i/(kT)}}{U(T)} = NkT\frac{d\ln U}{d\ln T} \tag{1.1.18}$$

For a free particle like that found in a monatomic gas, the partition function (see Aller[1] and section 11.1)* is

$$U(T) = \frac{(2\pi mkT)^{3/2}}{h^3} V \tag{1.1.19}$$

where V is the specific volume of the gas, m is the mass of the particle, and T is the kinetic temperature. Replacing $d\ln U/d\ln T$ in equation (1.1.18) by its value

---

* All text references may be found at the end of the chapter under References and Supplemental Reading.

obtained from equation (1.1.19), we get the familiar relation

$$E = \tfrac{3}{2}NkT \tag{1.1.20}$$

which is only correct if T is the kinetic temperature. Thus we arrive at a self-consistent solution if the parameter T is to be identified with the kinetic temperature.

The situation for a photon gas in the presence of material matter is somewhat simpler because the matter acts as a source and sink for photons. Now we can no longer apply the constraint $\delta N = 0$. This is equivalent to adding $\ln \alpha_2 = 0$ (that is, $\alpha_2 = 1$) to the equations of condition. If we let $\beta_2 = 1/(kT)$ as we did with the Maxwell-Boltzmann statistics, then the appropriate solution to the Bose-Einstein formula [equation (1.1.15)] becomes

$$\frac{N_i}{n} = \frac{1}{e^{h\nu/(kT)} - 1} \tag{1.1.21}$$

where the photon energy w_i has been replaced by $h\nu$. Since two photons in a volume h^3 can be distinguished by their state of polarization, the number of phase space compartments is

$$n = \frac{2}{h^3} \, dx_1 \, dx_2 \, dx_3 \, dp_1 \, dp_2 \, dp_3 \tag{1.1.22}$$

We can replace the rectangular form of the momentum volume $dp_1 \, dp_2 \, dp_3$ by its spherical counterpart $4\pi p^2 \, dp$, and remembering that the momentum of a photon is $h\nu/c$, we get

$$\frac{dN}{V} = \frac{8\pi \nu^2}{c^3} \frac{1}{e^{h\nu/(kT)} - 1} \, d\nu \tag{1.1.23}$$

Here we have have replaced N_i with dN. This assumes that the number of particles in any phase space volume is small compared to the total number of particles. Since the energy per unit volume dE_ν is just $h\nu \, dN/V$, we get the relation known as *Planck's law* or sometimes as the *blackbody law*:

$$dE_\nu = \frac{8\pi h}{c^3} \frac{\nu^3}{e^{h\nu/(kT)} - 1} \, d\nu \equiv \frac{4\pi}{c} B_\nu(T) \tag{1.1.24}$$

The parameter $B_\nu(T)$ is known as the *Planck function*. This, then, is the distribution law for photons which are in strict thermodynamic equilibrium. If we were to consider the Bose-Einstein result for particles and let the number of Heisenberg compartments be much larger than the number of particles in any volume, we would recover the result for Maxwell-Boltzmann statistics. This is further justification for using the Maxwell-Boltzmann result for ordinary gases.

f Thermodynamic Equilibrium—Strict and Local

Let us now consider a two-component gas made up of material particles and photons. In stars, as throughout the universe, photons outnumber material particles by a large margin and continually undergo interactions with matter. Indeed, it is the interplay between the photon gas and the matter which is the primary subject of this book. If both components of the gas are in statistical equilibrium, then we should expect the distribution of the photons to be given by Planck's law and the distribution of particle energies to be given by the Maxwell-Boltzmann statistics. In some cases, when the density of matter becomes very high and the various cells of phase space become filled, it may be necessary to use Fermi-Dirac statistics to describe some aspects of the matter. When both the photon and the material matter components of the gas are in statistical equilibrium with each other, we say that the gas is in strict thermodynamic equilibrium. If, for whatever reason, the photons depart from their statistical equilibrium (i.e., from Planck's law) but the material matter continues to follow Maxwell-Boltzmann statistics (i.e., to behave as if it were still in thermodynamic equilibrium), then we say that the gas (material component) is in local thermodynamic equilibrium (LTE).

1.2 Transport Phenomena

a Boltzmann Transport Equation

It is one thing to describe the behavior of matter and photons in equilibrium, but stars shine. Therefore energy must flow from the interior to the surface regions of the star and the details of the flow play a dominant role in determining the resultant structure and evolution of the star. We now turn to an extremely simple description of how this flow can be quantified; this treatment is due to Ludwig Boltzmann and should not be confused with the Boltzmann formula of Maxwell-Boltzmann statistics. Although the ideas of Boltzmann are conceptually simple, many of the most fundamental equations of theoretical physics are obtained from them. Basically the Boltzmann transport equation arises from considering what can happen to a collection of particles as they flow through a volume of phase space. Our prototypical volume of phase space was a six-dimensional "cube," which implies that it has five-dimensional "faces." The Boltzmann transport equation basically expresses the change in the phase density within a differential volume, in terms of the flow through these faces, and the creation or destruction of particles within that volume.

For the moment, let us call the six coordinates of this space x_i, remembering that the first three refer to the spatial coordinates and the last three refer to the momentum coordinates. If the "area" of one of the five-dimensional faces is A, the particle density is N/V, and the flow velocity is v, then the inflow of particles across that face in time dt is

$$v \frac{N}{V} A \, dt = \left[\frac{dx_i}{dt} f(x_i, t) \frac{dV}{dx_i} \right] dt \tag{1.2.1}$$

Similarly, the number of particles flowing out of the opposite face located dx_i away is

$$\left[\frac{dx_i}{dt} f(x_i + dx_i, t) \frac{dV}{dx_i} \right] dt = \text{number of outflowing particles} \tag{1.2.2}$$

The net change due to flow in and out of the six-dimensional volume is obtained by calculating the difference between the inflow and outflow and summing over all faces of the volume:

$$\sum_i \frac{dx_i}{dt} \left[\frac{f(x_i + dx_i, t) - f(x_i, t)}{dx_i} \right] dV \, dt = \sum_i \frac{dx_i}{dt} \frac{\partial f}{\partial x_i} dV \, dt \tag{1.2.3}$$

Note that the sense of equation (1.2.3) is such that if the inflow exceeds the outflow, the net flow is considered negative. Now this flow change must be equal to the negative time rate of change of the phase density (that is, df/dt). We can split the total time rate of change of the phase density into that part which represents changes due to the differential flow $\partial f/\partial t$ and that part which we call the *creation rate S*. Equating the flow divergence with the local temporal change in the phase density, we have

$$\sum_i \frac{dx_i}{dt} \frac{\partial f}{\partial x_i} dV \, dt = -\left(\frac{\partial f}{\partial t} - S \right) dV \, dt \tag{1.2.4}$$

Rewriting our phase space coordinates x_i in terms of the spatial and momentum coordinates and using the old notation of Isaac Newton to denote total differentiation with respect to time (i.e., the dot), we get

$$\frac{\partial f(x_i, p_i, t)}{\partial t} + \sum_{i=1}^{3} \left(\dot{x}_i \frac{\partial f}{\partial x_i} + \dot{p}_i \frac{\partial f}{\partial p_i} \right) = S \tag{1.2.5}$$

This is known as the *Boltzmann transport equation*, and it can be written in several different ways. In vector notation we get

$$\frac{\partial f}{\partial t} + \vec{v} \cdot \nabla f - \frac{1}{m} \nabla \Phi \cdot \nabla_v f = S \qquad (1.2.6)$$

Here the potential gradient $\nabla \Phi$ has replaced the momentum time derivative while ∇_v is a gradient with respect to velocity. The quantity m is the mass of a typical particle. It is also not unusual to find the Boltzmann transport equation written in terms of the total Stokes time derviative

$$\frac{D}{Dt} \equiv \frac{\partial}{\partial t} + \vec{v} \cdot \nabla \qquad (1.2.7)$$

If we take $\vec{v}$ to be a six-dimensional "velocity" and ∇ to be a six-dimensional gradient, then the Boltzmann transport equation takes the form

$$\frac{Df}{Dt} = S \qquad (1.2.8)$$

Although this form of the Boltzmann transport equation is extremely general, much can be learned from the solution of the homogeneous equation. This implies that $S = 0$ and that no particles are created or destroyed in phase space.

b Homogeneous Boltzmann Transport Equation and Liouville's Theorem

Remember that the right-hand side of the Boltzmann transport equation is a measure of the rate at which particles are created or destroyed in a phase space volume. Note that creation or destruction in phase space includes a good deal more than the conventional spatial creation or destruction of particles. To be sure, that type of change is included, but in addition processes which change a particle's position in momentum space may move a particle in or out of such a volume. The detailed nature of such processes will interest us later, but for the moment let us consider a common and important form of the Boltzmann transport equation, namely, that where the right-hand side is zero. This is known as the *homogeneous Boltzmann transport equation*. It is also better known as *Liouville's theorem of statistical mechanics*. In the literature of stellar dynamics, it is also occasionally referred to as *Jeans' theorem*[2], for Sir James Jeans was the first to explore its implications for stellar systems. By setting the right-hand side of the Boltzmann transport equation to zero, we have removed the effects of collisions from the system, with the result that the density of points in phase space is constant. Liouville's theorem is usually generalized to include sets or ensembles of particles. For this generalization, phase space is expanded to $6N$ dimensions, so that each particle of an ensemble has six position and momentum coordinates which are linearly independent of the coordinates of every other particle. This space is often called *configuration space*, since the entire ensemble of particles is represented by a point and its temporal history by a curve in this $6N$-dimensional space. Liouville's theorem holds here and implies that the density of points (ensembles) in configuration space is constant. This, in turn, can be used to demonstrate the determinism and uniqueness of new-tonian mechanics. If the configuration density is constant, it is impossible for two ensemble paths to cross, for to do so, one path would have to cross a volume element surrounding a point on the other path, thereby changing the density. If no two paths can cross, then it is impossible for any two ensembles to ever have exactly the same values of position and momentum for all their particles. Equi-valently, the initial conditions of an ensemble of particles uniquely specify its path in configuration space. This is not offered as a rigorous proof, only as a plausibility argument. More rigorous proofs can be found in most good books on classical mechanics.[3,4] Since Liouville's theorem deals with configuration space, it is some-times considered more fundamental than the Boltzmann transport equation; but for our purposes the expression containing the creation rate S will be required and therefore will prove more useful.

c Moments of the Boltzmann Transport Equation and Conservation Laws

By the *moment* of a function we mean the integral of some property of interest, weighted by its distribution function, over the space for which the distribution function is defined. Common examples of such moments can be found in statistics. The mean of a distribution function is simply the first moment of the distribution function, and the variance can be simply related to the second moment. In general, if the distribution function is analytic, all the information contained in the function is also contained in the moments of that function.

The complete solution to the Boltzmann transport equation is, in general, ex-tremely difficult and usually would contain much more information about the sys-tem than we wish to know. The process of integrating the function over its defined space to obtain a specific moment removes or averages out much of the information about the function. However, this process also usually yields equations which are much easier to solve. Thus we trade off information for the ability to solve the resulting equations, and we obtain some explicit properties of the system of interest. This is a standard "trick" of mathematical physics and one which is employed over and over throughout this book. Almost every instance of this type carries with it the name of some distinguished scientist or is identified with some fundamental conservation law, but the process of its formulation and its origin are basically the same.

We define the nth *moment* of a function f as

$$M_n[f(x)] = \int x^n f(x)\, dx \tag{1.2.9}$$

By multiplying the Boltzmann equation by powers of the position and velocity and integrating over the appropriate dimensions of phase space, we can generate equations relating the various moments of the phase density $f(\vec{x},\, \vec{v})$. In general, such a process always generates two moments of different order n, so that a succession of moment taking always generates one more moment than is contained in the resulting equations. Some additional property of the system will have to be invoked to relate the last generated higher moment to one of lower order, in order to close the system of equations and allow for a solution. To demonstrate this process, we show how the equation of continuity, the Euler-Lagrange equations of hydrodynamic flow, and the virial theorem can all be obtained from moments of the Boltzmann transport equation.

Continuity Equation and the Zeroth Velocity Moment   Although most moments, particularly in statistics, are normalized by the integral of the distribution function itself, we have chosen not to do so here because the integral of the phase density f over all velocity space has a particularly important physical meaning, namely, the local spatial density. Thus

$$\rho = m \int_{-\infty}^{+\infty} f(\vec{x},\, \vec{v})\, d\vec{v} \tag{1.2.10}$$

By $d\vec{v}$ we mean that the integration is to be carried out over all three velocity coordinates v_1, v_2, and v_3. A pedant might correctly observe that the velocity integrals should only range from $-c$ to $+c$, but for our purposes the newtonian view will suffice. Integration over the momentum space will properly preserve the limits. Now let us integrate the component form of equation (1.2.5) over all velocity space to generate an equation for the local density. Thus,

$$\int_{-\infty}^{+\infty} \left(\frac{\partial f}{\partial t} + \sum_{i=1}^{3} v_i \frac{\partial f}{\partial x_i} + \sum_{i=1}^{3} \dot{v}_i \frac{\partial f}{\partial v_i} \right) d\vec{v} = \int_{-\infty}^{+\infty} S\, d\vec{v} \tag{1.2.11}$$

Since the velocity and space coordinates are linearly independent, all time and space operators are independent of the velocity integrals. The integral of the creation rate S over all velocity space becomes simply the creation rate for particles in physical space, which we call $\Im$. By noting that the two summations in equation (1.2.11) are essentially scalar products, we can rewrite that moment and get

$$\frac{\partial}{\partial t} \int_{-\infty}^{+\infty} f\, d\vec{v} + \int_{-\infty}^{+\infty} \vec{v} \cdot \nabla f\, d\vec{v} + \int_{-\infty}^{+\infty} \dot{\vec{v}} \cdot \nabla_v f\, d\vec{v} = \Im \tag{1.2.12}$$

It is clear from the definition of ρ that the first term is the partial derivative of the local particle density. The second term can be modified by use of the vector identity

$$\vec{v} \cdot \nabla f = \nabla \cdot (f\vec{v}) - f\nabla \cdot \vec{v} \tag{1.2.13}$$

and by noting that $\nabla \cdot \vec{v} = 0$, since space and velocity coordinates are independent. If the particles move in response, to a central force, then we may relate their accelerations $\dot{\vec{v}}$ to the gradient of a potential which depends on only position and not velocity. The last term then takes the form $(\nabla\Phi/m) \cdot \int \nabla_v f \, d\vec{v}$. If we further require that $f(v)$ be bounded (i.e., that there be no particles with infinite velocity), then since the integral and gradient operators basically undo each other, the integral and hence the last term of equation (1.2.12) vanish, leaving

$$\frac{\partial n}{\partial t} + \nabla \cdot \int_{-\infty}^{+\infty} \vec{v} f \, d\vec{v} = \Im \tag{1.2.14}$$

The second term in equation (1.2.14) is the first velocity moment of the phase density and illustrates the manner by which higher moments are always generated by the moment analysis of the Boltzmann transport equation. However, the physical interpretation of this moment is clear. Except for a normalization scalar, the second term is a measure of the mean flow rate of the material. Thus, we can define a mean flow velocity $\vec{u}$ by

$$\vec{u} = \frac{\int \vec{v} f(v) \, dv}{\int f(v) \, dv} \tag{1.2.15}$$

which, upon multiplying by the particle mass, enables us to obtain the familiar form of the equation of continuity:

$$\frac{\partial \rho}{\partial t} + \nabla \cdot (\rho \vec{u}) = \Im m \tag{1.2.16}$$

This equation basically says that the explicit time variations of the density plus density changes resulting from the divergence of the flow is equal to the local creation or destruction of material $\Im$.

Euler-Lagrange Equations of Hydrodynamic Flow and First Velocity Moment of the Boltzmann Transport Equation The zeroth moment of the transport equation provided insight into the way in which matter is conserved in a flowing medium. Multiplying the Boltzmann transport equation by the velocity and integrating over all velocity space will produce momentum-like moments, and so we might expect that such operations will also produce an expression of the conservation of momentum. This is indeed the case. However, keep in mind that the velocity is a vector quantity, and so the moment analysis will produce a vector equation rather than the scalar equation, as was the case with the equation of continuity. Multiplying the Boltzmann transport equation by the local particle velocity $\vec{v}$, we get

$$\int_{-\infty}^{+\infty} \vec{v}\,\frac{\partial f}{\partial t}\,d\vec{v} + \int_{-\infty}^{+\infty} \vec{v}(\vec{v}\cdot\nabla f)\,d\vec{v} + \int_{-\infty}^{+\infty} \vec{v}(\dot{\vec{v}}\cdot\nabla_v f)\,d\vec{v} = \int_{-\infty}^{+\infty} \vec{v}S\,d\vec{v} \qquad (1.2.17)$$

We can make use of most of the tricks that were used in the derivation of the continuity equation (1.2.16). The first term can be expressed in terms of the mean flow velocity [equation (1.2.15)] while the second term can be expressed as

$$\int_{-\infty}^{+\infty} \vec{v}\nabla\cdot(\vec{v}f)\,d\vec{v} \qquad (1.2.18)$$

by using the vector identity given by equation (1.2.13). Since the quantity in parentheses of the third term in equation (1.2.17) is a scalar and since the particle accelerations depend on position only, we can move them and the vector scalar product outside the velocity integral and re-express them in terms of a potential, so the third term becomes

$$\frac{1}{m}\nabla\Phi\cdot\int_{-\infty}^{+\infty}(\nabla_v f)\vec{v}\,d\vec{v} \qquad (1.2.19)$$

The integrand of equation (1.2.19) is not a simple scalar or vector, but is the vector outer, or tensor, product of the velocity gradient of f with the vector velocity $\vec{v}$ itself. However, the vector identity given by equation (1.2.13) still applies if the scalar product is replaced with the vector outer product, so that the integrand in equation (1.2.19) becomes

$$(\nabla_v f)\vec{v} = \nabla_v(f\vec{v}) - f\nabla_v\vec{v} = \nabla_v(f\vec{v}) - \mathbf{1}f \qquad (1.2.20)$$

The quantity $\mathbf{1}$ is the unit tensor and has elements of the Kronecker delta δ_{ij} whose elements are 0 when $i \neq j$ and 1 when $i = j$. Again, as long as f is bounded, the integral over all velocity space involving the velocity gradient of f will be zero, and the first velocity moment of the Boltzmann transport equation becomes

$$\frac{\partial(n\vec{u})}{\partial t} + \int_{-\infty}^{+\infty} \vec{v}\nabla\cdot(\vec{v}f)\,d\vec{v} - n\nabla\Phi = \int_{-\infty}^{+\infty} S\vec{v}\,d\vec{v} \qquad (1.2.21)$$

Differentiating the first term and using the continuity equation (1.2.14) to eliminate $\partial n/\partial t$, we get

$$n \frac{\partial \vec{u}}{\partial t} - (\vec{u} \cdot \nabla n + n \nabla \cdot \vec{u})\vec{u} + \int_{-\infty}^{+\infty} \vec{v}\nabla \cdot (\vec{v}f)\, d\vec{v} + \rho\, \nabla\Phi$$

$$= \int_{-\infty}^{+\infty} \vec{v}S\, d\vec{v} - \int_{-\infty}^{+\infty} \vec{u}S\, d\vec{v} \qquad (1.2.22)$$

Since $\nabla \cdot \vec{v}$ is zero and the velocity and space coordinates are indpendent, we may rewrite the third term in terms of a velocity tensor defined as

$$\mathbf{u} = \frac{\displaystyle\int_{-\infty}^{+\infty} \vec{v}\vec{v}f(v)\, d\vec{v}}{\displaystyle\int_{-\infty}^{+\infty} f(v)\, d\vec{v}} \qquad (1.2.23)$$

Some rearranging and the use of a few vector identities lead to

$$\rho \frac{\partial \vec{u}}{\partial t} + \rho(\vec{u} \cdot \nabla)\vec{u} + \nabla \cdot [\rho(\mathbf{u} - \vec{u}\vec{u})] + \rho\, \nabla\Phi = \int_{-\infty}^{+\infty} mS(\vec{v} - \vec{v})\, d\vec{v} \qquad (1.2.24)$$

The quantity in brackets of the third term is sometimes called the *pressure tensor*. A density ρ times a velocity squared is an energy density, which has the units of pressure. We can rewrite that term so it has the form

$$\mathbf{P} = \frac{\displaystyle\rho \int_{-\infty}^{+\infty} f(v)(\vec{v} - \vec{u})(\vec{v} - \vec{u})\, dv}{\displaystyle\int_{-\infty}^{+\infty} f(v)\, d\vec{v}} \qquad (1.2.25)$$

which has the form of a moment of f. In this instance the moment is a tensor that more or less describes the difference between the local flow indicated by $\vec{v}$ and the mean flow $\vec{u}$. The form of the moment is that of a variance, and the tensor, in general, consists of nine components. Each component measures the net momentum transfer (or contribution to the local energy density) across a surface associated with that coordinate which results from the net flow coming from another coordinate. Thus the third term is simply the divergence of the pressure tensor, which is a vector quantity, and the first velocity moment of the Boltzmann transport equation becomes

$$\frac{\partial \vec{u}}{\partial t} + (\vec{u} \cdot \nabla)\vec{u} = -\nabla\Phi - \frac{1}{\rho} \nabla \cdot \mathbf{P} + \frac{1}{\rho} \int_{-\infty}^{+\infty} mS(\vec{v} - \vec{u})\, d\vec{v} \qquad (1.2.26)$$

This set of vector equations is known as the *Euler-Lagrange equations of hydro-dynamic flow*, and they are derived here in their most general form.

It is common to make some further assumptions concerning the flow to further simplify these flow equations. If we consider the common physical situation where there are many collisions in the gas, then there is a tendency to randomize the local velocity field $\vec{v}$ and thus to make $f(\vec{v}) = f(-\vec{v})$. Under these conditions, the pressure tensor becomes diagonal, all elements are equal, and its divergence can be written as the gradient of some scalar which we call the pressure P. In addition, the creation rate S in equation (1.2.26) which involves the effects of collisions will also become symmetric in velocity, which means that the entire integral over velocity space will vanish. This single assumption leads to the simpler and more familiar expression for hydrodynamic flow, namely

$$\frac{\partial \vec{u}}{\partial t} + (\vec{u} \cdot \nabla)\vec{u} = -\nabla\Phi - \frac{\nabla P}{\rho} \tag{1.2.27}$$

This assumption is necessary to close the moment analysis in that it provides a relationship between the pressure tensor and the scalar pressure P. From the definition of the pressure tensor, under the assumption of a nearly isotropic velocity field, P will be $P(\rho)$ and an expression known as an *equation of state* will exist. It is this additional equation that will complete the closure of the hydrodynamic flow equations and will allow for solutions. It is also worth remembering that if the mean flow velocity $\vec{u}$ is very large compared to the velocities produced by collisions, then the above assumption is invalid, no scalar equation of state will exist, and the full-blown equations of hydrodynamic flow given by equation (1.2.26) must be solved. In addition, a good deal of additional information about the system must be known so that a tensor equation of state can be found and the creation term can be evaluated.

It is worth making one further assumption regarding the flow equations. Consider the case where the flow is zero and the material is quiescent. The entire lef-hand side of equation (1.2.27) is now zero, and the assumption of an isotropic velocity field produced by random collisions holds exactly. The Euler-Lagrange equations of hydrodynamic flow now take the particularly simple form

$$\nabla P = -\rho \nabla\Phi \tag{1.2.28}$$

which is known as the *equation of hydrostatic equilibrium*. This equation is usually cited as an expression of the conservation of linear momentum. Thus the zeroth moment of the Boltzmann transport equation results in the conservation of matter, whereas the first velocity moment yields equations which represent the conservation of linear momentum. You should not be surprised that the second velocity moment will produce an expression for the conservation of energy. So far we have considered moment analysis involving velocity space alone. Later we shall see how moments taken over some dimensions of physical space can produce the diffusion approximation so important to the transfer of photons. As you might expect, moments taken over all physical space should yield "conservation laws" which apply to an entire system. There is one such example worth considering.

Boltzmann Transport Equation and the Virial Theorem   The virial theorem of classical mechanics has a long and venerable history which begins with the early work of Joseph Lagrange and Karl Jacobi. However, the theorem takes its name from work of Rudolf Clausius in the early phases of what we now call thermodynamics. Its most general expression and its relation to both subjects can be nicely seen by obtaining the virial theorem from the Boltzmann transport equation. Let us start with the Euler-Lagrange equations of hydrodynamic flow, which already represent the first velocity moment of the transport equation. These are vector equations, and so we may obtain a scalar result by taking the scalar product of a position vector with the flow equations and integrating over all space which contains the system. This effectively produces a second moment, albeit with mixed moments, of the transport equation. In the 1960s, S. Chandrasekhar and collaborators developed an entire series of virial-like equations by taking the vector outer (or tensor) product of a position vector with the Euler-Lagrange flow equations. This operation produced a series of tensor equations which they employed for the study of stellar structure. Expressions which Chandrasekhar termed "higher-order virial equations" were obtained by taking additional moments in r. However, the use of higher moments makes the relationship to the virial theorem somewhat obscure.

The origin of the position vector is important only in the interpretation of some of the terms which will arise in the expression. Remembering that the left-hand side of equation (1.2.27) is the total time derivative of the flow velocity $\vec{u}$, we see that this first spatial moment equation becomes

$$\int_V \rho\vec{r} \cdot \frac{d\vec{u}}{dt}\, dV + \int_V \rho\vec{r} \cdot \nabla\Phi\, dV + \int_V \vec{r} \cdot \nabla P\, dV = 0 \qquad (1.2.29)$$

With some generality

$$\int_V \rho\, \frac{dQ}{dt}\, dV = \frac{d}{dt}\int_V \rho Q\, dV \qquad (1.2.30)$$

Since $\vec{u}$ is just the time rate of change of position, we can rewrite $\vec{r} \cdot (d\vec{u}/dt)$ so that the first integral of equation (1.2.29) becomes

$$\int_V \rho\vec{r} \cdot \frac{d\vec{u}}{dt}\, dV = \frac{1}{2}\frac{d^2}{dt^2}\int_V \rho r^2\, dV - 2\mathbf{T} \qquad (1.2.31)$$

Here $\mathbf{T}$ is just the total kinetic energy due to the mass motions, as described by $\vec{u}$, of the system, and the integral can be interpreted as the moment of inertia about the *center*, or origin, of the coordinate frame which defines $\vec{r}$. The third integral in equation (1.2.29) can be rewritten by using the product law of differentiation and the divergence theorem:

$$\int_V \vec{r} \cdot \nabla P \, dV = \int_V \nabla \cdot (\vec{r}P) \, dV - \int_V P \nabla \cdot \vec{r} \, dV = \oint_s P_s \vec{r} \cdot \hat{n} \, dA - 3 \int_V P \, dV$$

$$(1.2.32)$$

It is also worth noting that $\nabla \cdot \vec{r} = 3$. We usually take the volume enclosing the object to be sufficiently large that $P_s = 0$. If we now make use of the ideal-gas law [which we derive in the next section along with the fact that the internal kinetic energy density of an ideal gas is $\epsilon = \frac{3}{2}\rho kT/(\mu m_h)$], we can replace the pressure P in the last integral of equation (1.2.32) with $\frac{2}{3}\epsilon$. The integral then yields twice the total internal kinetic energy of the system, and our moment equation becomes

$$\frac{1}{2}\frac{d^2 I}{dt^2} = 2(\mathbf{T} + \mathbf{U}) - \int_V \rho\vec{r} \cdot \nabla\Phi \, dV \qquad (1.2.33)$$

Here I is the moment of inertia about the origin of the coordinate system, and $\mathbf{U}$ is the total internal kinetic energy resulting from the random motion of the molecules of the gas. The last term on the right is known as the *virial of Clausius* whence the theorem takes its name. The units of that term are force times distance, so it is also an energy-like term and can be expressed in terms of the total potential energy of the system. Indeed, if the force law governing the particles of the system behaves as $1/r^2$, the virial of Clausius is just the total potential energy.[5] This leads to an expression sometimes called *Lagrange's identity* which was first developed in full generality by Karl Jacobi and is also called the *nonaveraged form* of the virial theorem:

$$\frac{1}{2}\frac{d^2 I}{dt^2} = 2(\mathbf{T} + \mathbf{U}) + \Omega \qquad (1.2.34)$$

If we consider a system in equilibrium or at least a long-term steady state, so that the time average of equation (1.2.34) removes the accelerative changes of the moment of inertia (that is, $\langle d^2 I/dt^2 \rangle = 0$), then we get the more familiar form of the virial theorem, namely,

$$2\langle \mathbf{T} \rangle + 2\langle \mathbf{U} \rangle + \langle \Omega \rangle = 0 \qquad (1.2.35)$$

It is worth mentioning that the use of the virial theorem in astronomy often replaces the time averages with ensemble averages over all available phase space. The theorem which permits this is known as the *ergodic theorem*, and all thermodynamics rests on it. Although such a replacement is legitimate for large systems consisting of many particles, such as a star, considerable care must be exercised in applying it to stellar or extragalactic systems having only a few members. However, the virial theorem itself has basically the form and origin of a conservation law, and when the conditions of the theorem's derivation hold, it must apply.

1.3 Equation of State for an Ideal Gas and Degenerate Matter

The formulation of the Boltzmann transport equation also provides an ideal setting for the formulation of the equation of state for a gas under wide-ranging conditions. The statistical distribution functions developed in Section 1.1 give us the distribution functions for particles which depend largely on how filled phase space happens to be. Those functions relate the particle energy and the kinetic temperature to the distribution of particles in phase space. This is exactly what is meant by $f(\vec{v})$. Thus we can calculate the expected relationship between the pressure as given by the pressure tensor and the state variables of the distribution function. The result is known as the *equation of state*.

As given in equation (1.2.24), the pressure tensor is $\rho(\mathbf{u} - \vec{u}\vec{u})$. If $f(\vec{v})$ is symmetric in $\vec{v}$, then $\vec{u}$ must be zero (or there must exist an inertial coordinate system in which $\vec{u}$ is zero), and the divergence of the pressure tensor can be replaced by the gradient of a scalar, which we call the *gas pressure*, and will be given by

$$P = \rho \,\frac{\displaystyle\int_{-\infty}^{+\infty} v^2 f(v)\, d\vec{v}}{\displaystyle\int_{-\infty}^{+\infty} f(v)\, d\vec{v}} \tag{1.3.1}$$

From Maxwell-Boltzmann statistics, the distribution function of particles, in terms of their velocity, as given by equation (1.1.16). If we regard the number of cells of phase space to be very large, we can replace N_i by dN and consider equation (1.1.16) to give the distribution function $f(\vec{v})$, so that

$$f(v) = \text{constant} \times e^{-mv^2/(2kT)} \tag{1.3.2}$$

Now, in general,

$$\int_0^{+\infty} x^{2n} e^{-ax^2}\, dx = \frac{1 \cdot 3 \cdot 5 \cdots (2n-1)}{2^{n+1} a^n}\left(\frac{\pi}{a}\right)^{1/2} \tag{1.3.3}$$

Substitution of equation (1.3.2) into equation (1.3.1) therefore yields

$$P = \frac{\rho kT}{m} = nkT \tag{1.3.4}$$

This is known as the *ideal-gas law*, and it is the appropriate equation of state for a gas obeying Maxwell-Boltzmann statistics. That is, we may confidently expect that this simple formula will provide the correct relation among P, T, and ρ as long as the cells of phase space do not become overly filled and quantum effects become important. If the density is increased without a corresponding increase in particle energy, a point will come when the available cells of phase space begin to fill up in accordance with the Pauli exclusion principle. As the most "popular" cells in phase space become filled, the particles will have to spill over into adjacent cells, producing a distortion in the distribution function (see Figure 1.5). When this happens, the gas is said to become *partially degenerate*. Figure 1.5 shows this effect and indicates a way to quantify the effect. We define a momentum p_0 as that momentum above which there are just enough particles to fill the remaining phase space cells below p_0. Thus

$$\int_0^{p_0} [N_{\text{max}} - N(p)]\, dp = \int_{p_0}^{\infty} N(p)\, dp \tag{1.3.5}$$

If we make the approximation that all the spaces in phase space are filled (i.e., a negligible number of particles exist with momentum above p_0), then the momentum distribution of the particles can be represented by a section of a sphere in

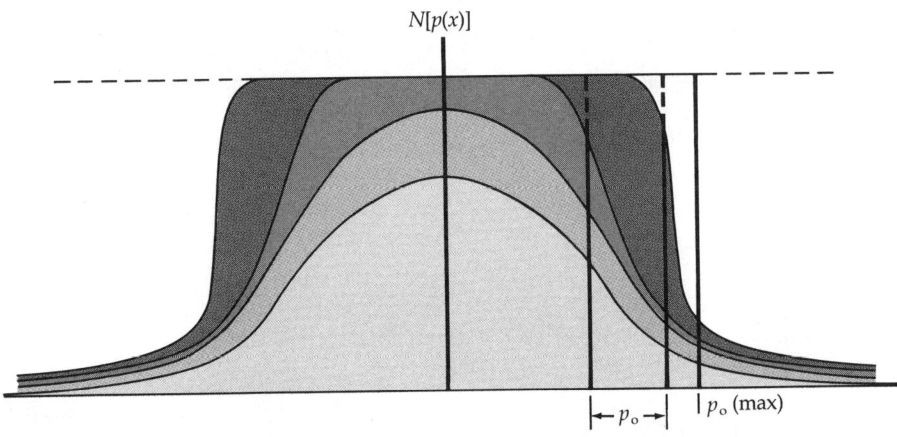

Figure 1.5 Momentum cross section of phase space with different particle densities. As the volume saturates, the distribution function departs further and further from maxwellian. The Fermi momentum p_0 represents that momentum such that the particles having greater momentum would just fill the momentum states below it.

momentum space so that

$$n(p)\,dp = 2\left(4\pi p^2\,\frac{dp}{h^3}\right) \tag{1.3.6}$$

The factor of 2 arises because the electron can have two spin states in a cell of size h^3. The number density of particles can then be given in terms of p_0 as

$$n = 2\int_0^{p_0}\frac{4\pi p^2}{h^3}\,dp = \frac{8\pi p_0^3}{3h^3} \tag{1.3.7}$$

We have already developed a relation for the scalar pressure in terms of the velocity distribution under the assumption of an isotropic velocity field in equation (1.3.1), and we need only to replace the velocity distribution $f(\vec{v})$ with a distribution function of momentum. However, we must remember that the integral in equation (1.3.1) is actually three integrals over each velocity coordinate which will all have the same value for an isotropic velocity field. The three integrals corresponding to the three components of velocity in equation (1.3.1) are equal for a spherical momentum space. Therefore one-third of the scalar form of equation (1.3.1) will represent the total contribution of the momentum to the pressure. Thus the pressure can be expressed in terms of the maximum momentum p_0, often known as the *Fermi momentum*, as

$$P = \frac{1}{3}\int_0^{p_0}\frac{p^2}{m}\,n(p)\,dp = \frac{1}{3}\int_0^{p_0}\frac{p^2}{m}\frac{8\pi p^2}{h^3}\,dp = \frac{8\pi p_0^5}{15mh^3} = \frac{h^3}{20m}\left(\frac{3}{\pi}\right)^{2/3}n^{5/3} \tag{1.3.8}$$

Using equation (1.3.7), we can eliminate p_0 and obtain a relationship between the pressure and the density. This, then, is the equation of state for totally degenerate matter, and since the electrons tend to become degenerate before any other particles, it is common to write the equation of state for electron degeneracy alone:

$$P = \left(\frac{h^2}{20m_e m_h}\right)\left(\frac{3}{\pi m_h}\right)^{2/3}\left(\frac{\rho}{\mu_e}\right)^{5/3} = 9.9\times10^{12}\left(\frac{\rho}{\mu_e}\right)^{5/3} \qquad \text{cgs} \tag{1.3.9}$$

Here m_e and m_h are the mass of the electron and hydrogen atom, respectively, while μ_e is the mean molecular weight of the free electrons.

If we consider a gas under extreme pressure, not only will the cells of phase space be filled, but also the maximum momentum will become very large. Although the mass m and momentum p both approach infinity as the particle energy increases, their ratio p/m does not. It remains finite and approaches the speed of light c. Since these particles also make the largest contribution to the pressure, we can estimate the effect of having a relativistically degenerate gas by replacing p/m by c in equation (1.3.8), and we get

$$P = \frac{1}{3}\int_0^{p_0} \frac{p^2}{m}\, n(p)\, dp = \frac{1}{3}\int_0^{p_0} \frac{8\pi p^3 c}{h^3}\, dp = \frac{2\pi p_0^4 c}{3h^3} \qquad (1.3.10)$$

which leads to an equation of state that depends on $\rho^{4/3}$ instead of $\rho^{5/3}$, as in the case of nonrelativistic degeneracy. Eliminating p_0, we obtain for the electron degeneracy

$$P = \left(\frac{hc}{8m_h}\right)\left(\frac{3}{\pi m_h}\right)^{1/3}\left(\frac{\rho}{\mu_e}\right)^{4/3} = 1.231 \times 10^{15}\left(\frac{\rho}{\mu_e}\right)^{4/3} \qquad \text{cgs} \quad (1.3.11)$$

The equations of state for degenerate matter that we have derived represent limiting conditions and are never exactly realized. In real situations we must consider the transition between the ideal-gas law and total degeneracy as well as the transition between nonrelativistic and completely relativistic degeneracy. One way to identify the range of state variables for which we can expect a transition zone is to equate the various equations of state and to solve for the range of state variables involved. Equating the ideal-gas law [equation (1.3.4)] with the equation for a totally degenerate gas [equation (1.3.9)], we can determine the range of density ρ_t and temperature T_t which lie in the transition zone between the two equations of state:

$$\frac{\rho_t}{\mu_e} = (2.40 \times 10^{-8})(T_t^{3/2}) \qquad \text{g/cm}^3 \qquad (1.3.12)$$

For a metal at 100 K, $\rho_t/\mu_e \approx 6 \times 10^{-5}$ g/cm^3, which implies that the electrons in such a conductor follow the degenerate equation of state and that virtually all the cells in phase space are full. This accounts for the high conductivity of metals, since the saturation of phase space cells implies that free electrons cannot scatter off the other particles in the metal (for in doing so they would have to move to a new cell in phase space, which is more than likely filled). Thus, they travel relatively unhindered through the conductor. In general, a totally degenerate gas proves to be an excellent conductor. For temperatures on the order of 10^7 K which prevail in the center of the sun, the transition densities occur at about 8×10^2 g/cm^3 which is significantly higher than we find in the sun. Thus, we may be assured that the ideal-gas law will be appropriate throughout the interior of the sun and most stars. However, white dwarfs do exceed the transition density for the temperatures we may expect in these stars. Therefore, we can expect a transition from the ideal-gas law which will prevail in the surface regions to total degeneracy in the interior. In this transition region the equation of state becomes more complex. A complete discussion of this region can be found in Cox and Giuli[6] and Chandrasekhar.[7] The basic philosophy is to write the equation of state in parametric form in terms of a degeneracy parameter ψ, where the equation of state becomes the ideal-gas law when $\psi \ll 0$ and the equation of state approaches the totally degenerate equation of state if $\psi \gg 0$. This parametric form can be written as

$$\frac{\rho}{\mu_e} = \left(\frac{4\pi}{h^3}\right)(2mkT)^{3/2} m_h F_{1/2}(\psi)$$

$$P_e = \left(\frac{8\pi}{3h^3}\right)(2mkT)^{3/2} kT F_{3/2}(\psi) \tag{1.3.13}$$

$$F_n(\psi) = \int_0^\infty \frac{x^n\, dx}{e^{x-\psi} + 1}$$

For the transition zone between nonrelativistic and relativistic degeneracy, Chandrasekhar[7] (p. 360) also gives a parametric equation of state in terms of the "maximum" momentum p_0 of the Fermi sea:

$$\frac{\rho}{\mu_e} = \left(\frac{8\pi}{3h^3}\right)(m_e c)^3 m_h x^3$$

$$P_e = \frac{\pi}{3h^3} m_e^4 c^5 f(x) \tag{1.3.14}$$

$$f(x) = x(2x^2 - 3)(x^2 + 1)^{1/2} + 3\sinh^{-1} x$$

As x approaches zero, the nonrelativistic equation of state is obtained whereas as x approaches ∞, the fully relativistic limit is obtained. In the rare case where the gas occupies both transition regions at the same time, the equation of state becomes quite complicated. Refer to Cox and Giuli[6] for a detailed description of this situation.

Before leaving this discussion of the equation of state and degenerate matter, we want to explore some consequences of the most notable aspect of the degenerate equation of state. Nowhere in either the nonrelativistic or the relativistic degenerate equation of state does the temperature appear. This complete lack of temperature dependence implies a unique relationship between the pressure and the density. Hydrostatic equilibrium [equation (1.2.28)] implies a relation among the pressure, mass, and radius. Since the mass, density, and radius are related by definition, these three relationships should allow us to find a unique relation between the mass of the configuration and its radius. Although a detailed investigation of the relation requires the solution of a differential equation coupled with some extremely nonlinear equations, we can get a sense of the mass-radius relation by considering the form of the equations that constrain the solution.

For spherical stars, hydrostatic equilibrium as expressed by equation (1.2.28) implies that

$$\frac{dP}{dr} \sim \frac{M(r)\rho}{r^2} \sim \frac{M(M/R^3)}{R^2} \tag{1.3.15}$$

Since we can also expect the pressure gradient to be proportional to P/R, the internal pressure in a star should vary as

$$P \sim \frac{M^2}{R^4} \tag{1.3.16}$$

For a totally degenerate gas, the equation of state requires that

$$P \sim \rho^{5/3} \sim \frac{M^{5/3}}{R^5} \tag{1.3.17}$$

Thus, we eliminate the pressure from these two expressions to get

$$R \sim M^{-1/3} \qquad \text{or} \qquad M \sim R^{-3} \tag{1.3.18}$$

We arrive at a curious result: As the mass of the configuration increases, the radius decreases. This situation, then, must prevail for white dwarfs. The more massive the white dwarf, the smaller its radius. In a situation where mass is added to a white dwarf, thereby causing its radius to decrease, the internal pressure must increase, which leads to an increase in the Fermi momentum p_0. Sooner or later the equation of state must change over to the fully relativistic equation of state. Here

$$P \sim \rho^{4/3} \sim \frac{M^{4/3}}{R^4} \tag{1.3.19}$$

If we again eliminate the pressure by using equation (1.3.16), then the radius also disappears and

$$M \sim \text{constant} \tag{1.3.20}$$

Thus, for a fully relativistic degenerate gas, there is a unique mass for which the configuration is stable. Should mass be added beyond this point, the star would

31

be forced into a state of unrestrained gravitational collapse. Much later we shall see that a further change in the equation of state, which occurs when the density approaches that of nuclear matter, can halt the collapse, allowing the formation of a neutron star. But for "normal" matter a limit is set by quantum mechanics, and this prevents the formation of white dwarfs with masses greater than about $1.4 M_\odot$. This is the limit found by S. Chandrasekhar in the late 1930s and for which he received the Nobel Prize in 1983. The configuration described by the fully relativistically degenerate equation of state is a strange one indeed, and we shall explore it in some detail later. For now, let us turn to the most basic assumptions that must be made for the study of stellar structure and to what they imply about the nature of stars.

Problems

1. Consider a standard deck of 52 playing cards dealt into four hands of 13 cards each. If a given suit distribution within a hand represents a macrostate while a specific set of cards within a suit represents a microstate, find (*a*) the number of possible macrostates for each hand, (*b*) the number of microstates allowed for each macrostate, and (*c*) the most probable macrostate.

2. Consider a space with three cells of size $2h^3$ and nine particles. Find the total number of macrostates, the total number of microstates, and the most probable macrostate, assuming the particles are (*a*) "Maxwellons," (*b*) fermions, and (*c*) bosons.

3. Given that

$$dN = B\left(\exp\frac{w}{kT} + \phi\right)^{-1} dp_x\, dp_y\, dp_z$$

and that $w = (p_x^2 + p_y^2 + p_z^2)/m$, find an expression for B in terms of N for the cases where $\phi = 0, \pm 1$.

4. Derive the equation of state for a Fermi gas from first principles.

5. Given that $f(x)$ is an analytic function in the interval $0 \leq x \leq \infty$, show that $f(x)$ can be represented in terms of the moments of the function $M_i[f(x)]$, where

$$M_i[f(x)] \equiv \int_0^\infty x^i f(x)\, dx$$

6. If the pressure tensor **P** has the form specified by equation (1.2.25), show that it can be rewritten as it appears in equation (1.2.24) (i.e., as the tensor operated on by the divergence operator in the third term on the left-hand side).

7. Show that the virial theorem holds in the form given by equation (1.2.35) even if the forces of interaction include velocity-dependent terms (i.e., such as Lorentz forces or viscous drag forces).

8. Show that the second velocity moment of the Boltzmann transport equation leads to an equation describing the conservation of energy.

References and Supplemental Reading

1. Aller, L. H., *The Atmospheres of the Sun and Stars*, 2d ed., Ronald Press, New York, 1963, pp. 104, 108.

2. Ogorodnikov, K. F., *Dynamics of Stellar Systems* (Trans.: J. B. Sykes, ed.: A. Beer), Macmillan, New York, 1965, p. 143.

3. Goldstein, H., *Classical Mechanics*, Addison-Wesley, Reading, Mass., 1959, p. 266.

4. Landau, L. D., and E. M. Lifshitz, *Mechanics* (Trans.: J. B. Sykes and J. S. Bell), Addison-Wesley, Reading, Mass., 1960, p. 147.

5. Collins, G. W., II, *The Virial Theorem in Stellar Astrophysics*, Pachart Publishing House, Tucson, Ariz., 1978, p. 14.

6. Cox, J. P., and R. T. Giuli, *Principles of Stellar Structure*, Gordon and Breach, New York, 1968, pp. 781, 812.

7. Chandrasekhar, S., *An Introduction to Stellar Structure* [1939], Dover, New York, 1957, p. 401.

The reader may wish to consult a number of supplemental references to further understand the material in this chapter. Since the approach is basically that of statistical mechanics, any good book on that subject should enhance the readers, understanding. Some examples are:

. Reif, F.: *Statistical Physics*, McGraw-Hill, New York, 1967.

. Akhiezer, A. I., and S. V. Peletminskii: *Methods of Statistical Physics*, Pergamon, New York, 1981.

. Andrews, F. C.: *Equilibrium Statistical Mechanics*, Wiley, New York, 1975.

. Mayer, J. E., and G. M. Mayer: *Statistical Mechanics*, Wiley, New York, 1977.

2

Basic Assumptions, Theorems, and Polytropes

. . .

2.1 Basic Assumptions

Any rational structure must have a beginning, a set of axioms, upon which to build. In addition to the known laws of physics, we have to assume a few things about stars to describe them. It is worth keeping these assumptions in mind for the day you encounter a situation in which the basic axioms of stellar structure no longer hold. We have already alluded to the fact that a self-gravitating plasma will assume a spherical shape. This fact can be rigorously demonstrated from the nature of an attractive central force, so it does not fall under the category of an axiom. However, it is a result which we shall use throughout most of this book. A less obvious axiom, but one which is essential for the construction of the stellar interior, is that the density is a monotonically decreasing function of the radius. Mathematically, this can be expressed as

$$\rho(r) \leq \langle \rho \rangle(r) \qquad \text{for } r > 0 \tag{2.1.1}$$

where

$$\langle \rho \rangle(r) \equiv \frac{M(r)}{4\pi r^3 / 3} \tag{2.1.2}$$

and $M(r)$ is the mass interior to a sphere of radius r and is just $\int 4\pi r^2 \rho \, dr$. In addition, we assume as a working hypothesis that the appropriate equation of state is the ideal-gas law. Although this is expressed here as an assumption, we shall shortly see that it is possible to estimate the conditions which exist inside a star and that they are fully compatible with the assumption.

It is a fairly simple matter to see that the free-fall time for a particle on the surface of the sun is about 20 min. This is roughly equivalent to the dynamical time scale, which is the time scale on which the sun will respond to departures from hydrostatic equilibrium. Most stars have dynamical time scales ranging from fractions of a second to several months, but in all cases this time is a small fraction of the typical evolutionary time scale. Thus, the assumption of hydrostatic equilibrium is an excellent one for virtually all aspects of stellar structure. In Chapter 1 we developed an expression for hydrostatic equilibrium [equation (1.2.28)], where the pressure gradient is proportional to the potential gradient and the local constant of proportionality is the density. For spherical stars, we may take advantage of the simple form of the gradient operator and the source equation for the gravitational potential to obtain a single expression relating the pressure gradient to $M(r)$ and ρ.

The source equation for the gravitational potential field is also known as *Poisson's equation*, and in general it is

$$\nabla^2 \Omega = 4\pi G\rho \tag{2.1.3}$$

which in spherical coordinates becomes

$$\frac{d}{dr}\left(r^2 \frac{d\Omega}{dr} \right) = 4\pi G\rho r^2 \tag{2.1.4}$$

Integrating this over r, we get

$$\frac{d\Omega}{dr} = \frac{G}{r^2} \int_0^r 4\pi r^2 \rho \, dr = \frac{GM(r)}{r^2} \tag{2.1.5}$$

Replacing the potential gradient from equation (1.2.28), we have

$$\frac{dP}{dr} = -\frac{GM(r)\rho(r)}{r^2} \tag{2.1.6}$$

This is the equation of hydrostatic equilibrium for spherical stars. Because of its generality and the fact that virtually no assumptions are required to obtain it, we can use its integral to place fairly narrow limits on the conditions that must prevail inside a star.

In equation (2.1.2), we introduced a new variable $M(r)$. Note that its invocation is equivalent to invoking a conservation law. The conservation of mass basically requires that the total mass interior to r be accounted for by summing over the density interior to r. Thus,

$$M(r) = \int_0^r 4\pi r^2 \rho \, dr \tag{2.1.7}$$

or its differential form

$$\frac{dM(r)}{dr} = 4\pi r^2 \rho \tag{2.1.8}$$

2.2 Integral Theorems from Hydrostatic Equilibrium

a Limits on State Variables

Following Chandrasekhar,[1] we wish to define a quantity $I_{\sigma,\nu}(r)$ which is effectively the σth moment of the mass distribution further weighted by $r^{-\nu}$. Specifically

$$I_{\sigma,\nu}(r) \equiv \frac{G}{4\pi} \int_0^r \frac{[M(r)]^\sigma}{r^\nu} \, dM(r) \tag{2.2.1}$$

There are quite a variety of physical quantities which can be related to $I_{\sigma,\nu}$. For example,

$$4\pi I_{11} = G \int_0^r M(r)\rho 4\pi r^2 \frac{dr}{r} \tag{2.2.2}$$

36   is just the absolute value of the total gravitational energy of the star.

We can use this integral quantity to place limits on physical quantities of interest if we replace ρ by $\langle \rho \rangle$ as defined by equation (2.1.2). Since

$$r^{\nu} = \left[\frac{M(r)}{\frac{4}{3}\pi \langle \rho \rangle(r)} \right]^{\nu/3} \tag{2.2.3}$$

we may rewrite $I_{\sigma,\nu}$ as

$$I_{\sigma,\nu}(r) = \frac{G}{4\pi} \int_0^r [M(r)]^{\sigma - \nu/3} \left(\frac{4\pi}{3} \right)^{\nu/3} \langle \rho \rangle(r)^{\nu/3}\, dM(r) \tag{2.2.4}$$

Now since our assumption of the monotonicity of ρ requires $\rho_c \geq \langle \rho \rangle(r) \geq \rho(r)$, we can obtain an inequality to set limits on $I_{\sigma,\nu}$. Namely,

$$\frac{G}{4\pi} \left(\frac{4\pi}{3} \right)^{\nu/3} \rho_c^{\nu/3} \frac{M(r)^{\sigma + 1 - \nu/3}}{\sigma + 1 - \nu/3} \geq I_{\sigma,\nu}(r) \geq \frac{G}{4\pi} \left(\frac{4\pi}{3} \right)^{\nu/3} \langle \rho \rangle(r) \frac{M(r)^{\sigma + 1 - \nu/3}}{\sigma + 1 - \nu/3} \tag{2.2.5}$$

Now let us relate $\langle P \rangle$, $\langle T \rangle$, and $\langle g \rangle$ to $I_{\sigma,\nu}$, where these quantities are defined as

$$\langle P \rangle \equiv \int_0^M P(r)\, \frac{dM(r)}{M}$$

$$\langle T \rangle \equiv \int_0^M T(r)\, \frac{dM(r)}{M} \tag{2.2.6}$$

$$\langle g \rangle \equiv \int_0^M g(r)\, \frac{dM(r)}{M}$$

Making use of the result that the surface pressure and temperature are effectively zero compared to their internal values, we can eliminate the temperature by using the ideal-gas law, integrate the first two members of equations (2.2.6) by parts, and eliminate the pressure gradient by utilizing hydrostatic equilibrium. We obtain

$$\langle P \rangle = \frac{I_{2,4}(R)}{M}$$

$$\langle T \rangle = \frac{4\pi \mu m_h}{3k} \frac{I_{1,1}(R)}{M} \tag{2.2.7}$$

$$\langle g \rangle = \frac{4\pi I_{1,2}(R)}{M}$$

The last of these expressions comes immediately from the definition of g. Applying the inequality [equation (2.2.5)], we can immediately obtain lower limits for these quantities of

$$\langle P \rangle \geq \frac{3GM^2}{20\pi R^4} = 5.4 \times 10^8 \left(\frac{M}{M_\odot}\right)^2 \left(\frac{R_\odot}{R}\right)^4 \qquad \text{atm}$$

$$\langle T \rangle \geq \frac{G\mu m_h M}{5kR} = 4.61 \times 10^6 \,\mu \left(\frac{M}{M_\odot}\right)\left(\frac{R_\odot}{R}\right) \qquad \text{K} \tag{2.2.8}$$

$$\langle g \rangle \geq \frac{3GM}{4R^2} = 2.05 \times 10^4 \left(\frac{M}{M_\odot}\right)\left(\frac{R_\odot}{R}\right)^2 \qquad \text{cm/s}^2$$

Since these theorems apply for any gas sphere in hydrostatic equilibrium where the ideal-gas law applies, we can use them for establishing the range of values to be expected in stars in general. In addition, it is possible to use the other half of the inequality to place upper limits on the values of these quantities at the center of the star.

b β^* Theorem and Effects of Radiation Pressure

We have consistently neglected radiation pressure throughout this discussion, and a skeptic could validly claim that this affects the results concerning the temperature limits. However, there is an additional theorem, also due to Chandrasekhar[1], (p. 73), which places limits on the effects of radiation pressure. This theorem is generally known as the β^* *theorem*. Let us define β as the ratio of the gas pressure to the total pressure, which includes the radiation pressure. The radiation pressure for a photon gas in equilibrium is just $P_r = aT^4/3$. Combining these definitions with the ideal-gas law, we can write

$$P_g = \beta P_T = \left[\frac{3}{a}\left(\frac{k}{\mu m_h}\right)^4 \frac{1-\beta}{\beta}\right]^{1/3} \rho^{4/3}$$

$$P_r = (1-\beta)P_T = \frac{aT^4}{3}$$

$$T = \left[\frac{3k\,(1-\beta)}{\mu a m_h \beta}\right]^{1/3} \rho^{1/3} \tag{2.2.9}$$

$$P_c = \frac{P_{g,c}}{\beta_c} = \frac{1}{\beta_c}\left[\frac{3}{a}\left(\frac{k}{\mu m_h}\right)^4 \frac{1-\beta_c}{\beta_c}\right]^{1/3} \rho_c^{4/3}$$

Using the integral theorems to place an upper limit on the central pressure, we get

$$P_c \leq \frac{1}{2} G \left(\frac{4\pi}{3}\right)^{1/3} \rho_c^{4/3} M^{2/3} \tag{2.2.10}$$

Equation (2.2.10), when combined with the last of equations (2.2.9) and solved for M, yields

$$M \geq \left(\frac{6}{\pi}\right)^{1/2} \left[\frac{1-\beta_c}{\beta_c^4}\left(\frac{k}{\mu m_h}\right)^4 \frac{3}{a}\right]^{1/2} G^{-3/2} \tag{2.2.11}$$

Now we define β^* to be the value of β which makes equation (2.2.11) an equality, and then we obtain the standard result that

$$\frac{1-\beta^*}{(\beta^*)^4} \geq \frac{1-\beta_c}{\beta_c^4} \tag{2.2.12}$$

Since $(1-\beta)/\beta^4$ is a monotone increasing function of $(1-\beta)$,

$$1-\beta^* \geq 1-\beta_c = \frac{P_{r,c}}{P_T} \tag{2.2.13}$$

Equation (2.2.11) can be solved directly for M in terms of β^*, and thus it places limits on the ratio of radiation pressure to total pressure for stars of a given mass. Chandrasekhar[1] (p. 75) provides the brief table of values shown in Table 2.1.

Table 2.1 Stellar Mass vs. $1 - \beta^*$

$1 - \beta^*$	$(M/M_\odot)\mu^2$
0.01	0.56
0.02	0.81
0.10	2.14
0.20	3.83
0.30	6.12
0.40	9.62
0.50	15.50
0.60	26.52
0.70	50.93
0.80	122.50
0.90	520.7
1.00	∞

As we shall see later, μ is typically of the order of unity (for example, μ is $\frac{1}{2}$ for pure hydrogen and 2 for pure iron). It is clear from Table 2.1 that by the time that radiation pressure accounts for half of the total pressure, we are dealing with a very massive star indeed. However, it is equally clear that the effects of radiation pressure must be included, and they can be expected to have a significant effect on the structure of massive stars.

2.3   Homology Transformations

The term *homology* has a wide usage, but in general it means "proportional to" and is denoted by the symbol $\sim$. One set is said to be *homologous* to another if the two can be put into a one-to-one correspondence. If every element of one set, say z_i, can be identified with every element of another set, say z_i', then $z_i \sim z_i'$ and the two sets are homologous. Thus a homology transformation is a mapping which relates the elements of one set to those of another. In astronomy, the term *homology* has been used almost exclusively to relate one stellar structure to another in a special way.

One can characterize the structure of a star by means of the five variables $P(r)$, $T(r)$, $M(r)$, $\mu(r)$, and $\rho(r)$ which are all dependent on the position coordinate r. In our development so far, we have produced three constraints on these variables— the ideal-gas law, hydrostatic equilibrium, and the definition of $M(r)$. Thus, specifying the transformation of any two of the five dependent variables and of the independent variable r specifies the remaining three. If the transformations can be written as simple proportionalities, then the two stars are said to be homologous to each other. For example, if

$$r' = C_1 r \qquad \rho'(r') = C_2 \rho(r) \qquad \xi'(r') = C_3 \xi(r) \tag{2.3.1}$$

then

$$\zeta'(r') = C_4 \zeta(r) \qquad \eta'(r') = C_5 \eta(r) \qquad \chi'(r') = C_6 \chi(r) \tag{2.3.2}$$

where ξ, ζ, η, and χ stand for any of the remaining structure variables. However, because of the constraints mentioned above, C_4, C_5, and C_6 are not linearly independent but are specified in terms of the remaining C's. Consider the definition of $M(r)$ and a homology transformation from $r \to r'$. Then

$$\frac{M'(r')}{M(r)} = \frac{\int_0^{r'} 4\pi (x')^2 \rho'(x')\,dx'}{\int_0^{r} 4\pi x^2 \rho(x)\,dx} = C_2 C_1^3 \tag{2.3.3}$$

In a similar manner, we can employ the equation of hydrostatic equilibrium to find the homology transformation for the pressure P, since

$$\frac{P'(r')}{P(r)} = \frac{\int_0^{r'} [GM'(x')\rho'(x')/x'^2]\,dx'}{\int_0^r [GM(x)\rho(x)/x^2]\,dx} = C_2^2 C_1^2 \tag{2.3.4}$$

If we take ξ to be the chemical composition μ, then the remaining structure variable is the temperature whose homology transformation is specified by the ideal-gas law as

$$\frac{P'(r')}{P(r)} = \frac{\rho'(r')kT'(r')/\mu'(r')}{\rho(r)kT(r)/\mu(r)} = C_2^2 C_1^2 \tag{2.3.5}$$

so that

$$\frac{T'(r')}{T(r)} = C_3 C_2 C_1^2 \tag{2.3.6}$$

Should we take ξ to be T, then the homology transform for μ is specified and is

$$\frac{\mu'(r')}{\mu(r)} = \frac{C_3}{C_1^2 C_2} \tag{2.3.7}$$

We can use the constraints specified by equations (2.3.3), (2.3.4), and (2.3.6) and the initial homology relations [equations (2.3.1)] to find how the structure variables transform in terms of observables such as the total mass M and radius R. Thus,

$$\frac{\rho'(r')}{\rho(r)} = \frac{M'}{M}\left(\frac{R}{R'}\right)^3$$

$$\frac{P'(r')}{P(r)} = \left(\frac{M'}{M}\right)^2 \left(\frac{R}{R'}\right)^4 \tag{2.3.8}$$

$$\frac{T'(r')}{T(r)} = \frac{\mu' M' R}{\mu M R'}$$

Since homology transformations essentially represent a linear scaling from one structure to another, it is not surprising that the dependence on mass and radius is the same as implied by the integral theorems [equations (2.2.8)].

The primary utility of homology transformations is that they provide a "feel" for how the physical structure variables change given a simple change in the defining parameters of the star, "all other things being equal." An intuitive feel for the behavior of the state variables P, T, and ρ which result from the scaling of the mass and radius is essential if one is to understand stellar evolution. Consider the homologous contraction of a homogeneous uniform density mass configuration. Here the total mass and composition remain constant, and we obtain a very specific homology transformation

$$\frac{\rho}{\rho_0} = \left(\frac{R_0}{R}\right)^3 \qquad \frac{P}{P_0} = \left(\frac{R_0}{R}\right)^4 \qquad \frac{T}{T_0} = \frac{R_0}{R} \qquad (2.3.9)$$

which is known as *Lane's law*[1] (p. 47) and has been thought to play a role in star formation. In addition, certain phases of stellar collapse have been shown to behave homologously. In these instances, the behavior of the state variables is predictable by simple homology transformations in spite of the complicated detailed physics surrounding these events.

2.4 Polytropes

We have progressed about as far as we can in setting conditions for stellar structure with the assumptions that we made. It is now necessary to add a constraint on the structure. Physically, the logical arenas to search for such constraints are energy production and energy flow, and we shall do so in later chapters. However, before we enter those somewhat complicated domains, consider the impact of a somewhat ad hoc relationship between the pressure and the density. This relationship has its origins in thermodynamics and results from the notion of polytropic change. This gives rise to the polytropic equation of state

$$P(r) = K\rho(r)^{(n+1)/n} \qquad (2.4.1)$$

where n is called the *polytropic index*. Clearly an equation of state of this form, when coupled with the equation of hydrostatic equilibrium, will provide a single relation for the run of pressure or density with position. The solution of this equation basically solves the fundamental problem of stellar structure insofar as the equation of state correctly represents the behavior of the stellar gas. Such solutions are called *polytropes* of a particular index n.

Many astrophysicists feel that the study of polytropes is of historical interest only. While it is true that the study of polytropes did develop early in the history of stellar structure, this is so because polytropes provide significant insight into the structure and evolution of stars. The motivation comes from the observation that ideal gases behave in a certain way when they change in an adiabatic manner. It is a generalization of this behavior which is characterized by the polytropic equation of state. Later we shall see that when convection is established in the interior of a star, it is so efficient that the resultant temperature gradient is that of an adiabatic gas responding to hydrostatic equilibrium. Such a configuration is a polytrope. We have already seen that the degenerate equations of state have the same form as the polytropic equation of state, and so we might properly expect that degenerate configurations will be well represented by polytropes. In addition, we shall find that in massive stars where the pressure is dominated by the pressure of radiation, the equation of state is essentially that of a photon gas in statistical equilibrium and that equation of state is also polytropic. The simple nature of polytropic structure and its correspondence to many physical stars provides a basis for incorporating additional effects (such as rotation) in a semianalytical manner and thereby offers insight into the nature of the effects in real stars. Thus, for providing insight into the structure and behavior of real stars, an understanding of polytropes is essential. However, even beyond the domain of stellar astrophysics, polytropes find many applications. Certain problems in stellar dynamics and galactic structure can be described by polytropes, and the polytropic equation of state has even been used to represent the density distribution of dark matter surrounding galaxies. But with the applications to stars in mind, let us consider the motivation for the polytropic equation of state.

a Polytropic Change and the Lane-Emden Equation

From basic thermodynamics we learn that the infinitesimal change in the heat of a gas dQ can be related to the change in the internal energy dU and the work done on the gas so that

$$dQ = dU + P\,dV = \frac{\partial U}{\partial T}\,dT + P\,dV \qquad (2.4.2)$$

The strange-looking derivative d is known as a *Pfaffian derivative*, and its most prominent property is that it is not an exact differential. A complete discussion of their mathematical properties is given by Chandrasekhar[1] (p. 17).

The ideal-gas law can be stated in its earliest form as $PV = RT$, which leads to

$$P\,dV + V\,dP = R\,dT \qquad (2.4.3)$$

where R is the gas constant and V is the specific volume (i.e., the volume per unit mass). Now let us define the specific heat at constant α, C_α, as

$$\left[\frac{dQ}{dT}\right]\bigg|_{\alpha=\text{const}} \equiv C_\alpha \tag{2.4.4}$$

Here the differentiation is done in such a way that α remains constant. Thus $(dU/dT)|_V$ is the specific heat C_V at constant volume. Using equation (2.4.3) to eliminate $P\,dV$ in equation (2.4.2), we get

$$C_P = C_V + R \tag{2.4.5}$$

where C_P is the specific heat at constant pressure.

   With this notion that $(dQ/dT)|_\alpha$ is the specific heat at constant α, we make the generalized definition of polytropic change to be

$$\frac{dQ}{dT} = C \tag{2.4.6}$$

where C is some constant. Using equations (2.4.2) and (2.4.3) and the definition of C, we can write

$$T^{C_V-C}V^{C_P-C_V} = \text{const} \tag{2.4.7}$$

Now for an ordinary gas it is common to define the ratio of specific heats C_P/C_V as γ. In that same spirit, we can define a polytropic gamma as

$$\gamma' = \frac{C_P - C}{C_V - C} \tag{2.4.8}$$

By use of the ideal-gas law, we can write

$$P = \text{const} \times V^{-\gamma'} = (\text{const})\left(\frac{k}{\mu m_h}\right)\rho^{\gamma'} = K\rho^{(n+1)/n} \tag{2.4.9}$$

Thus, we can relate the specific heat C associated with polytropic change to the polytropic index n to be found in the polytropic equation of state [equation (2.4.1)]. So

$$n = \frac{1}{\gamma' - 1} \tag{2.4.10}$$

If $C = 0$, then the general relation describes where the change in the internal energy is equal to the work *on* the gas [see equation (2.4.2)], which means the gas behaves adiabatically. If $C = \infty$, then the gas is isothermal.

The polytropic equation of state provides us with a highly specific relationship between P and ρ. However, hydrostatic equilibrium also provides us with a specific relationship between P and ρ, and we may use the two to eliminate the pressure P, thereby obtaining an equation in ρ alone which describes the run of density throughout the configuration. Differentiating equation (2.1.6) with respect to r and eliminating P by means of the polytropic equation of state, we get

$$\frac{d}{dr}\left[\frac{Kr^2(n+1)}{n\rho^{(n-1)/n}}\frac{d\rho}{dr}\right] = -4\pi r^2 G\rho \qquad (2.4.11)$$

This nonlinear second-order differential equation for the density distribution is subject to the boundary conditions $\rho(0) = \rho_c$ and $\rho(R) = 0$. Or to put it another way, the radius of the configuration is defined to be that value of r for which $\rho = 0$. The only free parameters in the equation are the polytropic index n and the parameter K, and any solution to such an equation is called a *polytrope*. The parameter K is related to the total mass of the configuration. In addition, the equation is generally known as the *Lane-Emden equation*. However, in this case, we have written it in physical variables.

During the nineteenth and early twentieth centuries, considerable effort was expended in the solution of this equation for various values of the polytropic index n. If one is going to investigate the general solution set of any equation, it is usually a good idea to express the equation in dimensionless form. This can be done to equation (2.4.11) by transformation to the so-called Emden variables given by

$$\rho = \lambda\theta^n \qquad r = \alpha\xi \qquad (2.4.12)$$

where

$$\alpha = \left[\frac{(n+1)K\lambda^{1/n-1}}{4\pi G}\right]^{1/2} \qquad (2.4.13)$$

Here λ is just a scaling parameter useful for keeping track of the units of ρ, and it plays no role in the resulting equation. It is clear that ξ is just a scaled, dimensionless radius while θ's meaning is more obscure. While θ is dimensionless by virtue of using λ to absorb the units of ρ, θ does vary as $\rho^{1/n}$ and is the normalized ratio of P/ρ. If we make the substitutions indicated by equation (2.4.12), we obtain the more familiar form of the Lane-Emden equation

$$\frac{1}{\xi^2}\frac{d}{d\xi}\left(\xi^2\frac{d\theta}{d\xi}\right) = -\theta^n \qquad (2.4.14)$$

By picking K and n we can transform any solutions of (2.4.14) and obtain the solution for the polytrope of a given mass M and index n in terms of the run of physical density with position. The nonlinear nature of the transformation has the advantage that the boundary conditions of the physical equation can easily be written as initial conditions at $\xi = 0$. The utility of λ now becomes clear because we can scale $\theta(0)$ to be 1, so that

$$\theta(0) = 1 \qquad \lambda = \rho_c \qquad \left.\frac{d\theta}{d\xi}\right|_{\xi=0} = 0 \qquad (2.4.15)$$

The last initial condition comes from hydrostatic equilibrium. As $r \to 0$, $M(r) \to 0$ as r^3 and $\rho \to \rho_c$. Thus it is clear from equation (2.1.6) that $dP/dr \to 0$ as well. This implies that $d\theta/d\xi \to 0$ as $\xi \to 0$.

In principle, we are now prepared to solve the Lane-Emden equation for any polytropic index n. Unfortunately, only three analytic solutions exist, and they are for $n = 0$, 1, and 5. None of these correspond to particularly interesting physical situations, but in the hopes of learning something about the general behavior of polytropic solutions we give them:

$$\theta_0(\xi) = 1 - \frac{\xi^2}{6} \qquad\qquad n = 0$$

$$\theta_1(\xi) = \frac{\sin \xi}{\xi} \qquad\qquad n = 1 \qquad (2.4.16)$$

$$\theta_5(\xi) = \left(1 + \frac{\xi^2}{3}\right)^{-1/2} \qquad\qquad n = 5$$

For $n = 0$, we see that the solution is monotonically decreasing toward the surface, which is physically reasonable. This is also true for $n = 1$ and $n = 5$, although the rate of decline is slower. Indeed, for the $n = 5$ case only, θ asymptotically approaches zero for arbitrarily large ξ. If we denote the value of ξ for which θ goes to zero as ξ_1, then

$$\xi_1 = \begin{cases} \sqrt{6} & n = 0 \\ \pi & n = 1 \\ \infty & n = 5 \end{cases} \qquad (2.4.17)$$

The value of r which corresponds to ξ_1 is clearly the radius R of the configuration. For other values of the polytropic index n it is possible to develop a series solution which is useful for starting many numerical methods for the solution. The first few terms in the solution are

$$\theta_n = 1 - \frac{1}{6}\,\xi^2 + \frac{n}{120}\,\xi^4 - \left(\frac{8n^2 - 5n}{15,120}\right)\xi^6 + \cdots + \qquad (2.4.18)$$

b Mass-Radius Relationship for Polytropes

For these solutions to be of any use to us, we must be able to relate them to a configuration having a specific mass and radius. We have already indicated how the radius is related to ξ_1 and α, which really means that the mass is related to n and K. Now let us turn to the relationship between the mass of the configuration and the parameters of the polytrope. By using the definition of $M(r)$ and the Lane-Emden equation, (2.4.14), to eliminate θ_n, we can write

$$M(\xi_1) = \int_0^{R/\alpha} 4\pi r^2 \rho \, dr = 4\pi\alpha^3\lambda \int_0^{\xi_1} \xi^2 \theta^n \, d\xi = -4\pi\alpha^3\rho_c\left(\xi_1^2 \frac{d\theta}{d\xi}\bigg|_{\xi_1}\right) \qquad (2.4.19)$$

for the total mass. Using $R = \alpha\xi_1$ to eliminate α from the expression for M, we can obtain a mass-radius relation for any polytrope:

$$GM^{(n-1)/n}R^{(3-n)/n} = -K(n+1)(4\pi)^{-1/n}\left(\xi_1^{(1+n)/(n-1)} \frac{d\theta}{d\xi}\bigg|_{\xi_1}\right) \qquad (2.4.20)$$

For a given configuration, equation (2.4.20) can be used to determine K since everything else on the right-hand side depends on only the polytropic index n. Thus, for a collection of polytropic model stars we can write the mass-radius relation as

$$M^{(n-1)/n}R^{(3-n)/n} = (\text{const})(n) \qquad (2.4.21)$$

c Homology Invariants

We can apply what we have learned about homology transformations to polytropes. In general, if $\theta_n(\xi)$ is a solution of the Lane-Emden equation, then $A^{2/(n-1)}\theta_n(A\xi)$ is also a solution (for a proof see Chandrasekhar[1], p. 101). Here A is an arbitrary constant, so $A\xi$ is clearly a homology transformation of ξ. This produces an entire family of solutions to the Lane-Emden equation, and it would be useful if we could obtain a set of solutions which contained all the homology solutions. To do this, we must find a set of variables which are invariant to homology transformations. Chandrasekhar[1] (p. 105) suggests the following variables

$$u \equiv \frac{d\ln[M(r)]}{d\ln r} = \frac{3\rho(r)}{\langle\rho(r)\rangle} = \frac{-\xi\theta^n}{d\theta/d\xi}$$

$$(n+1)v \equiv \frac{d\ln[P(r)]}{d\ln r} = -\frac{3}{2}\frac{GM(r)/r}{\frac{3}{2}kT/\mu m_h} = -(n+1)\xi\frac{d\theta/d\xi}{\theta}$$

$$(2.4.22)$$

as representing a suitable set of variables which are invariant to homology transformations. The physical interpretation of u is that it is 3 times the ratio of the local density to the local mean density, while $(n+1)v$ is simply 1.5 times the ratio of the local gravitational energy to the local internal energy. In general, these quantities will remain invariant to any change in the structure which can be described by a homology transformation. We can use these variables to rewrite the Lane-Emden equation so as to obtain all solutions which are homologous to each other:

$$\frac{u}{v}\frac{dv}{du} = -\frac{u+v-1}{u+nv-3}$$

$$(2.4.23)$$

Not all solutions to this equation are physically reasonable. For instance, at $\xi = 0$ we must require that $\theta(\xi)$ remain finite. One can show, by substituting into hydrostatic equilibrium as expressed in Emden variables, that $d\theta/d\xi = 0$ at $\xi = 0$. This requires that at the center of the polytrope the values $[u = 3, v = 0]$ set the initial conditions for the unique solution meeting the minimal requirements for being a physical solution. These solutions are known as the *E solutions*, and we have already given a series expansion for the θ_E solution in equation (2.4.18). By substituting this series into the equations for u and v and expanding by the binomial theorem, we obtain the following series solutions for u and v:

$$v = \frac{\xi^2}{3}\left(1 - \frac{3n-5}{30}\xi^2 + \frac{12n^2-39n+35}{1260}\xi^4 + \cdots +\right)$$

$$u = 3 - \frac{n\xi^2}{5} + \frac{(19n^2-25n)\xi^4}{1050}$$

$$(2.4.24)$$

$$-\frac{(472n^3-1275n^2+875n)\xi^6}{283{,}500} + \cdots +$$

As with the θ_E series, we may find the initial values for the numerical solution of the Lane-Emden equation and obtain the solution for a polytrope of any index which also satisfies hydrostatic equilibrium at its center. At the other end of the physical solution space, as $\xi \to \xi_1$, $\theta \to 0$, but the derivative of θ will remain finite. Thus as $u \to 0$, $v \to \infty$. The part of solution space which will be of physical

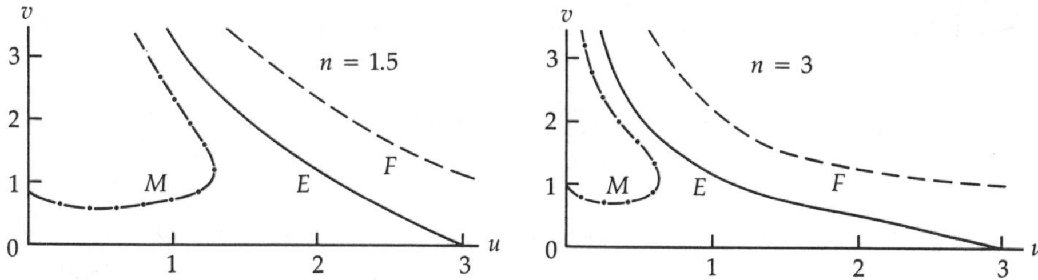

Figure 2.1 Solution for two common polytropes with physical interperations. The solid lines represent the *E* solutions which satisfy hydrostatic equilibrium at the origin. The dashed and dotted lines depict samples of the *F* and *M* solutions, respectively. The polytrope with $n = 1.5$ represents the solution for a star in convective equilibrium, while the $n = 3$ polytrope solution is what is expected for a star dominated by radiation pressure.

interest will then be limited to $u \geq 0$, $v \geq 0$. Figure 2.1 shows the solution set for two polytropic examples including the *E* solutions.

d   Isothermal Sphere

So far we have said nothing about what happens when the equation of state is essentially the ideal-gas law, but for various reasons the temperature remains constant throughout the configuration. Such situations can arise. For example, if the thermal conductivity is very high, the energy will be carried away rapidly from any point where an excess should develop. Such a configuration is known as an *isothermal sphere*, and it has a characteristic structure all its own. We already pointed out that an isothermal gas may be characterized by a polytropic $C = \infty$. A brief perusal of equations (2.4.8) and (2.4.10) will show that this leads to a polytropic index of $n = \infty$ and some problems with the Emden variables. Certainly the Lane-Emden equation in physical variables [equation (2.4.11)] is still valid since it involves only the hydrostatic equilibrium and the polytropic equation of state. However, we must investigate its value in the limit as $n \rightarrow \infty$. Happily, the equation is well behaved in that limit, and we get

$$\frac{d}{dr}\left(\frac{r^2}{\rho}\frac{d\rho}{dr}\right) = -\frac{4\pi r^2 G\rho}{K} \tag{2.4.25}$$

However, some care must be exercised in transforming to the dimensionless Emden variables since the earlier transformation will no longer work. The traditional transformation is

$$r = \alpha\xi \qquad \rho = \lambda e^{-\psi} \qquad \alpha = \left(\frac{K}{4\pi G\lambda}\right)^{1/2} \tag{2.4.26}$$

which leads to the Lane-Emden equation for the isothermal sphere

$$\frac{1}{\xi^2}\frac{d}{d\xi}\left(\xi^2\frac{d\psi}{d\xi}\right) = e^{-\psi} \tag{2.4.27}$$

The initial conditions for the corresponding E solution are $\psi(0) = 0$ and $d\psi/d\xi = 0$ at $\xi = 0$. All the homology theorems hold, and the homology invariant variables u and v have the same physical interpretation and initial values. In terms of these new Emden variables, they are

$$u = \frac{\xi e^{-\psi}}{d\psi/d\xi} \qquad v = \xi\frac{d\psi}{d\xi} \tag{2.4.28}$$

and the Lane-Emden equation in u and v is only slightly modified to account for the isothermal condition:

$$\frac{u}{v}\frac{dv}{du} = -\frac{u-1}{u+v-3} \tag{2.4.29}$$

The solution to this equation in the u-v plane is unique and is shown in Figure 2.2. In the vicinity of $\xi = 0$, ψ can be expressed as

$$\psi = \frac{\xi^2}{6} - \frac{\xi^4}{120} + \frac{\xi^6}{1890} - \frac{61\xi^8}{1{,}632{,}960} + \cdots + \tag{2.4.30}$$

which leads to the following expansions for the homology invariants u and v as

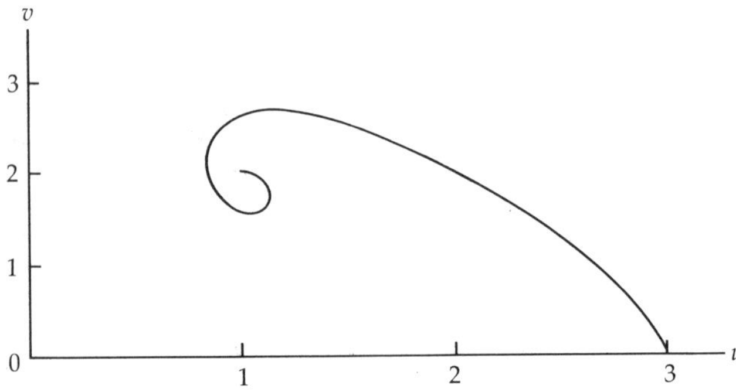

Figure 2.2 Solution for isothermal sphere in the uv plane. The solution is unique.

given by equation (2.4.28):

$$u = 3 - \frac{\xi^2}{5} + \frac{19\xi^4}{1050} - \frac{118\xi^6}{70,875} + \cdots +$$

$$v = \left(\frac{\xi^2}{3}\right)\left(1 - \frac{\xi^2}{10} + \frac{\xi^4}{105} - \cdots + \right) \tag{2.4.31}$$

The physical importance of the isothermal sphere is widespread, having applications from stellar cores to galaxy structure. So it is worthwhile to emphasize one curious aspect of the structure. Direct substitution of a density dependence with the form $\rho(r) \sim r^{-2}$ into equation (2.4.25) shows that such a density law will satisfy hydrostatic equilibrium at all points within an isothermal sphere. Thus, $\rho(r) \sim r^{-2}$ is often used to describe the radial density variation in spherically symmetric regions which are assumed to be isothermal.

e Fitting Polytropes Together

As we shall see later, many stars, including those on the main sequence, can be reasonably represented by a combination of polytropes where the local value of the polytropic index is chosen to reflect the physical constraints placed on the star by the mode of energy transport or possibly the equation of state. Thus, it is useful to understand what conditions must hold where the polytropes meet. Let us consider a simple star composed of a core and an envelope having different polytropic indices (see Figure 2.3).

Now let q be the fraction of the total mass in the core, n the polytropic index of the core, and m the total mass of the core. Physically, we must require that the pressure and density be continuous across the boundary. This implies that u and v are continuous across the boundary between the two polytropes. Since the initial conditions at the center of the core must be $u = 3$, $v = 0$, the core solution must be an E solution for the core index n. The envelope solution will not, in general, be an E solution; but as long as the central point ($u = 3$, $v = 0$) is not encountered, there is no violation of hydrostatic equilibrium by such a solution. Thus one can construct a reasonable model by proceeding outward along the core solution until the mass of the core is reached. This defines the fitting point in the u-v plane. One then searches for the F or M solutions which meet the core solution at the fitting point, to ensure continuity of P and ρ across the boundary. There will be many solutions corresponding to different values of the polytropic index of the envelope. Picking one such solution, one continues with this solution until ξ_1 is reached, at which point $M(\xi_1)$ should equal m/q. If it does not, then there is no

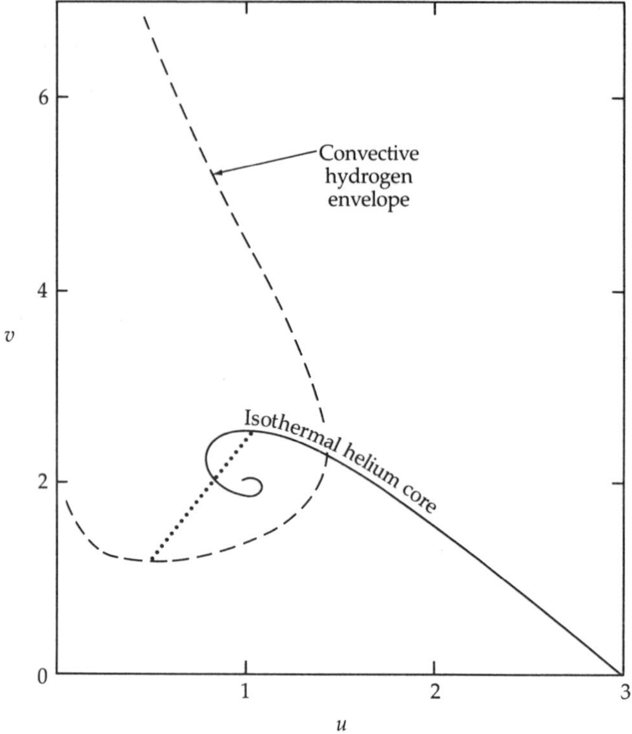

Figure 2.3 A model star composed of two polytropes. The outer convective hydrogen envelope can be represented by a polytrope of index $n = 1.5$, while the helium core is isothermal. The discontinuous change in u and v resulting from the change in chemical composition can be seen as a jump from the isothermal core solution toward the origin and the appropriate M solution for the envelope. Such a model can be expected to qualitatively represent the evolved phase of a red giant.

solution for that value of the polytropic index of the envelope and another solution at the fitting point should be chosen.

The techniques of J. L. Lagrange known as *variation of parameters* can be utilized to convert an error on the mass at ξ_1, $\delta M(\xi_1)$, to a correction in the polytropic index δn_e of the envelope solution. Any solution which satisfies the continuity conditions and the constraints set by the core mass and mass fraction is unique. In addition, it is possible to allow for a discontinuity in the chemical composition at the boundary by permitting a discontinuity in the density such that momentum conservation is maintained across the boundary. That is, pressure equilibrium must be maintained across the boundary. From the ideal-gas law, the ratio of the density in the envelope to that of the core is

$$\frac{\rho_e}{\rho_c} = \frac{\mu_e}{\mu_c} \tag{2.4.32}$$

This is equivalent to specifying a jump in u and $(n + 1)v$ by the ratio of the mean molecular weights of the core and envelope. Thus the fitting point, when it is reached, is displaced toward the origin in u and $(n + 1)v$ by the ratio of the mean molecular weights of the envelope and core. This displaced point in the u-v plane is the new point from which the solution is to be continued (see Figure 2.3). The solution is then completed as in the previous instance.

By making use of polytropic solutions, it is possible to represent stars with convective cores and radiative envelopes with some accuracy and to get a rough idea of the run of pressure, density, and temperature throughout the star. Polytropes are useful in determining the effects of the buildup of chemical discontinuities as a result of nuclear burning. As mentioned earlier, very massive stars are radiation-dominated and are quite accurately represented by polytropes of index $n = 3$ ($\gamma' = \frac{4}{3}$). Polytropes often can be used as an initial model which is then perturbed to approximate a given physical situation. For relatively little effort, polytropic models can provide substantial insight into the behavior of stars in response to various changes in physical conditions. We obtain this insight at a relatively low cost. To do significantly better, we must do much more. We will have to know, in some detail, how energy is transported throughout the star. But before we can do that, we must understand the detailed structure of the gas so that we can understand the properties which impede that flow of energy.

Problems

1. Use the integral theorems of Chandrasekhar to place limits on the central temperature of a star of given mass M.

2. Estimate the mass of a white dwarf at which the relativistic degenerate equation of state becomes essential for representing its structure.

3. Prove that all solutions to the Lane-Emden equation which remain finite at the origin ($\xi = 0$) must, of necessity, have

$$\left. \frac{d\theta}{d\xi} \right|_{\xi = 0} = 0$$

4. Show that the mass interior to ξ [that is, $M(\xi)$] in an isothermal sphere is given by

$$M(\xi) = 4\pi\alpha^3\lambda\xi^2 \frac{d\psi}{d\xi}$$

5. Find a series solution for the Lane-Emden equation in the vicinity of $\xi = 0$, subject to the boundary condition that $(d\theta/d\xi)|_{\xi = 0}$ be zero. This solution should have an accuracy of $O(\xi^{12})$.

6. Find a series solution for the isothermal sphere subject to the same conditions given in Problem 5.

7. Use the series solutions from Problems 5 and 6 to obtain corresponding series solutions for the homology invariants u and v.

8. Calculate a value for the free-fall time for an object on the surface of the sun to arrive at the center of the sun.

9. Show that the results of equation (2.3.8) are indeed correct. State clearly all assumptions you make during your derivation.

Reference and Supplemental Reading

1. Chandrasekhar, S. *An Introduction to the Study of Stellar Structure*, Dover, New York, 1939/1957.

For those who are interested in a further discussion of the integral theorems, some excellent articles are:

• Chandrasekhar, S.: *"An Integral Theorem on the Equilibrium of a Star,"* Ap. J. 87, 1938, pp. 535−552.

• Chandrasekhar, S.: *"The Opacity in the Interior of a Star,"* Ap. J. 86, 1937, pp. 78−83.

• Chandrasekhar, S.: *"The Pressure in the Interior of a Star,"* Ap. J. 85, 1937, pp. 372−379.

• Chandrasekhar, S.: *"The Pressure in the Interior of a Star,"* Mon. Not. R. astr. Soc. 96, 1936, pp. 644−647

• Milne, E. A.: *"The Pressure in the Interior of a Star,"* Mon. Not. R. astr. Soc. 96, 1936, pp. 179−184

For a complete discussion of polytropes and isothermal spheres, see any of these:

• Chandrasekhar, S.: *An Introduction to the Study of Stellar Structure*, Dover, New York, 1939/1957, chap. 4, p. 84.

• Eddington, A. S.: *The Internal Constitution of the Stars*, Dover, New York, 1926/1959, chap. 4, p. 79.

• Cox, J. P., and R. T. Giuli: *Principles of Stellar Structure*, Gordon and Breach, New York, 1968, chap. 12, p. 257.

An interesting example of the use of polytropes to explore the more complicated phenomenon of rotation can be found in

• Limber, D. N., and P. H. Roberts: "On Highly Rotating Polytropes V," *Ap. J.* 141, 1965, pp. 1439–1442.

• Geroyannis, V. S., and F. N. Valvi: "Numerical Implementation of a Perturbation Theory up to Third Order for Rotating Polytropic Stars: Parameters under Differential Rotation," *Ap. J.* 312, 1987, pp. 219–226.

A brief but useful account of the physical nature of polytropes may be found in

• Clayton, D. D.: *Principles of Stellar Evolution and Nucleosynthesis*, McGraw-Hill, New York, 1968, pp. 155–158.

3

Sources and Sinks of Energy

· · ·

We have come to that place in the study of stellar structure where we must be mindful of the flow of energy through the star. After all, stars do shine. So far, we have been able to learn much about the equilibrium structure of a star without considering that it is really a structure in a steady state, rather than one in perfect strict equilibrium. The basic reason that we have been able to ignore the flow of energy through the star is that, during a dynamical time, a very small fraction of the stored energy in the star escapes from the star. Although a star is not, strictly speaking, an equilibrium structure, it comes closer to being one than most any other object in the universe.

However, before delving into the actual movement of energy within the star, we must first identify the sources of that energy as well as the processes which impede its flow. This will also give us the chance to discuss the stores of energy within the star since these certainly represent a potential supply of flowing energy with which to generate the stellar luminosity.

3.1 "Energies" of Stars

One of the great mysteries of the late nineteenth and early twentieth centuries was the source of the energy required to sustain the luminosity of the sun. By then, the defining solar parameters of mass, radius, and luminosity were known with sufficient precision to attempt to relate them. For instance, it was clear that if the sun derived its energy from chemical processes typically yielding less that 10^{12} erg/g, it could shine no longer than about 10,000 years at its current luminosity. It is said that Lord Kelvin, in noting that the liberation of gravitational energy could only keep the sun shining for about 10 million years, found it necessary to reject Charles Darwin's theory of evolution because there would have been insufficient time for natural selection to provide the observed diversity of species.

a   Gravitational Energy

It is generally conceded that the sun has shone at roughly its present luminosity for at least the past 2 billion years and has been in existence for nearly 5 billion years. With this in mind, let us begin our study of the sources of stellar energy with an inventory of the stores of energy available to the sun. Perhaps the most obvious source of energy is that suggested by Lord Kelvin, namely, gravitation. From the integral theorems of Chapter 2, we may place a limit on the gravitational energy of the sun by remembering that $I_{1,1}(R)$ is related to the total gravitational potential energy. Thus, from equations (2.2.2) and (2.2.5)

$$\Omega \le -\frac{3}{5}\frac{GM^2}{R} \qquad (3.1.1)$$

The right-hand side of the inequality is the gravitational potential energy for a uniform density sphere, which provides a sensible upper limit for the energy. Remember that the gravitational energy is considered negative by convention, so a rather larger magnitude of energy may be available for a star that is more centrally concentrated than a uniform-density sphere. We may acquire a better estimate of the gravitational potential energy by using the results for a polytrope. Chandrasekhar[1] obtains the following result, due to Betti and Ritter, for the gravitational potential energy of a polytrope:

$$\Omega = -\frac{3}{5-n}\frac{GM^2}{R} \tag{3.1.2}$$

For a star in convective equilibrium (that is, $n = \frac{3}{2}$) the factor multiplying GM^2/R becomes $\frac{6}{7}$, or nearly unity. Note that for a polytrope of index 5, $\Omega \to -\infty$, implying an infinite central concentration of material. This is also one of the polytropes for which there exists an analytic solution and $\xi_1 = \infty$. Thus, one has the picture of a mass point surrounded by a massless envelope of infinite extent. Equation (3.1.2) also tells us that as the polytropic index increases, so does the central concentration.

It is not at all obvious that the total gravitational energy would be available to permit the star to shine. Some energy must be provided in the form of heat, to provide the pressure which supports the star. We may use the virial theorem [equation (1.2.35)] to estimate how much of the gravitational energy can be utilized by the luminosity. Consider a star with no mass motions, so that the macroscopic kinetic energy $\mathbf{T}$ in equation (1.2.35) is zero. Let us also assume that the equilibrium state is good enough that we can replace the time averages by the instantaneous values. Then the virial theorem becomes

$$2\mathbf{U} + \Omega = 0 \tag{3.1.3}$$

Remember that $\mathbf{U}$ is the total internal kinetic energy of the gas, which includes all motions of the particles making up the gas. Now we know from thermodynamics that not all the internal *kinetic* energy is available to do work, and it is therefore not counted in the internal energy of the gas. The internal kinetic energy density of a differential mass element of the gas is

$$d\mathbf{U} = \tfrac{3}{2}RT\,dm = \tfrac{3}{2}(C_P - C_V)T\,dm \tag{3.1.4}$$

where the relationship of the gas constant R to the specific heats was given in Chapter 2 [equation (2.4.5)]. However, from the definition of specific heats [equation (2.4.4)], the internal heat energy of a differential mass element is

$$dU = C_V T\,dm \tag{3.1.5}$$

Eliminating $T\,dm$ from equations (3.1.4) and (3.1.5) and integrating the energy densities of the entire star, we get

$$\mathbf{U} = \tfrac{3}{2}\langle\gamma - 1\rangle U \tag{3.1.6}$$

where U is the total internal heat energy or just the total internal energy. The quantity $\langle \gamma - 1 \rangle$ is the value of $\gamma - 1$ averaged over the star.

For simplicity, let us assume that γ is constant throughout the star. Then the virial theorem becomes

$$3(\gamma - 1)U + \Omega = 0 \qquad (3.1.7)$$

Remembering that the total energy $\mathbf{E}$ is the sum of the internal energy and the gravitational energy, we can express the virial theorem in the following ways:

$$U = \frac{-\Omega}{3(\gamma - 1)}$$

$$\mathbf{E} = -(3\gamma - 4)U \qquad (3.1.8)$$

$$\mathbf{E} = \frac{3\gamma - 4}{3(\gamma - 1)}\Omega$$

It is clear that for $\gamma > \frac{4}{3}$ (that is, $n < 3$) the total energy of the star will be negative. This simply says that the star is gravitationally bound and can be in equilibrium. So we can look for the physically reasonable polytropes to have indices less than or equal to 3. The case of $n = 3$ is an interesting one that we shall return to later, for it represents radiation-dominated gas. In the limit of complete radiation dominance, the total energy of the configuration will be zero.

b Rotational Energy

While utilizing the virial theorem to estimate the gravitational energy, we set the mass motions of the star to zero so that the macroscopic kinetic energy $\mathbf{T}$ was zero. However, stars do rotate, and we should not forget to count the rotational energy in the inventory of energies. We may place a reasonable upper limit on the magnitude of the rotational energy that we can expect by noting that (1) the moment of inertia of the star will always be less than that of a sphere of uniform density and (2) there is a limit to the angular velocity ω_c at which the star can rotate. Thus, for a centrally condensed star

$$\omega^2 \leq \frac{8GM}{27R_p^3} \qquad I_z < \tfrac{2}{5}MR^2 \qquad (3.1.9)$$

which implies that the rotational energy must be bounded by

$$E_{\mathrm{rot}} = \tfrac{1}{2}I_z\omega^2 < \frac{8}{135}\frac{GM^2}{R} \qquad (3.1.10)$$

Table 3.1 Mass Defect for Common Nuclear Fuels

Element	Atomic Weight (relative to ^{12}C)	Mass of H Used	Mass Loss by Fusion, %
H	1.00797	1.00797	—
^{4}He	4.0026	4.03188	0.726
^{12}C	12.000	12.0956	0.791
Fe	55.847	56.847	1.062

One may quibble that we have used the angular velocity limit for a centrally condensed star and the moment of inertia for a uniform-density star, but the fact remains that it is extremely difficult for a star to have more than about 10 percent of the magnitude of its gravitational energy stored in the form of rotational energy.

c Nuclear Energy

Of course, the ultimate upper limit for stored energy is the energy associated with the rest mass itself. It is also the common way of estimating the energy available from nuclear sources. Indeed, that fraction of the rest mass which becomes energy when four hydrogen atoms are converted to one helium atom provides the energy to sustain the solar luminosity. Table 3.1 is a short table giving the mass loss for a few common elements involved in nuclear fusion processes.

Clearly most of the energy to be gained from nuclear fusion occurs by the conversion of hydrogen to helium, and less than one-half of that energy can be obtained by all other fusion processes that carry helium to iron. Nevertheless, 0.7 percent of Mc^2 is a formidable supply of energy. Table 3.2 is a summary of the energy that one could consider as being available to the sun. All these entries are generous upper limits. For example, the sun rotates at less than 0.5 percent of its critical velocity, it was never composed of 100 percent hydrogen and will begin to change significantly when a fraction of the core hydrogen is consumed, and not all the gravitational energy could ever be converted to energy for release. In any

Table 3.2 Possible Sources of Solar Energy

Form of Energy	Amount, erg	$\%M_\odot$	Lifetime for constant $L_\odot$, years
Chemical (10^{12} erg/g)	1.98×10^{45}	10^{-7}	14,000
Rotational ($8\,\Omega/81$)	2.21×10^{47}	10^{-5}	1.4 million
Gravitational [$3GM^2/(5R)$]	2.24×10^{48}	10^{-4}	14 million
Nuclear ($0.0106M_\odot c^2$)	1.89×10^{52}	1	140 billion
Rest mass energy ($M_\odot c^2$)	1.78×10^{54}	100	1.4×10^{13}

event, only nuclear processes hold the promise of providing the solar luminosity for the time required to bring about agreement with the age of the solar system as derived from rocks and meteorites. However, the time scales of Table 3.2 are interesting because they provide an estimate of how long the various energy sources could be expected to maintain some sort of equilibrium configuration.

3.2 Time Scales

One of the most useful notions in stellar astrophysics for establishing an intuitive feel for the significance of various physical processes is the time required for those processes to make a significant change in the structure of the star. To enable us to estimate the relative importance of these processes, we shall estimate the time scales for several of them. In Chapter 2 we used the free-fall time of the sun to establish the fact that the sun can be considered to be in hydrostatic equilibrium. The statement was made that this time scale was essentially the same as the dynamical time scale. So let us now turn to estimating the time required for dynamical forces to change a star.

a   Dynamical Time Scale

The virial theorem of Chapter 1 [equation (1.2.34)] provides us with a ready way of estimating the dynamical time scale, for in the form given it must hold for all $1/r^2$ forces. Consider a star which is not in equilibrium because the internal energy is too low. As it enters the nonequilibrium condition, the star's kinetic energy will also be small. Thus, the virial theorem would require

$$\frac{d^2 I}{dt^2} \approx -\Omega \tag{3.2.1}$$

implying a rapid collapse. If we take as an average value for the accelerative change in the moment of inertia

$$\frac{d^2 I}{dt^2} \approx \frac{I}{\tau_d^2} \tag{3.2.2}$$

where τ_d is the dynamical time by definition, then we get

$$\tau_d^2 = \frac{-I}{\Omega} = \frac{\frac{2}{5}MR^2}{\frac{3}{5}GM^2/R} \tag{3.2.3}$$

or

$$\tau_d = \left(\frac{\frac{2}{3}R^3}{GM}\right)^{1/2} \tag{3.2.4}$$

Now we compare this to the free-fall time obtained by direct integration of

$$\frac{d^2r}{dt^2} = -\frac{GM(r)}{r^2} \tag{3.2.5}$$

remembering that since the star is "free-falling," $M(r)$ will always be the mass interior to r. Thus, a surface point will always be affected by the total mass M. With some attention to the boundary conditions [see equations (5.2.12) through (5.2.17)], direct integration yields a free-fall time of

$$\tau_f = \frac{\pi}{2}\left(\frac{R^3}{2GM}\right)^{1/2} = \left(\frac{3\pi}{32G\langle \rho_i\rangle}\right)^{1/2} \tag{3.2.6}$$

which is essentially the same (within about a factor of 1.4) as the dynamical time.

Although we considered a star having zero pressure in order to derive both those time scales, the situation would not be significantly different if some pressure did exist. While a collapse will cause an increase in the pressure, the virial theorem assures us that the gravitational energy will always exceed the internal energy of the gas unless there is a change in the equation of state resulting in a sudden increase in internal energy. However, for the interior of the star to adjust to the collapse, it is necessary for information regarding the collapse to be communicated throughout the star. This will be accomplished by pressure waves which travel at the speed of sound. The sound crossing time is

$$\tau_s = \frac{R}{\langle c_s\rangle} = R\left\langle\frac{\gamma P}{\rho}\right\rangle^{-1/2} = R\left\langle\left(\frac{\gamma kT}{\mu m_h}\right)^{-1/2}\right\rangle \tag{3.2.7}$$

For a monatomic gas $\gamma = \frac{5}{3}$. Hence

$$\langle c_s\rangle = \left(\frac{5k}{3\mu m_h}\right)^{1/2}\langle T^{1/2}\rangle \tag{3.2.8}$$

We may estimate the mean temperature for a uniform-density sphere from the integral theorems [equations (2.2.4) and (2.2.7)] and obtain

$$\tau_s = \left(\frac{3R^3}{GM}\right)^{1/2} \tag{3.2.9}$$

Although the sound crossing time is somewhat larger than the free-fall and dynamical time scales, they are all of the same order of magnitude, $\sqrt{R^3/(GM)}$. This is about 27 min for the sun. The similar magnitude for these times is to be expected since they have a common origin in dynamical phenomena. So we have finally justified our statement in Chapter 2 that any departure from hydrostatic equilibrium will be resolved in about 20 min. This short time scale is characteristic of the dynamical time scale; it is generally the shortest of all the time scales of importance in stars.

b Kelvin-Helmholtz (Thermal) Time Scale

Now we turn to some of considerations that led Lord Kelvin to reject the darwinian theory of evolution. These involve the gravitational heating of the sun. If you imagine the early phases of a star's existence, when the internal temperature is insufficient to ignite nuclear fusion, then you will have a physical picture of a cloud of gas which is slowly contracting and is thereby being heated. Ultimately some of the energy generated by this contraction will be released from the stellar surface in the form of photons. As long as the process is slow compared to the dynamical time scale for the object, the virial theorem in the form of equation (1.2.35) will hold and $\langle \mathbf{T} \rangle \approx 0$. Thus

$$\tfrac{1}{2}\langle \Omega \rangle = -\langle \mathbf{U} \rangle \tag{3.2.10}$$

which implies that one-half of the change in the gravitational energy will go into raising the internal kinetic energy of the gas. The other half is available to be radiated away. This was the mechanism that Lord Kelvin proposed was responsible for providing the solar luminosity, and he suggested a lifetime for such a mechanism to be simply the time required for the luminosity to result in a loss of energy equal to the present gravitational energy. If we estimate the latter by assuming that the star of interest is of uniform density, then

$$\tau_{\text{K-H}} = -\frac{\Omega}{L} = \frac{\frac{3}{5}GM^2}{RL} \tag{3.2.11}$$

This is known as the *Kelvin-Helmholtz gravitational contraction time*, and it is the same as the lifetime obtained from the gravitational energy given in the previous section. Since the star is simply cooling off and having its internal energy resupplied by gravitational contraction, some authors refer to this time scale as the thermal time scale. More properly, one could define the thermal time scale τ_{th} as the time required for the luminosity to result in an energy loss equal to the internal heat energy, and then one could relate that to the Kelvin-Helmholtz time by means of the virial theorem. That is,

$$\tau_{\text{th}} = \frac{\langle U \rangle}{L} = -\frac{\langle \Omega \rangle}{\langle 3(\gamma - 1) \rangle L} \approx \frac{\tau_{\text{K-H}}}{\langle 3(\gamma - 1) \rangle} \qquad (3.2.12)$$

Thus, we see that the two time scales are of the same order of magnitude, differing only by a factor of 2 for a monatomic gas. For the sun, both time scales are of the order of 10^{11} times longer than the dynamical time. In general, the thermal time scale is very much longer than the dynamical time scale. The thermal time scale is the time over which thermal instabilities will be resolved, and so they are always less important than dynamical instabilities.

c Nuclear (Evolutionary) Time Scale

In the beginning of this section we estimated the lifetime of the sun which could result from the dissipation of various sources of stored energy. By far the most successful at providing a long life was nuclear energy. The conversion of hydrogen to iron provided for a lifetime of some 140 billion years. However, in practice, when about 10 percent of the hydrogen is converted to helium in stars like the sun, major structural changes will begin to occur and the star will begin to evolve. We can define a time scale for these events in a manner analogous to other time scales as

$$\tau_n = \frac{K_n M c^2}{L} \qquad (3.2.13)$$

where K_n is just the fraction of the rest mass available to a particular nuclear process. While evolutionary changes often occur in one-tenth of the nuclear time scale, some stars show no significant change in less than $0.99\tau_n$. While in the terminal phases of some stars' lives the nuclear time scale becomes rather shorter than the thermal time scale and conceivably shorter than the dynamical time scale, for the type of stars we will be considering the nuclear time scale is usually very much longer than the other two. Certainly for main sequence stars we may observe that

$$\tau_d \approx \tau_f \approx \tau_s \ll \tau_{\text{KH}} \ll \tau_n \qquad (3.2.14)$$

It is important to understand that the time scales themselves may change with time. The nuclear time scale will depend on the nature of the available nuclear fuel. However, the time scales do indicate the time interval over which you may regard their respective processes as approximately constant. They are useful, for they are easy to estimate, and they indicate which processes within the star will be important in determining its structure at any given time.

3.3 Generation of Nuclear Energy

We have established that the most important source for energy in the sun results from nuclear processes. Therefore, it is time that we look more closely at the details of those processes with a view of quantifying the dependence of the energy generation rate on the local values of the state variables. During the last 50 years, great strides have been made in understanding the details of nuclear interactions. They have revealed themselves to be remarkably varied and complex. We do not attempt to delve into all these details; rather we sketch those processes of primary importance in determining the structure of the star during the majority of its lifetime. We leave it to others to describe the spectacular nuclear pyrotechnics which occur during the terminal phases of the evolution of massive stars. Indeed, the equilibrium processes that occur in the terminal phases of stellar evolution, giving rise to most of the heavier elements, are beyond the scope of this book. Nor do we attempt to develop a complete, detailed quantum theory of nuclear energy production. Those who thirst after that specific knowledge are referred to the excellent survey by Cox and Giuli[2] and other references at the end of this chapter. Instead, we concentrate on the physical principles which govern the production of energy by nuclear fusion.

a General Properties of the Nucleus

The notion that the atom can be viewed as being composed of a nucleus surrounded by a cloud of electrons which are confined to shells led to a very successful theory of atomic spectra. A very similar picture can be postulated for the nucleus itself, namely, that nucleons are arranged in shells within the nucleus and undergo transitions from one excited state (shell) to another subject to the same sort of selection rules that govern atomic transitions. The origin of the shell structure of any nucleus is that nucleons are fermions and therefore must obey the Pauli exclusion principle, just as the atomic electrons do. Thus, only two protons or two neutrons may occupy a specific cell in phase space (protons and neutrons have the same spin as electrons, so each species can have two of its kind in a quantum state characterized by the spatial quantum numbers).

However, the nucleons are much more tightly bound in the nucleus than the electrons in the atom. Whereas the typical ionization energy of an atom can be measured in tens to thousands of electronvolts, the typical binding energy of a nucleon in the nucleus is several million electronvolts. This large binding energy

65

and the Pauli exclusion principle can be used to explain the stability of the neutron in nuclei. Although free neutrons beta-decay to protons (and an electron and an electron antineutrino) with a half-life of about 10 min, neutrons appear to be stable when they are in nuclei. If neutrons did decay, the resulting proton would have to occupy one of the least tightly bound proton shells, which frequently costs more energy than is liberated by the beta decay of the neutron. Thus, unless the neutron decay can provide sufficient energy for the decay products to be ejected from the nucleus, the neutron must remain in the nucleus as a stable entity.

In general, for a nucleus to be stable, its mass must be less than the sum of the masses of any possible combination of its constituents. Thus, Li^5 is not stable, whereas He^4 is. A more detailed explanation of the reasons for the stability or instability of a particular nucleus requires a considerably more detailed discussion of nuclear interactions and nuclear structure than is consistent with the scope of this book. However, note that the instability of mass-5 nuclei posed one of the greatest barriers of the century to the understanding of the evolution of stars. The nuclear evolution beyond mass 5 was finally solved by Fred Hoyle, who showed that the triple-α process, which we consider later, could actually initiate synthesis of all the nuclei heavier than mass 12.

Before we turn to the specifics of nuclear energy production, it is worth saying something about notation. Consider the reaction where a particle a hits a nucleus X, producing a nucleus Y and another particle(s) b. In other words,

$$a + X \rightarrow Y + b \tag{3.3.1}$$

Such a reaction can be written $X(a, b)Y$. Usually for such a reaction to happen, it must be exothermic. That is, the rest energy of the initial constituents of the reaction must exceed that of the products:

$$E_n \equiv (m_a + M_X - m_b - M_Y)c^2 > 0 \tag{3.3.2}$$

b Bohr Picture of Nuclear Reactions

Although quantum mechanics formally describes the transition from the initial to the final state, it is convenient to break down the process and to say that a compound nucleus is formed by the collision and subsequently decays to the reaction products. With this assumption, a reaction can be viewed as consisting of two steps

$$a + X \rightarrow C^* \rightarrow Y + b \tag{3.3.3}$$

where C^* is the compound nucleus and the asterisk indicates that it is in an excited state. The compound nucleus can decay by various modes which have these convenient physical interpretations:

$$a + X \rightarrow C^* \rightarrow X + a \qquad \text{elastic scattering}$$

$$a + X \rightarrow C^* \rightarrow X^* + a \qquad \text{inelastic scattering}$$

$$a + X \rightarrow C^* \rightarrow Y + b \qquad \text{particle emission} \qquad (3.3.4)$$

$$a + X \rightarrow C^* \rightarrow C + \gamma \qquad \text{radiative capture}$$

Elastic scattering simply involves a particle "bouncing off" the nucleus in such a manner that the momentum and kinetic energy of both the constituents are conserved. However, inelastic scattering results in the nucleus being left in an excited state at the expense of the kinetic energy of the reactants. Particle emission is the process most often associated with nuclear reactions. The results of the interaction leave both reactants changed. Under certain conditions, the Bohr picture fails for these interactions since they proceed directly to the final state without the formation of a compound nucleus. In radiative capture, the compound nucleus decays from the excited state to a stable state by the emission of a photon.

The validity of this two-stage process, due to Neils Bohr, depends on the lifetime of the compound nucleus C^*. The duration of a nuclear collision can be characterized by the time it takes for the colliding particle to cross the nucleus. For typical nuclear radii and relative collision speeds of, say $0.1c$, this is about 10^{-21} s. If the lifetime of the compound nucleus is long compared to this crossing time, you may assume that the nucleons of the compound nucleus have undergone many "collisions" and that the interaction energy has been statistically redistributed among them. In short, the compound nucleus will have reached statistical equilibrium and reside in a well-defined state. In some sense, the compound nucleus can be said to exist. This effectively separates the details of the $C^* \rightarrow Y + b$ reaction from those of the $a + X \rightarrow C^*$ reaction. One might say that C^* will have "forgotten" about its birth.

More properly, the statistical equilibrium state of C^* is independent of the approach to that state. This was the case in Chapter 1 where we considered the establishment of statistical equilibrium for a variety of gases. It will also be the case when we consider the details of absorption and reemission of photons by atoms much later. Another way of stating this condition is to say that the average distance between collisions of a with the nucleons (the mean free path) is much less than the size of the nucleus. Experimentally, this appears to be true for collision energies below 50 MeV. Thus, if the energy is shared among more than a half dozen nucleons, any given nucleon will not have sufficient energy to exceed the binding energy and escape. The result is the formation of a stable nucleus by means of radiative capture.

By analogy to the photoexcitation of atoms, called *bound-bound transitions*, there exist resonances for nuclear reactions, particularly at low energy. A resonance is

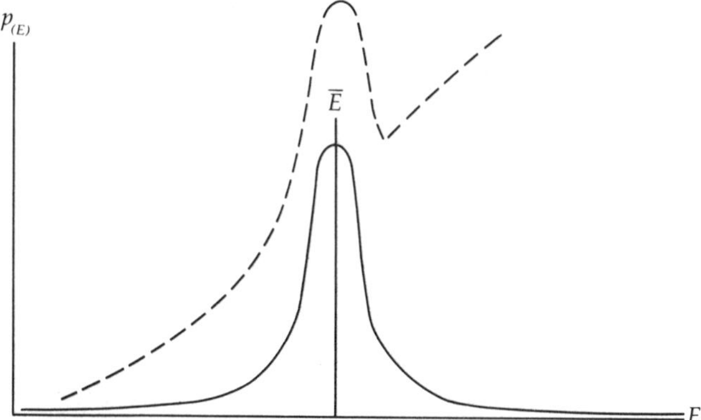

Figure 3.1 A typical damping, or dispersion, profile. A marked increase in the interaction probability occurs in the vicinity of the resonance energy $\bar{E}$. The width of the curve is characterized by the damping constant Γ.

an enhancement in the probability that a nuclear reaction will take place. Classically, one may view these as collision energies which excite particular nucleon shell transitions within the nucleus. These energies will be particularly favored for interactions and are known as the *resonance energies*. The probability density distribution with energy is characterized by a function known as a *damping, or dispersion, profile* whose form we will derive in some detail when we consider the formation of spectral lines in Chapter 13. All that need be understood is the general topological shape (see Figure 3.1) and the fact that the width of the probability maximum can be characterized by a width in energy usually denoted by Γ. As long as the resonance is a simple one and not blended with others, the energy at which the peak of the probability distribution occurs is known as the resonance energy.

c Nuclear Reaction Cross Sections

The words *cross section* have come to have a somewhat generic meaning in nuclear physics as a measure of the likelihood of a particular reaction taking place, in the sense that the larger the cross section, the greater the probability that the reaction will happen. The simplest way to visualize a reaction cross section is to consider the classical notion of a collision cross section. If you were to shoot a bullet through a swarm of hornets, the probability of hitting a particular hornet would be proportional to the cross-sectional area of the hornet as seen by the bullet. Of course, the cross-sectional area of the bullet will also play a role in determining the likelihood of hitting the hornet. The combined effect of these two cross-sectional areas is said to represent the geometric cross section of the collision. In a similar manner, one may interpret a nuclear reaction cross section as the "effective" geometric cross-sectional area of a collision between the particle and the nucleus. Remember that this is not a simple geometric cross section unless you are comfortable with the notion that the nucleus appears to have very different "sizes," as seen by the colliding particle, depending on the particle's energy.

In practice, the nuclear cross section will depend on all the quantities that govern the interaction between the colliding particles and the nucleons in the shell structure of the nucleus. The detailed calculation is usually very complicated, depending on the approximate wave function of the nucleus and the wave function of the colliding particle. A common approximation formula for nuclear cross sections, known as the *Breit-Wigner 1-level dispersion formula,* is

$$\sigma(a, b) = (2l + 1)\pi\lambdabar^2 \omega T_l(a) Y(E) S G(b) \tag{3.3.5}$$

where

$$\lambdabar = \frac{\lambda}{2\pi} = \frac{\hbar}{p} = \frac{\hbar}{\sqrt{2m_a E}}$$

$$L = \hbar[l(l + 1)]^{1/2} = \text{orbital angular momentum of } m_a \text{ about } X$$

$$T_l(a) = \text{transmission function of particle } a$$

$$\omega = \frac{2J + 1}{(2I_a + 1)(2I_x + 1)} \sim 1$$

$$I_a = \text{spin of particle } a$$

$$I_x = \text{nuclear spin of particle } X$$

$$\vec{J} = \vec{I}_a + \vec{I}_x \tag{3.3.6}$$

$$Y(E) \text{ allows for resonances}$$

$$S = 1 \text{ or } 2 \text{ depending on particle degeneracy}$$

$$G(b) \sim \frac{\Gamma(b)}{\Gamma} = \text{branching ratio for } b$$

$$\Gamma = \Gamma_\gamma + \sum_i \Gamma_i$$

$$\Gamma_\gamma = \text{damping constant for radiative capture (i.e., particle } b \text{ is a photon)}$$

$$\Gamma_i = \text{all other possible decay damping constants}$$

We make no attempt to derive this result. However, we do try to show that the result at least contains the right sort of terms and is reasonable. The term $\pi\lambdabar^2$ is essentially the geometric cross section of the colliding particle as it is related to

69

the particle's de Broglie wavelength. The angular momentum term $2l + 1$ is a measure of the impact parameter and the energy. As l increases, so does the impact parameter. For constant angular momentum, an increasing impact parameter will mean a decreasing collision energy, implying a net increase of the collision probability. However, as the impact parameter increases and the collision energy drops, the probability that the colliding particle will be able to overcome the coulomb barrier decreases drastically. Thus, we need be concerned only with $l = 0$ or 1. The term *transmission function* of particle a includes the probability that the particle will penetrate the coulomb barrier of the nucleus. The parameter ω allows for the spin-spin interactions of the nucleus and the particle and is of the order unity. Function $Y(E)$ includes the effects of resonances and from the dispersion curve in Figure 3.1 can clearly be a very strong function of collision energy E. The spin degeneracy parameter S is generally 1 except when a and X are the same kind of particle and also have zero spin; then $S = 2$. Finally, $G(b)$ is a measure of the probability that particle b will be created from the compound nucleus as opposed to some other possibility. Now that we have the nuclear reaction cross sections, we have to determine the rate at which collisions will occur. Then we will be able to find the energy produced by stellar material.

d Nuclear Reaction Rates

The reaction cross section of the previous section can be measured as a function of the collision energy (and some atomic constants) alone and therefore can be written as a function of the particle's velocity v relative to the target. By resurrecting the geometric interpretation of the cross section, the number of particles crossing an area (colliding with the target) per unit time is just $N\sigma(v)v$, where N is the density of colliding particles (see Figure 3.2).

Consider collisions between two different kinds of particles with a number density in phase space of dN_1 and dN_2. To obtain the number of collisions per second per unit volume, we must integrate over all available velocity space. That is, we must sum over the collision *between* particles so that the collision rate r is

$$r = \iint v\sigma(v) \, dN_1(\vec{v}_1) \, dN_2(\vec{v}_2) \tag{3.3.7}$$

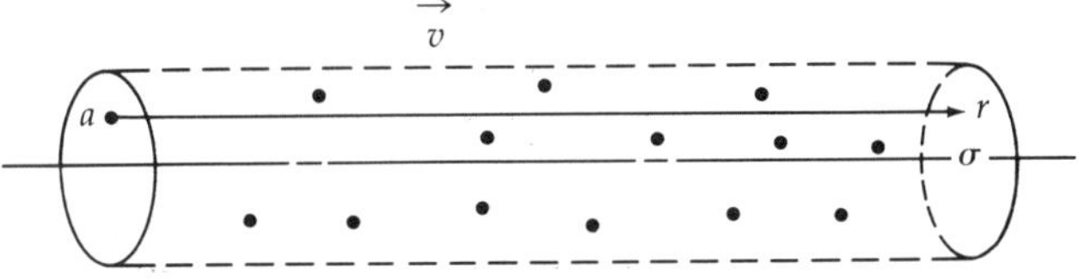

Figure 3.2 Schematic representation of a collision between particle a and a target with a geometric cross section σ.

Let us assume that the velocity distributions of both kinds of particles are given by maxwellian velocity distributions

$$dN = N\left(\frac{m}{2\pi kT}\right)^{3/2} e^{-mv^2/(2kT)}\, dv \tag{3.3.8}$$

so that equation (3.3.7) becomes

$$r = N_1 N_2 \left(\frac{m_1}{2\pi kT}\right)^{3/2} \left(\frac{m_2}{2\pi kT}\right)^{3/2} \int_{-\infty}^{+\infty} \int_{-\infty}^{+\infty} \exp\left(\frac{-m_1 v_1^2}{2kT} - \frac{m_2 v_2^2}{2kT}\right) v\sigma(v)\, d\vec{v}_1\, d\vec{v}_2 \tag{3.3.9}$$

If we transform to the center-of-mass coordinate system, assuming the velocity field is isotropic so that the triple integrals of equation (3.3.9) can be written as spherical "velocity volumes," then we can rewrite equation (3.3.9) in terms of the center-of-mass velocity v_0 and the relative velocity v as

$$r = N_1 N_2 \left(\frac{m_1}{2\pi kT}\right)^{3/2} \left(\frac{m_2}{2\pi kT}\right)^{3/2} \int_0^\infty \int_0^\infty \exp\left[\frac{-(m_1 + m_2)v_0^2 + \tilde{m} v^2}{2kT}\right]$$
$$\cdot\, v\sigma(v) 4\pi v_0^2 (4\pi v^2)\, dv_0\, dv \tag{3.3.10}$$

where

$$\tilde{m} = \frac{m_1 m_2}{m_1 + m_2}$$

$$\vec{v} = \vec{v}_1 - \vec{v}_2 \tag{3.3.11}$$

$$\vec{v}_0 = \frac{m_1 \vec{v}_1 + m_2 \vec{v}_2}{m_1 + m_2}$$

The integral over v_0 is analytic and is

$$\int_0^\infty e^{-(m_1 + m_2)v_0^2/(2kT)} 4\pi v_0^2\, dv_0 = \left(\frac{2\pi kT}{m_1 + m_2}\right)^{3/2} \tag{3.3.12}$$

which reduces equation (3.3.10) to

$$r = N_1 N_2 \left(\frac{\tilde{m}}{2\pi kT}\right)^{3/2} \int_0^\infty e^{-\tilde{m} v^2/(2kT)} v\sigma(v)(4\pi v^2)\, dv \tag{3.3.13}$$

Since the *relative* kinetic energy in the center-of-mass system is $E = \frac{1}{2}\tilde{m}v^2$, we can rewrite equation (3.3.13) in terms of an average reaction cross section $\langle \sigma(v) \cdot v \rangle$ so that

$$r = N_1 N_2 \langle \sigma(v)v \rangle \tag{3.3.14}$$

where

$$\langle \sigma(v)v \rangle = \frac{2}{\sqrt{\pi}} \left(\frac{1}{kT} \right)^{3/2} \int_0^\infty e^{-E/(kT)} \sigma(v)v E^{1/2}\, dE = \int_0^\infty n(E)v\sigma(v)\, dE \tag{3.3.15}$$

Thus $\langle \sigma(v) \cdot v \rangle$ is the "relative energy" weighted average of the collision probability of particle 1 with particle 2. When this average cross section is written, the explicit dependence on velocity is usually omitted, so that

$$\langle \sigma v \rangle \equiv \langle \sigma(v)v \rangle \tag{3.3.16}$$

If the collisions involve identical particles, then the number of *distinct* pairs of particles is $N(N-1)/2$, so the factor of $N_1 N_2$ in equation (3.3.14) is replaced by $N^2/2$. If we call the energy produced per reaction Q, we can write the energy produced per gram of stellar material as

$$\epsilon = \frac{r_{a,X}Q}{\rho} = \frac{N_a N_X \langle \sigma v \rangle Q}{\rho} \tag{3.3.17}$$

The number densities can be replaced with the more common fractional abundances by mass to get

$$\epsilon = \left[N_0^2 Q \langle \sigma v \rangle \left(\frac{X_X}{m_X} \right) \left(\frac{X_a}{m_a} \right) \right] \rho \tag{3.3.18}$$

where N_0 is Avogadro's number. Since $\langle \sigma v \rangle$ is a complicated function of temperature and must be obtained numerically, equation (3.3.18) is usually approximated numerically as

$$\epsilon \approx \epsilon_0 \rho T^\nu \tag{3.3.19}$$

where

$$\epsilon_0 = \frac{N_0^2 (X_a/m_a)(X_X/m_X)(Q\langle \sigma v \rangle|_{T_0})}{T_0^\nu} \tag{3.3.20}$$

Here ν itself is very weakly dependent on the temperature. Most of the important energy production mechanisms have this form. Equation (3.3.19) expresses the energy generated for a specific energy generation mechanism in terms of the state variables T and ρ. This is what we were after. Formulas such as these, where ϵ_0 has been determined, will enable us to determine the energy produced throughout the star in terms of the state variables. Before turning to the description of processes which impede the flow of this energy, let us consider a few of the specific nuclear reactions for which we have expressions of the type given by equation (3.3.20).

e Specific Nuclear Reactions

The nuclear reactions that provide the energy for main-sequence stars all revolve around the conversion of hydrogen to helium. However, this is accomplished in a variety of ways. We may divide these ways into two groups. The first is known as the proton-proton cycle (p-p cycle), and it begins with the conversion of two hydrogen atoms to deuterium. Several possibilities occur on the way to the production of ^{4}He. These alternate options are known as P2-P6 cycles. In addition to the proton-proton cycle, a series of nuclear reactions involving carbon, nitrogen, and oxygen also can lead to the conversion of hydrogen to helium with no net change in the abundance of C, N, and O. For this reason, it is known as the *CNO cycle*. These reactions and their side chains as given by Cox and Giuli[2] are shown in Table 3.3.

Besides the steps marked with asterisks, which denote reactions that occur by spontaneous decay and do not depend on local values of the state variables, the steps that are the important contributors to the energy supply have their contribution (their Q value) indicated. The energy of the neutrinos has not been included since they play no role in determining the structure of normal stars. When the p-p cycle dominates on the lower main sequence, most of the energy is produced by means of the P1 cycle. The neutrino produced in the fifth step of the P2 cycle is the high-energy neutrino which has been detected, but in unexpectedly low numbers, by the neutrino detection experiment of R. Davis in the Homestake Gold Mine. In general, the relative importance of the P1 cycle relative to P2 and P3 is determined by the helium abundance, since this governs the branching ratio at step 3 in the p-p cycle. If ^{4}He is absent, it will not be possible to make ^{7}Be by an α capture on ^{3}He.

Virtually all the energy of the CNO cycle is produced by step 6 as the production of ^{12}C from ^{15}N is strongly favored. However, all the higher chains close with only the net production of ^{4}He. The first stage of the P4 cycle is endothermic by 18 keV, so unless the density is high enough to produce a Fermi energy of 18 keV, the reaction does not take place. This requires a density of $\rho > 2 \times 10^4$ g/cm^3 and so will not be important in main sequence stars. Once ^{3}H is produced, it can

Table 3.3 Common Nuclear Reactions

Step	Q, MeV	Proton Cycles	Q, MeV	CNO Cycles	Step
1	1.18	$^1\mathrm{H}(p, \beta^+ v)^2\mathrm{H}$ ⎫	1.94	$^{12}\mathrm{C}(p, \gamma)^{13}\mathrm{N}$	1
2	5.49	$^2\mathrm{H}(p, \gamma)^3\mathrm{He}$ ⎬ × 2 ⎫ (P1)	1.51	$^{13}\mathrm{N} \rightarrow {}^{13}\mathrm{C} + \beta^+ + v$	2*
3	12.86	$^3\mathrm{He}(^3\mathrm{He}, 2p)^4\mathrm{He}$ ⎭	7.54	$^{13}\mathrm{C}(p, \gamma)^{14}\mathrm{N}$	3
or			7.29	$^{14}\mathrm{N}(p, \gamma)^{15}\mathrm{O}$	4
3	1.59	$^3\mathrm{He}(\alpha, \gamma)^7\mathrm{Be}$ (P2 & P3)	1.76	$^{15}\mathrm{O} \rightarrow {}^{15}\mathrm{N} + \beta^+ + v$	5*
4	0.06	$^7\mathrm{Be}(\beta^-, \bar{v})^7\mathrm{Li} \rightarrow {}^7\mathrm{Li}(p, \alpha)^4\mathrm{He}$ ⎱ (P2)	4.96	$^{15}\mathrm{N}(p, \alpha)^{12}\mathrm{C}$ $^{15}\mathrm{N}(p, \gamma)^{16}\mathrm{O}$	6
or		$Q = 17.35$ MeV ⎰			
4	0.13	$^7\mathrm{Be}(p, \gamma)^8\mathrm{B}$ ⎫		$^{16}\mathrm{O}(p, \gamma)^{17}\mathrm{F}$	7
5*	10.78	$^8\mathrm{B} \rightarrow {}^8\mathrm{Be} + \beta^+ + v$ ⎬ (P3)		$^{17}\mathrm{F} \rightarrow {}^{17}\mathrm{O} + \beta^+ + v$	8*
6*	0.09	$^8\mathrm{Be} \rightarrow 2{}^4\mathrm{He}$ ⎭		$^{17}\mathrm{O}(p, \alpha)^{14}\mathrm{N}$ $^{17}\mathrm{O}(p, \gamma)^{18}\mathrm{F}$	9
or				$^{18}\mathrm{F} \rightarrow {}^{18}\mathrm{O} + \beta^+ + v$	10*
3		$^3\mathrm{He}(\beta^-, v)^3\mathrm{H}$ ⟵ Endothermic by 18 KeV		$^{18}\mathrm{O}(p, \alpha)^{15}\mathrm{N}$ $^{18}\mathrm{O}(p, \gamma)^{19}\mathrm{F}$	11
4		$^3\mathrm{H}(p, \gamma)^4\mathrm{He}$ $^3\mathrm{H}(^3\mathrm{He}, np)^4\mathrm{He}$ $^3\mathrm{H}(^3\mathrm{H}, 2n)^4\mathrm{He}$ ⎬ (P4)		$^{19}\mathrm{F}(p, \alpha)^{16}\mathrm{O}$	12

Triple-α Process

Step		
1		$2(^4\mathrm{He}) + (\sim 100 \text{ KeV}) \rightarrow {}^8\mathrm{Be}^*$
2		$^8\mathrm{Be}^*(\alpha, {}^{12}\mathrm{C}^*)$
3		$^{12}\mathrm{C}^* \rightarrow {}^{12}\mathrm{C} + 2\gamma + 7.656$ MeV

* Reactions that occur by spontaneous decay and do not depend on the local values of the state variables.

be converted to ^{4}He by a variety of processes given in step 4. The last two are sometimes denoted P5 and P6, respectively, and are rare.

While the so-called triple-α process is not operative in main sequence stars, it does provide a major source of energy during the red giant phase of stellar evolution. The extreme temperature dependence of the triple-α process plays a crucial role in the formation of low-mass red giants, and we shall spend some time with it later. The ^{8}Be* is unstable and decays in an extremely short time. However, if during its existence it collides with another ^{4}He nucleus, ^{12}C can form, which is stable. The very short lifetime for ^{8}Be* basically accounts for the large temperature dependence since a very high collision frequency is required to make the process productive.

The exponent of the temperature dependence given in equation (3.3.20) and the constant ϵ_0 both vary slowly with temperature. This dependence, as given by Cox and Giuli[2] (p. 486), is shown in Table 3.4.

The temperature T_6 in Table 3.4 is given in units of 10^6. Thus $T_6 = 1$ is 10^6 K. It is a general property of these types of reaction rates that the temperature dependence "weakens" as the temperature increases. At the same time, the efficiency ϵ_0

Table 3.4 Temperature Dependence of υ and ϵ_0

	Proton-Proton			CNO Cycle			Triple-α Process	
T_6	ϵ_0 (cgs)	υ	ϵ_0 (cgs)	υ	T_8	ϵ_0 (cgs)	υ
10	7×10^{-2}	4.60	3×10^{-4}	22.9	0.8	2×10^{-12}	49
20	1	3.54	4.5×10^2	18	1.0	4×10^{-8}	41
40	9	2.72	3×10^7	14.1	2.0	15	19
80	43	2.08	2×10^{11}	11.1	3.0	6×10^3	12
100	—	—	2×10^{12}	10.2	4.0	10^5	7.9

increases. In general, the efficiency of the nuclear cycles rate is governed by the *slowest* process taking place. In the case of *p-p* cycles, this is always the production of deuterium given in step 1. For the CNO cycle, the limiting reaction rate depends on the temperature. At moderate temperatures, the production of ^{15}O (step 4) limits the rate at which the cycle can proceed. However, as the temperature increases, the reaction rates of all the capture processes increase, but the steps involving inverse β decay (particularly step 5), which do not depend on the state variables, do not and therefore limit the reaction rate. So there is an upper limit to the rate at which the CNO cycle can produce energy independent of the conditions which prevail in the star. However, at temperatures approaching a billion degrees, other reaction processes not indicated above will begin to dominate the energy generation and will circumvent even the beta-decay limitation.

We have now determined the various sources of energy that are available to a star so that it can shine. Clearly the only viable source of that energy results from nuclear fusion. The condition for the production of energy by nuclear processes can occur efficiently only under conditions that prevail near the center of the star. From there, energy must be carried to the surface in some manner in order for the star to shine. In the next chapter we investigate how this happens and describe the mechanisms that oppose the flow.

Problems

1. Using existing models or a current model interior program, find the expected solar neutrino flux (i.e., the flux of ^8B neutrinos) as a function of solar age from the zero-age model to the present.

2. What polytrope(s) would you use to describe the structure of the sun? How well do they match the standard solar model?

3. Consider a gas sphere that undergoes a pressure-free collapse. Let the free-fall time for material at the surface R be t_f. Find the mass distribution for an isothermal sphere and polytropes with indices n of 3 and 1.5 at (a) $t = 0.1t_f$, (b) $t = 0.5t_f$, and (c) $t = 0.8t_f$.

4. Use the virial theorem to find the fundamental radial pulsation period for a star where the equation of state is $P = K\rho^\gamma$. Find the behavior of this period as $\gamma \to \infty$.

5. Find the mass of a main sequence star for which the energy production by the *p-p* cycle equals that of the CNO cycle.

References and Supplemental Reading

1. Chandrasekhar, S., *An Introduction to the Study of Stellar Structure*, Dover, New York, 1939/1957, p. 101, eq. 90.

2. Cox, J. P., and R. T. Giuli, *Principles of Stellar Structure*, Gordon and Breach, New York, 1968, chap. 17, pp. 477–481.

We provided the bare minimum information regarding nuclear energy generation in this chapter. Further reading should be done in the following:

• Clayton, D. D.: *Principles of Stellar Evolution and Nucleosynthesis*, McGraw-Hill, New York, 1968, chaps. 4, 5, pp. 283–606.

• Rolfs, C., and W. S. Rodney: *Cauldrons in the Cosmos*, University of Chicago Press, Chicago, 1986.

• Rolfs, C., and H. P. Trautvetter: "Experimental Nuclear Astrophysics," *Ann. Rev. Nucl. Part. Sci.* 28, 1978, pp. 115–159.

• Bachall, J. N., W. F. Huebner, P. D. Parker, and R. K. Ulrich: *Rev. Mod. Phys.* 54, 1982, p. 767.

In addition, a good overview to the way in which the rates of energy generation interface with the equations of stellar structure is found in

• Schwarzschild, M.: *The Structure and Evolution of the Stars*, Princeton University Press, Princeton, N.J., 1958, pp. 73–88.

4

Flow of Energy through the Star and Construction of Stellar Models

· · ·

That the central temperatures of stars are higher than their surface temperatures can no longer be in doubt. The laws of thermodynamics thus ensure that energy will flow from the centers of stars to their surfaces. The physical processes that accomplish this will basically establish the temperature gradient within the star. This is the remaining relationship required for us to link the interior structure with that of the surface. The temperature gradient, along with the conservation laws of mass and energy, provide the three independent relationships necessary to relate the three state variables to the values they must have at the boundaries of the star.

Energy can move through a medium by essentially three ways, and they can each be characterized by the gas particles which carry the energy and the forces which resist these efforts. These mechanisms are *radiative transfer, convective transport,* and *conductive transfer* of energy. The efficiency of these processes is determined primarily by the amount of energy that can be carried by the particles, their number, and their speed. These variables set an upper limit to the rate of energy transport. In addition, the "opacity" of the material to the motion of the energy-carrying particles will also affect the efficiency. In the case of radiation, we have characterized this opacity by a collision cross section and the density. Another way to visualize this is via the notion of a mean free path. This is just the average distance between collisions experienced by the particles. In undergoing a collision, the particle will give up some of its energy, thereby losing its efficiency as a transporter of energy. We will see that, in general, there are large differences in the mean free paths for the particles that carry energy by these three mechanisms, and so one mechanism will usually dominate in the transfer of energy.

Before we can describe the radiative flow of energy, we must understand how the presence of matter impedes that flow. Thus we begin our discussion of the transport of energy by determining how the local radiative opacity depends on local values of the state variables. From the assumption of strict thermodynamic equilibrium (STE) we know that any impediment to the flow can be described in terms of a parameter that depends only on the temperature. However, to calculate that parameter, we have to investigate the detailed dependence of the opacity on frequency.

4.1 Ionization, Abundances, and Opacity of Stellar Material

We have now described the manner in which nuclear energy is produced in most stars, but before we can turn to the methods by which it flows out of the star, we must quantitatively discuss the processes which impede that flow. Each constituent of the gas will interact with the photons of the radiation field in a way that is characterized by the unique state of that particle. Thus, the type of atom, its state of ionization, and excitation will determine which photons it can absorb and emit. It is the combination of all the atoms, acting in consort, that produces the opacity of the gas. The details which make up this combination can be extremely complicated. However, several of the assumptions we have made, and justified, will make the task easier, and certainly the principles involved can be demonstrated by a few examples.

a   Ionization and the Mean Molecular Weight

Our first task is to ascertain how many of the different kinds of particles that make up the gas are present. To answer this question, we need to know not only the chemical makeup of the gas, but also the state of ionization of the atoms. We have already established that the temperatures encountered in the stellar interior are very high, so we might expect that most of the atoms will be fully ionized. While this is not exactly true, we will assume that it is the case. A more precise treatment of this problem will be given later when we consider the state of the gas in the stellar atmosphere, where the characteristic temperature is measured in thousands of degrees as opposed to the millions of degrees encountered in the interior.

We will find it convenient to divide the composition of the stellar material into three categories:

$$X = \text{mass fraction of gas which is hydrogen}$$

$$Y = \text{mass fraction of gas which is helium} \qquad (4.1.1)$$

$$Z = \text{mass fraction of gas which is everything else}$$

It is common in astronomy to refer to everything which is not hydrogen or helium as "metals." For complete ionization, the number of particles contributed to the gas *per element* is just

$$n_i = \frac{Z_i + 1}{A_i} \qquad (4.1.2)$$

where Z_i is the atomic number and A_i is the atomic weight of the element. Thus, the number of particles contributed by hydrogen will just be twice the hydrogen abundance, and for helium three-fourths the helium abundance. In general,

$$n_i \approx \frac{Z_i + 1}{2Z_i} \qquad (4.1.3)$$

The limit of equation (4.1.3) for the heavy metals is $\frac{1}{2}$. However, even at 10^7 K the inner shells of the heavy metals will not be completely ionized, and so $\frac{1}{2}$ will be an overestimate of the contribution to the particle number. This error is somewhat compensated by the 1 in the numerator of equation (4.1.3) for the light elements where it provides an underestimate of the particle contribution. Thus we take the total number of particles contributed by the metals to be $\frac{1}{2}Z$. The total number of particles in the gas from all sources is then

$$N = \frac{\rho}{m_h} \sum_i n_i x_i = \frac{\rho}{m_h} (2X + \tfrac{3}{4}Y + \tfrac{1}{2}Z) \qquad (4.1.4)$$

Since everything in the star is classed as either hydrogen, helium, or metals, $X + Y + Z = 1$ and we may eliminate the metal abundance from our count of the total number of particles, to get

$$N = \frac{1}{2}\left(\frac{\rho}{m_h}\right)\left(3X + \frac{1}{2}Y + 1\right) \tag{4.1.5}$$

Throughout the book we have introduced the symbol μ as the mean molecular weight of the gas without ever providing a clear definition for the quantity. It is clearly time to do so, and we can do it easily given the ideal-gas law and our expression for the total number of particles N. Remember

$$P = NkT = \frac{\rho kT}{\mu m_h} \tag{4.1.6}$$

The mean molecular weight μ must be defined so that this is a correct expression. Thus,

$$\mu = \frac{2}{1 + 3X + \frac{1}{2}Y} \tag{4.1.7}$$

b Opacity

In general, a photon can interact with atoms in three basic ways which result in the photon being absorbed:

1. Bound-bound absorption (atomic line absorption)

2. Bound-free absorption (photoionization)

3. Free-free absorption (bremsstrahlung)

For us to calculate the impeding effect of these processes on the flow of radiation, we must calculate the cross section for the processes to occur for each type of particle in the gas. For a particular type of atom, this parameter is known as the *atomic absorption coefficient* α_ν. The atomic absorption coefficient is then weighted by the abundance of the particle in order to produce the *mass* absorption coefficient κ_ν, which is the opacity *per gram* of stellar material at a particular frequency ν. Since we have assumed that the gas and photons are in STE, we can average the opacity coefficient over frequency to determine a mean opacity coefficient $\bar{\kappa}$, which will represent the average effect on the diffusion of energy through the star of the material itself. However, to calculate this mean, we must calculate κ_ν itself. As an example of how this is done, we shall consider the atomic absorption coefficient of hydrogen and "hydrogenlike" elements.

Classical View of Absorption   Imagine a classical electromagnetic wave encountering an atom. The time-varying electromagnetic field will cause the electron to be accelerated so that it oscillates at the same frequency as the wave. However, just as an accelerated charge radiates energy, to accelerate a charge requires energy, and in this instance it is the energy of the electromagnetic wave. If the electron is bound in an atom, the energy may just be sufficient to raise the electron to a higher orbit, and we say that a *bound-bound transition* has taken place. If the energy is sufficient to remove the electron and ionize the atom, we say that a *bound-free transition*, or *photoionization*, has taken place. Finally, if a free electron is passing an ion in an unbound orbit, it is possible for the electron to absorb the energy from an electromagnetic wave that happens to encounter this system. Energy and momentum are conserved among the two particles and the photon, with the result that the electron is moved to a different unbound orbit of higher energy relative to the ion. This is known as a *free-free absorption*.

Quantum Mechanical View of Absorption   In quantum mechanics the classical view of a finite cross section of an atom for electromagnetic radiation is replaced by the notion of a transition probability. That is, one calculates the probability that an electron will make a transition from some initial state to another state while in the presence of a photon. One calculates this probability in terms of the wave functions of the two states, and it usually involves a numerical integration of the wave functions over all space. Instead of becoming involved in the detail, we shall obtain a qualitative feeling for the behavior of this transition probability.

Within the framework of quantum mechanics, the probability that an electron in an atom will have a specific radial coordinate is

$$\langle \vec{r} \rangle = \int_V \Psi_i \vec{r} \Psi_i^* \, dV \tag{4.1.8}$$

where i denotes the particular quantum state of the electron (that is, n, j, l) and Ψ_i is the wave function for that state. In classical physics, the dipole moment $\vec{P}$ of a charge configuration is

$$\vec{P} = \int_v \vec{r} \rho_c(\vec{r}) \, dV \tag{4.1.9}$$

where ρ_c is the charge density. The quantum mechanical analog is

$$\langle \vec{P}_{ij} \rangle = e \int_v \Psi_i \vec{r} \Psi_j \, dV \tag{4.1.10}$$

where i denotes the initial state and j the final state.

Now, within the context of classical physics, the energy absorbed or radiated per unit time by a classical oscillating dipole is proportional to $\vec{P} \cdot \vec{P}$, and this result carries over to quantum mechanics. This classical power is

$$\mathbf{P} \propto v^4 P^2 \tag{4.1.11}$$

Since the absorbed power $\mathbf{P}$ is just the energy absorbed per second, the number of photons of energy hv that are absorbed each second is

$$N_p \propto v^3 P^2 \tag{4.1.12}$$

However, the number of photons absorbed per second will be proportional to the probability that one photon will be absorbed, which is proportional to the collision cross section. Thus, we can expect the atomic cross section to have a dependence on frequency given by

$$\alpha_v = \frac{\text{const}}{v^3 P^2} \tag{4.1.13}$$

In general, we can expect an atomic absorption coefficient to display the v^{-3} dependence while the constant of proportionality can be obtained by finding the dipole moment from equation (4.1.10). The result for the bound-free (b-f) absorption of hydrogen and hydrogenlike atoms is

$$\alpha_v^{\text{b-f}}(i, n) = \frac{64\pi^4 Z_i^4 m_e e^{10}}{3\sqrt{3}h^6 c} \frac{1}{n^5} S_{ni}^4 g_v^{\text{b-f}}(i, n)] \left(\frac{1}{v^3}\right)$$

$$\alpha_v^{\text{b-f}}(i, n) = (2.815 \times 10^{29}) \frac{Z_i^4 S_{ni}^4 g_v(i, n)}{n^5 v^3} \quad \text{cm}^2 \tag{4.1.14}$$

where

$$i = \text{state of ionization}$$

$$n = \text{principal quantum number of electron}$$

$$Z_i = \text{atomic number} \tag{4.1.15}$$

$$S_{ni} = \text{screening factor resulting from interior electrons}$$

$$g_v(i, n) = \text{gaunt factor} \approx 1$$

A similar expression can be developed for the free-free transitions of hydrogenlike atoms:

$$\alpha_v^{\text{f-f}}(i, p)\, dp = \frac{4\pi Z_i^2 e^6 S_{fi}^2}{3\sqrt{3}hcm_e^2 V(p)} g_v^{\text{f-f}} \left(\frac{1}{v^3}\right) dn_e(p) \tag{4.1.16}$$

Here, the atomic absorption coefficient depends on the momentum of the colliding electron. If one assumes that the momentum distribution can be obtained from Maxwell-Boltzmann statistics, then the atomic absorption coefficient for free-free transitions can be summed over all the colliding electrons and combined with that of the bound-free transitions to give a *mass* absorption coefficient for hydrogen that looks like

$$\kappa_v(\text{hydrogen}) = \frac{32\pi^2 e^6 R e^{-\chi_0/(kT)}}{3\sqrt{3}h^3 c m_h v^3} \left(\sum_{v_n < v}^{\infty} \frac{e^{-\epsilon_n/(kT)}}{n^3}\, \bar{g}_n + \frac{\bar{g}_{ff}\, kT}{2\chi_0} \right)$$

where

$\chi_0 = $ ionization potential of hydrogen

$R = $ gas constant

$\epsilon_n = $ excitation energy of v_nth energy level

$\bar{g}_n, \bar{g}_{ff} = $ gaunt factors averaged over frequency

(4.1.17)

The summation in equation (4.1.17) is to be carried out over all n such that $v_n < v$. That is, all series that are less energetic than the frequency v can contribute to the absorption coefficient. For us to use these results, they must be carried out for each element and combined, weighted by their relative abundances. This yields a frequency-dependent opacity per gram κ_v which can be further averaged over frequency to obtain the appropriate average effect of the material in impeding the flow of photons through matter. However, to describe the mean flow of radiation through the star, we want an estimate of the transparency or transmissivity of the material. This is clearly proportional to the inverse of the opacity. Hence we desire a reciprocal mean opacity. This frequency-averaged reciprocal mean is known as the *Rosseland mean* and is defined as follows:

$$\frac{1}{\bar{\kappa}_v} \equiv \frac{\int_0^{\infty} (1/\kappa_v)\, \partial B_v(T)/\partial T\, dv}{\int_0^{\infty} [\partial B_v(T)/\partial T]\, dv}$$

(4.1.18)

Here $B_v(T)$ is the Planck function, which is the statistical equilibrium distribution function for a photon gas in STE which we developed in Chapter 1 [equation (1.1.24)]. That such a mean should exist is plausible, since we are concerned with the flow of energy through the star, and as long as we assume that the gas and photons are in STE, we know how that energy must be distributed with wavelength. Thus, it would not be necessary to follow the detailed flow of photons in frequency space since we already have that information. That there should exist an average value of the opacity for that frequency distribution is guaranteed by the mean-value theorem of calculus. That the mean absorption coefficient should have the form given by equation (4.1.18) will be shown after we have developed a more complete theory of radiative transfer (see Section 10.4).

Approximate Opacity Formulas Although the generation of the mean opacity coefficient $\bar{\kappa}$ is essentially a numerical undertaking, the result is always a function of the state variables P, T, ρ, and μ. Before the advent of the monumental studies of Arthur Cox and others which produced numerical tables of opacities, much useful work in stellar interiors was done by means of expressions which give the approximate behavior of the opacity in terms of the state variables. The interest in these formulas is more than historical because they provide a method for predicting the behavior of the opacity in stars and a basis for understanding its relationship to the other state variables. If one is constructing a model of the interior of a star, such approximation formulas enable one to answer the question so central to any numerical calculation: Are these results reasonable? In general, these formulas all have the form

$$\bar{\kappa} = \bar{\kappa}_0 \rho^n T^{-s}$$

$$
\begin{array}{llll}
(a) \; n = 1 & s = 3.5 & \text{Kramers' law} \\
(b) \; n = 0.75 & s = 3.5 & \text{Schwarzschild's opacity} \\
(c) \; n = 0 & s = 0 & \text{electron scattering}
\end{array}
\tag{4.1.19}
$$

where $\bar{\kappa}_0$ depends on the chemical composition μ. Kramers' opacity is a particularly good representation of the opacity when it is dominated by free-free absorption, while the Schwarzschild opacity yields somewhat better results if bound-free opacity makes an important contribution. The last example of electron scattering requires some further explanation since it is not strictly a source of absorption.

Electron Scattering The scattering of photons at the energies encountered in the stellar interior is a fully conservative process in that the energy of the photon can be considered to be unchanged. However, its direction is changed, resulting in the photon describing a random walk through the star. This "walk" immensely lengthens the path taken by the photon and therefore increases its "stay" in the star. The longer the photon resides in the star, the greater its path and the greater its chances of being absorbed by an encounter with an atom. Thus, electron scattering, while not involved directly in the absorption of photons, does significantly contribute to the opacity of the gas. The photon flow is impeded by electron scattering, first, by redirecting the photon flow and, second, by lengthening the path and increasing the photon's chances of absorption.

As long as $hv \ll m_e c^2$, the electron will exhibit little or no recoil as a result of its collision with a photon, and the photon energy will be unchanged. This case is called *Thomson scattering*, and we can use the classical theory of electromagnetism to estimate its cross section. The energy radiated or absorbed per unit time by an oscillating free electron is

$$\frac{dE}{dt} = \frac{2e^4 E_0^2}{3m_e^2 c^3} \mathrm{Sin}^2\, \omega t \tag{4.1.20}$$

However, the power in an electromagnetic wave is given by the Poynting vector $\vec{S} = [c/(4\pi)](\vec{E} \times \vec{H})$. In vacuum $\vec{E} \perp \vec{H}$, and in gaussian units $|E| = |H|$. Therefore, the magnitude of the Poynting vector is

$$S = \frac{c}{4\pi}\, E_0^2\, \mathrm{Sin}^2\, \omega t \tag{4.1.21}$$

The ratio of the power absorbed from the wave to the power in the wave is $(dE/dt)/S$ and is the definition of a cross section. Thus, we can write the classical cross section for electron scattering as

$$\sigma_0 = \frac{8\pi}{3}\left(\frac{e^2}{m_e c^2}\right)^2 = 6.652 \times 10^{-25}\, \mathrm{cm}^2 \tag{4.1.22}$$

The quantity $e^2/(m_e c^2)$ is known as the *classical radius* of the electron and is roughly that radius for which the field energy of the electron is equal to its rest energy. The square of that radius yields a geometric cross section which is 1.5 times the classical cross section. This cross section is also known as the *Thomson cross section*.

Note that the Thomson cross section is not a function of frequency, which makes it particularly easy to incorporate in an expression for opacity. The symbol σ_e usually denotes the electron scattering coefficient *per gram* of stellar material, so that

$$\sigma_e = \frac{\sigma_0 n_e}{\rho} \tag{4.1.23}$$

However, in the limit where the total number of particles contributed to the particle density by metals is $\frac{1}{2}\mathbf{Z}$ and most of those are electrons, we get the electron density to be

$$n_e = \frac{\rho}{m_h}\left\langle\frac{Z}{A}\right\rangle \approx \frac{1}{2}\left(\frac{\rho}{m_h}\right)(1 + \mathbf{X}) \tag{4.1.24}$$

which yields an electron scattering cross section *per gram* of

$$\sigma_e \approx 0.2004(1 + \mathbf{X}) \text{ cm}^2/\text{g} \leq 0.4 \text{ cm}^2/\text{g} \tag{4.1.25}$$

Thus, there is a limit to the scattering coefficient per gram of stellar material of something less than 0.5 cm^2/g.

We have now shown how the energy generation rate and opacity per gram of stellar material can be related to the state variables P, T, ρ, and μ. You must not assume that this discussion is complete in every detail. The detailed calculation of these functions is extremely complex and has entertained some of the best minds of the twentieth century. What we have seen is some of the major physical principles which affect the outcome of such efforts. For the details of the modern values of these functions, you should consult the current literature as refinements continue. Nevertheless, from now on, we may assume that we have functions of the form

$$\epsilon = \epsilon(P, T, \rho, \mu) \qquad \kappa = \kappa(P, T, \rho, \mu) \tag{4.1.26}$$

at our disposal. Now we turn to the problem of describing the flow of energy through the star.

4.2   Radiative Transport and the Radiative Temperature Gradient

Although all forms of energy transport may be present at any given place in a star, we will see that their relative efficiency is such that generally only one form will be important for describing the flow of energy. The transport of energy by radiation is essentially the radiative diffusion of photons through the stellar material. It is the opacity of the material that opposes this flow. To establish the interplay between thermodynamics and radiative opacity, we assume that all the energy is flowing by this process.

a   Radiative Equilibrium

Since we are assuming that all the energy is flowing outward by means of radiative diffusion, the entire energy produced by the star within a sphere of radius r can be characterized by a local luminosity $L(r)$ which is entirely made up of photons. When this is the case, we may describe this flow of photons locally by defining the radiative flux as

$$F(r) = \frac{L(r)}{4\pi r^2} \tag{4.2.1}$$

When these conditions prevail, the entire flow of energy is carried by photons and the star is said to be in *radiative equilibrium.*

b Thermodynamic Equilibrium and Net Flux

In Chapter 1 we developed an elegant formalism to describe the flow of particles through space. In a later chapter we shall use this to produce an extremely general equation of radiative transfer which describes the flow in momentum space as well as physical space. But at this point, we are dealing with a gas in STE, and that fixes many properties of the gas. For example, we know that the phase density f that appears in the Boltzmann transport equation will be the Planck function, since we have shown f to be the equilibrium distribution function for photons in STE. We also know that while there is a net flow of photons, the energy involved in that flow must be small compared to the local energy density; otherwise, the photon gas could not be considered in equilibrium.

Another way of visualizing this is to observe that any system said to be in thermodynamic equilibrium cannot have temperature gradients. If it did, there would be a flow of energy driven by the temperature gradient. In a star we must have such a flow, or the star will not shine. What is important is the relative size of the temperature gradient through some volume for which the system is to be considered in equilibrium. In the case of the sun, this typical length would be the distance a photon travels before it encounters an atom. From the opacity calculations of Chapter 3 and our knowledge of the conditions within the sun, we would calculate that the mean free path for a photon in the center of the sun is less than a centimeter. Thus, as a measure of the extent to which STE is met in the sun, let us calculate

$$\left(\frac{\nabla T}{T}\right) l \approx \left(\frac{T_c/R}{T_c}\right) l \approx \frac{l}{R} \approx 10^{-11} \tag{4.2.2}$$

In other words, the change in the local temperature over a scale length appropriate for the photon gas is about 1 part in 10^{11}. There are few gaseous structures in the universe where the conditions for STE are met better than this. Small as this relative temperature gradient is, it drives the luminous flux of the sun, and so we must estimate its dependence on the state variables.

c Photon Transport and Radiative Gradient

Since we know so much about the nature of the photons in the star, we need not resort to the basic Boltzmann transport equation to describe how photons flow. Instead, consider the Euler-Lagrange equations of hydrodynamic flow. Since they were derived under fairly general conditions, they should be adequate to describe the flow of photons. Equation (1.2.27) provides a reasonably simple description of this process. But we are interested in a steady-state description, so all explicit time dependence in that equation must vanish. Thus, equation (1.2.27) becomes

$$(\vec{u} \cdot \nabla)\vec{u} = -\nabla\Phi - \left(\frac{1}{\rho}\right)\nabla P \tag{4.2.3}$$

However, in deriving equation (1.2.27), we averaged the local particle phase density over velocity space. For photons traveling at the velocity of light, this does not make much sense. Instead, the moment generation which led to the Euler-Lagrange equation of hydrodynamic flow should be carried out over momentum space, or photon frequency. Since this expression is for photons, P is the local radiation pressure due to photons and $\vec{u}$ is the mean flow velocity, or diffusion velocity, of those photons. However, in equation (1.2.27), ρ is the local mass density. For photons, this translates to the local energy density. In addition, the influence of gravity on the photons throughout the star can be estimated by the gravitational red shift that photons will experience which is

$$\frac{\Delta h\nu}{h\nu} \approx \frac{GM}{Rc^2} \approx 10^{-6} \tag{4.2.4}$$

Since the change in the photon energy resulting from moving through the gravitational potential is about 1 part in a million, we may safely neglect the influence of $\nabla\Phi$. In spherical coordinates, all spatial operators in equation (4.2.3) simply become derivatives with respect to the radial coordinate, so that equation (4.2.3) becomes

$$\vec{u}_r \cdot \left(\frac{d\vec{u}_r}{dr}\right)\hat{r} = -\left(\frac{1}{\rho_e}\right)\nabla P_r \tag{4.2.5}$$

where

$$\rho_e = \langle h\nu \rangle \int_0^\infty f(p)\frac{d\vec{p}}{c^2}$$

$$\tag{4.2.6}$$

$$\vec{u}_r = \left(\frac{\langle h\nu \rangle}{c^2}\right)^{-1} \frac{\int_0^\infty \vec{p}f(p)\,d\vec{p}}{\int_0^\infty f(p)\,d\vec{p}}$$

Here $\langle h\nu \rangle$ is the average photon energy, and $d\vec{p}$ indicates integration over all momentum coordinates which, in the absence of a strong potential gradient, can be represented by the differential spherical momentum volume $4\pi p^2\,dp$. We can then write equation (4.2.5) as

$$\nabla P_r = -\left(\int_0^\infty \vec{p}f(p)\,d\vec{p}\right)\cdot\left[\frac{d}{dr}\int_0^\infty \frac{\vec{p}f(p)\,d\vec{p}}{\rho_e}\right] \tag{4.2.7}$$

The first term on the right-hand side represents the net flow of momentum and can be related to the flow of radiant energy by

$$\int_0^\infty \vec{p}f(p)\,d\vec{p} = \frac{1}{c}\int_0^\infty hvf(p)\hat{r}(4\pi p^2)\,dp = \frac{\vec{F}(r)}{c} \tag{4.2.8}$$

and equation (4.2.7) becomes

$$\nabla P_r = \frac{-\vec{F}(r)}{c}\cdot\left\{\frac{d}{dr}\left[\frac{F(r)}{\rho_e c}\right]\right\} \tag{4.2.9}$$

The quantity $[F(r)/c\rho_e]$ is the fraction of photons which are participating in the net flow of energy. Thus the radial derivative represents the change in the fraction with r. The only reason for this fraction to change is the interaction of the flowing photons with matter. If we define the volume absorption coefficient α_v to be the "collision" cross section per unit volume, then the probability per unit length that a photon will be absorbed in passing through that volume is just α_v. However, the probability that one photon will be absorbed per unit length is equal to the fraction of n photons that will be absorbed in that same unit length. Thus, the second term on the right-hand side of equation (4.2.9) becomes

$$\frac{d}{dr}\left[\frac{F(r)/c}{\rho_e}\right] = \bar{\alpha}_v = \bar{\kappa}\rho \tag{4.2.10}$$

The radiation pressure gradient is now

$$\nabla P_r = -\frac{\bar{\kappa}\rho F(r)}{c} = \frac{-\bar{\kappa}\rho L(r)}{4\pi c r^2} \tag{4.2.11}$$

But in STE the radiation pressure depends on only a single parameter and is given by

$$P_r = \frac{aT^4}{3} \tag{4.2.12}$$

This implies we can write the radiation pressure gradient in terms of the temperature as

$$\frac{dP_r}{dr} = \frac{4}{3} aT^3 \frac{dT}{dr} \tag{4.2.13}$$

Equating this to the radiation pressure gradient from equation (4.2.11), we finally obtain an expression for the radiative temperature gradient:

$$\frac{dT}{dr} = -\frac{3\bar{\kappa}\rho L(r)}{16\pi acr^2 T^3} \tag{4.2.14}$$

This relationship specifies how the temperature must change if the energy is being carried by radiative diffusion and the specification is made in terms of the state variables and parameters that we have already determined characterize the problem.

d Conservation of Energy and the Luminosity

With the advent of the radiant flux $F(r)$, we have introduced a new variable to the problem. Relating the flux to the total luminosity [equation (4.2.1)] only transfers the source of the problem to the luminosity $L(r)$. That such a parameter is important should surprise no one, for the luminosity of a star is perhaps its most obvious characteristic. However, it is only with the transport of energy that we are faced with the internal energy, stored or produced, arriving at the surface and leaking into space. As far as the structure of the star is concerned, this is a "second-order" effect. It is only a small part of the internal energy that is lost during a dynamical time interval. However, for the proper understanding of the star as an object in steady state, it is a central condition which must be met, for in steady state the energy lost must be matched by the energy produced.

Fortunately, we have an additional fundamental constraint which we have not yet imposed that must be met by any physical system—the conservation of energy. This is completely analogous to the conservation of mass which we invoked in Chapter 2 [equations (2.1.7) and (2.1.8)], only now the total energy interior to r which must pass through r per unit time is called $L(r)$. Thus,

$$L(r) = \int_0^r 4\pi r^2 \rho \epsilon \, dr \tag{4.2.15}$$

The corresponding differential form is

$$\frac{dL(r)}{dr} = 4\pi r^2 \rho \epsilon \tag{4.2.16}$$

4.3 Convective Energy Transport

Our approach to the transport of energy by convection will be somewhat different from that for radiation. For radiation, we knew how much energy there was to carry—$L(r)/(4\pi r^2)$, and we set about finding the temperature gradient required to carry it. For convection, we will anticipate the answer by calculating the amount of energy that a superadiabatic temperature gradient will carry. For a wide range of parameters thought to prevail in the stellar interior, we shall discover that the adiabatic gradient is adequate to carry all the required energy. But first we must determine the adiabatic temperature gradient.

a *Adiabatic Temperature Gradient*

In Chapter 2 [equation (2.4.6)] we defined polytropic change in terms of a specific heat-like quantity C which is equal to the change in heat with respect to temperature. For an adiabatic change, the gas does no work on the surrounding medium, so that $C = 0$. The polytropic γ' as defined by equation (2.4.8) is

$$\gamma' = \gamma = \frac{C_P}{C_V} \tag{4.3.1}$$

Using equation (2.4.5) and the ideal-gas law, we have

$$\gamma = 1 + \frac{R}{C_V} = 1 + \frac{Nk}{\frac{3}{2}Nk} = \frac{5}{3} \tag{4.3.2}$$

or

$$n = \tfrac{3}{2} \tag{4.3.3}$$

where n is the appropriate polytropic index for an adiabatic gas.

Now the polytropic equation of state [equation (2.4.1)] and the ideal-gas law guarantee that

$$P = k\rho^{(n+1)/n}$$

$$T = \left(\frac{\mu m_h k}{k}\right)\rho^{1/n} \tag{4.3.4}$$

Forming the logarithmic derivative of P and T with respect to ρ, we get

$$\frac{1}{P}\frac{dP}{d\rho} = \frac{n+1}{n}\rho^{-1}$$

$$\frac{1}{T}\frac{dT}{d\rho} = \frac{1}{n}\rho^{-1} \tag{4.3.5}$$

Dividing these two equations yields

$$\frac{T}{P}\frac{dP}{dT} = n + 1 \tag{4.3.6}$$

which is known as the *polytropic temperature gradient* and for an adiabatic gas is just

$$\frac{d\ln P}{d\ln T} = 2.5 \tag{4.3.7}$$

b   Energy Carried by Convection

Imagine a small element of matter rising as a result of being somewhat hotter than its surroundings (see Figure 4.1). We can express the temperature difference between the gas element and its surroundings in terms of the external temperature gradient and the internal temperature gradient experienced by the small element as it rises. We assume that the element is behaving adiabatically, and so this internal gradient is the adiabatic gradient and the temperature difference is

$$\delta T = \left.\frac{dT}{dr}\right|_{ad}\delta r - \left.\frac{dT}{dr}\right|_{ext}\delta r \equiv \Delta\nabla T\,\delta r \tag{4.3.8}$$

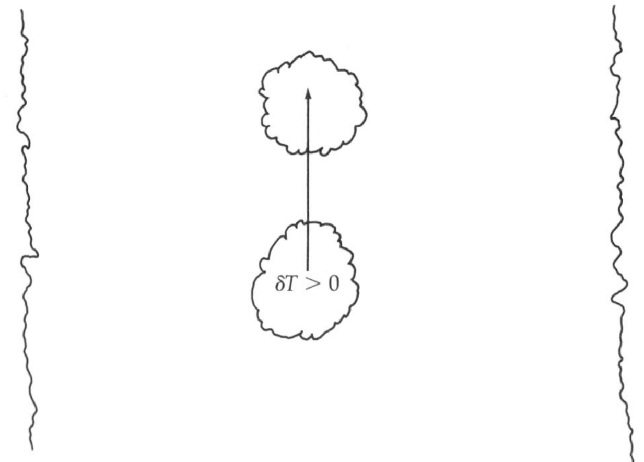

Figure 4.1 Schematic representation of a convective element of gas responding to a small temperature difference.

The flux of energy carried by this small convective element will be

$$F_{\text{conv}} = (\Delta \nabla T \, \delta r) \rho C_p v \qquad (4.3.9)$$

where v is the average velocity of the convective element which we must estimate.

The buoyant force experienced by the convective element will be determined solely by the density difference resulting from its slightly elevated temperature and is

$$F_b = \frac{Gm(r)}{r^2} \, \delta \rho \qquad (4.3.10)$$

Since the convective element rises adiabatically, pressure equilibrium will always be maintained during its ascent. Thus we can relate the variation in density to the variation in temperature by setting the variation of the ideal-gas law to zero:

$$\delta P = \frac{kT}{\mu m_h} \, \delta \rho + \frac{k\rho}{\mu m_h} \, \delta T = 0 \qquad (4.3.11)$$

We may use this and equation (4.3.8) to obtain the average buoyancy acting on the convective element. Initially, the buoyancy force is zero since the gradient difference does not produce a significant force until the element has traveled some distance. Thus, we take the average force to be one-half the maximum force, and we get

$$\langle F_b \rangle = -\frac{1}{2} \frac{GM(r)}{r^2} \frac{\rho}{T} \Delta \nabla T \, \delta r \qquad (4.3.12)$$

Now the buoyancy force will continuously accelerate the convective element, giving it a kinetic energy of $\frac{1}{2}\rho v^2$ which we can use to get an estimate of the convective velocity v. Thus,

$$\langle F_b \rangle \, \delta r = \tfrac{1}{2}\rho v^2 = \left[\frac{1}{2} \frac{GM(r)}{r^2} \frac{\rho}{T} (\Delta \nabla T \, \delta r) \right] \delta r \qquad (4.3.13)$$

which yields a convective velocity of

$$v_{\text{conv}} \approx \left[\frac{GM(r)}{Tr^2} \right]^{1/2} (\Delta \nabla T)^{1/2} \, \delta r \qquad (4.3.14)$$

We define

$$l = 2\,\delta r \tag{4.3.15}$$

This quantity l is known as the *mixing length* and is largely a free parameter of this theory of convection from which it takes its name. Typically it is taken to be of the order of a pressure scale height, and fortunately for the theory of stellar interiors, the results are not too sensitive to its exact value. In terms of the mixing length, the convective flux becomes

$$F_{\text{conv}} = C_P\rho \left[\frac{GM(r)}{Tr^2}\right]^{1/2} (\Delta\nabla T)^{3/2}\,\frac{l^2}{4} \tag{4.3.16}$$

Now all that remains is to estimate the difference in temperature gradients necessary to transport the energy of the star. We will require that the convection carry all the internal energy flowing through the star, so that

$$F_{\text{conv}} = \frac{L(r)}{4\pi r^2} \tag{4.3.17}$$

which yields the gradient difference of

$$\Delta\nabla T = \left[\frac{L^2(r)T}{C_P^2\rho^2 GM(r)\pi^2 l^4 r^2}\right]^{1/3} \tag{4.3.18}$$

To arrive at some estimate of the significance of this result, let us compare it to the adiabatic gradient. We use the adiabatic temperature gradient in equation (4.3.7) hydrostatic equilibrium [equation (1.2.28)], and the ideal-gas law to get

$$(\nabla T)_{\text{ad}} = \frac{dT}{dP}\frac{dP}{dr} = -\frac{\mu m_h GM(r)}{2.5kr^2} \tag{4.3.19}$$

Dividing equation (4.3.18) by the adiabatic gradient, we get

$$\frac{\Delta\nabla T}{(\nabla T)_{\text{ad}}} = \frac{5}{2}\left(\frac{L(r)}{C_P\pi\rho}\right)^{2/3} T^{1/3}\left[\frac{r}{GM(r)l}\right]^{4/3}\left(\frac{\mu m_h}{k}\right)^{-1} \tag{4.3.20}$$

For the sun, there is some evidence that a mixing length of about one-tenth of a solar radius is not implausible. Picking other values for the sun and trying to maximize equation (4.3.20), we have the following selection:

$$r = \frac{R}{10} \approx l \qquad\qquad M(r) = M_\odot$$

$$T = 10^7 \text{ K} \qquad\qquad \mu = 0.7$$

$$\rho \approx 1 \text{ g/cm}^3 \qquad\qquad L = L_\odot$$

$$C_P = 3.5 \qquad\qquad \frac{\Delta\nabla T}{(\nabla T)_{\text{ad}}} \approx 4 \times 10^{-3}$$

$$(4.3.21)$$

Thus, it would seem that the convective gradient will lie within a few tenths of a percent of the adiabatic gradient. This is the source of the statement in Chapter 2 that a polytrope of index $\frac{3}{2}$ represents convective stars quite well. Indeed, convection is so efficient that the adiabatic gradient will almost always suffice to describe convective stellar interiors. This is fortunate since the mixing length theory we have discussed here is admittedly rather crude. Unfortunately, this efficiency does not carry over into stellar atmospheres because the convective zones are bounded by the surface of the star, dropping the mixing lengths to numbers comparable to the photon mean free path, so that radiation competes effectively with convection regardless of the temperature gradient. For stellar interiors, the photon mean free path is measured in centimeters and the mixing length in fractions of a stellar radius. Thus convection, when established, will always be able to carry the stellar luminosity with a temperature gradient close to the adiabatic gradient.

4.4   Energy Transport by Conduction

a   Mean Free Path

Consider a simple monatomic gas where the kinetic energy per particle is $3kT/2$ so that the speed is

$$v = \left(\frac{3kT}{m}\right)^{1/2} \tag{4.4.1}$$

We will let the collisional cross section be just the geometric cross section, so that

$$\sigma = \pi(r_1 + r_2)^2 \tag{4.4.2}$$

where r_1 and r_2 are the radii of the two species of colliding particles. As we did with nuclear reaction rates, we get the collision frequency from the effective volume swept per unit time σv multiplied by the number density ρ/m. The time between collisions is just the reciprocal of the collision frequency, so the distance traveled between collisions is

$$l = v\left(\frac{\sigma v \rho}{m}\right)^{-1} = \frac{m}{\rho \sigma} \tag{4.4.3}$$

and is known as the mean free path for collisions.

b Heat Flow

The thermodynamic theory of heat says that the heat flux through a given area is proportional to the temperature gradient so that

$$F_{\text{cond}} = -K \nabla T \tag{4.4.4}$$

where Eddington[1] gives the conductivity K as

$$K = \frac{C_V \rho l v}{3} = \frac{C_V (kTm/3)^{1/2}}{\sigma} \tag{4.4.5}$$

If we compare the maximum luminosity obtainable with the conductive flux to the total solar luminosity, we have

$$\frac{L_{\text{cond}}}{L_\odot} = \frac{4\pi R_\odot^2 F_{\text{cond}}}{L_\odot} \approx 10^{-12} \tag{4.4.6}$$

With the gradient estimated as $T_c/R_\odot$, using the central temperature to make the conductivity as large as possible and taking the geometric cross section to be about 10^{-20} cm^2, we still fail to carry the solar luminosity by at least 5 orders of magnitude. Thus, conduction can play no significant role in the energy transport in the sun. Indeed, that is true for all normal stars. However, in white dwarf stars, where the electrons are degenerate, the mean free path of the electrons is comparable to the dimensions of the star itself. Then conduction becomes so important that the internal temperature distribution is essentially isothermal.

If we combine equations (4.2.14) and (4.2.1), we can write the radiative flux as

$$F_{\text{rad}} = \left(\frac{4ac}{3\bar{\kappa}\rho}\right) \nabla T(T^3) \tag{4.4.7}$$

which has the same form as equation (4.4.4). Thus we may define a conductive opacity from the conductivity, so that

$$\kappa_{\text{cond}} = \frac{4ac T^3}{3K\rho} \tag{4.4.8}$$

Then, if necessary, the conductive and radiative fluxes can be combined by augmenting the mean radiative opacity, so that

$$\frac{1}{\overline{\kappa}} = \frac{1}{\overline{\kappa}_{\text{rad}}} + \frac{1}{\kappa_{\text{cond}}} \tag{4.4.9}$$

4.5 Convective Stability

a Efficiency of Transport Mechanisms

We calculated the fluxes that can be transported by radiation, convection, and conduction, and we found that they produce rather different temperature gradients. However, we have seen from the integral theorems that the central temperature is set largely by the mass of the star, and in Chapter 3 we learned that the energy produced by nuclear processes will be a strong function of that temperature. Thus, virtually all the energy will be produced near the center and, in steady state, must make its way to the surface. In general, it would do this in the most efficient manner possible. That is, the mode of energy transport will be that which produces the smallest temperature gradient and the greatest luminosity. In short, the star will choose among the methods available to it and select that which allows it to leak away its energy as fast as possible.

To carry enough energy to support the luminosity of the sun, conductive transport would require an immense temperature gradient. This another way of saying that conduction is not important in the transport of energy. Convection will produce a temperature gradient which is nearly the adiabatic gradient and is fully capable of carrying all the energy necessary to sustain the solar luminosity. If we compare the radiative temperature gradient given in equation (4.2.14) and the adiabatic gradient as given in equation (4.3.19), we get

$$\frac{(\nabla T)_{\text{rad}}}{(\nabla T)_{\text{ad}}} = \frac{15\overline{\kappa}(r)\rho(r)L(r)k}{32\pi ac\mu m_h GM(r)T^3(r)} \sim 1 \tag{4.5.1}$$

From such an estimate the dominance of one mechanism over another is not obvious. Could both methods compete roughly equally? Or is it more likely that one method will prevail in part of the star, while the remainder will be the domain of   97 the other. We have continually suggested that the latter is the case, and now we shall see the reason for this assertion.

b Schwarzschild Stability Criterion

For convection to play any role whatever, convective elements must be formed, and the conditions must be such that the elements will rise and fall. The statistical distribution law says that particles exist with the full range of velocities, and it would be remarkable if the particles were so uniformly distributed that any given volume had *exactly* the same number of particles of each velocity. This would be a very special particle distribution and not at all a random one. A random distribution would require that on some scale some volumes have more high-speed particles than others and hence be considered to be hotter. In fact, an entire spectrum of such volumes will exist and can be viewed as perturbations to the mean temperature. Thus, the first of our conditions for convective transport will always be met. Temperature fluctuations will always exist. But will they result in elements that move? In developing an expression for the adiabatic gradient, we assumed that the convective element will expand adiabatically and so do no work on the surrounding medium. This is certainly the most efficient way the element can move, and it cannot be exactly met in practice. To move, the element must displace the material ahead of it. There must be some "viscous" drag on the element requiring the element to do "work" on the surrounding medium. So the adiabatic expansion of a convective element is clearly the "best it can do" in getting from one place to another. Let us see if we can quantify this argument.

Let us assume that the gas is an ideal gas, and for the reasons mentioned above we assume that the element will behave adiabatically. Under these conditions, we know that the element will follow a polytropic equation of state, namely,

$$P \propto \rho^{\gamma} \qquad \gamma = \tfrac{5}{3} \tag{4.5.2}$$

Now consider a volume element which is displaced upward and has state variables denoted with an asterisk, while the surrounding values are simply P, T, and ρ (see Figure 4.2).

If $\rho_2^* \geq \rho_2$, then the element will sink or will not have risen in the first place. Initially, we require the conditions at point 1 to be the same (we are displacing the element in an ad hoc manner). Thus,

$$P_1^* = P_1 \qquad T_1^* = T_1 \qquad \rho_1^* = \rho_1 \tag{4.5.3}$$

Adiabatic expansion of the element requires that pressure equilibrium be maintained throughout the displacement,

$$P_2^* = P_2 \tag{4.5.4}$$

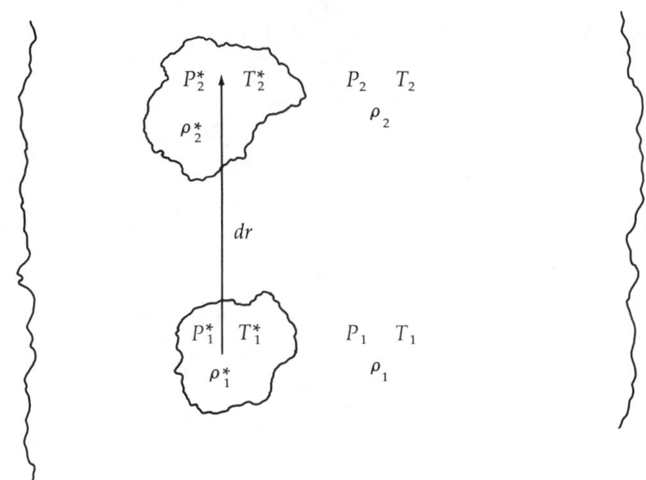

Figure 4.2 Schematic representation of a convective element with state variables denoted by an asterisk and surrounded by an ambient medium characterized by state variables P, T, and p. The element is initially at position 1 and is displaced through a distance dr to position 2.

We may express the conditions at point 2 in terms of a Taylor series and the conditions at point 1 so that

$$P_2^* = P_2 = P_1 + \frac{dP}{dr}\, dr + \cdots +$$

$$\rho_2 = \rho_1 + \frac{d\rho}{dr}\, dr + \cdots +$$

(4.5.5)

Using the equation of state, we may write

$$\rho_2^* = (\text{const})(P_2^*)^{1/\gamma} = (\text{const})(P_2)^{1/\gamma} = \rho_1\left(1 + \frac{1}{P_1}\frac{dP}{dr}\, dr + \cdots + \right)^{1/\gamma}$$

$$\approx \rho_1\left(1 + \frac{1}{\gamma P_1}\frac{dP}{dr}\, dr + \cdots + \right) \qquad (4.5.6)$$

If we take $\rho_2^* \geq \rho_2$ to be a condition for stability (i.e., the element will return to its initial position if displaced, then equations (4.5.5) and (4.5.6) require that

$$\frac{1}{\gamma P}\frac{dP}{dr} \geq \frac{1}{\rho}\frac{d\rho}{dr} \qquad (4.5.7)$$

The ideal-gas law requires that

$$\frac{1}{P}\frac{dP}{dr} = \frac{1}{\rho}\frac{d\rho}{dr} + \frac{1}{T}\frac{dT}{dr} \qquad (4.5.8)$$

which can be used to replace the density gradient in inequality (4.5.7) to get

$$\left(1 - \frac{1}{\gamma}\right)\frac{T}{P}\frac{dP}{dr} \leq \frac{dT}{dr} \qquad (4.5.9)$$

Dividing by dT/dr, we obtain the Schwarzschild stability criterion for a polytropic gas

$$\frac{d\ln P}{d\ln T} \geq \frac{\gamma}{\gamma - 1} = n + 1 \qquad (4.5.10)$$

which for a monatomic gas with $\gamma = \frac{5}{3}$ is just

$$\frac{d\ln P}{d\ln T} \geq 2.5 \qquad (4.5.11)$$

Thus, if the logarithmic derivative of pressure with respect to temperature is greater than or equal to 2.5, convection will not occur. In other words, if the actual temperature gradient is less than the adiabatic gradient, convection will not occur. This, then, is our means for deciding whether convection or radiation will be the dominant mode of energy transport. Should radiation be able to transport the energy with a temperature gradient less than the adiabatic gradient, no energy will be carried by convection, for the gas is stable against the thermal perturbations which must exist. However, if this is not the case, convection will be established; and it is so very efficient that it is capable of carrying all the energy with a temperature gradient that is just slightly superadiabatic. For most of stellar structure, we may regard energy transport as being bimodal; either radiation or convection will transport the energy, with the decision being made by equation (4.5.10). The Schwarzchild stability criterion has been shown to be quite general and will hold under the most varied of conditions, including those stars where general relativity must be included to describe their structure.

4.6 Equations of Stellar Structure

Having settled the mode of energy transport, we are in a position to describe the structure of a star in a steady-state condition. This is a good time to review briefly what we have done. The equations of stellar structure arise from conservation laws and relationships developed from the local microphysics. In Chapter 1, we posed the basic problem of stellar interiors to be the description of the variation of state variables P, T and ρ with position in the star. For spherical stars, this amounts to indicating their dependence on the radial coordinate r. In developing that description, we introduced additional variables and their relation to the state variables, so that by now our list of parameters has grown to nine members: $P(r)$, $T(r)$, $\rho(r)$, $M(r)$, $L(r)$, $\epsilon(r)$, $\bar{\kappa}(r)$, $\gamma(r)$, and $\mu(r)$. To specify these parameters, we have at our disposal three conservation laws and a transport equation in addition to three functional relationships derived from the microphysics. The function $\gamma(r)$ can also be specified by microphysics and is usually given by its adiabatic value. Only the variation of $\mu(r)$ needs to be specified *ab initio*. When we move to the stage of evolving the stellar models, the chemical composition will need to be specified for the initial model since the processes of nuclear energy generation will tell us how the composition changes with time. However, we must, at least initially, specify both the composition of the star and how it varies throughout the entire star. The use of a convective theory of transport which attempts to improve on the adiabatic gradient will also introduce another parameter, known as the *mixing length*, which must also be specified *ab initio*.

The constraints posed by the conservation laws take the form of differential equations whose solution is subject to a set of boundary conditions. Below is a summary of these differential equations and their origin:

$$(a)\quad \frac{dM(r)}{dr} = 4\pi r^2 \rho(r) \qquad\qquad \text{conservation of mass, eq. (2.1.8)}$$

$$(b)\quad \frac{dL(r)}{dr} = 4\pi r^2 \rho(r)\epsilon(r) \qquad\qquad \text{conservation of energy, eq. (4.2.16)}$$

$$(c)\quad \frac{dP(r)}{dr} = -\frac{GM(r)\rho(r)}{r^2} \qquad\qquad \text{conservation of momentum, eq. (2.1.6) (hydrostatic equilibrium)}$$

$$\frac{dT(r)}{dr} - \frac{3\bar{\kappa}(r)\rho(r)L(r)}{16\pi ac T^3(r)r^2} \qquad\qquad \text{radiative transport, eq. (4.2.4)}$$

$$(d)\quad \left(\frac{dT}{dr}\right)_{ad} = \frac{\mu m_h GM(r)}{(n+1)kr^2} \qquad\qquad \text{convective transport, eq. (4.3.19)}$$

$$(\Delta\nabla T) = \left[\frac{L^2(r)T(r)}{C_p^2\rho^2(r)GM(r)\pi^2 l^4 r^2}\right]^{1/3} \qquad \text{eq. (4.3.18)}$$

$$(4.6.1)$$

In addition to these differential equations we have the following relations from the microphysics:

$$\epsilon = \epsilon[T(r), \rho(r), \mu(r)] \qquad \text{nuclear energy production (Chap. 3)}$$

$$\bar{\kappa} = \bar{\kappa}[T(r), \rho(r), \mu(r)] \qquad \text{radiative opacity (Sec. 4.1)}$$

$$\gamma = \gamma[T(r), \rho(r), \mu(r)] = \tfrac{5}{3} \qquad \text{(Chap. 2)}$$

$$P = P[T(r), \rho(r), \mu(r)] \qquad \text{equation of state (Chap. 1)}$$

$$(4.6.2)$$

These eight relationships and the chemical composition completely specify the structure of the star. We now turn to describing methods by which their solution can be obtained.

4.7 Construction of a Model Stellar Interior

The construction of stellar models in steady state is essentially a numerical procedure which has been the subject of study of a large number of astrophysicists since the early 1950s and the pioneering work of Harm and Schwarzschild.[2] Basically two methods have been employed to solve the equations. The early work utilized a scheme described by Schwarzschild which amounts to a straightforward numerical integration of the differential equations of stellar structure. In the early 1960s, this procedure was superceded by a method due to L. Henyey which replaces the differential equations with a set of finite difference equations whose solution is carried out globally and enables one to include time-dependent phenomena in a natural way. However, since this method requires an initial solution which is usually obtained by the Schwarzschild procedure, we describe both methods.

a Boundary Conditions

Using the functional relations given by equations (4.6.2), we may reduce the problem of solving the structure equations to one of finding solutions for the four differential equations given in equations (4.6.1). These constitute a set of four nonlinear first-order differential equations in four unknowns. In general, such a system will have four constants of integration which must be specified to guarantee a solution. In principle, two of these constants are specified by requiring that the model be physically reasonable. These are

$$M(r) \to 0 \qquad \text{as } r \to 0$$

$$(4.7.1)$$

$$L(r) \to 0 \qquad \text{as } r \to 0$$

At the other end of the range of the independent variable,

$$M(r) \rightarrow M_* \qquad \text{as } r \rightarrow R_*$$
$$L(r) \rightarrow L_* \qquad \text{as } r \rightarrow R_* \tag{4.7.2}$$

However, five constants are specified by equations (4.7.1) and (4.7.2), if R_* is included as a parameter. Only four of these can be linearly independent. Thus, if one specifies M_* and R_*, the solution will specify L_*. Another aspect of the problem is that the constants are not all specified at the same boundary, and so it is not possible to treat the problem as an initial-value problem and to solve by straightforward numerical integration. Such problems are known as *two-point boundary-value problems*, and one must essentially guess the missing integration constants at one boundary, obtain the numerical solution complete to the other boundary, and adjust the guesses until the specified integration constants at the far boundary are obtained. A further problem arises from the fact that the equations of hydrostatic equilibrium and energy transport are numerically unstable as $r \rightarrow 0$ because the derivatives require the calculation of "0/0" at the origin. However, the problem can be recast as a double-eigenvalue problem with the fitting (solution adjustment) taking place in the interior but away from the boundary. This is essentially the Schwarzschild approach.

b Schwarzschild Variables and Method

When one is searching for the numerical solution to a physical problem, it is convenient to reexpress the problem in terms of a set of dimensionless variables whose range is known and conveniently limited. This is exactly what the Schwarzschild variables accomplish. Define the following set of dimensionless variables:

$$x = \frac{r}{R_*}$$

$$q = \frac{M(r)}{M_*} \qquad p = P\left(\frac{GM_*^2}{4\pi r_*^4}\right)^{-1} \tag{4.7.3}$$

$$f = \frac{L(r)}{L_*} \qquad t = T\left(\frac{m_h \mu GM_*}{kR_*}\right)^{-1}$$

Note that the first three variables are the fractional radius, mass, and luminosity, respectively, while the two at the right represent the pressure and temperature normalized by a constant which describes the way they vary homologously. In addition, let us assume that the opacity and energy generation rate can be approximated by

$$\bar{\kappa} = \kappa_0 \rho^n T^{-s} \qquad \epsilon = \epsilon_0 \rho^\lambda T^\nu \tag{4.7.4}$$

The differential equations of stellar structure then become

$$(a) \quad \frac{dp}{dx} = -\frac{qp}{tx^2} \qquad \text{hydrostatic equilibrium}$$

$$(b) \quad \frac{dq}{dx} = \frac{px^2}{t} \qquad \text{mass conservation}$$

$$(c) \quad \frac{df}{dx} = D(\lambda, v)x^2 p^{\lambda+1} t^{v-\lambda-1} \qquad \text{energy conservation} \qquad (4.7.5)$$

$$(d) \quad \frac{dt}{dx} = \frac{-C(n, s)fp^{n+1}}{x^2 t^{n+s+4}} \qquad \text{radiative equilibrium}$$

$$\frac{d\ln p}{d\ln t} = 2.5 \qquad \text{convective equilibrium}$$

which are subject to the boundary conditions

$$q(0) = f(0) = 0 \qquad q(1) = f(1) = 1 \qquad (4.7.6)$$

The parameters $C(n, s)$ and $D(\lambda, v)$ are the eigenvalues of the problem, and these values specify the type of star being considered. In physical variables they are

$$C(n, s) = \frac{3}{4(4\pi)^{n+2}ac} \left(\frac{k}{m_h G}\right)^{s+4} \frac{\kappa_0}{\mu^{s+4}} \frac{L_* R_*^{s-3n}}{M_*^{s-n+3}}$$

$$D(\lambda, v) = \frac{1}{(4\pi)^{\lambda}} \left(\frac{m_h G}{k}\right)^{v} (\epsilon_0 \mu^{v}) \frac{M_*^{v+\lambda+1}}{L_* R_*^{v+3\lambda}} \qquad (4.7.7)$$

Note that the ideal-gas law has been used to eliminate the density from the problem, and this many cause some problems with the solution at the surface where the pressure and temperature essentially go to zero, in addition to the numerical problems at the center when $x \rightarrow 0$. However, Schwarzschild shows that near the surface one may approximate the dimensionless pressure p and dimensionless temperature t by

$$t \approx \frac{n+1}{n+7.5} \frac{1-x}{x}$$

$$p \approx (\text{const})t^{(n+7.5)/(n+1)} \qquad \text{for radiative envelopes} \qquad (4.7.8)$$

If the star has a convective core, then all the energy is produced in a region where the structure is essentially specified by the adiabatic gradient and so the energy conservation equation [equation (4.7.5c)] is redundant. This means that the eigen-value $D(\lambda, v)$ is unspecified and the problem will be solved by determining $C(n, s)$ alone. Such a model is known as a *Cowling model*. The additional constraints on the solution are specified by the mass and size of the convective core (q_c and x_c). These are determined by the value of x for which $d(\ln p)/d(\ln t) < 2.5$, and the star becomes subject to the radiative temperature gradient. The stellar luminosity is then $L = L_c$, and for the envelope $f = 1$. While such a scheme works well for models with convective cores, numerical problems will generally occur at the center should it be in radiative equilibrium and the solution obtained numerically. However, a slightly different set of dimensionless variables can be defined where the pressure and temperature are scaled by their values at the center of the star. The differential equations of stellar structure become stable at $r = 0$ since the dimensionless pressure and temperature are both unity at the center by definition. One then integrates outward from the center with P_c and T_c as eigenvalues. The stellar mass, luminosity, and radius can be related to these new eigenvalues. That there are two distinct eigenvalues is demonstrated by the surface boundary condition that both the surface pressure and the surface temperature must vanish at the same value of x. Unfortunately the equations of stellar structure become numerically unstable near the surface for the same reasons that required the approximation of the solutions of equations (4.7.5) by equation (4.7.8). Although the errors in the model can be made small with the aid of modern computers, it is bad practice to numerically solve equations which are inherently unstable. For that reason, the usual procedure is to integrate from both the outside and the inside and to make the fit at the boundary between the core and envelope. The approximations near the surface are still present, but their effect on the solution is minimized. In actual practice, the fitting can be accomplished in the *UV* plane where the solutions are homologously invariant. The fitting procedure is similar to that described in Chapter 2.

Since Schwarzschild introduced this method of solution of the equations of stellar structure in the 1950s, many variants have been used by numerous investigators. In one form or another, all variants suffer from problems similar to those that plague the Schwarzschild procedure. In general, this approach to the numerical solution of two-point nonlinear boundary-value problems always suffers from the propagation of errors from one boundary to the other. The most serious of these errors are usually the truncation errors associated with the numerical integration scheme which tend to be systematic. However, this approach enabled the generation of stellar models which represented the steady-state aspect of stars for the first time. Although qualitative information about stellar evolution can be gained from polytropes (and we do so in Chapter 5), specific and detailed descriptions of stellar evolution require the generation of steady-state models. However, some aspects of stellar evolution happen on time scales which are very short compared to the thermal time scale, and in some instances short compared to the dynamical time scale. Often, substantive changes occur to the internal structure which produce only small changes at the surface. Thus, minor changes in the surface boundary conditions can reflect monumental changes in the internal structure of the star. In addition, we must include the time-dependent terms in the equations describing the conservation of momentum and energy. Specifically, if some of the generated energy does work on the star, causing it to expand as energy is liberated by contraction, then this energy must be included in the energy conservation equation relating the stellar luminosity to the sources of energy. This is usually accomplished by keeping track of the time rate of change of the entropy. The direct integration scheme does not readily lend itself to the inclusion of such terms. Such models are no longer merely steady-state models, and we will require more sophisticated tools to deal with them.

c   Henyey Relaxation Method for Construction of Stellar Models

To overcome some of the numerical instability problems described in the previous section, Louis Henyey et al.[3] developed a superior numerical scheme in the early 1960s. This method is the foundation for all modern stellar-model calculations. His approach was to transform the problem to a set of variables in which the non-linearity of the differential equations was minimized. The differential equations of stellar structure were then replaced with a set of finite difference equations whose solution could be carried out simultaneously over the entire model. This tended to reduce the effect of truncation error by spreading it more or less evenly across the model. Furthermore, the addition of time-dependent terms proved to be relatively easy to incorporate in the structure equations. We do not describe all the details that make this method so powerful, but only sketch the principles involved.

We begin by replacing the independent variable r with $M(r)$. Henyey noticed that the behavior of the equations was far more linear when the mass interior to r was used as the independent variable. The radial coordinate then becomes a dependent variable whose value must be found for any particular $M(r)$. If we make this transformation, the four differential equations of stellar structure become

$$\frac{\partial P}{\partial M(r)} = \frac{GM(r)}{4\pi r^4} \qquad \text{hydrostatic equilibrium}$$

$$\frac{\partial r}{\partial M(r)} = (4\pi r^2 \rho)^{-1} \qquad \text{conservation of mass}$$

$$\frac{\partial L}{\partial M(r)} = \epsilon(P,\, T,\, \mu) - T\frac{\partial S}{\partial t} \qquad \text{conservation of energy}$$

$$\frac{\partial \ln T}{\partial \ln P} = f(P,\, T,\, \mu) \qquad \text{energy transport}$$

$$(4.7.9)$$

Here we have explicitly included the time-dependent entropy term in the energy equation for purposes of example. In addition, we have written the energy transport term in a general manner which can accommodate either radiation or convection. Now we divide the star into $N - 1$ zones, starting with the center as the first point and ending at the surface or some outer point where the boundary conditions are known. By approximating the derivatives of equations (4.7.9) by the difference of the parameters at adjacent points, we get the following finite difference equations:

$$\frac{P_{i+1} - P_i}{M_{i+1} - M_i} = \frac{GM_{i+1/2}}{4\pi r_{i+1/2}^4}$$

$$\frac{r_{i+1} - r_i}{M_{i+1} - M_i} = (4\pi r_{i+1/2}^2)^{-1}$$

$$\frac{L_{i+1} - L_i}{M_{i+1} - M_i} = \epsilon_{i+1/2} - T_{i+1/2} \left.\frac{\partial S}{\partial t}\right|_{i+1/2} \tag{4.7.10}$$

$$\frac{T_{i+1} - T_i}{P_{i+1} - P_i} = \frac{T_{i+1/2}}{P_{i+1/2}} f(P_{i+1/2}, T_{i+1/2}, \rho_{i+1/2})$$

The subscript $i + \frac{1}{2}$ is used exclusively on the right-hand side of equations (4.7.10) to indicate that the value to be used is intermediate between the values at i and $i + 1$. It will turn out that we must have an initial guess of the model's structure to solve the finite difference equations. It is this guess which may supply the initial information for evaluating the parameters at the points $i + \frac{1}{2}$. Since the mass points M_i represent the independent variable of this problem, the four equations given in equations (4.7.10) contain eight unknowns. However, we have $N - 1$ systems of such equations with considerable overlap in unknowns among them. The situation at the outer zone will be handled somewhat differently since there is no $N + 1$ point. Thus if we count the total number of equations, we have $4N - 4$. But at each point there are only four unknowns, making the total number of unknowns of $4N$. The remaining four constraints are essentially the boundary conditions of the problem.

By analogy to the Schwarzschild problem, let us take the central boundary conditions to be $r_1 = L_1 = 0$, which removes two of the additional unknowns. Now if we choose two of the remaining unknowns at the surface, such as r_N and L_N, the problem is completely specified. Indeed, if we choose the surface pressure to be zero, then choosing a star of a particular mass and radius (and distribution of chemical composition) will specify the stellar configuration. One of the motivating notions that led Henyey to this type of technique was the ability to match a stellar interior to a model of the stellar atmosphere. This technique is ideally suited to do this. One simply takes the outer zone at that point where the physical parameters are known as the result of a separate study. In the second part of this book, we present a theory of stellar atmospheres which provides far more accurate surface boundary conditions than those of early investigators. In addition to improving the manner in which the surface boundary conditions are handled, it may be advisable to ignore the point at the center. A Taylor series expansion can be used to express the values of P_2, T_2, L_2, and r_2 in terms of the central temperature T_c and P_c. Because the system of equations is strongly diagonal, the solution is easier to come by if the central boundary conditions are expressed in this manner.

The Henyey approach shown in equations (4.7.10) represents the derivative of the structure equations by first-order finite differences. Thus the errors of the approximation are second order in those derivatives. This necessitates the use of the large number of zones to accurately represent the model, and it is this large number of zones that represents the primary computational burden in the construction of the model. Although increasing the order of the finite difference equations would improve the stability, it would also increase the density (i.e., the number of terms) of the resulting linear algebraic equations, slowing their solution and decreasing their stability. Budge[4] has shown that an improvement in the accuracy of the approximation can be achieved by using a Runga-Kutta fourth-order approximation for the derivatives without increasing the resulting linear equation density. Although there is some increase in the computational burden for obtaining the coefficients, this is more than offset by being able to reduce the number of zones in the model.

We still must solve these linear equations. It is not uncommon in the standard Henyey scheme to choose up to 500 zones in the star, which will yield some 2000 nonlinear equations in as many unknowns. Now it is clear why we need an initial solution. If we have a solution which is close to the correct one, we may express the correct solution in terms of the initial solution and a small linear correction to that solution. This will reduce the system of nonlinear equations to a linear system where the corrections are the unknowns. Such a scheme is known as a *Newton-Raphson iteration scheme*. Since the system is sparse (each equation contains only 8 of the 2000 unknowns) and the independent variable was chosen so as to make the equations somewhat linear, the iteration scheme is usually stable. However, the stability also depends on the quality of the initial solution. This is normally obtained by means of a Schwarzschild-type integration or a previously determined model.

It is clear that the Henyey method lends itself naturally to the problem of stellar evolution. In this case the initial model is a model calculated for an earlier time. Thus the procedure would be to start at some initial time with a Schwarzschild model, allow a small interval of time for the model to pass by, calculate the changes in the chemical composition resulting from nuclear processes, and modify the model accordingly. This serves as the initial first guess for the Henyey scheme, and a new model is produced. The effects of time are again allowed for, and the next Henyey model is constructed, etc. In this way an entire sequence of stellar models representing the life history of the star can be constructed. One generally starts the sequence when the star is well represented by a steady-state model, and the Schwarzschild solution gives an accurate description of the stellar structure. Such a time is the arrival on the main sequence and the accompanying onset of hydrogen burning. The resulting life history of the star is as good as the microphysics which goes into the solution and the quality of the computer and associated numerical techniques used to obtain the solution.

At this point, we have covered the fundamentals required to construct a model of the interior of a star. However, we should not leave the impression that such a model would reflect the accuracy of contemporary stellar interior models. There are many complications and refinements which should be treated and included to produce a model with modern sophistication. We have said nothing about the small departures of the equation of state from the ideal-gas law which occur at quite modest densities due to electron screening. Nor have we dealt with many of the vagaries of the theory of convection, such as semiconvection, convective overshoot, or mixing-length determination. These vagaries result largely from the primitive nature of the existing theory of convection, and while they do pose significant problems at certain points in a star's evolution, they do not affect the conceptual picture of stellar structure. It seems almost criminal not to devote more attention to the efforts of those who have labored to provide improved opacities and nuclear energy generation rates. But again, while these improve the details of the models and enhance our confidence in the predictions based on them, they do not conceptually change the basic physics upon which the models rest. While we have outlined the numerical procedures necessary to actually solve the structure equations, there is much cleverness and imaginative numerical analysis required to translate what we outlined to a computer program which will execute to completion in an acceptable time. Do not forget that the early models of Schwarzschild and Harm were calculated basically by hand, aided only by a desk calculator whose capabilities are far exceeded by even the cheapest pocket calculator of the present. It is no accident that the rapid advance of our knowledge of stellar structure parallels the explosive advance in our ability to carry out numerical calculations.

An understanding of the refinements of contemporary models is essential for any who would choose to do meaningful research in stellar interiors. It is not essential for those who would understand the results and their physical motivation, and it is to those people that this book is addressed. With the knowledge of the physical processes that determine the structure of stars, let us now turn to the crowning achievement of the study of stellar structure—the theory of stellar evolution.

Problems

1. Assume that there is a star in which the energy is generated uniformly throughout (that is, ϵ = constant). In addition, the opacity is constant (i.e., electron scattering). Further assume that the star is in radiative and hydrostatic equilibrium. Show that $1 - \beta$ is constant throughout the star and that it is a polytrope of index $n = 3$.

2. Suppose that in a star the only source of energy generation is radioactive decay, so the energy production per unit mass is constant and independent of density and temperature. Further suppose that the opacity is given by Kramers' law. Show that the structure of the star is described by a polytrope, and find the value · of the polytropic index n.

3. Compare the local value of the radiative gradient with the adiabatic gradient throughout the sun. Describe the regions of radiative and convective equilibrium in light of your results. What would you expect to be the effect on the radiative and convective zones of replacing 50% of the solar luminosity with an energy generation source which was more efficient?

4. For a sphere in radiative equilibrium and STE, show that the radiation pressure is one-third the energy density.

5. Since the convective temperature gradient differs systematically from the adiabatic gradient, it is possible that the cumulative effect is significant when it is integrated over the entire convective zone. Examine this effect in the sun, and decide whether it is significant.

6. Use a stellar interiors code or existing models to find the variation of the fractional ionization of hydrogen and helium with depth in the sun.

7. Use a stellar interiors code or existing models to find the fraction by mass and radius within which (*a*) 20%, (*b*) 50%, (*c*) 90%, and (*d*) 99% of the sun's energy is generated.

8. Repeat Problem 7 for a star of $10 M_\odot$.

9. Repeat Problem 7 but with $\mathbf{Z} = 10^{-8}$.

10. Determine the relative importance of bound-bound transitions, bound-free transitions, and electron scattering as opacity sources in the sun.

References and Supplemental Reading

1. Eddington, A. S., *The Internal Constitution of the Stars*, Dover, New York, 1926/1959, p. 276.

2. Harm, R., and M. Schwarzschild, "Numerical Integrations for the Stellar Interior," *Ap. J.*, Supp. No. 1, 1955, pp. 319–430.

3. Henyey, L. G., L. Wilets, K. H. Bohm, R. Lelevier, and R. D. Levee, "A Method for Automatic Computation of Stellar Evolution," *Ap. J.* 129, 1959, pp. 628–636.

4. Budge, K. G., "An Improved Method for Calculating Stellar Models," *Ap. J.* 312, 1987, pp. 217–218.

In addition to a very clear exposition of convection and the mixing-length theory, the clearest description of Schwarzschild variables and the Schwarzchild integration scheme can be found in

• Schwarzschild, M.: *Structure and Evolution of the Stars*, Princeton University Press, Princeton, N.J., 1958, chap. 3,

An exceptionally clear description of the Henyey method of integration of the equations of stellar structure is given by

• Kippenhahn, R., A. Weigert, and E. Hofmeister: "Methods for Calculating Stellar Evolution," *Methods in Computational Physics*, vol. 7 (Eds.: B. Adler, S. Fernbach, and M. Rotenberg), Academic, New York, 1967, pp. 129–190.

5

Theory of Stellar Evolution

One of the great triumphs of the twentieth century has been the detailed description of the life history of a star. We now understand with some confidence more than 90 percent of that life history. Problems still exist for the very early phases and the terminal phases of a star's life. These phases are very short, and the problems arise as much from the lack of observational data as from the difficulties encountered in the theoretical description. Nevertheless, continual progress is being made, and it would not be surprising if even these remaining problems were solved by the end of the century.

To avoid vagaries and descriptions which may later prove inaccurate, we concentrate on what is known with some certainty. Thus, we assume that stars can contract out of the interstellar medium, and generally we avoid most of the detailed description of the final, fatal collapse of massive stars. In addition, the fascinating field of the evolution of close binary stars, where the evolution of one member of the system influences the evolution of the other through mass exchange, will be left for another time. The evolution of so-called normal stars is our central concern.

Although the details of the theory of stellar evolution are complex, it is possible to gain some insight into the results expected of these calculations from some simple considerations. We have developed all the formalisms for calculating steady-state stellar models. However, those models could often be accurately represented by an equilibrium model composed of a polytrope or combinations of polytropes. We should then expect that the evolutionary history of a star could be approximately represented by a series of polytropic models. What is needed is to find the physical processes relating one of these models to another, thereby generating a sequence. Such a description is no replacement for model calculation for without the details, important aspects of stellar evolution such as lifetimes remain hidden. In addition, there are branching points in the life history of a star where the path taken depends on results of model calculations so specific that no general considerations will be able to anticipate them. However, a surprising amount of stellar evolution can be understood in terms of sequences of equilibrium models connected by some rather general notions concerning the efficiency of energy transfer. Descriptions of these models, and their relationship to one another, form the outline upon which we can hang the details of the model calculations.

In general, we trace the evolution of a star in terms of a model of that star's changing position on the Hertzsprung-Russell diagram. With that in mind, let us briefly review the range of parameters which define the internal structure of a star.

5.1   Ranges of Stellar Masses, Radii, and Luminosity

In Section 2.2b, we used the β^* theorem to show that as the mass of a star increases, the ratio of radiation pressure to total pressure also increases, so that by the time one reaches about $100M_\odot$ approximately 80 percent of the pressure will be supplied by the photons themselves. Although it is not obvious, at about this mass the outer layers can no longer remain in stable equilibrium, and the star will begin to shed its mass. Very few stars with masses above $100M_\odot$ are known to exist, and those that do show instabilities in their outer layers. At the other end of the mass scale, a mass of about $0.1M_\odot$ is required to produce core temperatures and densities sufficient to provide a significant amount of energy from nuclear processes. Thus, we can take the range of stellar masses to span roughly 3 powers of 10 with the sun somewhat below the geometric mean.

If we include the white dwarfs, which may be the size of the earth or less, the range of observed stellar radii is about 6 powers of 10, with the sun again near the geometric mean. The observed range of stellar surface temperatures is by far the smallest, being from about 2000 K for the coolest M star, to perhaps 50,000 K for some O-type stars. If the temperature range is combined with the range in radii, it is clear that we could expect a range as great as 17 powers of 10 in the luminosity. In practice, the largest stars do not have the highest temperatures, so that the range in luminosity is nearer 10 powers of 10. A reasonable range of these parameters is then

$$10^{-1}M_\odot \leq M_* \leq 10^2 M_\odot$$

$$10^{-3}R_\odot \leq R_* \leq 10^3 R_\odot \qquad (5.1.1)$$

$$10^{-4}L_\odot \leq L_* \leq 10^6 L_\odot$$

This, then, represents the ranges of the defining parameters of those objects we call normal stars. The theory of stellar evolution will tell us which parameters are related to various aspects of a star's life.

5.2   Evolution onto the Main Sequence

a   *Problems Concerning the Formation of Stars*

Since we began this discussion with the assumption that stars form by contraction from the interstellar medium, honesty requires that we describe several forces that mitigate against that contraction. For a star to form by gravitational contraction from the interstellar medium, all sources of energy which support the initial cloud must be dominated by gravity. For the typical interstellar cloud with sufficient mass to become a star, we shall see that not only is this not true of the collective sources of energy, but also it is not true of them individually.

Internal Thermal Energy   From the virial theorem as derived in Chapter 1 [equation (1.2.34)], the internal kinetic energy of the gas of the cloud must be less than one-half the gravitational energy in order for the moment of inertia to show any accelerative contraction. Thus for a uniform-density gas at a certain temperature T, the mass must be confined inside a sphere of a certain radius R_c. That radius can be found from

$$2\left(\frac{3\rho kT}{2\mu m_h}\right)\left(\frac{4\pi R_c^3}{3}\right) \leq \frac{GM^2}{R_c} \qquad (5.2.1)$$

or

$$R_c \leq \frac{GM\mu m_h}{3kT} \approx \frac{0.25(M/M_\odot)}{T} \quad \text{pc} \qquad (5.2.2)$$

This distance is sometimes known as the *Jeans length*, for it is the distance below which a gas cloud becomes gravitationally unstable to small fluctuations in density. For a solar mass of material with a typical interstellar temperature of 50 K, the cloud would have to be smaller than about 5×10^{-3} pc with a mean density of about 10^8 particles per cubic centimeter. This is many orders of magnitude greater than that found in the typical interstellar cloud, so it would seem unlikely that such stars should form.

Rotational Energy   The virial theorem can also be used to determine the effects of rotation on a collapsing cloud. Again, from Chapter 1, the rotational kinetic energy must be less than one-half the gravitational potential energy in order for the cloud to collapse. So

$$2(\tfrac{1}{2}I\omega^2) \leq \frac{GM^2}{R_c} \qquad (5.2.3)$$

which for a sphere of uniform density and constant angular velocity gives

$$R_c \leq \left(\frac{5GM}{2\omega^2}\right)^{1/3} \qquad (5.2.4)$$

The differential rotation of the galaxy implies that there must be a shear or velocity gradient which would impart a certain amount of rotation to any dynamical entity forming from the interstellar medium. For an Oort constant $A = 16$ km/s/kpc, this implies that

$$R_c \leq 0.9\left(\frac{M}{M_\odot}\right)^{1/3} \quad \text{pc} \qquad (5.2.5)$$

Thus, it would seem that to quell rotation, the initial mass of the sun must have been confined within a sphere of about 0.7 pc.

Magnetic Energy   A similar argument concerning the magnetic energy density M, where

$$M = \frac{H^2}{8\pi}\frac{4\pi R_c^3}{3} \qquad (5.2.6)$$

can be made by appealing to the virial theorem, with the result that

$$R_c \leq 0.37 \left(\frac{M}{M_\odot H_{\mu g}}\right)^{1/2} \quad \text{pc} \tag{5.2.7}$$

For a value of the ambient interstellar galactic magnetic field of 5 μg, we get

$$R_c \leq 0.17 \left(\frac{M}{M_\odot}\right)^{1/2} \quad \text{pc} \tag{5.2.8}$$

How are we to reconcile these impediments to gravitational contraction with the fact that stars exist? We can use the rotational and magnetic energies against one another. A moderate magnetic field of a spinning object will cause a great deal of angular momentum per unit mass to be lost by a star through the centripetal acceleration of a stellar wind. The resulting spin-down of the star will weaken the internal sources of the stellar magnetic field itself. Observations of extremely slow rotation among the magnetic A_p stars seem to suggest that this mechanism actually occurs. Clouds can be cooled by the formation of dust grains and molecules, as long as the material is shielded from the light of stars by other parts of the cloud. The high densities and low temperatures observed for some molecular clouds imply that this cooling, too, is occurring in the interstellar medium. However, unless some sort of phase transition occurs in the material, the thermal cooling time is so long that it is unlikely that the cloud will remain undisturbed for a sufficient time for the Jeans' condition to be reached. Thus it seems unlikely that the Jeans' condition can be met for low-mass clouds.

It is clear from equation (5.2.2) that $R_c \sim \sqrt{T/\rho}$, so for a given temperature the Jeans' length increases with decreasing density. However, the Jeans' mass increases as the cube of the Jeans' length. Thus, for a cloud of typical interstellar density to collapse, it must be of the order of $10^4 M_\odot$. It is thought that the contraction of these large clouds creates the conditions enabling smaller condensations within them to form protostars. The pressure that the large contracting cloud exerts on smaller internal perturbations of greater density may squeeze them down to within the Jeans' length, after which these internal condensations unstably contract to form the protostars of moderate mass. These are some of the arguments used to establish the conditions for gravitational contraction upon which all stellar formation depends; and since stars do form, something of this sort must happen.

b Contraction out of the Interstellar Medium

Since we have given some justification for the assumption that stars will form out of clouds of interstellar matter which have become unstable to gravitational collapse, let us consider the future of such a cloud.

Homologous Collapse   For simplicity, consider the cloud to be spherical and of uniform density. The equation of motion for a unit mass of material somewhere within the cloud is

$$\frac{d^2 r}{dt^2} = -\frac{GM(r)}{r^2} \tag{5.2.9}$$

If we assume that the material at the center does not move [that is, $v(0) = 0$], then the first integral of the equations of motion yields

$$v^2 = \frac{8\pi G\rho_0}{3} \int_0^r r\,dr = \frac{4\pi G\rho_0}{3}\, r^2 \tag{5.2.10}$$

or

$$v \propto r \tag{5.2.11}$$

This says that at any time the velocity of *collapse* is proportional to the radial coordinate. This is a self-similar velocity law like the Hubble law for the expansion of the universe, only in reverse. Thus, at any instant the cloud will look similar to the cloud at any other point in time, only smaller and with a higher density ρ_0. Thus, the density will remain constant throughout the cloud but steadily increase with time. Since the velocity is proportional to r, the collapse is homologous and we obtain Lane's law of Chapter 2 [equation (2.3.9)] which completely specifies the internal structure throughout the collapse.

One should not be left with the impression that this homologous collapse is uniform in time. It is not; rather, it proceeds in an accelerative fashion, resulting in a rapid compaction of the cloud. When the density increases to the point that internal collisions between particles produce a pressure sufficient to oppose gravity, the equations of motion become more complicated. Some of the energy produced by the collapse leaks away in the form of radiation from the surface of the cloud, and a temperature gradient is established. These processes destroy the self-similar, or homologous, nature of the collapse, and so we must include them in the equations of motion.

However simple and appealing this solution may be, it is a bit of a swindle. The mathematics is correct, and the assumption that $v(0) = 0$ may be quite reasonable. However, it is unlikely that most clouds are spherically symmetric and of uniform density. Density fluctuations must exist, and without the pressures of hydrostatic equilibrium to oppose the central force of gravity, there is no *a priori* reason to assume spherical symmetry. Normally this would seem like unnecessary quibbling with an otherwise elegant solution. Unfortunately, these perturbations are amplified by the collapse itself and destroy any possibility of the cloud maintaining a uniform density.

Nonhomologous Collapse Let us consider the same equations of motion as before, so that the first integral is given by

$$\int_{v_i^2}^{v^2} dv^2 = 2 \int_{r_i}^{r} \frac{4\pi}{3} G\rho_0 r \, dr = 2 \int_{r_i}^{r} \frac{GM(r) \, dr}{r^2} \tag{5.2.12}$$

Now we wish to follow the history of a point within which the mass is constant so that $M(r)$ is constant and

$$v^2(R) - v_i^2 = 2GM(R)\left(\frac{1}{R} - \frac{1}{R_i}\right) \tag{5.2.13}$$

The variable R has replaced r, and this introduces a minus sign into equation (5.2.13), since for a collapse $dr = -dR$. Equation (5.2.13) is just the energy integral, so there are no surprises here. Now we change variables so that

$$x = \frac{R}{R_i} \qquad \alpha = \frac{2GM(R_i)}{R_i^3} \qquad u = \frac{v}{R_i} \tag{5.2.14}$$

If we take the initial velocity of the cloud to be zero and the initial value of x to be 1, then equation (5.2.13) becomes

$$u^2 = \left(\frac{dx}{dt}\right)^2 = \frac{\alpha(1 - x)}{x} \qquad 0 \leq x \leq 1 \tag{5.2.15}$$

which can be integrated over time to give

$$t(x) = \alpha^{-1/2}\left[\frac{\pi}{2} - \mathrm{Sin}^{-1}\sqrt{x} + (x - x^2)^{1/2}\right] \tag{5.2.16}$$

Now α can be related to the mean density, so that we can rewrite equation (5.2.16) as

$$t\left(\frac{R}{R_i}\right) = \left[\frac{3}{8\pi G\overline{\rho_i(R_i)}}\right]^{1/2}\left\{\frac{\pi}{2} - \mathrm{Sin}^{-1}\left(\frac{R}{R_i}\right)^{1/2} + \left[\frac{R}{R_i} + \left(\frac{R}{R_i}\right)^2\right]^{1/2}\right\} \tag{5.2.17}$$

If $\overline{\rho_i(R_i)}$ is constant and not a function of R_i, then we recover the homologous contraction which is clearly not uniform in time. However, if the initial mean density is a decreasing function of R_i, then the collapse time of a sphere of material $M(R_i)$ is an increasing function of R_i. This means that initial concentrations of material will become more concentrated and any inhomogeneities in the density will grow unstably with time.

This is essentially the result found by Larson[1] in 1969. If the cloud is gravitationally confined within a sphere of the Jeans' length, the cloud will experience rapid core collapse until it becomes optically thick. If the outer regions contain dust, they will absorb the radiation produced by the core contraction and reradiate it in the infrared part of the spectrum. After the initial free-fall collapse of a $1M_\odot$ cloud, the inner core will be about 5 AU surrounded by an outer envelope about 20,000 AU. When the core temperature reaches about 2000 K, the H_2 molecules dissociate, thereby absorbing a significant amount of the internal energy. The loss of this energy initiates a second core collapse of about 10 percent of the mass, with the remainder following as a "heavy rain." After a time, sufficient matter has rained out of the cloud, and the cloud becomes relatively transparent to radiation and falls freely to the surface, producing a fully convective star. While this scenario seems relatively secure for low-mass stars (i.e., around $1M_\odot$), difficulties are encountered with the more massive stars. Opacities in the temperature range of 1500 to 3000 K make the evolutionary tracks somewhat uncertain. Indeed, there are some indications that massive stars follow a more homologous and orderly contraction to the state where they become fully convective.

Although this is the prevailing picture for the early phases of the evolution for low-mass protostars, there are some difficulties with it. Such stars would be shielded from observation by the infalling rain of material until quite late in their formation. Some suggest that since the entire configuration including the rain is hardly in a state of hydrostatic equilibrium, the arguments given above would not pertain until quite late in the star's formation, by which time the star may well have reached the main sequence. There seems to be little support in observation for this point of view, and the entire subject is still somewhat controversial.

Michael Disney[2] has pointed out that the details of the collapse from the interstellar medium depend critically on the ratio of the sound travel time to the free-fall time in the contracting protostar. Although this ratio is typically unity [equations (3.2.4), (3.2.6), and (3.2.9)], small departures from unity appear to matter. The free-fall time is basically the time during which the collapse takes place, and the sound travel time is the time required for the interior to sense the effects of pressure disturbances initiated at the boundary. Thus, if $\tau_s/\tau_f > 1$, the interior tends to be unaffected by the boundary pressure during the collapse. Any external pressure will then tend to compress the matter in the outer part of the collapsing cloud without affecting the interior regions, removing any density gradients that may exist in the perturbation and forcing the collapse to be more nearly homologous. This would reduce the effect of the rain and cause the protostar to collapse more as a unit. Any initial velocity resulting from the homologous collapse of the large cloud will only exacerbate the situation by significantly shortening the time required for the collapse. Thus the initial phases of star formation remain in some doubt and probably depend critically on the circumstances surrounding the initial conditions of the collapse of the larger cloud.

c Contraction onto the Main Sequence

Once the protostar has become opaque to radiation, the energy liberated by the gravitational collapse of the cloud cannot escape to interstellar space. The collapse will slow down dramatically, and the future contraction will be limited by the star's ability to transport and radiate the energy away into space. Initially, it was thought that such stars would be in radiative equilibrium and that the future of the star would be dictated by the process of radiative diffusion in the central regions of the star. Indeed, for most stars this is true for the phases just prior to nuclear ignition. However, Hayashi[3] showed that there would be a period after the central regions became opaque to radiation during which the star would be in convective equilibrium.

Hayashi Evolutionary Tracks In Chapter 4 we found that once convection is established, it is incredibly efficient at transporting energy. Thus, as long as there are no sources of energy other than gravitation, the future contraction will be limited by the star's ability to radiate energy into space rather than by its ability to transport energy to the surface. We have also learned that the structure of a fully convective star will essentially be that of a polytrope of index $n = 1.5$. We may combine these two properties of the star to approximately trace the path it must take on the Hertzsprung-Russell diagram.

With gravitation as the only source of energy and the contraction taking place on a time scale much longer than the dynamical time, the virial theorem allows one-half of the change in gravitational energy to appear as the luminosity and be radiated away into space. The other half will go into the internal energy of the star, increasing the internal temperature. Thus,

$$L = \frac{1}{2}\frac{d(GM^2/R)}{dt} = -\frac{\frac{1}{2}GM^2}{R^2}\frac{dR}{dt} \tag{5.2.18}$$

Since the luminosity is positive, dR/dt must be negative, which ensures that the star will contract. Since the luminosity is related to the surface parameters by

$$L = 4\pi R_*^2 \sigma T_e^4 \tag{5.2.19}$$

the change in the luminosity with respect to the radius will be

$$\frac{dL}{dR_*} = \frac{4L}{T_e}\frac{dT_e}{dR_*} + \frac{2L}{R_*} \qquad (5.2.20)$$

Equation (5.2.19) is essentially a definition of what we mean by the *effective temperature*. As long as the star remains in convective equilibrium, it will be a polytrope and the contraction will be a self-similar, and thus homologous, contraction.

Since the rate of stellar collapse is dictated by the photosphere's ability to radiate energy, we should expect the photospheric conditions to dictate the details of the collapse. Indeed, as we shall see in the last half of this book, the eigenvalues that determine the structure of a stellar atmosphere are the surface gravity and the effective temperature. So as long as the stellar luminosity is determined solely by the change in gravity, and the energy loss is dictated by the atmosphere, we might expect that the independent variable T_e to remain unchanged. However, it is necessary to show that such a sequence of models actually forms an evolutionary sequence. The extent to which this will be true depends on the radiative efficiency of the photosphere. This is largely determined by the opacity. At low temperatures the opacity will increase rapidly with temperature owing to the ionization of hydrogen. This implies that any homological increase of the polytropic boundary temperature at the base of the atmosphere will be met by an increase in the radiative opacity and a steepening of the resultant radiative gradient. This increase in the radiative opacity also forces the radiating surface farther away from the inner boundary, causing the effective temperature to remain unchanged. A much more sophisticated argument demonstrating this is given by Cox and Giuli.[4]

The star can effectively be viewed as a polytrope wrapped in a radiative blanket, with the changing size of the polytrope being dictated by the leakage through the blanket. The blanket is endowed with a positive feedback mechanism through its radiative opacity, so that the effective temperature remains essentially constant. The validity of this argument rests on the ability of convection to deliver the energy generated by the gravitational contraction efficiently to the photosphere to be radiated away. With this assumption, we should expect the effective temperature to remain very nearly constant as the star contracts. Thus dT_e/dR_* in equation (5.2.20) will be approximately zero, and we expect the star to move vertically down the Hertzsprung-Russell (H-R) diagram with the luminosity changing roughly as R_*^2 until the internal conditions within the star change. Thus for the Hayashi tracks

$$\frac{dT_e}{dR_*} = \frac{dT_e}{dL} = 0 \qquad \frac{d\ln L}{d\ln R_*} = +2 \qquad (5.2.21)$$

While the location of a specific track will depend on the atomic physics of the photosphere, the relative location of these tracks for stars of differing mass will be determined by the fact that the underlying star is a polytrope of index $n = \frac{3}{2}$. From the polytropic mass-radius relation developed in Chapter 2 [equation (2.4.21)] we see that

$$M^{1/3}R = \text{constant} \tag{5.2.22}$$

and that

$$\frac{d \ln R_*}{d \ln M} = -\frac{1}{3} \tag{5.2.23}$$

Equations (5.2.19) and (5.2.20) also imply that

$$\frac{dL}{dM} = \frac{2L}{R_*}\frac{dR_*}{dM} + \frac{4L}{T_e}\frac{dT_e}{dM} \tag{5.2.24}$$

If we inquire as to the spacing of the vertical Hayashi tracks in the H-R diagram, then we can look for the effective temperatures for stars of different mass but at the same luminosity. Thus, we can take the left-hand side of equation (5.2.24) to be zero and combine the right-hand side with equation (5.2.23) to get

$$\frac{d \ln T_e}{d \ln M} = +\frac{1}{6} \tag{5.2.25}$$

This extremely weak dependence of the effective temperature on mass means that we should expect all the Hayashi tracks for the majority of main sequence stars to be bunched on the right side of the H-R diagram. Since the star is assumed to be radiating as a blackbody of a given T_e and is in convective equilibrium, no other stellar configuration could lose its energy more efficiently. Thus no stars should lie to the right of the Hayashi track of the appropriate mass on the H-R diagram; this is known as the *Hayashi zone of avoidance*.

We may use arguments like these to describe the path of the star on the Hertzsprung-Russell diagram followed by a gravitationally contracting fully convective star (see Figure 5.1). As we suggested, this contraction will continue until conditions in the interior change as a result of continued contraction.

As the star moves down the Hayashi track, the internal temperature will increase in a homologous fashion so that $T \sim \mu M/R$. Hence we could expect the adiabatic gradient $\nabla T_{\text{ad}} \sim \mu M/R^2$. However, the radiative gradient is

$$\frac{dT}{dr} = -\frac{3\kappa\rho L(r)}{16\pi ac T^3 r^2} \sim \frac{\kappa(M/R^3)L}{(\mu M/R)^3 R^2} \sim \frac{\kappa L}{\mu^3 M^2 R^2} \tag{5.2.26}$$

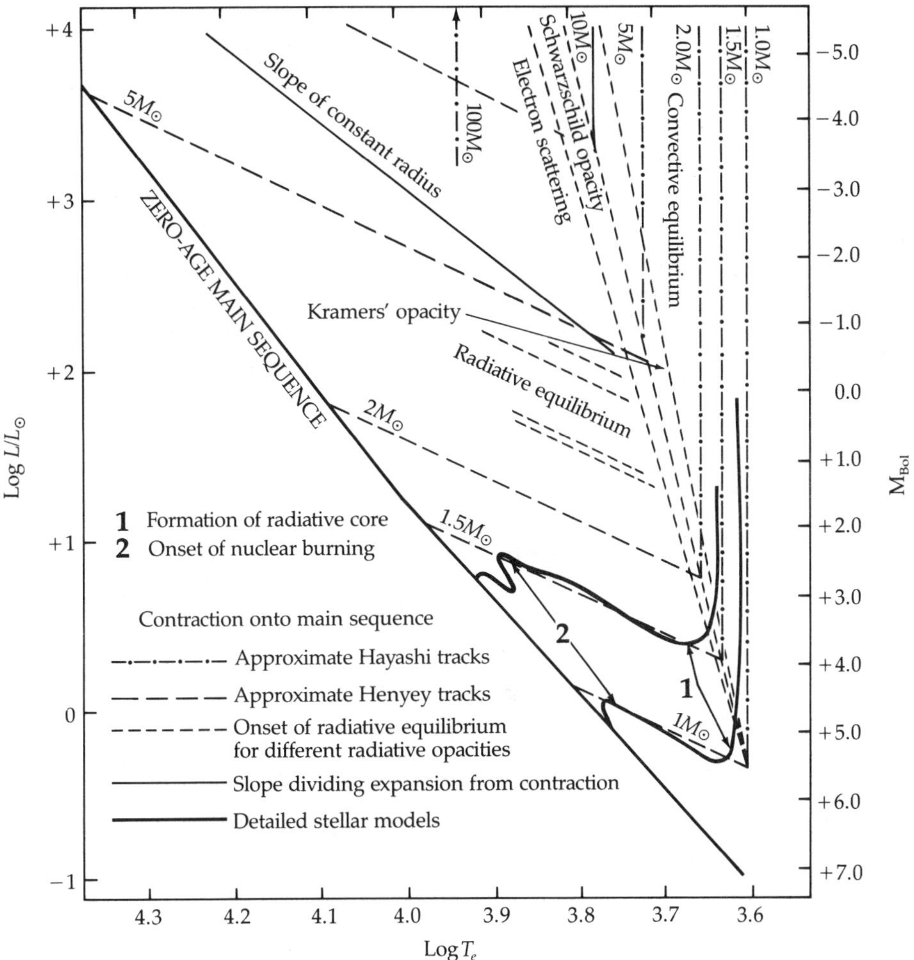

Figure 5.1 shows the schematic tracks for fully convective stars and radiative stars on their way
to the main sequence. The low dependence of the convective tracks on mass implies that most
contracting stars will occupy a rather narrow band on the right hand side of the H-R diagram.
The line of constant radius clearly indicates that stars on the Henyey tracks continue to contract.
The dashed lines indicate the transition from convective to radiative equilibrium for differing
opacity laws. The solid curves represent the computed evolutionary tracks for two stars of
differing mass.[5]

To find the homological behavior of the radiative opacity, we may use the approximate formulas [equation (4.1.19)] for Kramers-like opacity. Making use of the homology transformations for ρ and T, we can calculate the ratio of the adiabatic to
radiative gradient $\Re$ as

$$\Re \sim \frac{M^{s-n+3}}{L(R^{s-3n})} \tag{5.2.27}$$

As the star contracts down the Hayashi track, $\mathfrak{R}$ will steadily increase. At some point, depending on the dominant source of opacity, the adiabatic gradient will exceed the radiative gradient, and convection will cease. This will not happen globally all at once; rather, a radiative core will form that propagates outward until the entire star is radiative. At that point the mode of collapse will change because the primary barrier to energy loss will move from the photosphere to the interior and the diffusion of radiant energy. Since all the models on the Hayashi tracks are convective polytropes, we might expect this point to happen at the same value of $\mathfrak{R}$ for stars of differing mass. If this is the case, then, remembering that for stars on the Hayashi tracks $L \sim R^2$, we may use equation (5.2.27) to find that the locus of points of constant $\mathfrak{R}$ lies along a line such that

$$\frac{d \ln L}{d \ln M} = 2 \left(\frac{s - n + 3}{s - 3n + 2} \right) \tag{5.2.28}$$

Remembering that for electron scattering $n = s = 0$, while for Kramers' opacity $s = \frac{7}{2}$ and n is 1 or 0.75, depending on the relative dominance of free-free to bound-free opacity, we can obtain the appropriate mass luminosity law for the dominant source of opacity at the point of transition from convective to radiative equilibrium. Combining this with equation (5.2.25), we find that the locus of points in the H-R diagram will be described by

$$\frac{d \ln L}{d \ln T_e} = \begin{cases} +18 & \text{electron scattering} \\ +21.24 & \text{Schwarzschild opacity for bound-bound opacity} \\ +26.4 & \text{Kramers' opacity for free-free opacity} \end{cases} \tag{5.2.29}$$

For the very massive stars, radiation pressure may play an important role toward the end of the Hayashi contraction phase, so that the onset of radiative equilibrium occurs sooner, increasing the value on the right-hand side of equation (5.2.28) slightly. But for stars with a mass less than about $3 M_\odot$, equations (5.2.25) and (5.2.29) will describe their *relative* positions on the H-R diagram with some accuracy.

Henyey Evolutionary Tracks　After sufficient time has passed for the adiabatic gradient to exceed the radiative gradient, convection ceases and the main barrier to energy loss is no longer the ability of the photosphere to radiate energy into space. Rather the radiative opacity of the core slows the leakage of energy generated by gravitational contraction, and the atmosphere no longer provides the primary barrier to the loss of energy. Further contraction now proceeds on the Kelvin-Helmholtz time scale. As the star continues to shine, the gravitational energy continues to become more negative, and to balance it, in accord with the virial theorem, the internal energy continues to rise. This results in a slow but steady increase in the temperature gradient which results in a steady increase in the luminosity as the radiative flux increases. This increased luminosity combined with the ever-declining radius produces a sharply rising surface temperature as the photosphere attempts to accommodate the increased luminosity. This will yield tracks on the H-R diagram which move sharply to the left while rising slightly (see Figure 5.1). For the reasons mentioned above, the beginning of these tracks will be along a series of points which move upward and to the left for stars of greater mass.

We may quantify this by asking how the luminosity changes in time. We differentiate equation (5.2.18) and obtain

$$\frac{dL}{dt} = -\frac{1}{2}\frac{d^2\Omega}{dt^2} = -\frac{aGM^2}{R^2}\left[-\frac{2}{R}\left(\frac{dR}{dt}\right)^2 + \frac{d^2R}{dt^2}\right] \qquad (5.2.30)$$

The parameter a is simply a measure of the central condensation of the model, which we require to be independent of time. This requirement is satisfied if the contraction is homologous. If we further invoke the virial theorem and require that the contraction proceed so as to keep the second derivative of the moment of inertia equal to zero, then

$$\frac{d^2I}{dt^2} = \frac{d^2}{dt^2}\left(\alpha MR^2\right) = 0 \Rightarrow \left(\frac{dR}{dt}\right)^2 + R\frac{d^2R}{dt^2} = 0 \qquad (5.2.31)$$

Using this to replace the second derivative in equation (5.2.30), we get

$$\frac{d\ln L}{d\ln R} = -3 \qquad (5.2.32)$$

Multiplying equation (5.2.24) by $(R/L)(dM/dR)$ and combining with the above result, we get

$$\frac{d\ln T_e}{d\ln R} = -\frac{5}{4} \qquad \frac{d\ln L}{d\ln T_e} = +\frac{12}{5} \qquad (5.2.33)$$

Thus we can expect the star to move upward and to the left on the H-R diagram with a slope of 2.4. This path will eventually carry it to the main sequence where nuclear burning will set in as a consequence of the steady increase in the central temperature. Since the effect of onset of nuclear burning will be similar for a wide range of main sequence stars, we can expect the luminosity distribution of the Henyey tracks with mass to be reflected on the main sequence. As a result, we might expect that equation (5.2.28) will reflect the main sequence mass-luminosity

125

relation. Indeed, Harris, Strand, and Worley[6] give empirical values of 2.76 for the exponent on the mass of the mass-luminosity relation for the lower main sequence and 4 for the upper main sequence. The values obtained from equation (5.2.28) corresponding to different types of opacity are

$$\frac{d\ln L}{d\ln M} = \begin{cases} +3 & \text{electron scattering} \\ +3.54 & \text{Schwarzschild's opacity} \\ +4.4 & \text{Kramers' opacity} \end{cases} \qquad (5.2.34)$$

Although the proper evolutionary tracks for a star contracting to the main sequence require more exact modeling than can be done with polytropes, the overall effects can be estimated by considering sequences of equilibrium configurations linked by a description of those physical processes which limit the energy flow from the star. To dramatize this, we included in Figure 5.1 some calculated evolutionary tracks for two stars. The salient features of the pre-main sequence evolution are reasonably described by the curves in spite of the crude assumptions involved.

5.3 Structure and Evolution of Main Sequence Stars

When the track of a gravitationally contracting star intersects the main sequence, the star has reached a point in its life when it will be stable for an extended period of time. This is ensured observationally by the fact that about 90 percent of all stars reside on or very near the main sequence and so must be involved in the utilization of their most prolific and efficient source of energy—the fusion of hydrogen into helium. So we may take the intersection of the Henyey track with the main sequence as an indication that hydrogen ignition has begun in the stellar core.

Actually nuclear processes begin somewhat before the main sequence is encountered. The first constituents of the star to undergo nuclear fusion are deuterium and lithium which require conditions substantially below that of hydrogen for their ignition. However, their abundance is sufficiently low that they provide little more than a stabilizing effect on the star as it proceeds along its Henyey track, causing the star to hook onto the main sequence.

For the sun, about a million years is required for the equilibrium abundances of the proton-proton cycle to be established with sufficient accuracy for their use in the energy generation schemes. At this point, the star can be said to have arrived at the *zero-age main sequence*. As the name implies, this is generally taken as the beginning point of stellar evolution calculations, as the onset of nuclear burning makes the details of the prior evolution largely irrelevant to the subsequent evolution. In some real sense, the star forgets where it came from.

Since it is fairly obvious that the effects of stellar evolution during the main sequence phase will result in little movement on the H-R diagram, we need to understand more of the structure of the interior to appreciate these effects. Therefore, we begin by describing the structure to be expected for the hydrogen-burning models that describe the main sequence. The structure of main sequence stars can be readily broken into two distinct groups: those that occupy the upper half of the main sequence and those that occupy the lower main sequence.

a   Lower Main Sequence Stars

We define the *lower main sequence* to be those stars with masses less than about 2 solar masses. For these stars, after the trace elements with low ignition temperatures have been exhausted and hydrogen fusion has begun, the equilibrium structure is established in about a million years. The mass of these stars is insufficient to produce a central temperature high enough to initiate the CNO cycle, so the primary source of energy is the proton-proton cycle. Models indicate that in the sun 98 percent of the energy is supplied by the proton-proton cycle. The relatively low dependence on temperature of the proton-proton cycle implies that the energy generation will be less concentrated toward the center than would be the case with the CNO cycle. This and the modest central temperature imply that a temperature gradient less than the adiabatic gradient is all that is required to carry the energy produced by the *p-p* cycle. Thus, these stars have a central core which is in radiative equilibrium. However, in the sun, the adiabatic gradient is never far from the actual temperature gradient, and it would take a very little increase in the temperature gradient to cause the core to become unstable to convection. Indeed, the conditions for convective instability are met in the outer regions of these stars, resulting in the formation of a convective envelope. In the sun, this point is reached at about $0.75R_\odot$, so that about 98.8 percent of the mass is included in the radiative core. Ultimately, the situation is reversed near the surface, as it must be, for the energy leaves the surface of the star by radiation.

The existence of the radiative core in stars of the lower main sequence has a significant effect on the subsequent evolution of the star. The ^{4}He, which is the end product of hydrogen burning, remains in the locale in which it is produced. However, since the production rate is strongly dependent on temperature, the helium abundance increases more rapidly as one approaches the center of the star. The helium must be supported against its own gravity while it contributes nothing to the support of the remainder of the star. As a result, the internal temperature will increase to maintain the luminosity in the face of decreasing hydrogen abundance and the increasing mass of the particles (i.e., the ^{4}He). This is why the temperature scales with the mean molecular weight μ [see equation (2.3.8)]. Thus, we should expect stars like the sun to slowly increase in brightness, as the internal temperature rises, during their main sequence lifetime. Indeed, the standard solar model indicates that the solar luminosity has increased by about 40 percent since its arrival on the zero-age main sequence.

Toward the end of the star's main sequence life, the helium abundance will rise to the point where a core of helium, surrounded by a hydrogen-burning shell, will form in the center of the star. The support of this isothermal helium core is eventually helped by the Pauli exclusion principle. In Chapter 1, we outlined the equation of state to be expected for a gas where all the available h^3 volumes of phase space were filled. Because of their lower mass, this condition will be experienced first by the electrons. The degenerate equation of state does not contain the temperature and therefore permits the existence of an electron pressure capable of assisting in the support of the helium core; this equation is independent of the conditions existing in the hydrogen-burning shell. Thus as the core builds, we could expect its structure to shift from that of an isothermal sphere, described in Chapter 2, to that of a polytrope with a $\gamma = \frac{5}{3}$, as would be dictated by the electron pressure of a fully degenerate gas. This change from an isothermal sphere to a polytrope will dictate the mass distribution, for the pressure of the ions becomes small compared to that of the electrons. However, because of the high conductivity of a degenerate gas, the configuration will remain isothermal since any energy surplus can immediately be transported to a region of energy deficit by electron conduction. Thus, the region is still known as the isothermal helium core, even though the pressure equilibrium is dictated by the electron pressure behaving as a polytropic gas with a γ of $\frac{5}{3}$.

Therefore, the main sequence lifetime of a low-mass star consists of a steady energy output from hydrogen burning in an environment of steadily increasing helium. On a nuclear time scale, the helium abundance increases preferentially in the most central regions, causing the temperature to rise, which results in a slow increase in the luminosity throughout the main sequence lifetime of the star. After about 10 percent of the radiative core mass has been consumed, an isothermal helium core begins to form and structural changes begin to occur very rapidly. This signals the end of the main sequence lifetime.

b Upper Main Sequence Stars

The situation regarding the stellar structure for stars of more than 2 solar masses is nearly reversed from that of the lower main sequence. For stars on the main sequence, the observed mass-radius relation is approximately

$$M \propto R^{4/3} \tag{5.3.1}$$

However, from the homology relations in Chapter 2 [equations (2.3.8)], we know that

$$T \propto \frac{M}{R} \tag{5.3.2}$$

Therefore, for stars along the main sequence, we expect the central temperature to increase slowly as we proceed up the main sequence in accord with

$$T_c \propto R^{1/3} \propto M^{1/4} \qquad\qquad (5.3.3)$$

This slow rise in the central temperature will result in a greater fraction of the energy being produced by the more temperature-sensitive CNO cycle. Thus, by the time one reaches stars of greater than about 2 solar masses, the CNO cycle will be the dominant source of energy production. The much larger temperature sensitivity of the CNO cycle as compared to the *p-p* cycle means that the region of energy production will be rather more centrally concentrated than in stars of less mass. This, in turn, requires a steeper temperature gradient in order to transport the energy to the outer parts of the star. Since in the sun the radiative gradient was already quite close to the adiabatic gradient, this small increase is sufficient to cause the inner regions to become convectively unstable, and a substantial convective core will be established. However, in the outer parts of the star, the declining density causes the product of $\bar{\kappa}\rho$, which appears in the radiative gradient [equation (4.2.14)], to reduce the radiative gradient below that of the adiabatic gradient, and so convection stops. Thus, we have a star composed of a convective core surrounded by an envelope in radiative equilibrium. This role reversal for the core and envelope has a profound effect on the evolution of the star.

The presence of a convective core ensures that the inner regions of the star will be well mixed. As helium is produced from the burning of hydrogen, it is mixed thoroughly throughout the entire core. Thus, we do not have a buildup of a helium core that increases in helium abundance toward the center in these stars. Instead, the entire convective core is available as a fuel source for energy production at the center of the star. For this reason, energy production is remarkably steady in these stars until the entire convective core is nearly exhausted of hydrogen. Even as exhaustion approaches, the extreme temperature dependence of the CNO cycle implies that deficits produced by the declining availability of hydrogen fuel can be made up by modest increases in the temperature and hence minor changes in the structure of the star. Indeed, it is not until more than 99 percent of the convective core mass has been converted to helium that truly significant changes occur in the structure of the star and the star can be said to be leaving the main sequence.

5.4 Post Main Sequence Evolution

The evolution of stars off the main sequence represents the response of the star to a depletion of the available fuel supply, and it can be qualitatively understood by examining the response of the core and envelope to the attempts of the nuclear burning regions to adjust to the diminution of the available hydrogen. In stars of the lower main sequence, the hydrogen-burning shell begins to move into a region of declining density, resulting in a decrease in available hydrogen. For stars of the upper main sequence, the situation is somewhat different. The convective nature of the core ensures the existence of mass motions, which continue to bring hydrogen into the central regions for hydrogen burning until the entire core is depleted. These two rather different approaches to hydrogen exhaustion produce somewhat different evolutionary futures for the two kinds of stars, so we examine them separately.

a Evolution off the Lower Main Sequence

The development of a helium core, which signals the onset of post main sequence evolution, is surrounded by a thin hydrogen-burning shell. The hydrogen burning continues in a shell around the helium core which steadily grows outward, in mass, through the star. However, the helium core must be supported against its own gravity as well as support the weight of the remaining star, and its energy sources are all on the outside. As a result, it is impossible for the hydrogen-burning shell to establish a temperature gradient within the helium core. Only gravitational contraction of the core will result in the release of energy inside the helium core, and except for this source of energy the helium core must be isothermal, with its temperature set by the burning of hydrogen surrounding it. But the rate of hydrogen burning is dictated largely by the mass of material lying above the burning zone, because this is the material that must be kept in equilibrium. As the mass of the isothermal helium core increases, the equilibrium temperature of the core will also rise, and this demand can be met only by a slow contraction of the helium core. The slow increase in the core temperature triggers a steady increase in the stellar luminosity.

As the isothermal core grows through the addition of He from the hydrogen-burning shell, the core temperature must rise in order for it to remain in equilibrium and support the outer layers of the star. Since an isothermal sphere is a unique polytropic configuration, it seems reasonable that there would be a limit to the amount of overlying material that such an isothermal core could support. This limit is known as the *Chandrasekhar-Schönberg (C-S) limit.* The limit will depend solely on the mass fraction of the isothermal core and the mean molecular weights of the core and envelope. Should the core exceed this limiting mass fraction, it must contract to provide the temperature and pressure gradients necessary to support the remainder of the star as well as itself.

Chandrasekhar-Schönberg Limit A detailed evaluation of the Chandrasekhar-Schönberg limit requires matching the isothermal core solution to the pressure required to support the overlying stellar mass. The maximum mass fraction that an isothermal core can have is

$$q_{C\text{-}S} \approx 0.37 \left(\frac{\mu_o}{\mu_i}\right)^2 \tag{5.4.1}$$

where μ_o and μ_i are the mean molecular weight of the outer region and core, respectively. Although the specific calculation of $q_{C\text{-}S}$ requires detailed consideration of the isothermal core solution, we can provide an argument for the plausibility of such a limit by considering the virial theorem for the core alone:

$$3(\gamma - 1)U_c + \Omega_c - 3P\left(\frac{4\pi r_c^3}{3}\right) = 0 \tag{5.4.2}$$

where

$$U_c = \frac{kTM_c}{\mu_i m_h}$$

$$\Omega_c \approx -\frac{GM_c^2}{r_c} \tag{5.4.2a}$$

and r_c is the radius of the isothermal core. The third term on the left-hand side arises because we cannot take the volume integrals, which yield the global theorem, over a surface where the pressure is zero. Thus, we must include a "surface" term which is effectively the surface pressure times the enclosed volume of the core. We solve this expression for the pressure at the boundary of the core to obtain

$$P(r_c) = \frac{1}{4\pi r_c^3}\left[\frac{3(\gamma - 1)kTM_c}{\mu_i m_h} - \frac{GM_c^2}{r_c}\right] \tag{5.4.3}$$

Now we wish to find the maximum core radius which will provide sufficient pressure to support the remaining star. We can find a maximum pressure by differentiating equation (5.4.3) with respect to r_c and finding that value of r_c for which the pressure gradient is zero. Certainly any core which yields a zero surface pressure gradient is the largest physically reasonable core. This calculation results in a maximum r_c given by

$$\frac{1}{r_c} = \frac{9(\gamma - 1)kT}{4\mu_i m_h GM_c} \tag{5.4.4}$$

Substitution into equation (5.4.3) yields the maximum surface pressure attainable at the surface of the core

$$P_m(r_c) = 3\left[\frac{9(\gamma - 1)}{4}\right]^4 \left(\frac{kT}{\mu_i m_h}\right)^4 G^{-3}M_c^{-2} \tag{5.4.5}$$

Remember that the homologous behavior of the temperature allows us to write

$$T = (\text{const})\left(\frac{\mu_o M}{R}\right) \tag{5.4.6}$$

So we can eliminate the temperature from (5.4.5) and express the result with a term which is homologous to the pressure of the envelope. Thus,

$$P_m(r_c) = (\text{const})\left(\frac{T^4}{\mu_i^4 M_c^2}\right) = (\text{const})\left(\frac{M^2}{R^4}\right)\left(\frac{\mu_o}{\mu_i}\right)^4\left(\frac{M}{M_c}\right)^2 \tag{5.4.7}$$

The term M^2/R^4 is homologous to the pressure of the envelope at the surface of the core. The ratio of this to the maximum attainable core pressure must be less than unity for the core to be able to support the envelope, so

$$q_{\text{C-S}} \equiv \frac{M_c}{M} = (\text{const})\left(\frac{P_e}{P_m}\right)^{1/2}\left(\frac{\mu_o}{\mu_i}\right)^2 \leq (\text{const})\left(\frac{\mu_o}{\mu_i}\right)^2 \tag{5.4.8}$$

where the constant is the same as in equation (5.4.1). Thus, we see that the isothermal core can, at most, support about 37 percent of the mass of the star; but if the core is primarily helium, this limit is reduced to about 10 percent.

Degenerate Core   Only for stars near the upper end of our range ($M \approx M_\odot$) will the mass of the core approach the Chandrasekhar-Schönberg limit without becoming degenerate and the core undergoing further gravitational contraction. For stars with masses $M \leq 1.3 M_\odot$, the slowly developing isothermal core will be degenerate from a point in its development when the core mass is well below the Chandrasekhar-Schönberg limit. Under these conditions, that limit does not apply because the added pressure of the degenerate electron gas is sufficient to support nearly any additional mass. Thus, the isothermal helium cores of lower main sequence stars can increase to virtually any mass below the Chandrasekhar degeneracy limit. As mass is added to the core, we can expect the core to contract according to the mass-radius law for degenerate configurations that we derived in Chapter 1 [equation (1.3.18)]. This law, following differentiation with respect to time, indicates that the core will shrink on the same time scale that mass is added to it, and that is the nuclear time scale.

Progress to Red Giant Phase   In terms of its physical size, this isothermal degenerate helium core is never very large. Thus the post main sequence evolution of a low-mass star can be viewed as the processing of stellar material through the burning zone, with the resultant helium being packed into a very small volume of systematically higher mean molecular weight. The declining density just above the helium core will lead to an increase in temperature, in order for the nuclear energy generation mechanisms to supply the energy required to support the star. However, an increase in the central temperature would lead to an increase in the temperature gradient and an increase in the luminosity. The increased luminosity, in turn, causes the outer envelope of the star to expand, decreasing the temperature gradient. Equilibrium is established at a higher shell temperature and somewhat greater luminosity and temperature gradient. The result is that the star moves upward and very slightly to the right on the H-R diagram. The process continues until the temperature gradient exceeds the adiabatic gradient. Then the entire outer envelope becomes convective. The increase in physical size of the envelope lowers the surface temperature and thereby increases the radiative opacity in the outer layers. This further decreases the efficiency of radiative transport and hastens the formation of the outer convection zone. The outer envelope is now well approximated by a polytrope of index $n = 1.5$, and the conditions for the Hayashi tracks become operative.

The star now approximately follows the track of a fully convective star, only now in reverse. The continual decline of the available hydrogen supply in the shell burning region, which becomes extremely thin, leads to a steady increase in the shell temperature and accompanying rise in luminosity. With the outer convection zone behaving as a good polytrope and efficiently carrying the energy to the surface, the energy loss is again limited by the photosphere, and the star expands rapidly to accommodate the increased energy flow. The star now moves nearly vertically up the giant branch as a red giant.

Helium Flash   As the temperature of the hydrogen-burning shell increases and the degenerate core builds in mass, the temperature eventually reaches approximately 10^8 K. This is about the ignition temperature of helium via the triple-α processes. Under normal conditions, the burning of helium could begin in a measured way, which would allow for an orderly transition of nuclear energy generation processes. However, the core is degenerate, so the electron pressure is only weakly dependent on temperature. Indeed, the limiting equation of state for total degeneracy does not contain the temperature at all. Thus helium burning sets in with unrelenting ferocity.

With its extreme dependence on temperature, the triple-α process initiates a thermal runaway, which is limited only by the eventual removal of the degeneracy from the core. The complete equation of state for a partially degenerate gas does, indeed, contain the temperature, and at a sufficiently high temperature the equation of state will revert to the ideal-gas law. When this occurs, the core rapidly expands, cools, and reaches equilibrium, with helium continuing to burn to carbon in its center.

The response of the core to this entire process is so swift that the total energy produced is a small fraction of the stored energy of the star. In addition, the site for the production of the energy is sufficiently far removed from the outer boundary that energy is diffused smoothly throughout the star and never makes a noticeable change in the star's appearance.

The duration of the flash is so much shorter a time than the dynamical time scale for the entire star that one would expect all manifestation of the flash would be damped out by the overlying star and remain hidden from the observer. However, detailed hydrodynamical calculations[7] indicate that the pressure pulse resulting from the rapid expansion of the core arrives at the surface with a velocity well in excess of the escape velocity. This result may well produce a one-time mass loss of the order of 30 percent, which would affect the subsequent evolution.

Terminal Phases of Low-Mass Evolution   Initially, after helium ignition, the hydrogen-burning shell continues to supply about 90 percent of the required support energy. However, now an orderly transition of energy mechanisms can take place, resulting in the transfer from hydrogen burning to helium burning over an extended time. The star will move somewhat down the giant branch and out on the horizontal branch, from near the peak of the giant branch where the helium flash took place. Meanwhile the helium core is in convective equilibrium, with the convection zone extending almost to the hydrogen shell. The reexpansion of the core is responsible for the contraction of the outer envelope, causing the star to move out onto the horizontal branch. It appears likely that after helium burning has ceased and the resultant carbon core is contracting, the outer envelope becomes unstable to radiation pressure and lifts off the star, forming a planetary nebula and leaving the hot core, which now relieved of its outer burden simply cools. If the mass is below the Chandrasekhar limiting mass for carbon white dwarfs, the star continues to cool to the virtually immortal state of a white dwarf.

Structure and Evolution of White Dwarfs   We have already discussed much that is relevant to the description of this abundant stellar component of the galaxy, and we return to the subject in Section 6.4. In Section 1.3 we derived the equation of state appropriate for a relativistic and a nonrelativistic degenerate gas and found them to be polytropic. In Section 2.4 we developed the mass-radius relation for polytropes in general, which provides the approximate results appropriate for white dwarfs. In Chapter 6 we will see how the relativistic equation of state and the theory of general relativity lead to an upper limit of the mass that one can expect to find for white dwarfs. However, some description of the white dwarfs formed by the evolution of low-mass stars and their subsequent fate is appropriate.

There are basically two approaches to the theory of white dwarfs. The first is to observe that a relativistically degenerate gas will behave as a polytrope and to explore the implications of that result. The second is to investigate the detailed physics that specifies the equation of state and to create models based on the results. Cox and Giuli[8] and references therein provide an excellent example of the latter. Our approach will be much nearer the former.

The ejection of a planetary nebula during the later phases of the evolution of a low-mass star leaves a hot degenerate core of carbon and oxygen exposed to the interstellar medium. While such a core may range in mass from about 0.1 of a solar mass to perhaps a little over 1 solar mass, its future will be remarkably independent of its mass. While the actual run of the state variables will pass through regions of degeneracy through partial degeneracy to a nondegenerate surface layer, the basic properties of the star can be understood by treating the stars as polytropes.

From observation we know that the white dwarf remains of stellar evolution are about a solar mass confined to a volume of planetary dimensions and thus will have a mean density of the order of $\rho = 10^6$ g/cm^3. If we assume that the gas is fully ionized, then the typical energy of an electron will be about 0.1 MeV for a fully degenerate gas. If the stellar core were at 10^7 K, the typical ion would have an energy of about 1 keV. Since energy densities are like pressures, even if the number densities were the same for the electrons and ions, the pressure of the electrons would dominate. In fact, since the typical ion produces many electrons, the dominance of the electron pressure is even greater. Thus the structure will be largely determined by the electron pressure, and the ions may be largely ignored. However, Hamada and Salpeter[9] have shown that at densities around 10^8 g/cm^3 the Fermi energy of the electron "sea" becomes so high that inverse beta decay becomes likely and some of the electrons disappear into the protons of the nuclei, causing the limiting mass to be somewhat reduced over what would be expected for a purely degenerate gas. Also, the thermal energy of the ions is lost, permitting the star to shine.

Since these stars are largely degenerate, most of the momentum states in phase space are full, and an electron that is perturbed from its place in phase space has to travel quite a distance before it can find an empty place. This implies that the mean free path of electrons will be very long in spite of the high densities. Such electrons play essentially the same role as the conduction electrons of a conductor, so that the electrical and thermal conductivity will be very high in a degenerate gas. This crowding of the states in phase space also results in the reduction in radiative opacity since it is difficult for a photon to move an electron from one state to another. As a result, it is very difficult for temperature gradients to exist within a fully degenerate configuration. However, in the outer regions of the white dwarf where the gas becomes partially degenerate, the opacity rapidly rises and the conductivity drops, giving rise to a steep temperature gradient with the result that the energy flow to the surface is seriously impeded. Thus L. H. Aller has likened a white dwarf to a metal ball wrapped in an insulating blanket. Since the structure of a polytrope is stable and independent of the temperature, the evolutionary history of a white dwarf largely revolves on the details of its cooling.

In 1952, Leon Mestel[10] took basically this classical approach to the cooling of white dwarfs and found that the cooling curve $d(\ln L)/d(\ln t)$ was approximately constant and independent of time. Iben and Tutukov,[11] using a much more detailed analysis and equation of state, found virtually the same result which they regarded as occurring through a series of accidents. Their results give $d(\ln L)/d(\ln t) \approx (-1.4, -1.6)$ for $5 < \log t_{\mathrm{yr}} < 9.4$. It is true that a considerable number of effects complicate the simple picture of a polytrope wrapped in a blanket.

For example, while we may neglect the ions for an excellent approximation of the description of the white dwarf structure, the contribution of the ion pressure will make the star slightly larger than one would expect from the polytropic approximation. Because of the extreme concentration of the star, a small contraction produces a considerable release of gravitational energy, which is then to be radiated away. This extends the cooling time significantly over that which would be expected simply for a cooling polytrope. Toward the end of the cooling curve, a series of odd things happen to the equation of state for the white dwarf. As the interior regions cool, they undergo a series of phase transitions, first to a liquid state and then to a crystalline phase. Each of these transitions results in a "heat of liquefaction or crystallization" being released and increasing the luminosity temporarily. Problems of the final cooling remain in the understanding of the low-temperature–high-density opacities that will determine the flow of radiation in this final descent of the white dwarf to a cool cinder, called a *black dwarf*, in thermal equilibrium with the ambient radiation field of the galaxy. The question is of considerable interest since such objects could be detected only by their gravitational effect and could bear on the question of the "missing mass."

b Evolution away from the Upper Main Sequence

The evolution of the more massive stars that inhabit the upper main sequence is driven by the same processes that govern the evolution of the lower main sequence, namely, the enhaustion of hydrogen fuel. However, the processes are quite different. The exhaustion of the convective core leads to the production of a helium center, as in the lower main sequence stars, but now the core will have to contend with the Chandrasekhar-Schönberg limit.

Nature of the Massive Helium Core   In massive stars, as the hydrogen is depleted in the helium core, the temperature rises rapidly, to produce the energy necessary to accommodate the demands of stellar structure. For stars with masses greater than about $7M_{\odot}$, the resulting helium core will be greater than the Chandrasekhar-Schönberg limit; and to make up for the energy deficit caused by the failing hydrogen burning, the core will have to contract. Since the contraction must maintain a temperature gradient, the contraction will proceed rather faster than would be expected for an isothermal core. However, the steep temperature gradient established by the terminal phases of core hydrogen burning will be relaxed, because the energy generated by gravitational contraction will not be as centrally concentrated as it was from hydrogen burning. This drop in temperature gradient will cause convection to cease, yielding a core in radiative equilibrium supplying the required stellar energy by contraction.

The contraction of virtually any polytrope will result in an increase in internal temperature. This is really a consequence of the virial theorem. However, the cessation of hydrogen burning in the core and the resultant decrease in the temperature gradient imply a change in the overall structure of the core, and thus the polytropic analogy is somewhat strained. The decline in the temperature gradient actually implies that the core will suffer a reduction of its internal energy while the boundary temperature increases. This loss of internal energy goes into the support of the outer envelope. Thus, both this energy and the energy generated by contraction are available for the support of the outer layers of the star.

The increase in the boundary temperature of the core will result in a slow expansion and cooling of the radiative envelope for the same reason described for low-mass stars. After a suitable rise in the boundary temperature, hydrogen is re-ignited in a shell surrounding the helium core. This results in a marked change in the temperature gradient of the hydrogen-burning shell at the core boundary. The localization of the energy production in such a small region steepens the temperature gradient to the point where the temperature gradient exceeds the adiabatic gradient, driving the entire envelope into convection. The outer envelope now rapidly transfers the energy to the surface, which again becomes the limiting barrier to its escape. The star moves rapidly toward the giant branch as a star with a helium core surrounded by a hydrogen-burning shell and covered with a deep convective envelope. This envelope is so deep that it reaches into the region where nuclear processing has taken place, dredging up some of this material to the surface. The result of this process is evident in the atmospheric spectra of some late-type supergiants.

Stars with masses less than about $5M_\odot$ will end their hydrogen burning with a helium core below that of the Chandrasekhar-Schönberg limit and will contend with a slowly contracting isothermal core right through the ignition of a hydrogen-burning shell. The future of such a star is mirrored in the behavior of lower-mass stars except that helium ignition takes place before the core becomes significantly degenerate. The result is that these stars experience no helium flash, and the transition to helium burning is orderly.

Ignition of the Massive Helium Core   In both cases described above, the contraction of the helium core proceeds while the hydrogen shell is burning. In the more massive stars, where the core is above the Chandrasekhar-Schönberg limit, the star must do so to maintain a temperature gradient for its own support. In the less massive case, the core grows slowly as a result of the processing of the hydrogen in the energy-generating shell. The added mass results in a slow core contraction. In both instances, the decreasing density in the hydrogen-burning shell necessitates a rise in the temperature required for energy generation. After the increasing temperature gradient caused by this increasing shell temperature has forced the envelope to become fully convective, the convective envelope continues to expand, for the energy escape is again limited by the radiative efficiency of the photosphere.

Eventually the central temperature reaches 10^8 K, and helium ignition takes place. The ignition has a dramatic effect on the core but does not exhibit the explosive nature of the helium flash. The core undergoes a rapid expansion and, because of the huge temperature dependence of the triple-α process, becomes unstable to convection. This produces an expanded convective helium core surrounded by a hydrogen-burning shell. This shell supplies more than 90 percent of the energy required to maintain the luminosity. The rapid core expansion is accompanied by a contraction of the outer envelope with a corresponding increase in the surface temperature. The star moves off to the left in the H-R diagram, maintaining about the same total luminosity. The hydrogen-burning shell continues to supply the majority of the energy throughout the helium-burning phase which proceeds in much the same manner as the main sequence core hydrogen-burning phase, but on a much shorter time scale.

Terminal Phases of Evolution of Massive Stars   The end phase of the evolution of massive stars is still somewhat murky and the subject of active research. Initially, the helium burning continues in the core until the core becomes largely carbon. At the point of helium exhaustion, the core again gravitationally contracts, rising in temperature, until helium is ignited in a shell source around the core. For stars with a mass between 3 and 7 solar masses, there is some evidence that the carbon core which develops is degenerate and that ignition, when it occurs, occurs explosively, perhaps producing a supernova. For stars of more than 10 solar masses, carbon burning can take place in a nonviolent manner, producing cores of neon, oxygen, and finally silicon, each surrounded by a shell source of the previous core material which continues to provide some energy to the star. The results of silicon burning yield elements of the iron group for which further nuclear reactions are endothermic, and so this burning will not only fail to contribute energy for the support of the star but also rob it of energy. In addition, the densities become high enough that the electrons are forced into the protons of the nuclei by means of inverse β decay. This effect has been called *neutronization* of the core. Both mechanisms produce a significant number of neutrinos which also do not take part in the support of the star against gravity and which can be viewed as "cooling" mechanisms for the core.

This rapid cooling precipitates a rapid collapse of the core followed by the entire star. The in-fall velocity soon exceeds the speed of sound, resulting in the formation of a shock wave, and interior densities may become large enough for the material to become opaque to neutrinos. Under some conditions the endothermic nuclear reactions may bring about the disintegration of the iron-group elements into ^{4}He. Which process dominates for stars of which mass is not at all clear. The shock wave formed by the in-fall may "bounce," or the increased neutrino opacity may provide sufficient energy and momentum to ensure that a large fraction of the star will be blown into the interstellar medium.

The remnants, if any, could be a neutron star or a black hole. Although the details of the terminal phases of massive stars remain somewhat unclear, it is virtually certain that these phases are likely to end with the production of a supernova and the subsequent enrichment of the interstellar medium by heavy elements.

c Effect of Mass Loss on the Evolution of Stars

Throughout our discussion of stellar evolution we have assumed that the mass of the evolving star remains essentially constant. Certainly there is a reduction in mass from the nuclear production of energy, but this can never exceed 0.7 percent and therefore can be safely neglected. However, as we shall see in Section 16.3, stars may exhibit rather large winds emanating from their atmospheres. In some cases these winds may result in mass-loss rates exceeding $10^{-5}M_\odot$ per year and so could lead to a significant reduction in the mass of the star during its nuclear lifetime.

Ever since the observation by Armin Deutch[12] that the red supergiant α Herculis was losing mass faster than about $10^{-8}M_\odot$ per year, interest in the effects of mass loss on the evolution of a star has been high. Generally one would expect that a star slowly losing mass would follow the classical evolutionary track represented by models of stars having successively lower and lower mass. Thus, for a main sequence star that is losing mass at a significant rate, the evolutionary track on the H-R diagram would not rise like the constant mass models but move steadily to the right and perhaps down until the asymptotic giant branch is reached. In addition, the lifetime would be significantly enhanced as a result of the reduced stellar mass. Such were the conclusions reached by Massevitch[13] and collaborators[14] in the late 1950s. However, the evolutionary tracks of $1-2M_\odot$ models failed to fit the H-R diagrams of globular clusters, and in the absence of evidence for mass loss from main sequence stars their work was largely ignored. However, during the 1960s and 1970s it became clear that virtually all early type stars exhibited significant stellar winds, and it was likely that their evolution from the main sequence was affected.[13,14] In general, the more massive the star, the greater the fractional mass-loss rate. This result would explain why stars in the range of a few solar masses were unaffected and why the constant mass models fit the globular cluster H-R diagrams relatively well. However, from studies of the ratio of blue to red supergiants in the Milky Way and other galaxies, Humphreys and Davidson[15] concluded that stars more massive than $50M_\odot$ never make it to the red supergiant phase but remain confined to the left-hand side of the H-R diagram throughout their lives. Lamers[16] found that this did not appear to be the case for stars in our galaxy, and he concluded that a $100M_\odot$ could loose only 15 percent of its mass during its main sequence lifetime.

The impact of such mass loss on the subsequent evolution of the star seems to depend on an accurate knowledge of the mass-loss rate during evolution, which in turn rests on the specifics of the origin of stellar winds. Using an empirically inspired mass-loss rate, Brunish and Truran[17] found that mass loss affected the evolution of stars less than $30M_\odot$ more than it did that of the massive stars. However, Sreenivasan and Wilson[18,19] found that to include rotation and a theoretically motivated origin for the mass-loss rate resulted in a much more complicated evolutionary history. By adjusting the amount of rotation present in the initial star, they were able to match most of the observations. However, it is fair to say that as of the present, much remains poorly understood about the specific role played by mass loss in the massive stars. It also appears that a proper understanding will require models that correctly couple the atmosphere to the interior and include rotation in a physically self-consistent way.

5.5 Summary and Recapitulation

In this chapter we have sketched the evolution of normal stars from their contraction out of the interstellar medium to their probable fate. We have not discussed many topics and details that are important to the detailed understanding of stellar evolution, and some important problems remain unsolved. For the evolution of individual stars, an area of singular importance that was acknowledged only in passing concerns the origin of the elements. The production of the elements through nuculosynthesis was suggested by Burbidge, Burbidge, Fowler and Hoyle,[20] and the early view of the important processes are reviewed by Bashkin.[21] The manifestation of these elements in the outer layers of the star and the internal processes by which they got there are covered by Wallerstein.[22] In addition, we have said nothing about the fascinating topic of the evolution of close binary stars where the futures of the components are linked through the process of mass exchange. Nor have we touched on the tricky processes by which a white dwarf cools off. We also avoided the effects of rotation and magnetic fields on the evolution of stars along with the details of the dynamic collapse of stars, and these should be regarded as fertile areas for research. However, we did delineate major events in the lives of normal stars. Specifically, we used the efficiency of energy transport, the temperature sensitivity of the nuclear reactions, and the radiative ability of the photosphere to indicate the probable direction that the evolution of stars will take. Simple arguments of efficiency lead to a remarkably accurate description of the early phases of stellar evolution to the main sequence. Post main sequence evolution is made more complicated by the zonal nature of the star, complications to the equation of state, and the existence of multiple energy sources. Nevertheless, we can see the basic conservation laws of physics at work during the latter phases of stellar evolution and can get a feel for the important processes at work. We close this discussion with another view of the interplay between the core and outer envelope along with a detailed look at the evolution of a $5M_\odot$ star.

a   Core Contraction—Envelope Expansion: Simple Reasons

For years there has been some debate over why the envelope of an evolving star expands when the core contracts, for many people find the result counterintuitive. A number of explanations have been suggested, and objections have been made to almost all, of being simplistic or incomplete. Some have regarded the question as being so complicated that it is not useful to search for a single cause, and in response to the question of envelope expansion they simply say, "It happens because my computer tells me it does." This is no answer at all, for it offers no insight into the physical phenomena that result in the particular behavior exhibited by the star. Certainly the physical situation which leads to the expansion of the envelope during core contraction is complicated, and simple answers, in some sense, will always be incomplete. However, we should make the effort to identify the important processes at work which dominate the result.

It would be useful if we could clarify the question a little. A star goes through several different phases as it evolves from the main sequence to the giant branch of the H-R diagram, and all phases result in an expansion of the envelope accompanying some contraction of the core. However, the structure of the core and that of the envelope differ widely in these various phases, as do the magnitude and time scale for the resulting core contraction—envelope expansion. We attempted to make plausible the expansion of the convective envelope which accompanies the temperature increase of the hydrogen-burning shell, resulting from the contraction of the radiative helium-rich core, by appealing to the behavior of convective polytropes. Although such envelopes will not be the complete polytropes of the Hayashi tracks, it seems reasonable that the stars will approach the tracks in their general behavior, for the same principles that result in the Hayashi tracks are operative in the expansion of the convective envelope. However, in envelope expansion, the processes are reversed and less perfectly followed, since only the envelope is involved. The general expansion of the radiative zone overlying the helium core which initiates the departure of the star from the main sequence cannot be described in the same manner. The expansion would be represented by a partial polytrope having an index that varies in time. Indeed, for the lower main sequence stars, that zone is surrounded by a convective region that, together with the radiative zone, makes up the envelope outside the hydrogen-burning shell. This compound envelope undergoes the expansion.

With such a great variety of situations leading to envelope expansion, does it even make sense to look for a common cause? If we found one, it would be necessarily vague about details since it must apply in a wide variety of circumstances. Should this common cause exist, it must result from a very general principle in order for it to apply in these many diverse circumstances. Let us consider two very general principles to see if they can provide a qualitative indication as to how the entire star will behave should the central regions, embodying the majority of the mass, contract. First, the conservation of energy will require that the total energy of the star be written as

$$E = \langle \Omega \rangle + \langle U \rangle - \int_0^t L \, dt + \int_0^t \epsilon \, dt \qquad (5.4.9)$$

For any time scale that is less than the Kelvin-Helmholtz time, the magnitude of the integrals will be less than either the gravitational or internal energy, because the Kelvin-Helmholtz time is essentially the time required for the star's luminosity to lose an amount of energy equal to the internal energy. Since the integrals appear with opposite sign and are approximately equal, and since we have included all sources of energy available to the star explicitly, we may write

$$E \approx \langle \Omega \rangle + \langle U \rangle \approx \text{constant} \qquad (5.4.10)$$

The contractions and expansions of interest occur on time scales very much longer than the dynamic time scale, so we can be certain that the time-averaged form of the virial theorem will apply. Thus

$$\langle \Omega \rangle + 2\langle U \rangle = 0 \qquad (5.4.11)$$

The combination of equations (5.4.10) and (5.4.11) requires that the gravitational and potential energy, separately, be constant. Now as long as we average over a dynamical time, equation (5.4.11) will be valid, while equation (5.4.10) becomes more exact for shorter times t since the integrals of equation (5.4.9) will contribute less. Thus we may drop the averages and expect that for any time greater than the dynamical time but shorter than the Kelvin-Helmholtz time

$$\Omega \approx \text{constant} \approx \Omega_c + \Omega_e \qquad (5.4.12)$$

Now, for upper main sequence stars, the mass of the core substantially exceeds that of the envelope, so

$$\Omega \approx \frac{GM_c^2}{R_c} + \frac{GM_cM_e}{R_*} \tag{5.4.13}$$

If, for simplicity, we further hold the masses of the core and envelope constant during the core contraction, we have

$$\frac{dR_*}{dR_c} \approx -\left(\frac{M_c}{M_e}\right)\left(\frac{R_*}{R_c}\right)^2 \ll -1 \tag{5.4.14}$$

The sign of equation (5.4.14) indicates that we should expect the observed radius of the star to increase for any decrease in the core radius, and the magnitude of the right-hand side implies that a very large amplification of the change in core size would be seen in the stellar radius. One can argue that the assumptions are only approximately true or that the time scales involved occasionally approach the Kelvin-Helmholtz time, but that will affect only the degree of the change, not the sign. Indeed, detailed evolutionary model calculations throughout the period of evolution from the main sequence to the giant branch indicate that the total gravitational and internal energies are indeed constant to about 10 percent. The accuracy for shorter times is considerably better. For lower main sequence stars, the mass of the core is less than that of the envelope. Nevertheless, a result similar to equation (5.4.14), with the same sign, is obtained although the magnitude of the derivative is not as large.

The nature of this argument is so general that we may expect any action of the core to be oppositely reflected in the behavior of the envelope regardless of the relative structure. Thus, we can understand the global response of the star to the initial contraction of the core when the overlying layers are in radiative equilibrium as well as the subsequent rapid expansion to and up the giant branch when the outer envelope is fully convective. In addition, contraction of the stellar envelope following the core expansion accompanying the helium flash, which leads to its position on the horizontal branch, can also be qualitatively understood. In general, whenever the core contracts, we may expect the envelope to expand, and *vice versa*. Detailed model calculations confirm that this is the case.

b Calculated Evolution of a $5M_\odot$ Star

In this final section we look at the specific track on the H-R diagram made by a $5M_\odot$ star as determined by models made by Icko Iben.[23] This is best presented in the form of a figure and is shown in Figure 5.2. Similar calculations have been done for representative stellar masses all along the main sequence, so the evolutionary tracks of all stars on the main sequence are well known. The basic nature of the theory of stellar evolution can be confirmed by comparing the location of a collection of stars of differing mass but similar physical age with the H-R diagrams of clusters of stars formed about the same time. A reasonable picture is obtained for a large variety of clusters with widely ranging ages. It would be presumptuous to attribute this picture to chance. While much remains to be done to illuminate the details of certain aspects of the theory of stellar evolution, the basic picture seems secure.

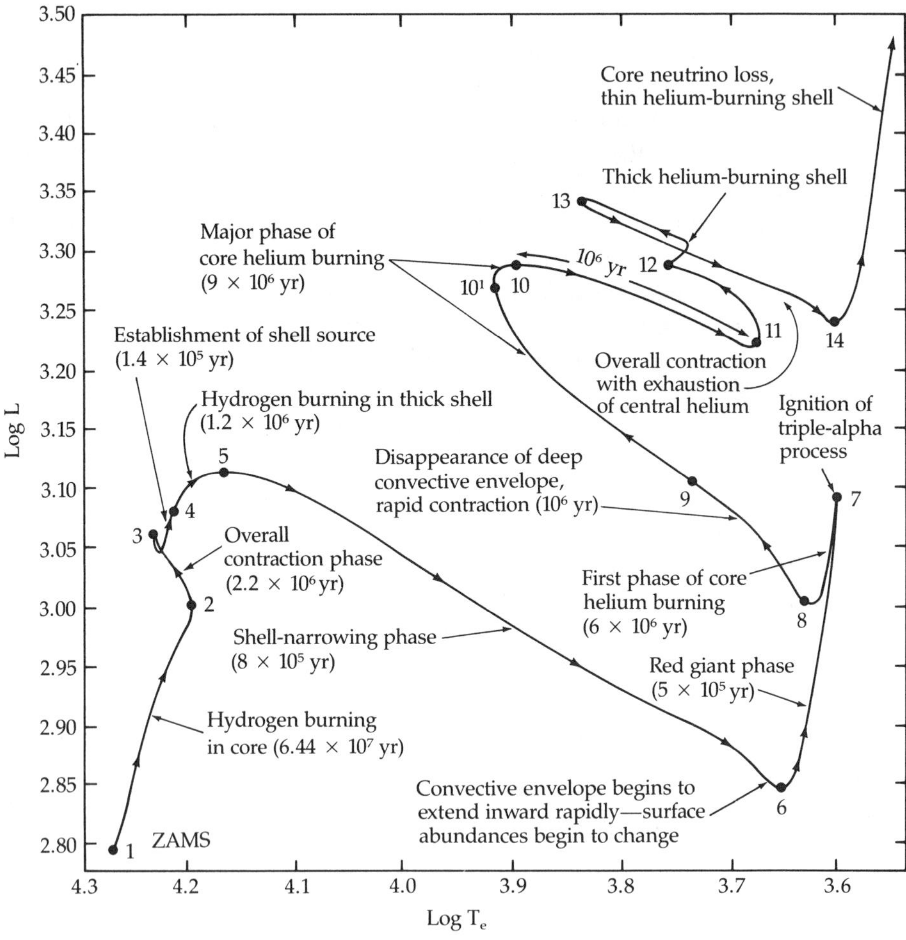

Figure 5.2 delineates the evolution of a $5M_\odot$ star from its arrival on the main sequence through its demise at the onset on carbon burning.[23] The labeled points are points of interest discussed in the chapter and their duration, place in the stellar lifetime, and the significant physical process taking place are given in Table 5.1.

Table 5.1 History of a $5M_\odot$ Star

Point	Duration	Elapsed Time	
Location	*yr*	*yr*	Primary Physical Activity
(1–2)	6.4×10^7	6.40×10^7	H burning core
(2–3)	2.2×10^6	6.62×10^7	Core exhaustion and contraction
(3–4)	1.4×10^5	6.63×10^7	Establishment of hydrogen-burning shell
(4–5)	1.2×10^6	6.75×10^7	H shell thickens
(5–6)	8.0×10^5	6.83×10^7	H exhaustion, envelope expansion to convection
(6–7)	5.0×10^5	6.88×10^7	Core contraction, envelope expansion
(7–8)	6.0×10^6	7.48×10^7	He ignition and burn, envelope contraction, and core expansion
(8–9)	1.0×10^7	8.48×10^7	Primary He burning phase
(9–10)	1.0×10^6	8.58×10^7	He core grows, envelope expands
(10–11)	$<10^5$	8.58×10^7	Core contraction, He shell ignition
(11–12)	$<10^4$	8.58×10^7	He exhaustion before C ignition

Problems

1. Find the fraction by mass and radius inside of which 20, 50, and 99 percent of the sun's energy is generated. Compare the results with a star of the same chemical composition but 10 times the mass.

2. Determine the mass for which stars with the chemical composition of the sun derive equal amounts of energy from the CNO and *p-p* cycles.

3. Determine the relative importance of free-free and bound-free absorption and electron scattering as opacity sources in the sun.

4. Calculate the evolutionary tracks for a $1M_\odot$ and $10M_\odot$ star.

5. Choose a representative set of models from the evolutionary calculations in Problem 4.
    (*a*) Calculate the moment of inertia, gravitational and internal energies of the core and envelope, and the total energy of the star.
    (*b*) Determine the extent to which the conditions in Section 5.4c are met during the evolution of the star.

6. Compute the Henyey track for a $1M_\odot$ star, and compare it with that of a polytrope of index $n = 3$. Would you recommend that the comparison be made with a polytrope of some different index? If so, why?

7. Compute the evolutionary track for the sun from early on the Hayashi track as far as you can. At what point do you feel the models no longer represent the actual future of the sun, and why?

8. Discuss the evolution of a $5M_\odot$ star as it leaves the main sequence. Detail specifically the conditions that exist immediately before and after the onset of hydrogen shell burning.

9. Consider a star composed of an isothermal helium core and a convective hydrogen envelope. Suppose that the mass in the core remains constant with time but that the core contracts. By constructing a model of appropriate polytropes, show what happens to the envelope and comment on the external appearance of the star.

References and Supplemental Reading

1. Larson, R. B., "The Dynamics of a Collapsing Proto-star," *Mon. Not. R. Astr. Soc.* 145, 1969, pp. 271−295.

2. Disney, M. J., "Boundary and Initial Conditions in Protostar Calculations," *Mon. Not. R. Astr. Soc.* 175, 1976, pp. 323−333.

3. Hayashi, C., "Evolution of Protostars," *Ann. Rev. Astr. and Astrophys.* 4, 1966, pp. 171−192.

4. Cox, J. P., and R. T. Giuli, *Principles of Stellar Structure*, Gordon and Breach, New York, 1968, chap. 26, pp. 743−750.

5. Iben, I., Jr., "Normal Stellar Evolution," *Stellar Evolution* (Eds.: H-Y Chiu and A. Muriel), M.I.T. Press, Cambridge, Mass., 1972, pp. 44, 53.

6. Harris, D. L., II, K. Aa. Strand, and C. E. Worley, "Empirical Data on Stellar Masses, Luminosities, and Radii," *Basic Astronomical Data* (Ed.: K. Aa. Strand), *Stars and Stellar Systems*, vol. 3, University of Chicago Press, Chicago, 1963, p. 285.

7. Deupree, R. G., and P. W. Cole, "Mass Loss During the Core Helium Flash" *Ap. J.* 249, 1981, pp. L35−L38.

8. Cox, J. P., and R. T. Giuli, *Principles of Stellar Structure*, Gordon and Breach, New York, 1968, chap. 26, pp. 874−943.

9. Hamada, T., and E. E. Salpeter, "Models for Zero Temperature Stars," *Ap. J.* 134, 1961, pp. 683−698.

10. Mestel, L., "On the Theory of White Dwarf Stars, I The Energy Sources of White Dwarfs," *Mon. Not. R. Astr. Soc.* 112, 1952, pp. 583−597. Particularly see sec. 4, "The Rate of Cooling of a White Dwarf," pp. 590−593.

11. Iben, Jr., I., and A. V. Tutukov, "Cooling of Low-Mass Carbon-Oxygen Dwarfs from the Planetary Nucleus Stage through the Crystallization Stage," *Ap. J.* 282, 1984, pp. 615–630.

12. Deutch, A. J., "The Circumstellar Envelope of Alpha Herculis" *Ap. J.* 123, 1956, pp. 210–227.

13. Massevitch, A. G., "The Evolution of Stars in the χ and h Persei Double Cluster," *Soviet Astronomy* 1, 1957, pp. 177–182.

14. Ruben, G., and A. G. Massevitch, "An Investigation of Evolutionary Paths for a Homogeneous Stellar Model with a Convective Core," *Soviet Astronomy* 1, 1957, pp. 705–718.

15. Humphreys, R. M., and K. Davidson, "Studies of Luminous Stars in Nearby Galaxies III. Comments on the Most Massive Stars in the Milky Way and the Large Magellanic Cloud," *Ap. J.* 232, 1979, pp. 409–420.

16. Lamers, H. J. G. L. M., "Mass Loss from O and B Stars," *Ap. J.* 245, 1981, pp. 593–608.

17. Brunish, W. M., and J. W. Truran, "The Evolution of Massive Stars I. The Influence of Mass Loss on Population I Stars," *Ap. J.* 256, 1982, pp. 247–258.

18. Sreenivasan, S. R., and W. J. F. Wilson, "The Evolution of Massive Stars Losing Mass and Angular Momentum: Supergiants," *Ap. J.* 290, 1985, pp. 653–659.

19. _______ "The Evolution of Massive Stars Losing Mass and Angular Momentum: Rotational Mixing in Early-Type Stars," *Ap. J.* 292, 1985, pp. 506–510.

20. Burbidge, E. M., G. R. Burbidge, W. A. Fowler, and F. Hoyle, "Synthesis of the Elements in Stars," *Rev. of Mod. Phy.* 29, 1957, pp. 547–650.

21. Bashkin, S., "The Origin of the Elements," *Stellar Structure*, (Eds.: L. H. Aller and D. B. McLaughlin), University of Chicago Press, Chicago, 1965, pp. 1–60.

22. Wallerstein, G., "Mixing in Stars," *Sci.* 240, 1988, pp. 1743–1750.

23. Iben, Jr., I. "Stellar Evolution within and off the Main-Sequence," *Ann. Rev. Astr. and Astrophys.* 5, 1967, pp. 571–626.

A review of the traditional collapse can be found in

• Hayashi, C.: "Evolution of Protostars," *Ann. Rev. Astr. and Astrophys.* 4, 1966, pp. 171–192.

A comprehensive review of the problems involved in nonhomologous collapse can be found in

• Larson, R.: "Processes in Collapsing Interstellar Clouds," *Annual Review of Astronomy and Astrophysics*, (Eds.: L. Goldberg, D. Layzer, and J. Phillips), Annual Reviews, Palo Alto, Calif., 1973, pp. 219–238.

An excellent classical description of the early phases of the contraction of proto-stars can be found in

• Bodenheimer, P.: "Stellar Evolution toward the Main Sequence," *Rep. Prog. Phys.* 35, 1972, pp. 1–54.

A more recent description of the hydrodynamic problems of collapsing stars is contained in

• Tohline, J. E.: "Hydrodynamic Collapse," *Fund. Cosmic Phys.* 8, 1982, pp. 1–82.

One of the best physical reviews of main sequence evolution and the phases leading up to the red giant phases is still

• Schwarzschild, M.: *Structure and Evolution of the Stars*, Princeton University Press, Princeton, N.J., 1958, chaps. 5, 6, pp. 165–227.

For the evolution of a normal $5M_\odot$ star, the description by

• Iben, Jr., I.: "Stellar Evolution within and off the Main Sequence," *Ann. Rev. Astr. and Astrophys.* 5, 1967, pp. 571–626.

or

• Iben, Jr., I.: "Normal Stellar Evolution," *Stellar Evolution* (Eds.: H-Y Chiu and A. Muriel), M.I.T. Press, Cambridge, Mass., 1972, pp. 1–106.

is still regarded as basically correct. However, the broadest based and most contemporary survey of the results of stellar evolution calculations remains

• Cox, J. P., and R. T. Giuli: *Principles of Stellar Structure*, Gordon and Breach, New York, 1968, chap. 26, pp. 944–1028.

An extremely literate presentation of the early theory of white dwarfs can be found in this fine review article:

• Mestel, L.: "The Theory of White Dwarfs," *Stellar Structure* (Eds.: L. H. Aller and D. B. Mclaughlin), *Stars and Stellar Systems*, vol. 5, University of Chicago Press, Chicago, 1965, pp. 297–325.

6

Relativistic Stellar Structure

· · ·

In the next two chapters we consider some specific problems which lie outside the realm of normal stellar structure. In the past several decades, it has become increasingly clear that a large number of stars require some further subtleties of physics for their proper description. Two areas that we shall consider involve the initial assumption of spherical symmetry and the assumption that the gravitational field can be described by the newtonian theory of gravity with sufficient accuracy to properly represent the star. In this chapter, we investigate some of the ramifications of the general theory of relativity for highly condensed objects and supermassive stars.

Although the application of the general theory of relativity to astronomical problems has a long and venerable history dating back to Einstein himself, it was not until the discovery of pulsars in the 1960s that a great deal of interest was directed toward the impact of the theory on stellar structure. To be sure, the pioneering theoretical work was done 30 years earlier and can be traced to Landau[1] in 1932. The fundamental work of Oppenheimer[2,3] and collaborators still provides the fundamental basis for most models requiring general relativity for their representation. But it was the discovery that neutron stars actually existed and were probably the result of the dynamical collapse of a supernova that led to the construction of modern models that represent our contemporary view of these objects.

It is not our intent to provide a complete description of the general theory of relativity so that the readers can understand all the ramifications for stellar structure implied by that theory. For that, the readers are referred to *Gravitation* by Misner, Thorne, and Wheeler.[4] Rather, let us outline the origin of the fundamental equations of relativistic stellar structure and the results of their applications to some simple objects, without the rigors of their complete derivation. The intent here is to provide some physical insight into the role played by general relativity in a variety of objects for which that role is important.

6.1 Field Equations of the General Theory of Relativity

The general theory of relativity is a classical field theory of gravitation in which all variables are assumed to be continuous and are uniquely specified. Thus, the Heisenberg uncertainty principle and quantum mechanics play no direct role in the theory. Although it is traditional to present general relativity in a system of units where $c = h = G = 1$, we adopt the nontraditional notion of generally maintaining the physical constants in the expressions in the hopes that the physical interpretation of the various terms may be clearer to the readers. However, we adopt the Einstein summation convention where repeated indices are summation indices for this section, to avoid the host of summation signs that would otherwise accompany the tensor calculus.

The basic philosophy of general relativity is to relate the geometry of space-time, which determines the motion of matter, to the density of matter-energy, known as the *stress energy tensor*. This relation is accomplished through the Einstein field equations. The geometry of space-time is dictated by the metric tensor, which defines the properties of that geometry and basically describes how travel in one coordinate involves another coordinate, so that

$$ds^2 = \mathbf{g}_{\mu\nu}\, dx^\mu\, dx^\nu \tag{6.1.1}$$

The elements of the metric tensor are dimensionless; for ordinary euclidean space they are all unity if $\mu = \nu$ and zero otherwise. If one were doing geometry on a deformed rubber sheet, this would not necessarily be true. In general, the distance traveled, expressed in terms of any set of local coordinates, will depend on the orientation of those coordinates on the rubber sheet. The coefficients that "weight" the role played by each coordinate in determining the distance according to equation (6.1.1), for all directions traveled, are the elements of the metric tensor. Now the field equations relate second derivatives of the metric tensor to the properties of the local matter-energy density, expressed in terms of the stress-energy tensor. Specifically the Einstein field equations are

$$\mathbf{G}_{\mu\nu} = \frac{8\pi G}{c^2}\,\mathbf{T}_{\mu\nu} \tag{6.1.2}$$

Here $\mathbf{G}_{\mu\nu}$ is known as the *Einstein tensor*, and $\mathbf{T}_{\mu\nu}$ is the stress energy tensor in physical units (say, grams per cubic centimeter). The quantity G/c^2 is a very small number in any common system of units, which shows that the departure from euclidean space is small unless the stress-energy is exceptionally large. The specific relation of the metric tensor to the Einstein tensor is extremely complicated and for completeness is given below.

Define

$$\Gamma_{\beta\mu\nu} \equiv \frac{1}{2}\left(\frac{\partial \mathbf{g}_{\beta\nu}}{\partial x^\mu} + \frac{\partial \mathbf{g}_{\beta\mu}}{\partial x^\nu} + \frac{\partial \mathbf{g}_{\mu\nu}}{\partial x^\beta}\right) \tag{6.1.3}$$

and

$$\Gamma^{\alpha}{}_{\mu\nu} \equiv \mathbf{g}^{\alpha\beta}\Gamma_{\beta\mu\nu} \tag{6.1.4}$$

where $\mathbf{g}^{\alpha\beta}$ is the matrix inverse of $\mathbf{g}_{\alpha\beta}$. The symbol $\Gamma_{\beta\mu\nu}$ is known as the *Christoffel symbol*. The Christoffel symbols and their derivatives can be combined to produce the Riemann curvature tensor

$$\mathbf{R}^{\alpha}{}_{\beta\gamma\delta} = \frac{\partial \Gamma^{\alpha}{}_{\beta\delta}}{\partial x^\gamma} - \frac{\partial \Gamma^{\alpha}{}_{\beta\gamma}}{\partial x^\delta} + \Gamma^{\alpha}{}_{\mu\gamma}\Gamma^{\mu}{}_{\beta\delta} - \Gamma^{\alpha}{}_{\mu\delta}\Gamma^{\mu}{}_{\beta\gamma} \tag{6.1.5}$$

which, when it is summed over two of its indices, produces the *Ricci tensor*

$$\mathbf{R}_{\mu\nu} = R^{\alpha}{}_{\mu\alpha\nu} \tag{6.1.6}$$

This can be further summed (contracted) over the remaining two indices to yield a quantity known as the *scalar curvature*

$$R = \mathbf{R}^{\mu}{}_{\mu} \tag{6.1.7}$$

Finally, the Einstein tensor can be expressed in terms of the Ricci tensor, the scalar curvature, and the metric tensor itself as

$$\mathbf{G}_{\mu\nu} = \mathbf{R}_{\mu\nu} - \mathbf{g}_{\mu\nu}R \qquad (6.1.8)$$

For a given arbitrary metric, the calculations implied by equations (6.1.4) through (6.1.8) are extremely tedious but conceptually simple. Since the metric tensor depends on only the geometry and since the operations described in forming the Riemann and Ricci tensors and scalar curvature are essentially geometric, nothing but geometry appears in the Einstein tensor. Hence the saying "The left-hand side of the Einstein field equations is geometry, while the right-hand side is physics."

6.2 Oppenheimer-Volkoff Equation of Hydrostatic Equilibrium

a Schwarzschild Metric

For reasons that are obvious by now, much of the initial progress in general relativity was made by considering highly symmetric metrics which simplify the Einstein tensor. So let us consider the most general metric which exhibits spherical symmetry. This is certainly consistent with our original assumption of spherical stars. If we take the usual spherical coordinates r, θ, ϕ, and let t represent the time coordinate, then the distance between two points in this spherical metric can be written as

$$ds^2 = -e^{\lambda(r)}\,dr^2 - r^2\,d\theta^2 - r^2(\mathrm{Sin}^2\,\theta)\,d\phi^2 + \frac{e^{\alpha(r)}\,dt^2}{c^2} \qquad (6.2.1)$$

where $\lambda(r)$ and $\alpha(r)$ are arbitrary functions of the radial coordinate r. We must also make some assumptions about the physics of the star in question. This amounts to specifying the stress energy tensor.

Consistent with our assumption of spherical symmetry, let us assume that the material of the star has an equation of state which exhibits no transverse strains, so that all the off-diagonal elements of the stress energy tensor are zero and the first three spatial elements are equal to the matter equivalent of the energy density. The fourth diagonal component is just the matter density, so

$$\mathbf{T}^{11} = \mathbf{T}^{22} = \mathbf{T}^{33} = -\frac{P}{3c^2} \qquad \mathbf{T}^{44} = \rho \qquad (6.2.2)$$

This is equivalent to saying that the equation of state has the familiar form

$$P = P(\rho) \qquad (6.2.3)$$

Now if we take the metric tensor specified by equation (6.2.1), go through the operations specified by equations (6.1.2) through (6.1.8), and sum over the three spatial indices because of the spherical symmetry, then the Einstein field equations become

$$
e^{-\lambda}\left(\frac{\alpha'}{r} + \frac{1}{r^2}\right) - \frac{1}{r^2} = \frac{8\pi G}{c^2}\frac{P}{c^2}
$$

$$
e^{-\lambda}\left(\frac{\lambda'}{r} - \frac{1}{r^2}\right) + \frac{1}{r^2} = \frac{8\pi G}{c^2}\rho
$$

(6.2.4)

Here the ′ denotes differentiation with respect to the radial coordinate r. This solution must hold through all space, including that outside the star where $P = \rho = 0$. If we take the boundary of the star to be where $r = R$, then for $r > R$ we get the Schwarzschild metric equations

$$
e^{-\lambda(r)}\left(\frac{1}{r}\frac{d\alpha(r)}{dr} + \frac{1}{r^2}\right) - \frac{1}{r^2} = 0
$$

$$
e^{-\lambda(r)}\left(\frac{1}{r}\frac{d\lambda(r)}{dr} - \frac{1}{r^2}\right) + \frac{1}{r^2} = 0
$$

(6.2.5)

which have solutions

$$
e^{-\lambda(r)} = 1 + \frac{A}{r} \qquad e^{-\alpha(r)} = B\left(1 + \frac{A}{r}\right)
$$

(6.2.6)

where A and B are arbitrary constants of integration for the differential equations and are to be determined from the boundary conditions. At large values of r, we require that the metric go over to the spherical metric of euclidean flat space, so that

$$
\operatorname*{Lim}_{r \to \infty} e^{\lambda(r)} = \operatorname*{Lim}_{r \to \infty} e^{\alpha(r)} = 1
$$

(6.2.7)

and $B = 1$. A line integral around the object must yield a temporal period and distance consistent with Kepler's third law, meaning that A is related to the newtonian mass of the object. Specifically,

$$
A = -\frac{2GM}{c^2}
$$

(6.2.8)

which has the units of a length and is known as the *Schwarzschild radius.*

b Gravitational Potential and Hydrostatic Equilibrium

Since

$$e^{\alpha(r)} \simeq 1 + \alpha(r) \simeq 1 + \frac{2GM}{c^2 r} \qquad (6.2.9)$$

we know that

$$\alpha(r) = \frac{2\Omega}{c^2} \qquad (6.2.10)$$

where Ω is the newtonian potential at large distances. The parameter $\alpha(r)$ then plays the role of a potential throughout the entire Schwarzschild metric. So we can solve the first of equations (6.2.4) for its spatial derivative and get

$$\frac{d\Omega}{dr} = \frac{G[M(r) + 4\pi r^3 P/c^2]}{r[r - 2GM(r)/c^2]} \qquad (6.2.11)$$

This is quite reminiscent of the newtonian potential gradient, except (1) that the mass has been augmented by a term representing the local "mass" density attributable to the kinetic energy of the matter producing the pressure and (2) that the radial coordinate has been modified to account for the space curvature. Now even in a noneuclidean metric we have the reasonable result

$$\nabla P = -\tilde{\rho}\nabla\Omega \qquad (6.2.12)$$

where $\tilde{\rho}$ is the total local mass density, so the matter density ρ must be increased by P/c^2 to include the mass of the kinetic energy of the gas. [For a rigorous proof of this see Misner, Thorne, and Wheeler[4] (p. 601)]. Combining equations (6.2.11) and (6.2.12), we get

$$\frac{dP}{dr} = -\frac{G(\rho + P/c^2)[M(r) + 4\pi r^3 P/c^2]}{r[r - 2GM(r)/c^2]} \qquad (6.2.13)$$

This is known as the *Oppenheimer-Volkoff equation of hydrostatic equilibrium*, and along with the equation of state it determines the structure of a relativistic star.

6.3 Equations of Relativistic Stellar Structure and Their Solutions

In many respects the construction of stellar models for relativistic stars is easier than that for newtonian models. The reasons can be found in the very conditions which make consideration of general relativity important. Except in the case of supermassive stars, when gravity has been able to compress matter to such an extent that general relativity is necessary to describe the metric of the space occupied by the star, all forms of energy generation which might provide opposition to gravity have ceased. Because of the high degree of compaction, the material generally has a high conductivity and is isothermal, so its cooling rate is limited only by the ability of the surface to radiate energy. In addition, the high density leads to equations of state in which the kinetic energy of the gas is relatively unimportant in determining the state of the gas. The pressure is determined by internuclear forces and thus depends on only the density. In a way, the messy detailed physics of a low-density gas, which depends on its chemical composition and internal energy, has been "squeezed" out of it and replaced by a simpler environment where gravity rules supreme. To be sure, the equation of state of nuclear matter is still an area of intense research interest. But progress in this area is limited as much by our inability to test the results of theoretical predictions as by the theoretical difficulties themselves.

a   A Comparison of Structure Equations

To see the sort of simplification that results from the effects of extreme gravity, let us compare the equations of stellar structure in the newtonian limit and the relativistic limit.

NonrelativisticRelativistic

(a)
$$\frac{dM(r)}{dr} = 4\pi r^2 \rho \qquad \text{Conservation of mass} \qquad \frac{dM(r)}{dr} = 4\pi r^2 \rho$$

(b)
$$\frac{d\Omega}{dr} = \frac{GM(r)}{r^2} \qquad\qquad \left\{ \frac{d\Omega}{dr} = \frac{G[M(r) + 4\pi r^3 P/c^2]}{r[r - 2GM(r)/c^2]} \right.$$

Hydrostatic equilibrium

(c)
$$\frac{dP}{dr} = -\frac{\rho\, d\Omega}{dr} \qquad\qquad \left. \frac{dP}{dr} = -\left(\rho + \frac{P}{c^2}\right)\frac{d\Omega}{dr} \right.$$

(d)
$$\frac{dL(r)}{dr} = 4\pi r^2 \rho \epsilon \qquad \text{Conservation of energy} \qquad \epsilon = 0$$

(e)
$$P = \frac{\rho k T}{\mu m_h} \qquad\qquad \text{Equation of state} \qquad\qquad P = P(\rho)$$

(f)
$$\frac{dT}{dr} = f(P, T, \rho) \qquad\qquad\qquad \text{The equation of state does not depend on } T$$

(g)
$$\kappa = \kappa(P, T, \rho) \qquad\qquad\qquad \kappa \text{ is irrelevant if } \epsilon = 0$$

$$(6.3.1)$$

For relativistic stellar models, we need only solve equations (6.3.1*a*) through (6.3.1*c*) and (6.3.1*e*) subject to certain boundary conditions. Combining equations (6.3.1*b*) and (6.3.1*c*), we have just three equations in three unknowns—$M(r)$, P, and ρ. Two of the equations are first-order differential equations requiring two constants of integration. One additional eigenvalue of the problem is required because we must specify the type (mass) of star we wish to make.

Thus,

$$P(R) = 0 \qquad M(R) = M \qquad M(0) = 0 \tag{6.3.2}$$

For the eigenvalue, we might just as well have specified the central pressure for that would lead to a specific star and would make the problem an initial-value problem. We can gain some insight into the effects of general relativity by looking at a concrete example.

b A Simple Model

The reduction of the equation of state to the form $P = P(\rho)$ is reminiscent of the polytropic equation of state. For polytropes, the combination of the equation of state with hydrostatic equilibrium led to the Lane-Emden equation, which specified the entire structure of the star subject to certain reasonable boundary conditions. To be sure, we could write a similar "relativistic" Lane-Emden equation for relativistic polytropes, but instead we take a different approach. Let us consider a situation where the constraint presented by the equation of state is replaced by a direct constraint on the density. While this does not result in a polytropic equation of state, it is illustrative and analytic, allowing for the solution to be obtained in closed form. Assume the density to be constant, so that

$$\rho(r) = \rho_0 = \text{constant} \tag{6.3.3}$$

The first of the two remaining equations of stellar structure then has the direct solution

$$M(r) = \frac{4\pi r^3 \rho_0}{3} \tag{6.3.4}$$

while the Oppenheimer-Volkoff equation of hydrostatic equilibrium becomes

$$\frac{dP(r)}{dr} = -\frac{4\pi G r \rho_0^2 [1 + P/(\rho_0 c^2)][1 + 3P/(\rho_0 c^2)]}{3[1 - 8\pi G \rho_0 r^2/(3c^2)]} \tag{6.3.5}$$

This equation has an analytic solution which can be obtained by direct, albeit somewhat messy, integration. We can facilitate the integration by introducing the variables

$$y = \frac{P}{\rho_0} \qquad \gamma = \frac{8\pi G \rho_0}{3c^2} = \frac{2GM}{R^3 c^2} \tag{6.3.6}$$

and rewrite the equation for hydrostatic equilibrium as

$$\frac{dy}{dr} = -\frac{1}{2}\gamma c^2 \frac{(1 + y/c^2)(1 + 3y/c^2)r}{1 - \gamma r^2} \tag{6.3.7}$$

which is subject to the boundary condition $y(R) = 0$. With zero as a value for the surface pressure, the solution of equation (6.3.7) is

$$y = c^2 \frac{(1 - \gamma r^2)^{1/2} - (1 - \gamma R^2)^{1/2}}{3(1 - \gamma R^2)^{1/2} - (1 - \gamma r^2)^{1/2}} \tag{6.3.8}$$

In terms of physical variables this is

$$P(r) = \rho_0 c^2 \frac{[1 - 2GMr^2/(R^3 c^2)]^{1/2} - [1 - 2GM/(Rc^2)]^{1/2}}{3[1 - 2GM/(Rc^2)]^{1/2} - [1 - 2GMr^2/(R^3 c^2)]^{1/2}} \tag{6.3.9}$$

Now we evaluate equation (6.3.9) for the central pressure by letting r go to zero. Then

$$P_c = \rho_0 c^2 \frac{1 - [1 - 2GM/(Rc^2)]^{1/2}}{3[1 - 2GM/(Rc^2)]^{1/2} - 1} \tag{6.3.10}$$

As the central pressure rises, the star will shrink, reflecting the larger effects of gravity, so that

$$\lim_{P_c \to \infty} R = \frac{9}{8}\frac{2GM}{c^2} = \frac{9}{8}R_s \tag{6.3.11}$$

where R_s is the Schwarzschild radius. This implies that the smallest stable radius for such an object would be slightly larger than its Schwarzschild radius. A more reasonable limit on the central pressure would be to limit the speed of sound to be less than or equal to the speed of light. A sound speed in excess of the speed of light would suggest conditions where the gas would violate the principle of causality. Namely, sound waves could propagate signals faster than the velocity of light. Since P/ρ_0 is the square of the local sound speed, consider

$$\lim_{P_c \to c^2\rho_0} R = \tfrac{4}{3}R_s \tag{6.3.12}$$

This lower value for the central pressure yields a somewhat larger minimum radius. Since any reasonable equation of state will require that the density monotonically decrease outward and since causality will always dictate that the sound speed be less than the speed of light, we conclude that any stable star must have a radius R such that

$$R \geq \tfrac{4}{3}R_s \tag{6.3.13}$$

In reality, this is an extreme lower limit, and neutron stars tend to be rather larger and of the order of 4 or 5 Schwarzschild radii. Nevertheless, neutron stars still represent stellar configurations in which the general theory of relativity plays a dominant role.

c Neutron Star Structure

The larger size of actual neutron stars, compared to the above limit, results from detailed consideration of the physics that specifies the actual equation of state. Although this is still an active area of research and is likely to be so for some time, we will consider the results of an early equation of state given by Salpeter.[5,6] He shows that we can write a parametric equation of state in the following way:

$$P = \tfrac{1}{3}K\left(\operatorname{Sinh} t - 8\operatorname{Sinh}\frac{t}{2} + 3t \right) \tag{6.3.14}$$

$$\rho = K(\operatorname{Sinh} t - t)$$

where

$$K = \frac{\pi \mu_0^4 c^5}{4h^3}$$

$$t = 4\operatorname{Log}\left\{ \frac{\hat{p}}{\mu_0 c} + \left[1 + \left(\frac{\hat{p}}{\mu_0 c}\right)^2 \right]^{1/2} \right\} \tag{6.3.15}$$

and $\hat{p}$ is the maximum Fermi momentum and may depend weakly on the temperature. The relationship between the mass and central density is shown in Figure 6.1. If one includes the energy losses from neutrinos due to inverse beta decay, there

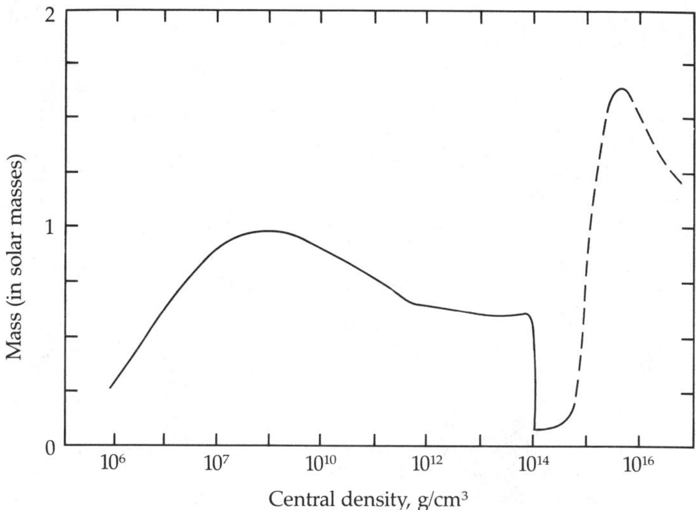

Figure 6.1 shows the variation of the mass of a degenerate object central density. The large drop in the stable mass at a density of about 10^{14} gm/cm^3 represents the transition from the electron degenerate equation of state to the neutron degenerate equation of state.

exists a local maximum for the mass at around 1 solar mass. More recent modifications to the equation of state show a second maximum occurring at slightly more than 2 solar masses. Considerations of causality set an absolute upper limit for neutron stars at about 5 solar masses. So there exists a mass limit for neutron stars, as there does for white dwarfs, and it is probably about 2.5 solar masses. However, unlike the Chandrasekhar limit, this mass limit arises because of the effects of the general theory of relativity. As we shall see in the next section, this relativistic mass limit is also applicable to white dwarfs.

We have not said anything about the formidable problems posed by the formulation of an equation of state for material that is unavailable for experiment. To provide some insight into the types of complications presented by the equation of state, we show in Figure 6.2 the structure of a neutron star as deduced by Rudermann.[7]

The equation of state for the central regions of such a star still remains in doubt as the Fermi energy reaches the level for the formation of hyperons. Some people have speculated that one might reach densities sufficient to yield a "quark soup." Whatever the details of the equation of state, they matter less and less as one approaches the critical mass. Gravity begins to snuff out the importance of the local microphysics. By the time one reaches a configuration that has contracted within its Schwarzschild radius, only the macroscopic properties of total mass, angular momentum, and charge can be detected by an outside observer. Although

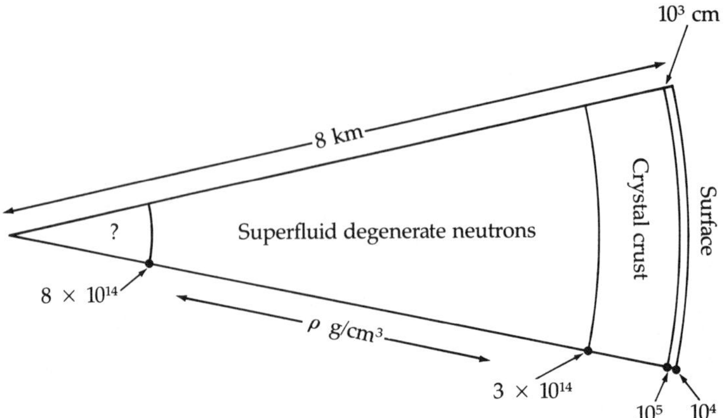

Figure 6.2 shows a section of the internal structure of a neutron star. The formation of crystal structure in the outer layers of the neutron star greatly complicates its equation of state. Its structure may be testable by observing the shape changes of rapidly rotating pulsars as revealed by discontinuous changes in their spin rates as they slow down.

this ultimate result occurs only when the object has reached the Schwarzschild radius, aspects of its approach are manifest in the insensitivity of the global structure to the equation of state as the limiting radius is approached. This has the happy result for astronomy that the mass limit for neutron stars can comfortably be set at around $2\frac{1}{2}M_\odot$ regardless of the vagaries of the equation of state. It has an unhappy consequence for physics in that neutron stars will prove a difficult laboratory for testing the details of the equation of state for high-density matter.

6.4   Relativistic Polytrope of Index 3

In Chapter 2, we remarked that the equation of state for a totally relativistic degenerate gas was a polytrope of index 3. In addition, we noted that an object dominated by radiation pressure would also be a polytrope of index $n = 3$. In the first category we find the extreme white dwarfs, those nearing the Chandrasekhar degeneracy limit. In the second category we find stars of very great mass where, from the β^* theorem, we can expect the total pressure to be very nearly that of the pressure from photons. It is somewhat curious that such different types of stars should have their structures given by the same equilibrium model. However, both types are dominated by relativistic (in the sense of the special theory of relativity) particles, and this aspect of the gas is characterized by a polytrope of index $n = 3$.

Our approach to the study of these objects will be a little different from our previous discussions of stellar structure. Rather than concentrate on the internal properties and physics of these objects, we consider only their global properties, such as mass, radius, and internal energy. This will be sufficient to understand their stability and evolutionary history. The ideal vehicle for such an investigation is the virial theorem.

a   Virial Theorem for Relativistic Stars

The virial theorem for relativistic particles differs somewhat from that derived in Chapter 1. The effect of special relativity on the "mass" or momentum of such particles increases the gravitational energy required to confine the particles as the internal energy increases (see Collins[8]). Thus, for stable configurations, instead of

$$2\mathbf{T} + \Omega = 0 \qquad (6.4.1)$$

we get

$$\mathbf{T} + \Omega = 0 \qquad (6.4.2)$$

which specifies the total energy of the configuration as

$$E = \mathbf{T} + \Omega = 0 \qquad (6.4.3)$$

This is sometimes called the *binding energy* because it is the energy required to disperse the configuration throughout space. Thus polytropes of index $n = 3$ are neutrally stable since it would take no work at all to disperse them and as such these polytropes represent a limiting condition that can never be reached. To investigate the fate of objects approaching such a limit, it is necessary to look at the behavior of those conditions that lead to small departures from the limit. One of those conditions is the distortion of the metric of space caused by the matter-energy itself and so well described by the general theory of relativity.

Phenomenologically, we may view the effects of general relativity as increasing the effective "force of gravity." Thus, as we approach the limiting state of the relativistic polytrope, we would expect the effects of general relativity to cause the configuration to become unstable to collapse. So it is general relativity which sets the limit for the masses of white dwarfs, not the Pauli exclusion principle, just as general relativity set the limit for the masses of neutron stars. We could also expect such an effect for supermassive stars dominated by photon pressure.

To quantify these effects, we have to appeal to the virial theorem in a non-euclidean metric. Rather than rederive the Boltzmann transport equation for the Schwarzschild metric, we obtain the relativistic Euler-Lagrange equations of hydrodynamic flow and take the appropriate spatial moments. We skip directly to the result of Fricke[9]

$$\frac{1}{2}\frac{d^2 I_r}{dt^2} = 3\int_V P\, dV + \Omega$$

$$+ \frac{G}{c^2}\left[\int_V \frac{M(r)(P + \rho\dot{r}^2)}{r}\, dV - 3\int_V \frac{GM^2(r)\rho}{r^2}\, dV\right] \quad (6.4.4)$$

Here I_r is the moment of inertia defined about the center of the Schwarzschild metric. The effects of general relativity are largely contained in the third term in brackets, which is multiplied by G/c^2 and contains the additions to the potential energy of the kinetic energy of the gas particles (as represented by the pressure) and the kinetic energy of mass motions of the configuration (as represented by $\dot{r}^2\rho$). The physical interpretation of the second term in the brackets is more obscure. For want of a better description, it can be viewed as a self-interaction term arising from the nonlinear nature of the general theory of relativity. Except for the relativistic term, equation (6.4.4) is very similar to its newtonian counterpart in Chapter 1 [equation (1.2.34)]. The effect of the internal energy is included in the term $3\int_V P\, dV$. Since we will be considering stars that are near equilibrium, we take the total kinetic energy of mass motions to be zero. The $\dot{r}^2\rho$ term was included in the relativistic term to emphasize its relativistic role.

A common technique in stellar astrophysics is to perform a variational analysis of the virial theorem as expressed by equation (6.4.4), but such a process is quite lengthy. Instead, we estimate the effects of general relativity by determining the magnitude of the relativistic terms as $\gamma \to \frac{4}{3}$. Obviously if the left-hand side of equation (6.4.4) becomes negative, the star will begin to acceleratively contract and will become unstable. Thus we investigate the conditions where the star is just in equilibrium. Replacing $3\int_V P\, dV$ with its equivalent in terms of the internal energy [see equation (5.4.2)], we get

$$3(\gamma - 1)U + \Omega = -\frac{G}{c^2}\int_V \frac{M(r)P}{r}\, dV + \frac{3G^2}{c^2}\int_V \frac{M^2(r)\rho}{r^2}\, dV \quad (6.4.5)$$

Now, since $\gamma = \frac{4}{3}$ is a limiting condition, let

$$\gamma = \tfrac{4}{3} + \epsilon \qquad n = 3(1 - 3\epsilon) \quad (6.4.6)$$

where ϵ is positive. We may use the variational relation between the internal and potential energies

$$\delta U = -3(\gamma - 1)^{-1}\,\delta\Omega \quad (6.4.7)$$

(see Chandrasekhar[10]), and we get

$$\lim_{\gamma \to 4/3 + \epsilon} [3(\gamma - 1)U + \Omega] = (1 + 3\epsilon)(U_0 + \delta U) + \Omega_0 + \delta\Omega = -3\Omega_0 \qquad (6.4.8)$$

Here the subscript zero denotes the value of quantities when $\gamma = \frac{4}{3}$, and $U_0 = -\Omega_0$ for that value of γ, so the virial theorem becomes

$$3\epsilon\Omega_0 = \frac{G}{c^2} \int_V \frac{M(r)P}{r} \, dV - \frac{3G^2}{c^2} \int_V \frac{M^2(r)\rho}{r^2} \, dV \qquad (6.4.9)$$

We may now estimate the magnitude of the relativistic terms on the right-hand side as follows. Consider the first term where

$$\frac{G}{c^2} \int_V \frac{M(r)P}{r} \, dV = \frac{G}{c^2} \left(\overline{\frac{M}{r}} \right) \left(\frac{-\Omega_0}{3} \right) \approx \frac{GM}{Rc^2} \left(\frac{-\Omega_0}{3} \right) = -\frac{1}{6} \left(\frac{R_s}{R} \right) \Omega_0 \qquad (6.4.10)$$

Here we have taken the pressure-weighted mean of M/R to be M/R, and R_s is the Schwarzschild radius for the star. The second term can be dealt with in a similar manner, so

$$\frac{3G^2}{c^2} \int_V \frac{M^2(r)\rho}{r^2} \, dV = \frac{3G^2}{c^2} \overline{\left(\frac{M}{r} \right)^2} M \approx \frac{3GM^2}{2R} \frac{R_s}{R} = -\frac{R_s}{R} \Omega_0 \qquad (6.4.11)$$

Again, we have replaced the mean of M/R by M/R. Since the means of the two terms are not of precisely the same form, we expect this approach to yield only approximate results. Indeed, the central concentration of the polytrope will ensure that both terms are underestimates of the relativistic effects. Even worse, the mean-square of $M(r)/r$ in equation (6.4.11) will yield an even larger error than that of equation (6.4.10). Since the terms differ in sign, the combined effect could be quite large. However, we may be sure that the result will be a lower limit of the effects of general relativity, and the approximations do demonstrate the physical nature of the terms. With this large caveat, we shall proceed.

Substituting into equation (6.4.9), we get

$$\frac{R}{R_s} \approx \frac{5}{18\epsilon} \qquad (6.4.12)$$

Now all that remains to be done is to investigate how $\gamma \to \frac{4}{3}$ in terms of the defining parameters of the star (M, L, R), and we will be able to estimate when the effects of general relativity become important.

b   Minimum Radius for White Dwarfs

We have indicated that the effects of general relativity should bring about the collapse of a white dwarf as it approaches the Chandrasekhar limiting mass. If we can characterize the approach of γ to $\frac{4}{3}$ in terms of the properties of the star, we will know how close to the limiting mass this occurs. As $\gamma \to \frac{4}{3}$, the degeneracy parameter in the parametric degenerate equation of state approaches infinity. Carefully expanding $f(x)$ of equation (1.3.14) and determining its behavior as $x \to \infty$, we get

$$\operatorname*{Lim}_{x \to \infty} f(x) = 2(x^4 - x^2) \tag{6.4.13}$$

From the polytropic equation of state

$$\gamma = \frac{d(\ln P)}{d(\ln \rho)} \tag{6.4.14}$$

Evaluating the right-hand side from the parametric equation of state [equation (1.3.14)] and the result for $f(x)$ given by equation (6.4.13), we can combine it with the definition of ϵ from equation (6.4.6) to get

$$\epsilon \approx \frac{2x^{-2}}{3} \tag{6.4.15}$$

If we neglect the effects of inverse beta decay in removing electrons from the gas, we can write the density in terms of the electron density and, with the aid of equation (1.3.14), in terms of the degeneracy parameter x:

$$\rho = \frac{8\pi c^3 m_e^3 \mu_e m_h x^3}{3h^3} \tag{6.4.16}$$

If we approximate the density by its mean value, we can solve for the average square degeneracy parameter for which we can expect collapse:

$$\bar{x}^2 = \left(\frac{9h^3}{32\pi^2 c^3 m_e^3 \mu_e m_h} \right)^{2/3} \left(\frac{M}{R^3} \right)^{2/3}$$
$$= (7.12 \times 10^6)\mu_e^{2/3} \left(\frac{M_\odot}{M} \right)^{4/3} \left(\frac{R_s}{R} \right)^2 \tag{6.4.17}$$

Combining this with equations (6.4.15) and (6.4.12), we obtain an estimate for the manner in which the minimum stable radius of a white dwarf depends on mass as the limiting mass is approached:

$$\frac{R}{R_s} \approx 144\mu_e^{-2/9}\left(\frac{M_\odot}{M}\right)^{4/9} \qquad (6.4.18)$$

A more precise calculation involving a proper evaluation of the relativistic integrals and evaluation of the average internal degeneracy by Chandrasekhar and Trooper[11] yields a value of 246 Schwarzschild radii for the minimum radius of a white dwarf, instead of about 100 given by equation (6.4.18). We can then combine this with the mass-radius relation for white dwarfs to find the actual value of the mass for which the star will become unstable to general relativity. This is about 98 percent of the value given by the Chandrasekhar limit, so for all practical purposes the degeneracy limit gives the appropriate value for the maximum mass of a white dwarf.

However, massive white dwarfs do not exist because general relativity brings about their collapse as the star approaches the Chandrasekhar limit. This point is far more dramatic in the case of neutron stars. Here the relativistic terms bring about collapse long before the entire star becomes relativistically degenerate. A relativistically degenerate neutron gas has much more kinetic energy per gram than a relativistically degenerate electron gas, since a relativistic particle must have a kinetic energy greater than its rest energy, by definition. To contain such a gas, the gravitational forces must be correspondingly larger, which implies a greater importance for general relativity. Indeed, to confine a fully relativistically degenerate configuration, it would be necessary to restrict it to a volume bounded by its Schwarzschild radius. This is not to say that the cores of neutron stars cannot be relativistically degenerate. Indeed they can, but the core is contained by the weight of the nonrelativistically degenerate layers above as well as its own self-gravity.

c Minimum Radius for Supermassive Stars

Since the early 1960s, supermassive stars have piqued the interest of some. It was thought that such objects might provide the power source for quasars. While their existence might be ephemeral, if supermassive stars were formed in sufficient numbers, their great luminosity might provide a solution to one of the foremost problems of the second half of the twentieth century. However, truly massive stars are subject to the same sort of instability as we investigated for white dwarfs. Indeed, the problem was first discussed by W. Fowler[12,13] for the supermassive stars and later extended by Chandrasekhar and Trooper[11] to white dwarfs. Finally the problem was rediscussed by Fricke,[9] and the effect of the metric on the relativistic integrals was correctly included.

By now you should not be surprised that such an instability exists, because we know that massive stars are dominated by radiation pressure and can be well represented by polytropes of index $n = 3$. For supermassive stars, the departure from being a perfect relativistic polytrope results from some of the total energy being provided by the kinetic energy of the gas particles. To quantify the instability, we may proceed as we did with the white dwarf analysis. If we write the equilibrium virial theorem and split the $3\int PdV$ term into a sum of the gas pressure and the radiation pressure, then we get

$$0 = 3 \int_V \beta P\, dV + 3 \int_V (1 - \beta) P\, dV + \Omega + \frac{G}{c^2} \int_V \frac{M(r)P}{r}\, dV$$

$$- \frac{3G^2}{c^2} \int_V \frac{M^2(r)\rho}{r^2}\, dV \tag{6.4.19}$$

However, the total energy of the configuration is

$$E = \frac{3}{2} \int_V \beta P\, dV + 3 \int_V (1 - \beta) P\, dV + \Omega + \frac{3G}{c^2} \int_V \frac{M(r)P}{r}\, dV$$

$$- \frac{3G^2}{2c^2} \int_V \frac{M^2(r)\rho}{r^2}\, dV \tag{6.4.20}$$

Subtracting equation (6.4.19) from (6.4.20), we get

$$E = -\frac{3}{2} \int_V \beta P\, dV + \frac{2G}{c^2} \int_V \frac{M(r)P}{r}\, dV + \frac{3G^2}{2c^2} \int_V \frac{M^2(r)\rho}{r^2}\, dV \approx 0 \tag{6.4.21}$$

When the total energy of the configuration becomes zero, we will have reached its minimum stable radius. Making the same approximations for the relativistic integrals that were made for the white dwarf analysis, we get

$$\frac{R_0}{R_s} \approx \frac{5}{3\beta} \tag{6.4.22}$$

From equation (2.2.11) we saw that the central value of beta, β_c, was bounded by the mass, so that

$$\beta_c \propto M^{-1/2} \tag{6.4.23}$$

Using the constant of proportionality implied by equation (2.2.11) and combining with equation (6.4.22), we get

$$\frac{R_0}{R_s} \simeq 0.18 \left(\frac{M}{M_\odot} \right)^{1/2} \tag{6.4.24}$$

Thus a supermassive star of $10^8 M_\odot$ will become unstable at about 1800 Schwarzschild radii or about 14 AU. In units of the Schwarzschild radius, this result is rather larger than that for white dwarfs. This can be qualitatively understood by considering the nature of the relativistic particles providing the majority of the internal pressure in the two cases. The energy of the typical photon providing the radiation pressure for a supermassive star is far less than the energy of a typical degenerate electron whose degenerate pressure provides the support in a white dwarf. Thus a weaker gravitational field will be required to confine the photons as compared to the electron. This implies that as the total energy approaches zero, the mass required to confine the photons can be spread out over a larger volume, when measured in units of the Schwarzschild radius, than is the case for the electrons. This argument implies that neutron stars should be much closer to their Schwarzschild radius in size, which is indeed the case.

Perhaps the most surprising aspect of both these analyses is that general relativity can make a significant difference for structures that are many hundreds of times the dimensions that we usually associate with general relativity.

6.5 Fate of Supermassive Stars

The relativistic polytrope can be used to set minimum sizes for both white dwarfs and very massive stars. However, supermassive stars are steady-state structures and will evolve, while white dwarfs are equilibrium structures and will remain stable unless they are changed by outside sources. Let us now see what can be said about the evolution of the supermassive stars.

a Eddington Luminosity

Sir Arthur Stanley Eddington observed that radiation and gravitation both obey inverse-square laws and so there would be instances when the two forces could be in balance irrespective of distance. Thus there should exist a maximum luminosity for a star of a given mass, where the force of radiation on the surface material would exactly balance the force of gravity. If we balance the gravitational acceleration against the radiative pressure gradient [equation (4.2.11)] for electron scattering, we can write

$$\nabla P_r = -\frac{\sigma_e \rho L(r)}{4\pi c R^2} = -\frac{GM\rho}{R^2} \tag{6.5.1}$$

Therefore, any object that has a luminosity greater than

$$L_{\mathbf{Edd}} = \frac{4\pi G c}{\sigma_e} M \approx (1.3 \times 10^{38}) \left(\frac{M}{M_\odot}\right) \quad \text{erg/s} \tag{6.5.2}$$

will be forced into instability by its own radiation pressure. This effectively provides a mass-luminosity relationship for supermassive stars since these radiation-dominated configurations will radiate near their limit.

b Equilibrium Mass-Radius Relation

If we now assume that the star can reach a steady-state that represents a near-equilibrium state on a dynamical time, then the energy production must equal the energy lost through the luminosity. Eugene Cappriotti[14] has evaluated the luminosity integral and gets

$$ L = \int_V \rho \epsilon \, dV \simeq (2.9 \times 10^{-67}) \left(\frac{M^{8.625}}{R^{16.25}} \right) \qquad \text{erg/s} \qquad (6.5.3) $$

We can assume that massive stars will derive the nuclear energy needed to maintain their equilibrium from the CNO cycle, can evaluate ϵ as indicated in Chapter 3 [equation (3.3.19)], and can evaluate the central term of equation (6.5.3) to obtain the approximate relation on the right. Assuming that the stars will indeed radiate at the Eddington luminosity, we can use equation (6.5.2) to find

$$ R_e \approx (1.7 \times 10^{11}) \left(\frac{M}{M_\odot} \right)^{0.47} \qquad \text{cm} \qquad (6.5.4) $$

Thus we have a relation between the mass and radius for any supermassive star that would reach equilibrium through the production of nuclear energy. However, we have yet to show that the star can reach that equilibrium state.

c Limiting Masses for Supermassive Stars

Let us rewrite equation (6.4.19) for the total energy, taking care to express the relativistic integrals as dimensionless integrals by making use of the homology relations for pressure and density:

$$ E = -\frac{1}{2}\bar{\beta}\Omega + \frac{2G^2M^3}{R^2c^2} \int_0^1 \left[\frac{M(r)}{M} \frac{R}{r} \right]^2 \frac{P}{P_c} \frac{\rho}{\rho_c} \frac{dM(r)}{M} $$

$$ - \frac{9G^2M^3}{2R^2c^2} \int_0^1 \left[\frac{M(r)}{M} \right]^2 \frac{dM(r)}{M} \qquad (6.5.5) $$

We must be very careful in evaluating these integrals, for any polytrope in euclidean space as the radial coordinate used to obtain those integrals is defined by the Schwarzschild metric (see Fricke,[9] p. 942). We must do so here since we will not be content to find a crude result for the mass limits. Replacing β by its limiting value given by the β^* theorem and evaluating the relativistic integrals for a polytrope of index $n = 3$, we obtain

$$E = -\frac{27GM_\odot^2}{4R_\odot}\left(\frac{M}{M_\odot}\right)^{3/2}\frac{R_\odot}{R} + 5.07\frac{G^2M_\odot^2}{R_\odot^2 c^2}\left[\left(\frac{M}{M_\odot}\right)^{3/2}\frac{R_\odot}{R}\right]^2 \tag{6.5.6}$$

If we now seek the radial value for which $E = 0$, we get

$$R_0 \approx 0.756\left(\frac{GM_\odot}{c^2}\right)\left(\frac{M}{M_\odot}\right)^{3/2} = (1.1 \times 10^5)\left(\frac{M}{M_\odot}\right)^{3/2} \quad \text{cm}$$

$$= 0.37R_s\left(\frac{M}{M_\odot}\right)^{1/2} \tag{6.5.7}$$

Comparing this result with equation (6.4.24), we see that our crude approximation of the relativistic integrals was low by about a factor of two. Equation (6.5.6) is quadratic in $M^{3/2}/R$ and will become positive at small R. Figure 6.3 shows the

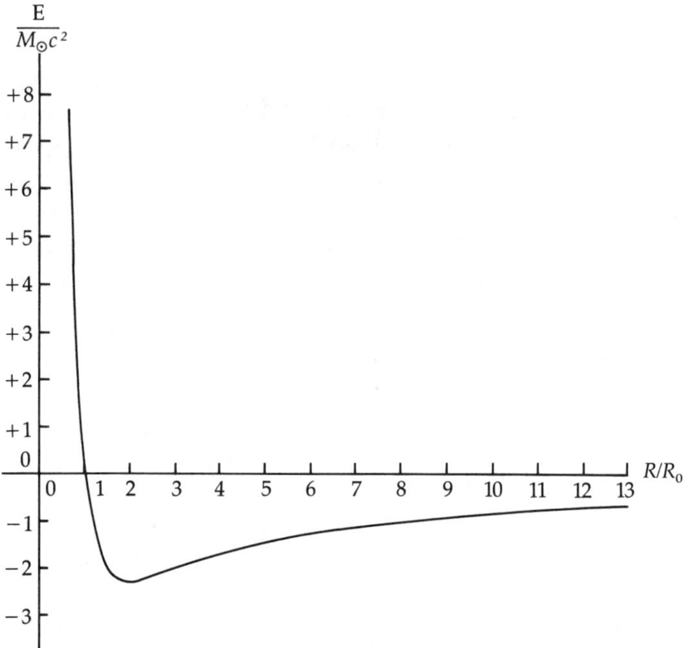

Figure 6.3 shows the variation of the binding energy in units of the rest energy of the sun as a function of the radius in units of the minimum stable radius. It is clear that a minimum (most negative) binding energy exists and that the minimum is a specific value for all supermassive stars.

dependence of the binding energy on the radius. Differentiation of equation (6.5.6) shows that the greatest (most negative) binding energy will occur at

$$R_m = 2R_0 \qquad (6.5.8)$$

and this corresponds to an energy of

$$E_m = -2.25 M_\odot c^2 \qquad (6.5.9)$$

This energy is a constant because of the quadratic nature of the energy equation. The relativistic terms simply vary as the next power of $M^{3/2}/R$, compared to the newtonian terms. Hence the minimum will depend only on physical constants.

A star that is contracting toward its equilibrium position may reach equilibrium for any radial value that is greater than or equal to R_m, providing an energy source exists to replace the energy lost to space. We have already found the equilibrium radius for energy produced by the CNO cycle [equation (6.5.4)]. Combining that with the minimum energy radius, we find

$$\frac{M}{M_\odot} \le 5.2 \times 10^5 \qquad (6.5.10)$$

Thus any star with a mass less than about half a million solar masses can come to equilibrium burning hydrogen via the CNO cycle, albeit with a short lifetime. More massive stars are destined to continue to contract. Of course, more massive stars will produce nuclear energy at an ever-increasing rate as their central temperatures rise. However, the rate of energy production cannot increase without bound. This is suggested by the declining exponents of the temperature dependence shown in Table 3.4. The nuclear reactions that involve β decay set a limit on how fast the CNO cycle can run, and β decay is independent of temperature. So there is a maximum rate at which energy can be produced by the CNO cycle operating in these stars.

If the nuclear energy produced is sufficient to bring the total energy above the binding energy curve, the star will explode. However, should the energy not be produced at a rate sufficient to catch the binding energy that is rising due to the relativistic collapse, the star will continue an unrestrained collapse to the Schwarzschild radius and become a black hole. Which scenario is played out will depend on the star's mass. For these stars, the temperature gradient will be above the adiabatic gradient, so convection will exist. However, the only energy transportable by convection is the kinetic energy of the gas, which is an insignificant fraction of the total energy. Therefore, unlike normal main sequence stars, although it is present, convection will be a very inefficient vehicle for the transport of energy. This is why the star remains with a structure of a polytrope of index $n = 3$ in the presence of convection. The pressure support that determines the density distribution comes entirely from radiation and is not governed by the mode of energy transport. We saw a similar situation for degenerate white dwarfs. The equation of state indicated that their structure would be that of a polytrope of index $n = 1.5$ (for nonrelativistic degeneracy) and yet the star would be isothermal due to the long mean free path of the degenerate electrons. However, the structure is not that of an isothermal sphere because the pressure support comes almost entirely from the degenerate electron gas and is largely independent of the energy and temperature distribution of the ions.

The star will radiate at the Eddington luminosity, and that will set the time scale for collapse. Remember that the total energy of these stars is small compared to the gravitational energy. So most of the energy derived from gravitational contraction must go into supporting the star, and very little is available to supply the Eddington luminosity. This can be seen from the relativistic virial theorem [equation (6.4.2)], which indicates that any change in the gravitational energy is taken up by the kinetic energy. Relativistic particles (in this case, photons) are much more difficult to bind by gravitation than ordinary matter; thus little of the gravitational energy resulting from collapse will be available to let the star shine. The collapse will proceed very quickly on a time scale that is much nearer to the dynamical time scale than the Kelvin-Helmholtz time scale. The onset of nuclear reactions will slow the collapse, but will not stop it for the massive stars.

A dynamical analysis by Appenzeller and Fricke[15,16] (see also Fricke[9]) shows that stars more massive than about $7.5 \times 10^5 M_\odot$ will undergo collapse to a black hole. Here the collapse proceeds so quickly and the gravity is so powerful that the nuclear reactions, being limited by β decay at the resulting high temperatures, do not have the time to produce sufficient energy to arrest the collapse. For less massive stars, this is not the case. Stars in the narrow range of $5 \times 10^5 M_\odot \leq M \leq 7.5 \times 10^5 M_\odot$ will undergo explosive nuclear energy generation, resulting in the probable destruction of the star.

Nothing has been said about the role of chemical composition in the evolution of these stars. Clearly, if there is no carbon present, the CNO cycle is not available for the stabilization of the star. Model calculations show that the triple-α process cannot stop the collapse. For stars with low metal abundance, only the proton-proton cycle is available as an energy source. This has the effect of lowering the value of the maximum stable mass. Surprisingly, there is no range at which an explosion occurs. If the star cannot stabilize before reaching R_m, it will continue in a state of unrestrained gravitational collapse to a black hole. Thus, it seems unlikely that stars more massive than about a half million solar masses could exist. In addition, it seems unlikely that black holes exist with masses greater than a few solar masses and less than half a million solar masses. If they do, they must form by accretion and not as a single entity.

Problems

1. Describe the physical conditions that correspond to polytropes of different indices, and discuss which stars meet these conditions.

2. What modifications must be made to the classical equation of hydrostatic equilibrium to obtain the Oppenheimer-Volkoff equation of hydrostatic equilibrium?

3. Find the mass-radius law for supermassive stars generating energy by means of the proton-proton cycle. Assume that the metal abundance is very small.

4. Determine the mass corresponding to a white dwarf at the limit of stability to general relativity.

5. Evaluate the relativistic integrals in equation (6.4.4) for a polytrope of index $n = 3$. Be careful, for the euclidean metric appropriate for the polytropic tables is not the same as the Schwarzschild metric of the equation (see Fricke,[9] p. 941).

6. Use the results of Problem 5 to reevaluate the minimum radius for white dwarfs.

7. Assuming that a neutron star can be represented by a polytrope with $\gamma = \frac{3}{2}$, find the minimum radius for a neutron star for which it is stable against general relativity. To what mass does this correspond?

References and Supplemental Reading

1. Landau, L., "On the Theory of Stars," *Physik Zeitz. Sowjetunion* 1, 1932, pp. 285–288.

2. Oppenheimer, J. R., and H. Snyder, "On Continued Gravitational Contraction," *Phys. Rev.* 56, 1939, pp. 455–459.

3. Oppenheimer, J. R., and G. Volkoff, "On Massive Neutron Cores," *Phys. Rev.* 55, 1939, pp. 374–381.

4. Misner, C. W., K. S. Thorne, and J. A. Wheeler, *Gravitation*, Freeman, New York, 1970.

5. Salpeter, E. E., "Matter at High Densities," *Ann. Rev. Phys.* 11, 1960, pp. 393–417.

6. Salpeter, E., "Energy and Pressure of a Zero Temperature Plasma," *Ap. J.* 134, 1961, pp. 683–698.

7. Rudermann, M., "Pulsars: Structure and Dynamics," *Ann. Rev. Astro. and Astrophys.* 10, 1972, pp. 427–476.

8. Collins, G. W., II, *The Virial Theorem in Stellar Astrophysics* Pachart Publishing House, Tucson, Ariz., 1978, pp. 34–37.

9. Fricke, K. J., "Dynamical Phases of Supermassive Stars," *Ap. J.* 183, 1973, pp. 941–958.

10. Chandrasekhar, S., *An Introduction to the Study of Stellar Structure*, Dover, New York, 1939/1957, p. 53, eq. 102.

11. Chandrasekhar, S., and R. F. Trooper, "The Dynamical Instability of the White-Dwarf Configuration Approaching the Limiting Mass," *Ap. J.* 139, 1964, pp. 1396–1397.

12. Fowler, W. A., "Massive Stars, Relativistic Polytropes, and Gravitational Radiation," *Rev. Mod. Phys.*, 36, 1964, pp. 545–555, 1104 errata.

13. Fowler, W. A., "The Stability of Supermassive Stars," *Ap. J.* 144, 1966, pp. 180–200.

14. Cappriotti, E. R., private communication, 1986.

15. Appenzeller, I., and K. Fricke, "Hydrodynamic Model Calculations for Supermassive Stars, I: The Collapse of a Nonrotating $0.75 \times 10^6 M_\odot$ Star," *Astr. and Ap.* 18, 1972, pp. 10–15.

16. Appenzeller, I., and K. Fricke, "Hydrodynamic Model Calculations for Supermassive Stars, II: The Collapse and Explosion of a Nonrotating $5.2 \times 10^5 M_\odot$ Star," *Astr. and Ap.* 21, 1972, pp. 285–290.

For a more detailed view of the internal structure of neutron stars, see

• Baym, G., H. Bethe, and C. J. Pethick: "Neutron Star Matter," *Nuc. Phys.* A175, 1971, pp. 225–271.

No introduction to the structure of degenerate objects would be complete without a reading of

• Hamada, T., and E. E. Salpeter: "Models for Zero-Temperature Stars," *Ap. J.* 134, 1961, pp. 683–698.

I am indebted to E. R. Cappriotti for introducing me to the finer points of supermassive stars, and most of the material in Sections 6.4b, 6.5b, and 6.5c was developed directly from his notes on the subject. Those interested in the historical development of supermassive stars should read

• Hoyle, F., and W. A. Fowler: "On the Nature of Strong Radio Sources," *Mon. Not. R. astr. Soc.* 125, 1963, pp. 169–176.

• Faulkner, J., and J. R. Gribbin: "Stability and Radial Vibration Periods of the Hamada-Salpeter White Dwarf Models," *Nature* 218, 1966, pp. 734–736.

While there are many other contributions to the subject that I have not included, these will acquaint the readers with the important topics and the flavor of the subject.

7

Structure of Distorted Stars

· · ·

Throughout this book we have assumed that stars are spherical. This reduces the problem of stellar structure to one dimension, greatly simplifying its description. Unfortunately, many stars are not spherical, but are distorted by their own rotation or the presence of a nearby companion. Not only does this add geometrical complications to the mathematical representation of the equations of stellar structure, but also new physical phenomena, such as global circulation currents, may result. Major problems are created for the observational comparison with theory in that the appearance of a star will now depend on its orientation with respect to the observer. Some quantities, such as the total luminosity and the stellar effective temperature, are no longer accessible to observation. With these problems in mind, we consider some approaches to developing a theoretical framework for the structure of distorted stars.

The removal of spherical symmetry, by increasing the number of dimensions required for the description of the star's structure, will considerably increase the number of equations to be solved to obtain that structure. Rather than develop those equations in detail, we indicate how they are obtained and the basic procedures for their solution. We consider only those cases that exhibit axial symmetry so that the number of dimensions is increased by 1. This is sufficient to illustrate most problems generated by distortion without raising the complexity to an unacceptable level. It also provides a framework for the description of a significant number of additional stars.

7.1 Classical Distortion: The Structure Equations

The loss of spherical symmetry will change the familiar equations of stellar structure to vector form. Before developing the specific equations for axial distortion, let us consider the general form of these equations. In Chapter 6 we compared the relativistic equations of stellar structure to the classical spherical equations. In a similar manner, let us begin our discussion of distortion with a comparison of the classical spherical equations with their counterparts for distorted stars.

a A Comparison of Structure Equations

Below is a summary of the equations of stellar structure for spherically symmetric stars and stars which suffer a general distortion.

Spherical		Nonspherical
(a) $\dfrac{dM(r)}{dr} = 4\pi r^2 \rho$	mass conservation	$\nabla^2 \Omega = 4\pi G \rho$ Poisson's eq.
(b) $\dfrac{dL(r)}{dr} = 4\pi r^2 \rho \epsilon$	energy conservation	$\nabla \cdot \vec{F} = \rho\epsilon - T\dfrac{\partial S}{\partial t}$
(c) $\dfrac{dT(r)}{dr} = -\dfrac{3\bar{\kappa}\rho L(r)}{16\pi ac T^3 r^2}$	radiative equilibrium	$\nabla T = -\dfrac{3\bar{\kappa}\rho}{4ac T^3}\vec{F}$
(d) $\dfrac{dP(r)}{dr} = -\dfrac{GM(r)\rho}{r^2}$	hydrostatic equilibrium	$\nabla P = -\rho\,\nabla\Omega + \rho\vec{D}$
$\kappa = \kappa(P,\,T,\,\rho)$	opacity	$\kappa = \kappa(P,\,T,\,\rho)$
(e) $\epsilon = \epsilon(P,\,T,\,\rho)$	energy generation	$\epsilon = \epsilon(P,\,T,\,\rho)$
$P = P(T,\,\rho,\,\mu)$	equation of state	$P = P(T,\,\rho,\,\mu)$

$$(7.1.1)$$

The variable $M(r)$ that is so useful for spherical structure is replaced by the potential, given here as the gravitational potential. In principle, the potential could contain a contribution from other physical phenomena such as magnetism or rotation. Poisson's equation is a second-order partial differential equation and replaces the first-order total differential equation for spherical structure. So the price we pay for the loss of spherical symmetry is immediately obvious. While the conservation of energy equation remains a scalar equation, as it should, it now involves a vector quantity, the radiative flux, and an additional term that anticipates some results from later in the chapter. The quantity S is the entropy of the gas, and in Chapter 4 [see equation (4.6.9)] we saw that this term had to be included when the models were changing rapidly in time. In this case, the term is required to describe the flow of energy due to mass motions resulting from the distortion itself. Both radiative and hydrostatic equilibrium become vector equations where we have explicitly indicated the presence of a perturbing force by the vector $\vec{D}$ which, should it be derivable from a potential, could be included directly in the potential. This perturbing force is assumed to be known. The quantities such as κ and ϵ, which depend on the local microphysics, presumably will not be directly affected by the presence of a macroscopic perturbing force. A possible exception could be the case of distortion by a magnetic field where the local field would contribute to the total pressure and, in extreme cases, could affect the opacity.

b Structure Equations for Cylindrical Symmetry

To minimize the complexity, we consider cases resulting in the loss of only one symmetry coordinate, and we deal with those systems exhibiting axial symmetry. This is clearly appropriate for rapidly rotating stars as well as stars distorted by the presence of a companion. In addition, we shall see that it also is appropriate for the distortion introduced by an ordered magnetic field that itself exhibits axial symmetry.

To specifically see the effects that result from a distortion force, we have to express that force in some appropriate coordinate system. The distortion force was represented in the structure equations, (7.1.1), by the vector $\vec{D}$ in the equation of hydrostatic equilibrium. For axial symmetry, cylindrical and spherical polar coordinates both form suitable coordinate systems for this description (see Figure 7.1). We express the components of the perturbing force in terms of Legendre polynomials of the polar angle θ. Once the perturbing force has been characterized, we shall indicate, in the next section, how the solution of the structure equations proceeds.

The Legendre polynomials form an orthogonal set of polynomials over a finite, defined range. Specifically, let

$$\mu = \mathrm{Cos}\,\theta \qquad (7.1.2)$$

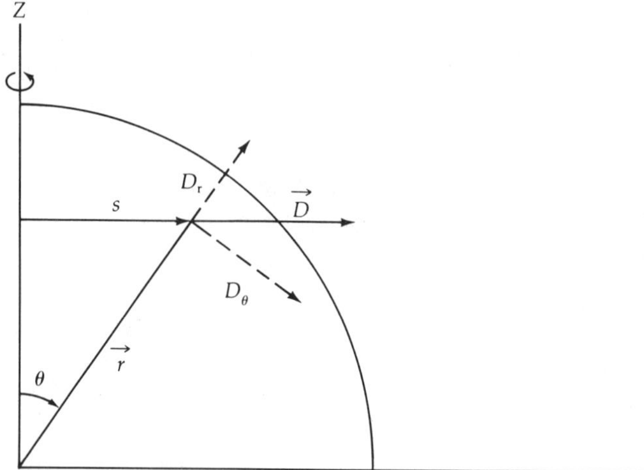

Figure 7.1 Suitable coordinate systems to describe an axially symmetric perturbing force $\vec{D}$ with components D_r and D_θ. We have chosen to illustrate rotational distortion so that Z represents the spin axis. However, we could have illustrated gravitational distortion by an external object, in which case Z would lie along the line of centers of the system and $\vec{D}$ would point to the center of the other object.

Then the Legendre polynomials form an orthonormal set in the interval $-1 \leq \mu \leq +1$ subject to the normalization condition

$$\int_{-1}^{+1} P_n(\mu)P_m(\mu)\,d\mu = \delta_{m,n}\left(\frac{2}{2n+1}\right) \qquad (7.1.3)$$

Here $\delta_{m,n}$ is the Kronecker delta which is 1 if $m = n$ and 0 otherwise. Various members of the set of Legendre polynomials can be generated from the recursion relation

$$P_{m+1}(\mu) = \frac{2m+1}{m+1}\,\mu P_m(\mu) - \frac{m}{m+1}\,P_{m-1}(\mu) \qquad (7.1.4)$$

where the first three members of the set are

$$P_0(\mu) = 1$$

$$P_1(\mu) = \mu = \operatorname{Cos}\theta \qquad (7.1.5)$$

$$P_2(\mu) = \frac{3}{2}\,\mu^2 - \frac{1}{2} = 1 - \frac{3}{2}\operatorname{Sin}^2\theta$$

Before we can specify the effects of the perturbing force in detail, we must indicate its nature. So let us turn to some simple examples of distorting forces and their effects on the structure equations.

Rigid Rotation   For our first example, we consider the case where the star is rotating as a rigid body. This yields a simple expression for the magnitude of the distorting force produced by the local centripetal acceleration, which is

$$|D| = \omega^2 r \operatorname{Sin} \theta \tag{7.1.6}$$

where ω is the angular velocity of the star and is assumed to be constant. The components of the acceleration are then

$$D_r = \omega^2 r \operatorname{Sin}^2 \theta \qquad D_\theta = \omega^2 r \operatorname{Sin} \theta \operatorname{Cos} \theta \tag{7.1.7}$$

which can be expressed in terms of Legendre polynomials and their derivatives as

$$D_r = \frac{2}{3} \omega^2 r[1 - P_2(\mu)] \qquad D_\theta = -\frac{1}{3} \omega^2 r \frac{\partial P_2(\mu)}{\partial \theta} \tag{7.1.8}$$

Due to the axial symmetry, $D_\phi = 0$ and it is a simple matter to show that the curl of D, $\nabla \times \vec{D}$, is 0 so that $\vec{D}$ is derivable from a scalar potential by

$$\vec{D} = \nabla \Lambda \tag{7.1.9}$$

where

$$\Lambda = -\tfrac{1}{2}\omega^2 r^2 \operatorname{Sin}^2 \theta \tag{7.1.10}$$

Thus, the components of the perturbing force and the rotational potential can be expressed in terms of the Legendre polynomials as

$$D_r = A(r) + B(r)P_2(\mu)$$

$$D_\theta = C(r) \frac{\partial P_2(\mu)}{\partial \theta} \tag{7.1.11}$$

$$\Lambda = Q_0(r) + Q_1(r)P_2(\mu)$$

Although the above relations are correct for $\omega = $ constant, it is worth considering the functional dependence of ω for which it is true in general. Consider the nature of centripetal acceleration in a cylindrical coordinate system where the radial coordinate is denoted by s. The components of $\vec{D}$ are

$$D_z = D_\phi = 0 \qquad D_s = \omega^2 s \hat{s} \tag{7.1.12}$$

In order for the rotational force to be derivable from a scalar potential, its curl must be zero. The cylindrical components of the curl are

$$(\nabla \times \vec{D})_s = \frac{1}{s}\frac{\partial D_z}{\partial \phi} - \frac{\partial D_\phi}{\partial s} = 0$$

$$(\nabla \times \vec{D})_z = \frac{\partial D_s}{\partial \phi} = 0 \tag{7.1.13}$$

$$(\nabla \times \vec{D})_\phi = \frac{\partial D_s}{\partial z} - \frac{\partial D_z}{\partial s} = \frac{\partial(\omega^2 s)}{\partial z}$$

The radial component is identically zero, so we may suspect that if the object exhibits axial symmetry, ω cannot be a function of ϕ. In this case, the z component of the curl would also be zero. Thus, the condition that the rotational force be derivable from a scalar potential boils down to the ϕ component of the curl being zero, so that

$$\frac{\partial(s\omega^2)}{\partial z} = 0 \tag{7.1.14}$$

Thus,

$$\omega \neq \omega(z) \tag{7.1.15}$$

so the angular velocity must be constant on cylinders. It can be shown that if the perturbing force is not derivable from a potential, then no equilibrium solution of the structure equations exists. This is sometimes called the *Taylor-Proudman theorem*,[1] and it basically guarantees that if the star has reached an equilibrium angular momentum distribution, the angular velocity will be constant on cylinders.

Gravitational Distortion by an External Point Mass   Now let us return to the spherical polar coordinates that we used to obtain the components of the rotational force. The force will be directed toward an external point mass located along the z axis at a distance d from the center of the star (see Figure 7.2).

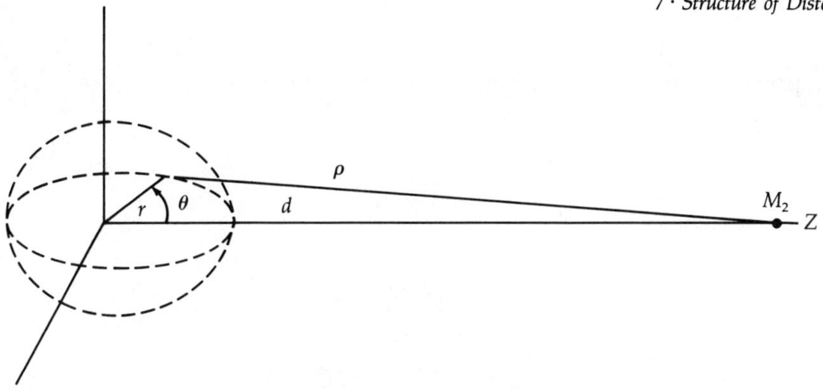

Figure 7.2 Type of distortion to be expected from the presence of a companion such as in a close binary system. For simplification, the rotational distortion is considered negligible. The distance from any point in the star to the perturbing mass is denoted ρ.

Now the perturbing potential of the point mass M_2 is

$$\Omega_2 = \frac{GM_2}{\rho} \tag{7.1.16}$$

where

$$\rho^2 = d^2 + r^2 - 2rd\,\mathrm{Cos}\,\theta = d^2\left[1 + \left(\frac{r}{d}\right)^2 - 2\frac{r}{d}\,\mathrm{Cos}\,\theta\right] \tag{7.1.17}$$

So the potential can be written in terms of our coordinates and the stellar separation as

$$\Omega_2 = \frac{GM_2}{d}\left[1 - 2\frac{r}{d}\,\mathrm{Cos}\,\theta + \left(\frac{r}{d}\right)^2\right]^{-1/2} \tag{7.1.18}$$

Equation (7.1.18) is rather nonlinear in the θ coordinate, so to express the potential in term of Legendre polynomials, we can make use of the "generating function" (see Arfken[2] for the development of this generating function) for the Legendre polynomials

$$(1 - 2\alpha\mu + \alpha^2)^{-1/2} = \sum_{i=0}^{\infty} P_i(\mu)\alpha^i \tag{7.1.19}$$

So the perturbing potential becomes

$$\Omega_2 = \frac{GM_2}{d}\sum_{i=0}^{\infty}\left(\frac{r}{d}\right)^i P_i(\mathrm{Cos}\,\theta) \tag{7.1.20}$$

Since the perturbing force is conservative, we may obtain it from

$$\vec{D} = +\nabla\Omega_2 \tag{7.1.21}$$

which has components

$$D_r = \frac{GM_2}{d} \sum_{i=1}^{\infty} \left(\frac{i}{d}\right)\left(\frac{r}{d}\right)^{i-1} P_i(\text{Cos }\theta)$$

$$D_\theta = \frac{GM_2}{d} \sum_{i=0}^{\infty} \left(\frac{1}{r}\right)\left(\frac{r}{d}\right)^{i} \frac{\partial P_i(\text{Cos }\theta)}{\partial\theta} \tag{7.1.22}$$

$$D_\phi = 0$$

So far, the only approximation that we have made is that the perturbing potential is that of a point mass. To simplify the remaining discussion, we assume that the point mass is distant compared to the size of the object so that

$$D_r = \frac{GM_2}{d}\left[\frac{\text{Cos }\theta}{d} + \frac{2r}{d^2}P_2(\text{Cos }\theta) + \cdots + \right]$$

$$D_\theta = \frac{GM_2}{d}\left[\frac{-\text{Sin }\theta}{d} + \frac{r}{d^2}\frac{\partial P_2(\text{Cos }\theta)}{\partial\theta} + \cdots + \right] \tag{7.1.23}$$

Note that the zeroth-order terms of the components can be added vectorially to give

$$\vec{D}_0 = \frac{GM_2}{d^2}\hat{d} \tag{7.1.24}$$

This is just the gravitational force that is balanced by the acceleration resulting from the orbital motion of the system, and so this force can be made to vanish by going to a rotating coordinate system. In such a system, the components of the perturbing force will just be the first-order terms, so that

$$\tilde{D}_r = \frac{2GM_2r}{d^3}P_2(\text{Cos }\theta) \qquad \tilde{D}_\theta = \frac{GM_2r}{d^3}\frac{\partial P_2(\text{Cos }\theta)}{\partial\theta} \tag{7.1.25}$$

These components of the perturbing force have the same form as those of rotation [see equation (7.1.11)], and any method which is applicable to the solution of the structure equations for rotational distortion will also be applicable to the problem of gravitational distortion.

Distortion Resulting from a Toroidal Magnetic Field Consider the Lorentz force of an internal magnetic field on the material of the star:

$$\vec{f} = \frac{\vec{j} \times \vec{B}}{c} = -(4\pi c)^{-1}[\vec{B} \times (\nabla \times \vec{B})] \tag{7.1.26}$$

The perturbing acceleration due to this force will be

$$\vec{D} = \frac{\vec{f}}{\rho} = (4\pi \rho c)^{-1}[\vec{B} \times (\nabla \times \vec{B})] \tag{7.1.27}$$

Now assume a special, but not implausible, geometry for the internal stellar magnetic field. Specifically, let us choose a toroidal field that exhibits pure axial symmetry so that

$$\vec{B} = \psi(r)(\text{Sin } \theta)(\hat{\phi}) \tag{7.1.28}$$

where $\psi(r)$ contains the arbitrary, but presumed known, variation of the field with the radial coordinate r. Since the field only has a ϕ component, the vector part of equation (7.1.27) is

$$\vec{B} \times (\nabla \times \vec{B}) = -\hat{r}[B_\phi(\nabla \times \vec{B})_\theta] + \hat{\theta}[B_\phi(\nabla \times \vec{B})_r] \tag{7.1.29}$$

The θ and r components of the curl of $\vec{B}$ are

$$(\nabla \times \vec{B})_\theta = -r^{-1}\frac{\partial(rB_\phi)}{\partial r} = -\frac{\left[B_\phi + r\,\text{Sin }\theta\left(\dfrac{\partial \psi}{\partial r}\right)\right]}{r}$$

$$(\nabla \times \vec{B})_r = (r\,\text{Sin }\theta)^{-1}[\psi(r)\,\text{Sin }\theta\,\text{Cos }\theta + B_\phi\,\text{Cos }\theta] = \frac{2\psi(r)\,\text{Cos }\theta}{r} \tag{7.1.30}$$

which yields for the vector components of the perturbing field

$$D_r = (4\pi\rho c)^{-1}\left[\frac{\psi^2(r)}{r} - \psi(r)\frac{\partial \psi}{\partial r}\right]\text{Sin}^2\,\theta = \tilde{A}(r) + \tilde{B}(r)P_2\,(\text{Cos }\theta)$$

$$D_\theta = (4\pi\rho c)^{-1}\frac{\psi^2(r)}{r}\frac{\partial P_2(\text{Cos }\theta)}{\partial \theta} = \tilde{C}(r)\frac{\partial P_2(\text{Cos }\theta)}{\partial \theta} \tag{7.1.31}$$

Again, these components have the same form as those of rotational distortion.   183

Thus we can expect to be able to solve a wide variety of distortion problems by considering the single case of an axis-symmetric perturbing force of the form given in equations (7.1.11), (7.1.25), and (7.1.31). We now consider some aspects of the solution of such problems.

7.2 Solution of Structure Equations for a Perturbing Force

The equations given by equations (7.1.1), and which arise from the perturbations discussed in Section 7.1, are partial differential equations and must be solved numerically. The numerical solution of partial differential equations constitutes a major area of study in its own right and is beyond the scope of this book. So we leave the numerical methods required for the actual solution to others and another time. Instead, we concentrate on the conditions required for the equations to have a solution and some of the implications of those solutions.

Since the perturbing forces derived in Section 7.1 are all conservative forces (that is, $\nabla \times \vec{D} = 0$), they are all derivable from some scalar potential which we can call Λ. This can be added to the gravitational potential so that we have a generalized potential to enter into the structure equations which we can call

$$\Phi(r, \theta, \phi) = \Omega + \Lambda \tag{7.2.1}$$

Since all the forces exhibited axial symmetry, there will be no explicit dependence of the generalized potential on ϕ. There will be sets of values of θ and r, for which Φ is constant. For the unperturbed gravitational potential alone, these would be spheres of a given radius. For the generalized potential, they will be surfaces that exhibit axial symmetry. Such surfaces are known as *level surfaces* since a particle placed on one would feel no forces that would move it along the surface. Thus, if $\hat{n}$ represents a normal to such a surface, the gradient of the potential can be expressed as

$$\nabla\Phi = \hat{n}\,\frac{d\Phi}{dn} \tag{7.2.2}$$

As long as the chemical composition is constant, the state variables will be constant on level surfaces. This is sometimes known as *Poincaré's theorem* which we prove for rotation in the next section. However, the result is entirely reasonable. The values of the state variables change in response to forces acting on the gas. Since the potential is constant on a level surface and its gradient is always normal to the surface, there are no forces along the surface to produce such differences.

If we take the state variables to be constant along level surfaces of constant potential, we can expect the variables to have the same functional dependence on the coordinates as the potential itself. Thus, from the form of Λ given by equation (7.1.11), the state variables, and those parameters that depend directly on them, can be written as

$$P(r, \theta) = P_0(r) + P_2(r)P_2(\cos \theta)$$

$$\rho(r, \theta) = \rho_0(r) + \rho_2(r)P_2(\cos \theta)$$

$$\epsilon(r, \theta) = \epsilon_0(r) + \epsilon_2(r)P_2(\cos \theta) \qquad (7.2.3)$$

$$\kappa(r, \theta) = \kappa_0(r) + \kappa_2(r)P_2(\cos \theta)$$

$$\Omega(r, \theta) = \Omega_0(r) + \Omega_2(r)P_2(\cos \theta)$$

The gravitational potential must also be written with a θ dependence, because the perturbing force will rearrange the matter density so that the potential is no longer spherically symmetric.

We now regard equations (7.2.3) as perturbative equations in the traditional sense in that the terms with subscript 2 will be considered to be small compared to the terms with subscript 0.

a   Perturbed Equation of Hydrostatic Equilibrium

Substituting the perturbed form of the structure variables given by equation (7.2.3) into the equation of hydrostatic equilibrium [equation (7.1.1.*d*)], we get

$$\begin{aligned}
\nabla P = \nabla[P_0(r) + P_2(r)P_2(\cos \theta)] &= -\rho\,\nabla\Phi = -\rho\,\nabla\Omega + \rho\vec{D} \\
&= -\{\rho_0(r)\,\nabla\Omega_0(r) + \rho_0(r)\nabla[\Omega_2(r)P_2(\cos \theta)] \\
&\quad + \rho_2(r)P_2(\cos \theta)\,\nabla\Omega_0(r)\} + \rho_0(r)\vec{D} \\
&\quad + \rho_2(r)P_2^2(\cos \theta)\,\nabla\Omega_2(r) + \rho_2(r)P_2(\cos \theta)\vec{D}
\end{aligned} \qquad (7.2.4)$$

The terms on the last line of equation (7.2.4) are small "second-order" terms by comparison to the other terms, so, in the tradition of perturbative analysis, we will ignore them. Since the equations must hold for all values of θ, the r component of the gradient yields two distinct equations and the θ component yields one equation. These are basically the zeroth- and second-order equations from the two components of the gradient. However, in general, there will be no zeroth-order θ equation, since the unperturbed state is spherically symmetric. Remembering the form for $\vec{D}$

185

from equation (7.1.11), we see that the partial differential equations for hydrostatic equilibrium are

$$\frac{\partial P_0(r)}{\partial r} = -\rho_0(r)\frac{\partial \Omega_0(r)}{\partial r} + \rho_0(r)A(r)$$

$$\frac{\partial P_2(r)}{\partial r} = -\rho_0(r)\frac{\partial \Omega_2(r)}{\partial r} - \rho_2(r)\frac{\partial \Omega_0(r)}{\partial r} + \rho_0(r)B(r) \qquad (7.2.5)$$

$$P_2(r) = -\rho_0(r)\Omega_2(r) + \rho_0(r)C(r)$$

b Number of Perturbative Equations versus Number of Unknowns

The number of independent partial differential equations generated by the vector equation of hydrostatic equilibrium is 3. In general, the vector equations of stellar structure will yield three such independent equations while the scalar equations will produce only two, since there is no θ component. In Table 7.1 we summarize the number of equations we can expect from each of the structure equations.

Each of the perturbed variables P, T, and Ω will produce a first- and second-order unknown function of r for a total of six unknowns. The perturbations in the density ρ are not linearly independent since they are related to those of P and T by the equation of state. A similar situation exists for the opacity $\bar{\kappa}$ and energy generation ϵ. However, the radiative flux is a vector quantity and will yield two unknown perturbed quantities, F_{0r} and F_{2r}, from the r component and one, $F_{1\theta}$, from the θ component. Thus the total number of unknowns in the problem is 9, and the problem is overdetermined and has no solution. This implies that we have left some physics out of the problem.

In counting the unknowns resulting from perturbing equations (7.1.1), we implicitly assumed that there were no mass motions present in the star, with the result that $\partial S/\partial t$ in equation (7.1.1b) was taken to be zero. If we assume that a stationary state exists, then we can represent the local time rate of change of entropy by a

Table 7.1 **The Number of Independent Scalar Structure Equations**

Equation	Zeroth-order r	Second-order r	Second-order θ	Total
(7.1.1a)	1	1		2
(7.1.1b)	1	1		2
(7.1.1c)	1	1	1	3
(7.1.1d)	1	1	1	3
Total	4	4	2	10

velocity times an entropy gradient, so equation (7.1.1b) becomes

$$\mathbf{V} \cdot \vec{F} = \rho\epsilon - \vec{v} \cdot T\,\mathbf{V}S \tag{7.2.6}$$

Thus, we have added a velocity with three components, each of which will have two perturbed parameters. However, in general $\vec{v}_0$ will be zero, since we assume no circulation currents in the unperturbed model. In addition, $\mathbf{V}S$ will exhibit axial symmetry and have no ϕ component. Thus the $v_{2\phi}$ perturbed parameter will be orthogonal to $\mathbf{V}S$ and not appear in the final equations. This leaves us with 11 unknowns and 10 equations. However, we have not included the fact that mass conservation must be involved with any transport of matter, and modifying the conservation of mass equation to include mass motions will provide one more equation, completing the specification of the problem.

7.3 Von Zeipel's Theorem and Eddington-Sweet Circulation Currents

For a solution to exist for the structure of a distorted star, we had to invoke mass motions in the star itself. This result was essentially obtained by von Zeipel[3] in the middle 1920s. At that time, the source of stellar energy was unknown, and von Zeipel set about to place constraints on the energy generation within a distorted star and in so doing produced one of the most misunderstood theorems of stellar astrophysics. The theorem is essentially a proof by contradiction that stars cannot simultaneously satisfy radiative and hydrostatic equilibrium if the stars are distorted. The normal version of the theorem is given for rigidly rotating stars, and this is the version quoted by Eddington.[4] However, in the original publication, the version developed for rigid rotation is followed immediately by a version appropriate for tidally distorted stars.[5] Thus, clearly the theorem results from the induced distortion itself and is independent of the details that produce the distortion. We describe the version for rotation here, but keep in mind that is it the distortion that is important, not the mechanism by which that distortion is generated.

a   Von Zeipel's Theorem

As originally stated by von Zeipel[3] in 1924, this theorem says that for a rigidly rotating star in hydrostatic and radiative equilibrium, the rate of energy generation is given by

$$\epsilon = (\text{const})\left(1 - \frac{\omega^2}{2\pi G\rho}\right) \tag{7.3.1}$$

In light of what we now know about stars, this is an absurd result, because it requires that the energy generation rate become negative near the surface as the density goes to zero. As is the case when any theorem yields an absurd result, one must challenge the assumptions. To see where the trouble is likely to be, let us sketch von Zeipel's argument.

The equation of hydrostatic equilibrium

$$\nabla P = \rho\, \nabla \Phi \tag{7.3.2}$$

indicates that the potential gradient is related to the pressure gradient by the scalar density ρ. Hence, both vectors point in the same direction, and we can describe the change in pressure as a proportional change in potential so that

$$dP = \rho\, d\Phi \tag{7.3.3}$$

From this it is clear that the pressure must be constant on a level surface where $d\Phi = 0$. This is equivalent to saying that the pressure can be written as a function of the potential Φ alone. If the pressure is a function of Φ alone, then the scalar ρ, relating the potential and pressure gradients, must also be a function of Φ alone. Or

$$\rho = \frac{dP(\Phi)}{d\Phi} = \rho(\Phi) \tag{7.3.4}$$

As long as the chemical composition μ is constant, or at least not varying over an equipotential (level) surface, the ideal-gas law guarantees that the temperature will also be a function of Φ alone:

$$T = \frac{P(\Phi)\mu m_h}{k\rho(\Phi)} = T(\Phi) \tag{7.3.5}$$

This is what we stated in Section 7.2 to be Poincaré's theorem.

Now the radiative temperature gradient which arises from radiative equilibrium requires that

$$\vec{F} = \frac{4acT^3}{3\bar{\kappa}\rho}\,\hat{n} = -\left(\frac{c}{\bar{\kappa}\rho}\right)\nabla\left(\frac{aT^4}{3}\right) = -\frac{c}{\bar{\kappa}\rho}\frac{dP_r}{dn}\,\hat{n} \tag{7.3.6}$$

which, expressed in terms of the potential gradient, becomes

$$\vec{F} = -\frac{c}{\bar{\kappa}\rho}\frac{dP_r}{d\Phi}\frac{d\Phi}{dn}\,\hat{n} = -\frac{c}{\bar{\kappa}\rho}\frac{dP_r}{d\Phi}\nabla\Phi \tag{7.3.7}$$

However, since $\bar{\kappa}$, ρ, and T are all state variables or functions of them, they are all functions of Φ alone and

$$\vec{F} = f(\Phi)\,\nabla\Phi \tag{7.3.8}$$

But $\nabla\Phi$ is just the local gravity, and it is most certainly not a function of the potential alone or constant on level surfaces. Indeed, for a critically rotating star, the gravity varies from the mass-gravity at the pole to zero at the equator, where the mas -gravity is balanced by the centripetal acceleration. Thus, equation (7.3.8) basically says that in the presence of the radiative temperature gradient,

$$\left|\vec{F}\right| = (\text{const})\left|\vec{g}\right| \tag{7.3.9}$$

which is sometimes known as *von Zeipel's law of gravity darkening.*

If we further consider radiative equilibrium in the absence of mass motions, we can write

$$\nabla \cdot \vec{F} = \rho\epsilon = \nabla \cdot [f(\Phi)\,\nabla\Phi] = f(\Phi)\,\nabla^2\Phi + \nabla f(\Phi) \cdot \nabla\Phi \tag{7.3.10}$$

For a star in rigid rotation, $\nabla^2\Phi$ will depend on only the density and some constants and so will be a function of Φ alone. The left-hand side of equation (7.3.10) will depend on only the state variables and must also be a function of Φ alone. But, again, the gravity $\nabla\Phi$ is not a function of Φ alone, so

$$\nabla f(\Phi) = 0 \tag{7.3.11}$$

Therefore, evaluating $\nabla^2\Phi$ by means of equations (7.1.10) and (7.2.1), we get

$$\epsilon = (\text{const})\frac{\nabla^2\Phi}{\rho} = (\text{const})\left(1 - \frac{\omega^2}{2\pi G\rho}\right) \tag{7.3.12}$$

which is von Zeipel's result.

The absurdity of equation (7.3.12) results primarily from the assumption that the effects of mass motions are not present in equation (7.3.10). The addition of mass motions removes the exclusive dependence of radiative equilibrium on the potential, and the remainder of the argument falls apart, releasing the constraint on ϵ. The gradient of $f(\Phi)$ is no longer zero and allows for the variation of ϵ with radius that we know must exist. However, as we shall see, small amounts of energy are all that is required to be carried by the currents of the mass motions. Thus the radiative gradient will still be basically the temperature gradient that is operative in the radiative zones of the star. The result given in equation (7.3.9) will still be largely correct, and we may expect the radiative flux to be redistributed in accordance with the local value of the gravity. Therefore, particularly for the rapidly rotating upper main sequence stars with radiative envelopes, we may expect that their surface will not be uniformly bright, but will become darker with decreasing local gravity. While it is true that the conditions of radiative equilibrium become rather different in the stellar atmosphere as the photons begin to escape into outer space, the thickness of the atmosphere compared to the depth of the radiative envelope is minuscule, and whatever variation of radiative flux has been established at the base of the atmosphere will be largely reflected in the flux emerging from the star. So von Zeipel's theorem, while telling us nothing about the energy generation within the star, does tell us quite a lot about the manner in which the radiation leaves the star.

b Eddington-Sweet Circulation Currents

We have seen that radiative and hydrostatic equilibrium cannot be simultaneously satisfied in a distorted star and that the failure of these conditions results in the mass motion of material carrying energy to make up the deficit produced by the departure from spherical geometry. That the energy transfer is accomplished by means of the physical motion of material seems ensured. There simply is no other mechanism to effect the transfer. Radiation has been accounted for, conduction is ineffective, and the environment is stable against classical convection. These arguments persuaded Eddington[4] (p. 286) to suggest the existence of such currents which were later quantified by Sweet.[6] Let us now estimate the speed of these currents and determine the amount of energy they may carry. The currents will be quite slow since even in the most distorted of stars, the local departure of the energy flux from spherical symmetry is quite small. Even so, any mass motion could be important if it transports material throughout the star on a nuclear time scale. The possibility would then exist for a resupply of nuclear fuel, and that could upset some of our stellar evolution calculations.

Conservation of Energy and Circulation Velocity   The distortion of a star will force a departure from radiative equilibrium and a change in the divergence of the radiative flux from that expected for spherical stars. We argued earlier that the change in the divergence will be brought about by the additional nonradial transport of energy by mass motions, as expressed by the second term on the right-hand side of equation (7.2.6). Thus, to estimate the velocity of those motions, we must estimate the entropy gradient that distortion will establish.

From thermodynamics remember that the entropy can be expressed in terms of the state variables of an ideal gas as

$$S = C_P \ln T - n\mathrm{R} \ln P + S_0 \tag{7.3.13}$$

Therefore, the general energy source term in equations (7.1.1b) can be written as

$$T\frac{\partial S}{\partial t} = C_P\frac{\partial T}{\partial t} - \frac{\partial P}{\partial t} \tag{7.3.14}$$

and the entropy gradient of equation (7.2.6) becomes

$$T\nabla S = C_P\nabla T - \nabla P \tag{7.3.15}$$

Now the temperature and pressure gradients are both normal to equipotential surfaces, so the vector nature of equation (7.3.15) is unimportant and it must hold for the magnitude of the individual terms. Therefore,

$$T|\nabla S| = C_P\rho|\nabla P|\left(\frac{1}{\rho}\frac{|\nabla T|}{|\nabla P|} - \frac{1}{\rho C_P}\right) \tag{7.3.16}$$

Equation (7.3.16), when combined with the ideal-gas law and the fact that the temperature and pressure gradients point in the same direction, enables equation (7.3.15) to be written as

$$T|\nabla S| = C_P\rho\left(\frac{T}{P}\right)|\nabla P|\left(\frac{P}{\rho T}\frac{dT}{dP} - \frac{P}{C_P\rho T}\right) \tag{7.3.17}$$

Now the adiabatic gradient can be expressed as

$$\left.\frac{dT}{dP}\right|_{\text{ad}} = \frac{T}{P(n+1)} = \frac{P/(kT)}{\rho C_P} \tag{7.3.18}$$

Since for the zeroth-order values of these gradients

$$\left(\frac{dT}{dP}\right)_0 = \frac{(dT/dr)_0}{(dP/dr)_0} \tag{7.3.19}$$

we can write the zeroth-order value for the entropy gradient as

$$T|\nabla S|_0 = \left[|\nabla P|\left(\frac{dT}{dP}\right)_{\text{ad}}^{-1}\left(\frac{dT}{dP} - \left.\frac{dT}{dP}\right|_{\text{ad}}\right)\right]_0 = \left(|\nabla P|\frac{\Delta\nabla T}{\nabla T|_{\text{ad}}}\right)_0 \tag{7.3.20}$$

In the equilibrium model, there are no mass motions, the velocity in equation (7.2.6) is already a second-order term, and so to estimate its value we need only keep zeroth-order terms in the entropy gradient. The zeroth-order pressure gradient is just

$$\nabla P \simeq \nabla P_0 = \frac{dP_0}{dr} = \frac{GM_0(r)\rho_0(r)}{r^2} = -g_0\rho_0(r) \qquad (7.3.21)$$

Combining this with equations (7.2.6) and (7.3.20), we can write the perturbed equation for energy conservation as

$$\nabla \cdot \vec{F} = \rho\epsilon - \vec{g}_0 \cdot \vec{v}_2\rho_0(r)\frac{(\Delta\nabla T)_0}{(\nabla T|_{ad})_0}$$

$$= \rho_0\epsilon_0 - g_0 v_{2,0}(r)\rho_0(r)\frac{(\Delta\nabla T)_0}{(\nabla T|_{ad})_0} \qquad (7.3.22)$$

Now, from von Zeipel's gravity darkening law [equations (7.3.6) and (7.3.7)], we have

$$\vec{F} = -\left(\frac{4acT^3}{3\bar{\kappa}\rho}\frac{dT}{d\Phi}\right)\vec{g} \qquad (7.3.23)$$

which means that we can write the divergence of the flux as

$$\nabla \cdot \vec{F} = \nabla\Phi \cdot \frac{d}{d\Phi}\left(-\frac{4acT^3}{3\bar{\kappa}\rho}\frac{dT}{d\Phi}\nabla\Phi\right)$$

$$= -\nabla\Phi \cdot \nabla\Phi\left[\frac{d}{d\Phi}\left(\frac{4acT^3}{3\bar{\kappa}\rho}\frac{dT}{d\Phi}\right)\right] - \frac{4acT^3}{3\bar{\kappa}\rho}\frac{dT}{d\Phi}\nabla \cdot \nabla\Phi \qquad (7.3.24)$$

But since the radiative flux and gravity are vectors pointing in the same direction,

$$\nabla \cdot \vec{F} = |\vec{g}|^2\left[\frac{d(F/g)}{d\Phi} + \frac{F}{g}\cdot\nabla^2\Phi\right] \qquad (7.3.25)$$

For rotation we can obtain the generalized potential from equations (7.1.1a) and (7.1.10). Expressing the rotational potential in cylindrical coordinates, we get

$$\nabla^2\Phi = 4\pi G\rho - s^{-1}\frac{\partial}{\partial s}\left[s\frac{\partial}{\partial s}\left(\frac{1}{2}\omega^2 s^2\right)\right] = 4\pi G\rho - s^{-1}\frac{\partial}{\partial s}(\omega^2 s^2) \qquad (7.3.26)$$

Equation (7.3.24) for the perturbed flux divergence can be broken into its perturbed components so that

$$\nabla \cdot \vec{F}_0 = \frac{d(F_0/g_0)}{d\Phi} g_0^2 + \frac{F_0}{g_0} \left\{ 4\pi G\rho - \left[\frac{1}{s} \frac{d}{ds} (\omega^2 s^2) \right]_0 \right\}$$

$$\nabla \cdot \vec{F}_2 = \frac{d(F_0/g_0)}{d\Phi} 2g_0 g_2 - \frac{F_0}{g_0} \left[\frac{1}{s} \frac{d}{ds} (\omega^2 s^2) \right]_2 \tag{7.3.27}$$

Since the zeroth-order flux-to-gravity ratio can be written as

$$\frac{F_0}{g_0} = \frac{L(r)/(4\pi r^2)}{GM(r)/r^2} = \frac{L(r)}{4\pi GM(r)} \tag{7.3.28}$$

its derivative with respect to the generalized potential is

$$\frac{d(F_0/g_0)}{d\Phi} = -\frac{1}{g_0^2} \frac{\rho_0 L(r)}{M(r)} \tag{7.3.29}$$

The luminosity can be written in terms of an average energy generation rate $\bar{\epsilon}$ and $M(r)$ so that

$$L(r) = \int_0^r 4\pi r^2 \rho \epsilon \, dr \equiv \bar{\epsilon} M(r) \tag{7.3.30}$$

which yields

$$\nabla \cdot \vec{F}_2 = -\bar{\epsilon}\rho_0 \frac{2g_2}{g_0} - \frac{L(r)}{4\pi GM(r)} \left[\frac{1}{s} \frac{d}{ds} (\omega^2 s^2) \right]_2 \tag{7.3.31}$$

If we further assume that the distortion is small so that $g_2/g_0 \ll 1$, then equation (7.3.31) can be combined with (7.3.22) to give the velocity for the induced circulation currents as

$$v_c \equiv v_{2,0}(r) = \frac{(\nabla T|_{ad})_0}{(\Delta \nabla T)_0} \frac{L(r)}{4\pi g_0 \rho GM(r)} \left[s^{-1} \frac{d(\omega^2 s^2)}{ds} \right]_2 \tag{7.3.32}$$

Eddington-Sweet Time Scale and Mixing   If we take reasonable values for the parameters in equation (7.3.32), namely,

$$g_0 = \frac{GM}{R^2} \qquad L(r) = L \qquad \omega = \text{const} = w \left(\frac{8GM}{27R^3} \right)^{1/2} \tag{7.3.33}$$

then we can rewrite the circulation velocity as

$$v_c = \frac{(\nabla T|_{ad})_0}{(\Delta \nabla T)_0} \frac{\bar{\rho}}{\rho} \frac{16R^2L}{81GM^2} w^2 \tag{7.3.34}$$

Here we have introduced the fractional angular rotational velocity w, which is just ω normalized by the critical angular velocity where the effective equatorial gravity is zero for a centrally condensed star (Roche model). In addition, if we introduce a time scale such as the Kelvin-Helmholtz time scale [equation (3.2.11)], we can further reduce the expression for the circulation velocity to some dimensionless ratios multiplied by R/τ_{K-H} and get

$$v_c \approx \frac{(\nabla T|_{ad})_0}{(\Delta \nabla T)_0} \frac{\bar{\rho}}{\rho} \frac{16w^2}{135} \frac{R}{\tau_{K-H}} \tag{7.3.35}$$

If we introduce the Eddington-Sweet time scale as the time required for the circulation currents to carry the material a distance R, then

$$\tau_{E-S} = \tau_{K-H} \frac{135}{16w^2} \frac{\rho}{\bar{\rho}} \frac{(\Delta \nabla T)_0}{(\nabla T|_{ad})_0} \tag{7.3.36}$$

The sun is a rather slowly rotating star, and if the angular velocity of the core is displayed on the surface, there is little rotational distortion, so the Eddington-Sweet currents should be small. Certainly the core density will exceed the mean density, and we have already indicated that the radiative core of the sun is barely stable against convection. Thus, the following values for the solar radiative core should provide fair estimates of the internal conditions necessary for the evaluation of the circulation currents:

$$w_\odot^2 \sim 10^{-5} \qquad \frac{\rho(core)}{\bar{\rho}} > 1 \qquad \frac{(\Delta \nabla T)_0}{(\nabla T|_{ad})_0} \sim 1 \tag{7.3.37}$$

These values, when substituted into equation (7.3.35), yield

$$\tau_{E-S} \geq 10^5 \tau_{K-H} > \tau_n \tag{7.3.38}$$

So the material in the core takes much longer than the nuclear time scale to circulate, and we would not expect the core of the sun, or solar-type stars that are slowly rotating, to be mixed. So the stellar evolution scenarios we developed for lower main sequence stars in Chapter 5 remain intact.

The situation is less clear for upper main sequence stars. Here the envelope has a density that is lower than the mean density, is in radiative equilibrium, and is in danger of supplying fuel to the core. In addition, many of these stars rotate very rapidly so that we might expect much larger circulation current velocities. Reasonable values of the parameters in equation (7.3.35) for rapidly rotating B stars are

$$w_B^2 \approx 1 \qquad \frac{\rho(\text{env})}{\bar{\rho}} < 1 \qquad \frac{(\Delta \nabla T)_0}{(\nabla T|_{\text{ad}})_0} \sim 1 \qquad (7.3.39)$$

which lead to

$$\tau_{\text{E-S}}(B \text{ star}) \sim \tau_{\text{K-H}} < \tau_n \qquad (7.3.40)$$

On the basis of this analysis we would have to conclude that there is an excellent possibility that rapidly rotating stars on the upper main sequence may be mixed thoroughly throughout and their main sequence life times may be prolonged. However, we have not dealt with the formation of the helium core itself and the effects caused by the change in chemical composition.

7.4   Rotational Stability and Mixing

A complete discussion of the stability of a rotating star is quite complicated and beyond the scope of this book. However, we consider some of the important effects on the stability of rotating stars. The usual approach to the subject of stability involves finding the spectrum of perturbations for which the equations of motion are stable (i.e., the perturbations will damp out with time). A related approach is to use the virial theorem,[7] which after all is just a spatial moment of the equations of motion. Various physical processes may occur and give rise to an instability:

1. Buoyancy forces that result from thermal stratification

2. Perturbations that may grow in the presence of an angular momentum gradient

3. Instabilities in the presence of a magnetic field

4. Shear instabilities producing flows both parallel and perpendicular to the local gravity field

5. Failure of the equipotential surfaces being surfaces of constant temperature and pressure [that is, $\omega \neq \omega(s)$]

6. Development of a molecular weight gradient as a result of nuclear evolution

7. Diffusion of heat, angular momentum, and the mean molecular weight

Of all these effects, probably the most important for the stability of rotating stars is the various shear instabilities.

a Shear Instabilities

The existence of a velocity gradient implies the presence of particle interactions resulting from changes in the macroscopic velocity field. These interactions result from the collisions that are the product of the differential stream motion of the gas, and the severity of these collisions is usually characterized by the viscosity v of the material. The viscosity will try to remove the velocity gradient. However, if the shear is too great, the velocity field will break up into turbulent flow. The conditions of the flow can be characterized by a dimensionless number known as the *Reynolds number* R_e, which for rotating stars is

$$R_e \approx \frac{\omega R_*^2}{v} \tag{7.4.1}$$

Should this number exceed a critical value, known as the *critical Reynolds number*, which is about 10^3, the flow will break up into turbulent eddies and the smooth macroscopic motion will become chaotic.

It is useful to break the notion of shear motion into two limiting cases. Motion along the equipotential surfaces will be unopposed by gravity and any of the phenomena that arise from the gravity field. Thus, perturbations that produce horizontal shear can grow unopposed except by the dissipative forces that arise from the viscosity of the gas. However, shear instabilities that arise from motions perpendicular to the equipotential surfaces must overcome forces caused by the temperature and perhaps molecular weight gradients. Thus, the star will be much more stable against vertical shear instabilities, and the time scales for their respective growths will be quite different. For the horizontal shear instabilities the time scale is dominated by the viscosity, while for vertical shear instabilities the time scale for development is essentially the thermal, or Kelvin-Helmholtz, time scale. Thus,

$$t_h \approx \frac{R_*^2}{v} \qquad t_v \approx t_{\text{K-H}} \tag{7.4.2}$$

The nature of the viscosity of stellar material has long been a subject of heated debate. If one calculates the viscosity simply on the basis of the collisional interaction of the atoms of the gas, one will obtain an extremely small number and an associated growth time scale which is long compared to the nuclear time scale for the star. However, if the flow becomes turbulent, then the dominant collisions occur, not between atoms, but between turbulent elements, giving rise to a "turbulent viscosity" which is many orders of magnitude greater than the kinematic viscosity of the atoms themselves. Unfortunately, the theory of turbulent flow is not sufficiently developed to yield reliable values for the turbulent viscosity, so we must rely on empirical values for systems with dimensions vastly smaller than those of stars. Nevertheless, the prevailing opinion seems to be that turbulent viscosity will be many orders of magnitude greater than kinematic viscosity, and so shear instabilities will be of considerable importance in bringing about the redistribution of angular momentum within the star.

From the arguments in Section 7.3 [equation (7.3.40)] it seemed likely that the Eddington-Sweet circulation currents could redistribute material and angular momentum on a time scale comparable to the Kelvin-Helmholtz time for a rapidly rotating star. This is the same order of magnitude as the time scale for the development of the vertical shear instabilities. However, it is rather greater than the time scale for the horizontal shear instabilities, should they result from turbulent flow. So these horizontal shear instabilities would appear to be the dominant phenomenon that redistributes the material and angular momentum in the most rapidly rotating stars. This would lead to a steady-state rotation law where the angular velocity was constant on equipotential surfaces and had a condition for stability of the form

$$\frac{\partial^2(\omega s^2)^2}{\partial(\cos\theta)^2} \neq 0 \tag{7.4.3}$$

The most plausible rotation law that would satisfy these constraints is rigid rotation, and this may well be the only equilibrium law for rapidly rotating stars. However, many questions must be answered before it can be determined if this law actually exists in these stars.

b Chemical Composition Gradient and Suppression of Mixing

While discussing von Zeipel's theorem and developing Poincaré's theorem [equation (7.3.5)], we saw that the temperature would be constant on an equipotential surface only if the star were homogeneous, so that the chemical composition μ did not appear on the right-hand side of equation (7.3.5). In an evolving star, the chemical composition is continually changing as a result of nuclear processes. Thus, for the early-type stars, we expect the convective core to change its chemical composition on a nuclear time scale, causing μ to increase with time. This will lead to a discontinuity in the chemical composition at the core-envelope interface. Now imagine a blob of helium displaced upward by the circulation currents into the less dense hydrogen envelope. The forces of hydrostatic equilibrium will tend to restore the higher-density helium to the core, while the circulation currents will try to mix the helium higher in the hydrogen envelope. Fricke and Kippenhahn[8] have shown that ratio of the circulation velocity to the restoring velocity induced by hydrostatic equilibrium is given by

$$\frac{v_c}{v_\mu} \sim \frac{0.3w^2}{\Delta\mu/\mu_c} \tag{7.4.4}$$

Since the greatest value of w which is allowed is unity, and since a pure helium core will produce $\Delta\mu/\mu_c \sim 0.5$, we would expect the core-envelope interface to be stable against any vertical motion that would allow mixing. For the typical B star where $w \sim 0.4$, a reasonably small gradient in the chemical composition will stabilize the star against rotationally driven mixing, so we may expect the stellar evolution scenarios for the upper main sequence stars described in Chapter 5 to remain correct.

c Additional Types of Instabilities

Conditions that can lead to instability in a rotating star seem so numerous that some physicists have despaired from finding any angular momentum distribution that is stable for the lifetime of the star, and it may well be true that no such distribution exists. The number and type of instabilities that can occur are indeed legion. However, what is relevant for the theory of stellar evolution is the time scale for the development of these instabilities and what they do to the star. For rapidly rotating stars, shear instabilities are likely to occur and lead to a rotation law where the angular velocity is constant on equipotential surfaces. There are additional constraints on the rotation law. Should an outward displacement that conserves angular momentum produce a perturbation that has a greater angular velocity than the local velocity field, the perturbation will be dynamically unstable and will grow on the dynamical time scale. This basically geometric instability is sometimes called the *Solberg-Høiland instability*, and it constrains the angular momentum per unit mass so that

$$\left.\frac{\partial(\omega s^2)^2}{\partial s}\right|_{\rho\,=\,\text{const}} > 0 \tag{7.4.5}$$

Thus, angular velocity laws that decrease faster than s^{-2} will be dynamically unstable. A similar criterion holds for the Goldreich-Schubert-Fricke instability. However, the time scale for its development is very much longer because this instability basically arises from the removal of buoyancy stabilization of the temperature gradient by thermal diffusion. If we add to the angular velocity constraints the notion that the rotation law should be derivable from a potential [equation (7.1.13)], then the constraints on the angular velocity distribution become

$$\frac{\partial(\omega s^2)^2}{\partial s} > 0 \qquad \frac{\partial(\omega s^2)^2}{\partial z} = 0 \tag{7.4.6}$$

The notion that the rotation law should be conservative is largely based on personal prejudice and will be wrong if dissipative forces like those arising from viscosity are present. Under these conditions the criterion for stability becomes

$$\left\{\frac{\partial[\ln(\omega s^2)^2]}{\partial z}\right\}^2 \geq \frac{v}{kT}\frac{g_{\text{eff}}}{\omega^2(s)} \tag{7.4.7}$$

Since the quantity $v/(kT)$ is usually quite small for stars [i.e., of the order of 10^{-6} (cgs)], the Goldreich-Schubert-Fricke instability is unimportant except in cases of slow rotation and long nuclear time scales. Thus, this instability has been applied to the sun with some interesting results. However, it can be easily stabilized by a molecular weight gradient such as that described by equation (7.4.4).

Under conditions of rapid rotation, one might expect non-axis-symmetric motions to occur that can separate surfaces of constant pressure from equipotential surfaces. Instabilities resulting from such situations are generally referred to as *baroclinic instabilities*. These and other types of diffusive instabilities we leave to others to discuss.

Problems

1. Discuss the problems you would encounter in describing the structure of a very rapidly rotating magnetic neutron star. Specifically discuss how you would propose calculating a model of the structure, and list the assumptions you would make.

2. Show that

$$\frac{v_c}{v_\mu} \sim \frac{0.3w^2}{\Delta\mu/\mu_c}$$

and clearly state the assumptions you would make.

3. Indicate how the conservation of mass equation should be modified to accommodate the flow of matter resulting from the Eddington-Sweet currents.

References and Supplemental Reading

1. Lamb, H. *Hydrodynamics*, 5th ed., Cambridge University Press, New York, 1924, pp. 216–217.

2. Arfken, G. *Mathematical Methods for Physicists*, Academic, New York, 1970, pp. 534–538.

3. von Zeipel, H. "The Radiative Equilibrium of a Rotating System of Gaseous Masses" *Mon. Not. R. astr. Soc.* 84, 1924, pp. 665–701.

4. Eddington, A. S. *The Internal Constitution of the Stars* [1926], Dover, New York, 1959, pp. 282–283.

5. von Zeipel, H. "The Radiative Equilibrium of a Double-Star System with Nearly Spherical Components," *Mon. Not. R. astr. Soc.* 84, 1924, pp. 702–719.

6. Sweet, P. A. "The Importance of Rotation in Stellar Evolution," *Mon. Not. R. astr. Soc.* 110, 1950, pp. 548–558.

7. Collins, G. W., II, *The Virial Theorem in Stellar Astrophysics*, Pachart, Tucson, Ariz., 1978, Chap. 3, pp. 61–102.

8. Fricke, K. J. and R. Kippenhahn, "Evolution of Rotating Stars," *Annual Review of Astronomy and Astrophysics* (Ed.: L. Goldberg), vol. 10, Annual Review, Palo Alto, Calif., 1972, pp. 45–72.

During the last quarter of a century, much has been done regarding the structure of distorted stars. A useful historical review through the early 1970s can be found in

• Roxburgh, I. W.: "Rotation and Stellar Interiors," *Stellar Rotation* (Ed.: A. Slettebak), D. Reidel, Dordrecht, Holland, 1970, pp. 9–19.

However, the most comprehensive review of the problems relating to the structure of rotating stars is

• Toussel, J. L.: *The Theory of Rotating Stars*, Princeton University Press, Princeton, N.J., 1978.

For the fundamental literature on distorted polytropes, see

• Chandrasekhar, S.: "The Equilibrium of Distorted Polytropes, I (The Rotational Problem)," *Mon. Not. R. astr. Soc.* 93, 1933, pp. 390–405.

• Chandrasekhar, S.: "The Equilibrium of Distorted Polytropes, II (The Tidal Problem)," *Mon. Not. R. astr. Soc.* 93, 1933, pp. 449–471.

More recent work on this subject can be found in Limber and Roberts (1965) and Geroyannis and Valvi (1987) (see References and Supplemental Reading in Chapter 2).

8

Stellar Pulsation
and Oscillation

· · ·

That some stars vary in brightness has been known from time immemorial. That this variation is the result of intrinsic changes in the star itself has been known for less than 100 years, and the causes of those variations have been understood for less than 30 years. It is not a simple matter to distinguish the light variations resulting from eclipses by a companion from those caused by physical changes in the star itself. However, for the cepheid variables (named for the prototype example δ Cephei), the treatment of the light variations as if they resulted from eclipses by a companion leads to some absurd results. If this were the case, the orbit of the companion would have to be highly elliptical with the semimajor axes pointed toward the earth. This somewhat Ptolemaic view suggests that the binary hypothesis is incorrect but is far from conclusive. Further analysis shows that the sum of the radii for the two hypothetical stars would exceed their separation. Such a situation is even less likely. However, the coup de grace is administered to the binary hypothesis when one considers the radial velocity curve produced by the star throughout a period of light variation. Here one finds that maximum radial velocity occurs near minimum light, which is nearly the reverse of that required by the binary hypothesis. Thus, one is left with almost no other option than to conclude that this class of stars is intrinsically varying.

While a number of other types of stars were seen to exhibit similar behavior, considerable interest was generated by the cepheid variables themselves when it was found that a relationship exists between the period of variability and the absolute luminosity at maximum brightness. This period-luminosity relationship was found to hold true, in one form or another, for the intrinsic variables. While it is true that the RR Lyrae stars have basically the same absolute magnitude regardless of their period, this can be viewed as a period-luminosity relation since the very determination of the RR Lyrae nature of a star establishes its luminosity. Thus, variable stars could be used as "standard candles" for the purposes of distance determination. Although it is not necessary to understand the physical processes giving rise to the variation of light in these stars in order to use them to determine distances to associated objects, the lack of such understanding may result in large errors of interpretation. Such was the case with the cepheid variables, and the result was an error in the distances to galaxies of about a factor of 2. Although this error was uncovered before the nature of stellar pulsation was understood, there can be little doubt that such an understanding would have signaled the danger much earlier and possibly prevented much confusion.

The detailed nature of the origin of pulsation for the intrinsic variables was not understood until the computational power existed for the accurate construction of model stars. Then, for the two most important classes of intrinsic variables, it was found that the mechanism producing the light variations resulted from oscillations in the ionization equilibrium in the outer layers of the star and had nothing to do with oscillations in the energy generation process itself. However, to construct these models, it is necessary to have some idea about the nature of the stars themselves. Here their variations provide the clues required. From the stability of the periods of these pulsating stars, it is clear that the oscillations are resonant oscillations, and their periods will tell us much about the stars involved. We have already established the conceptual foundation for this determination in the virial theorem.

8.1 Linear Adiabatic Radial Oscillations

Just as the pitch or oscillation frequency of an orchestral chime conveys information about its length, so the oscillation period of a pulsating variable should indicate its size. The low-pitch, or long-period, chimes or bells are characteristically larger than higher-pitch ones. So it is with the most common of the pulsating stars. The RR Lyrae stars having pulsational periods of less than a day are both hotter and less luminous, and therefore smaller, than the longer-period cepheid variables. The cepheid variables, in turn, are significantly hotter and smaller than the even longer-period Mira variables. Only for the cataclysmic variables, where the pulsations are probably mediated by an outside source, does this size-period relationship fail. This is as it should be, for the pulsation of a star, being a hydrodynamic phenomenon, should take place on the hydrodynamic time scale roughly equal to the sound crossing time. Let us now consider how we may quantify this notion.

Begin by assuming that the pulsations take place in an adiabatic manner. By this we mean that the energy associated with the motion of the pulsation does no net work on the gas of the star and is therefore conserved from one oscillation to the next. Thus an adiabatic oscillation can proceed forever once it is established. This lack of sharing of energy with the star means that driving and damping terms that must be present in any real situation are assumed to be negligible. Thus we cannot hope to determine anything about the evolutionary history of such oscillations in real stars. The origin or fundamental cause of such oscillations will not be found by such analysis. However, we can learn something about the eigenfrequencies for those oscillations and the stellar parameters upon which they depend.

The history of the theory of stellar pulsation can be traced to Eddington,[1,2] but a conceptually simple picture was introduced by Ledoux[3] and is nicely described by Ledoux and Walraven.[4] This approach involves the virial theorem and so focuses on the global properties of the stars and their relation to the periods of pulsation. This approach was developed by Chandrasekhar[5] to deal with some extremely complicated problems and is summarized by Collins.[6]

a   Stellar Oscillations and the Variational Virial Theorem

Consider a spherical star in equilibrium. Now allow the moment of inertia, the internal energy, and the gravitational potential energy to vary in such a manner that the virial theorem always holds. Then the virial theorem as given by equation (1.2.34) can be written as

$$\frac{1}{2}\frac{d^2(\delta I)}{dt^2} = 2\,\delta U + \delta \Omega \qquad (8.1.1)$$

The conservation of mass requires that the mass interior to some radial distance r remain constant throughout the oscillation, so that

$$M(r) = M(r + \delta r) = M(r_0) = \int_0^{r_0} 4\pi r_0^2 \rho \, dr \qquad (8.1.2)$$

This is virtually a definition of what is meant by $M(r)$. The variations of the moment of inertia and gravitational potential energy can simply be calculated from their definitions as

$$\delta I = 2 \int_0^{I_0} \frac{\delta r}{r_0} \, dI_0 \qquad \delta \Omega = - \int_0^{\Omega_0} \frac{\delta r}{r_0} \, d\Omega_0 \qquad (8.1.3)$$

To calculate the effect of the oscillations on the total internal energy, we have to consider how the gas responds to the radial motion of material. In terms of the state variables, the total internal energy is

$$2U = 3 \int_V P \, dV = 3 \int_0^M \frac{P}{\rho} \, dM(r) \qquad (8.1.4)$$

For adiabatic pulsations

$$\frac{\delta P}{P_0} = \gamma \, \frac{\delta \rho}{\rho_0} \qquad (8.1.5)$$

Now let us assume that the pulsations can be characterized by

$$\xi \equiv \frac{\delta r}{r_0} \qquad (8.1.6)$$

and make the sensible requirement that ξ remain finite at the origin.

The conservation of mass [equation (8.1.2)] requires that

$$\delta[M(r)] = 0 = \int_0^r 8\pi r_0 \rho \, dr \, dr + \int_0^r 4\pi r_0^2 \, \delta\rho_0 \, dr + \int_0^r 4\pi r_0^2 \rho_0 \, d(\delta r) \quad (8.1.7)$$

while the definition of ξ requires

$$d\xi = \frac{d(\delta r)}{r_0} - \frac{\delta r \, dr}{r_0^2} \qquad (8.1.8)$$

Assuming the variations are small and by keeping only the first-order terms, we get

$$\int_0^r 4\pi r_0^2 \, \delta\rho \, dr = - \int_0^r 4\pi \rho_0 r_0^2 \left(3\xi + r_0 \frac{d\xi}{dr_0} \right) dr \qquad (8.1.9)$$

Since this integral must hold for all values of r, the integrands must be equal, or

$$\frac{\delta\rho}{\rho_0} = -\left(3\xi + r_0\frac{d\xi}{dr_0}\right) \tag{8.1.10}$$

When this is combined with equation (8.1.5), the variation of the internal energy as given by equation (8.1.4) becomes

$$2\,\delta U = -3\int_0^M \frac{P_0}{\rho_0}(\gamma - 1)\left(3\xi + r_0\frac{d\xi}{dr_0}\right)dM(r) \tag{8.1.11}$$

This may be combined with equations (8.1.1) and (8.1.3) to give the constraint on radial oscillations implied by the virial theorem. However, to arrive at some sensible result, we have to make some further assumption about the nature of the oscillation. Let us assume that the amplitude variation is linear with position, so that the pulsation varies homologously within the star and is simply periodic in time and

$$\xi = \xi_0 e^{i\sigma t} \tag{8.1.12}$$

Substitution of this form of the variation into the variational forms of the moment of inertia, internal and potential energies, and subsequently into the variational virial theorem [equation (8.1.1)] gives

$$\sigma^2 = -\frac{\langle 3\gamma - 4\rangle\Omega_0}{I_0} \tag{8.1.13}$$

where $\langle\gamma\rangle$ is an average throughout the star weighted by the gravitational potential. For a homogeneous uniform star that behaves as a perfect gas throughout, this becomes

$$\sigma^2 = \frac{9GM}{4R_0^3} = 3\pi G\bar{\rho}_0 \tag{8.1.14}$$

If we compare this to the free-fall time given by equation (3.2.6), we see that the pulsation period, which is just $2\pi/\sigma$, is of the same order. Specifically,

$$P_p(\text{sec.}) = \frac{8\sqrt{2}\tau_f}{3} = \left(\frac{4\pi}{3G\bar{\rho}}\right)^{1/2} \approx (8\times10^3)(\bar{\rho})^{-1/2}\text{g/cm}^3 \tag{8.1.15}$$

Since pulsation is a dynamic phenomenon, we should not be surprised that it takes place on a dynamical time scale and its period is slightly longer than the free-fall time. For cepheid variables, observationally determined values for the mean density (that is, 10^{-3} g/cm^3 $> \bar{\rho} > 10^{-6}$ g/cm^3) imply that the characteristic pulsation periods should lie in the interval $0.3d < P_p < 90d$ which, conveniently, is observed for these stars. Thus these stars can be understood as undergoing radial oscillations.

b   Effect of Magnetic Fields and Rotation on Radial Oscillations

The impact of rotational motion or the presence of a strong magnetic field can be significant to the characteristic periods of pulsation for a star. Detailed predictions are difficult, for they depend on knowledge of the internal angular momentum and magnetic field distribution. However, the virial theorem gives us some insight into the nature of such effects. We need only calculate the variational behavior for the magnetic energy density and angular momentum (see Collins[6]) to find that the pulsational frequency, under the assumptions made in obtaining equation (8.1.13), is

$$\sigma^2 = -\langle 3\gamma - 4\rangle(\Omega_0 + M_0) + \langle 5 - 3\gamma\rangle\omega_0\mathscr{L}_0 \qquad (8.1.16)$$

Here M_0 is the total internal magnetic energy of the equilibrium configuration while $\mathscr{L}_0$ is the total angular momentum. It is clear that the effect of rotation on stars where $\frac{4}{3} < \gamma < \frac{5}{3}$ will cause a decrease in the pulsational period, while the presence of a strong magnetic field will cause the period to increase.

c   Stability and the Variational Virial Theorem

If we eliminate the total internal energy in favor of the total energy E, the virial theorem for static stars as given by equation (1.2.34) can be written as

$$\frac{1}{2}\frac{d^2 I}{dt^2} = 2E - \Omega > 2E \qquad (8.1.17)$$

Thus, if the total energy of any configuration is greater than zero, the moment of inertia will increase without bound and the system can be said to be *dynamically unstable*. This condition is sometimes called *Jacobi's stability criterion* and as stated is a sufficient condition for a system to be dynamically unstable. We may extend this condition to the variational virial theorem by noting that if $\sigma^2 < 0$, then the perturbations have the form

$$\xi = \xi_0 e^{\pm 2\pi t/t_0} \qquad (8.1.18)$$

Since we may expect the full spectrum of perturbations to be present in any configuration, the $\pm$ sign does not matter, for some perturbation will grow exponentially without bound and the object will be unstable. A quick inspection of equations (8.1.13) and (8.1.16) shows that they represent a set of sufficient conditions for a star to be dynamically unstable. In the absence of magnetic fields and rotation, equation (8.1.13) shows that a *necessary* condition for a star to be stable is that $\gamma > \frac{4}{3}$, which is consistent with what we learned in Chapter 6 about the relativistic polytrope. In the absence of rotation, equation (8.1.16) implies that the magnitude of the gravitational energy must exceed that of the magnetic energy if the star is to remain stable. In Chapter 7 we saw that the presence of a magnetic field would tend to distort the star, and this would seem to be a destabilizing process. This is not necessarily the case for rotation since equation (8.1.16) indicates that for $\frac{4}{3} < \gamma < \frac{5}{3}$ rotation actually seems to help stabilize the star. This occurs because as an oscillation takes place in a rotating star, it is necessary for the pulsating material to conserve angular momentum. For reasonable values of γ, this removes energy from the gas, thereby enhancing the stability of the motion of the perturbation.

The oscillations described so far represent only the fundamental or lowest frequency of oscillation that one could expect. It is this fundamental mode that is limited by the global characteristics of the star. However, it is possible for the star to oscillate at higher frequencies. Under these conditions, the oscillations will take the form of standing waves, with nodes at the surface and the center of the star and possibly elsewhere. However, to show this, it is necessary to consider the internal structure in greater detail than that afforded by the virial theorem.

d   Linear Adiabatic Wave Equation

To find higher-order modes of oscillation, it is necessary to track the motion of the gas within the star. This can be done by considering the equations of motion for the gas. In Chapter 1 we developed the Euler-Lagrange equations of hydrodynamic flow [equation (1.2.27)]. These can be use to develop equations of motion for small-amplitude oscillations. Remember that

$$\frac{\partial \vec{u}}{\partial t} + (\vec{u} \cdot \nabla)\vec{u} = -\nabla\Omega + \frac{\nabla P}{\rho} \tag{8.1.19}$$

Since $\vec{u}$ represents the motion of the gas during the pulsation, we can assume it to be small. Under these conditions, the second term of equation (8.1.19) will be second-order, and we may write the equations of motion as

$$r_0 \frac{d^2\xi}{dt^2} = -\nabla\Omega + \frac{\nabla P}{\rho} \tag{8.1.20}$$

where

$$\vec{u} = \frac{d(r\xi)}{dt} = r_0 \frac{d\xi}{dt} \qquad P = P_0 + \delta P \tag{8.1.21}$$

$$\Omega = \Omega_0 + \delta\Omega \qquad \rho = \rho_0 + \delta\rho$$

The subscript 0 refers to the equilibrium configuration, so hydrostatic equilibrium requires that $\rho\nabla\Omega_0 = -\nabla P_0$. The variational form of the equations of motion becomes

$$r_0 \frac{d^2\xi}{dt^2} = -\nabla\delta\Omega + \frac{\nabla\delta P}{\rho_0} - \frac{\delta\rho\nabla P}{\rho_0^2} \tag{8.1.22}$$

Assuming that the variation has the form given by equation (8.1.12) and that the variation of the density is given by equation (8.1.10), we can use the fact that

$$\nabla\delta\Omega = \xi\nabla\Omega_0 = -\frac{\xi\nabla P_0}{\rho_0} \tag{8.1.23}$$

and some algebra to write

$$\frac{d}{dr}\left(\gamma_0 r^4 \frac{d\xi_0}{dr}\right) + \xi_0\left\{\sigma^2\rho r^4 + r^3 \frac{d}{dr}[(3\gamma - 4)P_0]\right\} = 0 \tag{8.1.24}$$

This is known as the *linear adiabatic wave equation* for $\xi(r)$ (see Cox[7]); since all the coefficients are real, σ^2 must also be real, and pure standing waves are possible. Clearly this is a linear homogeneous second-order differential equation in the displacement ξ_0, so we can expect some ambiguity in the solution. This ambiguity takes the form of an amplitude that can be scaled. That is, if $\xi_0(r)$ represents a solution to the wave equation, so does $A\xi_0(r)$. Since the spatial variation must vanish at the origin, the boundary condition at the center is

$$\left.\frac{d\xi_0}{dr}\right|_{r=0} = 0 \tag{8.1.25}$$

The appropriate boundary condition for the surface is rather more difficult to obtain and is given by Cox[7] (pp. 77–80) as

$$\left.\frac{\delta P}{P_0}\right|_{r=R} = -\left(\frac{\sigma^2 R^3}{GM} + 4\right)\xi_0 = -\gamma\left(3\xi_0 + R_0 \left.\frac{d\xi_0}{dr}\right|_{r=R}\right)$$

$$= -\gamma \left.\frac{\delta\rho}{\rho_0}\right|_{r=R} \tag{8.1.26}$$

These conditions provide the two constants required for the integration of the wave equation, but the actual solution takes the form of a two-point boundary-value problem. Actually with the equilibrium structure known from models, the problem is to find the eigenvalue σ^2 which satisfies the wave equation subject to the boundary conditions. Thus, in principle, we can find the entire spectrum of allowed adiabatic oscillations for a particular star. Since the solutions are pure standing waves, the state variables oscillate locally about the equilibrium values, passing through them twice each cycle. However, to understand the origin of these oscillations, we must describe the nonadiabatic processes which drive and damp them.

8.2 Linear Nonadiabatic Radial Oscillations

The nonadiabatic processes that give rise to stellar pulsations are basically thermodynamic, so we should expect the time scale for their development to be roughly the thermal or Kelvin time scale. Since this is generally much longer than dynamical time scale of the resonant oscillation period, we might anticipate the energies involved in the nonadiabatic processes to be significantly less than those of the pulsational motions themselves. Small as these effects are, they are responsible for the origin of the pulsations.

Throughout the book we have generally treated the gas in stars as an ideal gas. Its thermodynamic properties could thus be represented by the parameter γ, which is just the ratio of the specific heats of the gas. Now that we will be dealing with nonadiabatic processes, we should provide a more complete description of how a gas behaves when it departs from being an ideal gas.

a Adiabatic Exponents

We will define several quantities that describe the change of the state variables of the gas with respect to one another when the gas is subject to an adiabatic change. It may seem a little odd that an adiabatic change is used to characterize the nonadiabatic behavior of a gas, but for the defined parameters to properly describe the behavior of the gas alone, it is necessary to separate the external environment of the gas from the gas itself. Hence we consider externally imposed changes to the gas that do no net work on the gas. If the gas is an ideal gas, then the single parameter γ is sufficient to describe all the changes of the state variables with respect to each other. If the gas is not an ideal gas, then it is necessary to describe the change of each state variable with respect to the others. Thus we define

$$\Gamma_1 \equiv \left(\frac{d\ln P}{d\ln \rho}\right)_{\text{ad}} \qquad \Gamma_3 - 1 \equiv \left(\frac{d\ln T}{d\ln \rho}\right)_{\text{ad}} \qquad \frac{\Gamma_2 - 1}{\Gamma_2} \equiv \left(\frac{d\ln T}{d\ln P}\right)_{\text{ad}} \tag{8.2.1}$$

Since there are three state variables, there can be only two linearly independent changes of one with respect to another. That is clear from the definition of Γ_2. Since the Γ's are defined in terms of logarithmic derivatives, they appear as exponents in the actual relations between the state variables themselves. Since the Γ's describe the response of the gas to an adiabatic change, they are known as the *adiabatic exponents* of the gas. Under the conditions where the thermodynamic pressure is entirely due to an ideal gas, $\Gamma_1 = \gamma$.

As one proceeds inward through the outer envelopes of stars, there exist regions where the dominant elements that make up the star (i.e., hydrogen and helium) change from being largely neutral to being largely ionized. This change usually makes for a relatively small change in temperature and so is physically a relatively narrow zone in the envelope. A gas that undergoes a change in its ionization state in response to a small change in temperature finds itself with more degrees of freedom than an ideal gas. This can be described as a change in the values of the adiabatic exponents, and it will play an important role in understanding stellar pulsation.

b   Nonadiabatic Effects and Pulsational Stability

In Section 8.1 we investigated the dynamical stability of a star to small pulsation. However, even where stable adiabatic pulsations are possible, the presence of nonadiabatic terms may force oscillations to grow or damp out in time. Thus we must investigate how the presence of nonadiabatic effects will affect the stability of stellar oscillations. We know that the thermodynamic nature of the nonadiabatic terms will cause them to effect the oscillations on a thermal time scale, but it would be useful to have a method of estimating what that time scale might be and to know whether the oscillations will grow or die out. To do this, we estimate the rate at which energy is transferred from the kinetic and potential energies associated with the oscillatory motion to the stellar gas by the nonadiabatic processes.

Let us begin by defining the nature of the perturbation to have the form

$$\xi = \xi_0 e^{i\omega t} \qquad \omega = \sigma + i\eta \tag{8.2.2}$$

The pulsational frequency is complex so as to represent the dissipational losses through the quantity η as well as the pure oscillatory motion having a frequency σ. The linear analysis of a simple harmonic oscillator shows that

$$\frac{W}{\Psi} = -\frac{4\pi\eta}{\sigma} = -2P_p\eta \tag{8.2.3}$$

where W is the total work done on the oscillator by the restoring forces over a complete cycle while Ψ is the total oscillation energy (potential and kinetic). We may rewrite this for η and replace W/P_p by the average rate of energy loss $\langle dW/dt\rangle$. We get

$$\eta = \frac{-\frac{1}{2}(W/P_p)}{\Psi} = \frac{-\frac{1}{2}\langle dW/dt\rangle}{\Psi} \tag{8.2.4}$$

The instantaneous work done on the star by the oscillation can be represented as the sum of the gravitational and pressure forces times a differential displacement divided by a differential time and integrated over the entire mass of the star, so

$$\frac{dW}{dt} = \int_0^M \left(g - \frac{1}{\rho}\frac{\partial P}{\partial r}\right)\frac{dr}{dt}\,dM(r) + \frac{d}{dt}\left(\int_0^M \frac{GM(r)\,dM(r)}{r}\right)$$

$$ - \int_0^M 4\pi r^2 \frac{\partial P}{\partial M(r)}\,\dot{r}\,dM(r) \tag{8.2.5}$$

Integrating the second term on the right by parts and using the condition that the surface pressure is zero, we get

$$\frac{dW}{dt} = -\frac{d}{dt}\int_0^M \Omega(r)\,dM(r) + \int_0^M P\frac{\partial}{\partial t}\left(\frac{1}{\rho}\right)dM(r) \tag{8.2.6}$$

Now, by averaging over one full pulsational period, the gravitational forces vanish, since they are fully conservative, so that

$$\left\langle\frac{dW}{dt}\right\rangle = \frac{1}{P_p}\int_0^{P_p}\frac{dW}{dt}\,dt = \frac{1}{P_p}\int_0^{P_p}\int_0^M P\frac{\partial(1/\rho)}{\partial t}\,dM(r)\,dt \tag{8.2.7}$$

However, since the heat increase (dq/dt) over a complete cycle of the pulsation can be expressed in terms of the pressure and density changes,

$$\int_0^{P_p}\frac{dq}{dt}\,dt = \int_0^{P_p} P\frac{\partial(1/\rho)}{\partial t}\,dt \tag{8.2.8}$$

which means that the average energy transfer can be written as

$$\left\langle\frac{dW}{dt}\right\rangle = \frac{1}{P_p}\int_0^{P_p}\int_0^M \frac{dq}{dt}\,dM(r)\,dt \tag{8.2.9}$$

Since the pulsation represents a closed cycle, the star behaves as if it were a Carnot engine. If $\langle dW/dt\rangle > 0$, then the pressure forces are doing positive work on the star and some source of that work must be found. Indeed, each mass shell may be treated separately by calculating the $\oint P\,dV$ forces for that shell. So the star may be viewed as a sum of Carnot engines, some of which feed energy into the pulsation and others which remove it. If the sum of all the Carnot engines produces $\langle dW/dt \rangle > 0$, the star is unstable and the oscillations will grow on a thermal time scale τ_{th} [see equation (3.2.12)] which we can estimate from the inverse of η. Thus

$$\frac{\tau_{th}}{P_p} \equiv \frac{1}{\eta P_p} = \frac{-2\Psi}{\int_0^{P_p} \int_0^M (dq/dt)\,dM(r)\,dt} \tag{8.2.10}$$

Cox[7] (see p. 117) finds that τ_{th}/P_p ranges from the order of 1 to about 1000 for common variables extending from the Mira variables to the RR Lyrae stars, respectively. This would imply that the pulsations in these stars ought to damp out in less than 1000 periods due to thermal losses. Since this is clearly not the case, we must find some sort of driving mechanism.

c   Constructing Pulsational Models

While it is possible to develop a nonadiabatic wave equation (see Cox,[7] pp. 72, 73) similar to equation (8.1.24), we forgo doing so here. Instead let us consider a few aspects of the actual construction of a pulsating model. In principle, one has a complete equilibrium model which provides the run of state variables throughout the star at equilibrium. It is against this background that one formulates the problem of how those state variables will vary with position and time. By assuming a variation of the form given in equation (8.2.2) for each of the state variables, one has the four dependent variables $\delta r/r_0$, $\delta\rho/\rho_0$, $\delta T/T_0$, and dL/L_0, with $M(r)$ as the independent variable. The situation is similar to that encountered in Chapter 4 when we arranged the variables in the same manner to utilize the Henyey scheme for the construction of equilibrium models. Just as the construction of those models involved the solution of a two-point boundary-value problem for a particular eigenvalue, so this problem involves finding the complex eigenfrequency ω.

It is clear that unlike the linear adiabatic wave equation, the linear nonadiabatic wave equation will be complex. Hence the four dependent variables will be complex, with the real part describing the oscillatory motion and the imaginary part describing the damping of the solution. The central boundary conditions can make use of the fact that all variations must vanish at the center. Indeed, many early models[8,9] simply required the solution to go over to the adiabatic solution in the deep interior of the star. However, the surface boundary conditions are rather more tricky. Perhaps the simplest condition that can be employed at the surface is the radiative condition

$$\left.\frac{\delta L}{L_0}\right|_R = 2\left.\frac{\delta r}{r_0}\right|_R + 4\left.\frac{\delta T}{T_0}\right|_R \tag{8.2.11}$$

In addition to the difficulties of posing the correct boundary conditions, significant numerical problems are encountered in the actual solution. These problems were overcome by Castor[10] and Iben,[11] so reliable pulsational models can now be obtained.

d   Pulsational Behavior of Stars

We already indicated that the helium and hydrogen ionization zones might be expected to play a significant role in determining the actual nature of pulsating stars, since they represent zones where the value of Γ_3 varies with position. However, such zones exist in all stars, but not all stars exhibit radial oscillations. Indeed, further thought about the role of ionization zones would lead one to believe that the destabilizing effect on a radial oscillation entering the ionization zone from below brought about by the rapidly declining value of Γ_3 would be offset by the damping effect of a rising Γ_3 at the top of the zone. For the majority of stars, this appears to be the case. Something else must be operative, and to understand its nature, we must consider the nature of nonadiabatic forces near the surface of the star.

Consider the part of the stellar envelope that is relatively near the surface. This region is what Cox[7] (see pp. 140, 141) has called the *transition zone*. As one moves up through the stellar envelope toward the surface, the thermal cooling time steadily decreases. The nearer one is to the surface, the less time is required for energy to diffuse to the surface and escape. Thus the nonadiabatic effects will be more pronounced since pulsational energy that appears as thermal energy will be quickly radiated away. Changes in luminosity generated deeper in the star will become "frozen" in the star and not travel with the wave. When this transition zone coincides with an ionization zone, the potential exists for the nonadiabatic effects of ionization to also be frozen in. Thus, if the ionization zone lies at the right depth, the driving effects caused by the decline in Γ_3 upon entering the zone will not be reversed upon leaving, because the luminosity changes so induced become frozen in.

Since the conditions for ionization are set by atomic physics, we expect that stars with the appropriate effective temperatures and gravities to produce an ionization zone at the correct diffusion depth corresponding to the transition zone will exhibit radial pulsations. There is indeed a region on the H-R diagram known as the *instability strip* which encompasses the cepheid variables, W Virginis stars, δ Scuti stars, RR Lyrae stars, and dwarf cepheids, where this condition appears to be met for the He II ionization zone. The He II ionization zone seems to be the dominant zone for most of the common variable stars. The H and He I ionization zones lie so close together that they may be virtually considered as one. They invariably lie above the He II ionization zone in a region where the luminosity changes have already been frozen in and so play little role in the hot stars. However, this zone is suspected of introducing a phase shift between the luminosity variations and the radial velocity variation from that which is expected. Normally one would expect the time of maximum compression to be the time of maximum luminosity. However, the luminosity maximum lags somewhat, and this may result from its being delayed in the hydrogen ionization zone. In late-type variables where the He II zone is so deep as to be below the transition zone, the hydrogen ionization zone may be the primary driver of the pulsations. However, for these stars the situation becomes complicated by the large geometric extent of the atmosphere and the complex nature of the opacity.

We have seen how the onset of He II ionization at the appropriate place in the star can cause conditions to exist that will drive pulsations simply by a rapid decline of Γ_3. This cause of pulsation is called the γ *mechanism*. However, there is another contributor to the destabilizing effect of ionization. In most of the stellar interior, an increase in temperature is accompanied by a decline in opacity. Certainly Kramers' law [equation (4.1.19)] implies this. Thus any temperature increase accompanying the compression of a passing pulsational perturbation would cause a decline in the local opacity and a release of the radiation trapped there. This would tend to stabilize the region against pulsations by removing the pulsational energy in the form of radiation. However, at low temperatures a rise in temperature results in an increase in opacity. This effect was largely responsible for the nearly vertical tracks of collapsing convective protostars (see Section 5.2c). If the opacity increases with increasing temperature, the opacity will tend to trap the energy at that point and prevent the energy from diffusing away. This is a destabilizing effect that tends to feed the pulsations, and it has been called the κ *mechanism*. Of these two, the γ mechanism seems to be the more important for the pulsation of cepheid variables.

When the oscillations become large, the linear theory we have been describing becomes invalid. However, the complications introduced by these nonlinearities are well beyond the scope of this book, and we leave them to others to discuss.

8.3　Nonradial Oscillations

So far we have considered only those oscillations that involve the radial coordinate. While these oscillations seem sufficient to explain the majority of known pulsating stars, other less dramatic phenomena result from more complicated oscillations. Indeed, one would expect that most pulsational energy would appear in the fundamental radial mode of oscillation, and it is precisely those modes involving the modulation of the greatest amount of energy that can be most easily detected. However, the detection of short-period oscillations of low amplitude in the sun suggests that more complicated types of oscillations can occur. Their importance to the structural models of the sun and their probable detection in some early-type stars require that we spend a little time discussing them. However, the subject is too broad and many of the results are too uncertain to do more than sketch the nature of the problem. To give the greatest insight into the nature of the problem, we concentrate on the adiabatic oscillations. The true cause of the oscillations lies in nonadiabatic theory, as it did for radial oscillations, and the results are still rather uncertain. In addition, the theory for oscillations among stars that are not spherically symmetric is still in its infancy. It was clear from Chapter 6 that the loss of spherical symmetry resulted in a substantial increase in the complexity of the theoretical description. No less is to be expected from pulsation theory. From the small amount of energy involved in the present cases of nonradial oscillations, it will be appropriate to use perturbation theory and to assume that the amplitudes of the oscillations are small.

a Nature and Form of Oscillations

Just as there existed a wave equation for radial oscillations, so there will be a wave equation for nonradial oscillations. However, instead of being a scalar equation in the radial coordinate r alone, it will be a vector equation whose solution will represent the behavior of the displacement vector $\delta\vec{r}$ and the associated variations of the state variables in the various dimensions that define the star. Since our problem will be an adiabatic one, the solutions will be undamped waves propagating not only along the radial coordinate but also over the surface coordinates. In Chapter 7 we saw that it was possible to represent the polar angle variation in terms of a series of Legendre polynomials. This was a special case of a much more general representation of the angular variation of solutions to a wide range of important physical equations. Laplace's equation, the Schrödinger equation, and the wave equation of classical electrodynamics are only a few of the equations whose solutions can be described in terms of spherical harmonics. Spherical harmonics are basically the product of the elements of two sets of functions. One set of functions describes the solution variation in the polar angle and is represented by the orthonormal Legendre polynomials. The orthogonal functions that describe the behavior in the azimuthal coordinate are just $e^{im\phi}$. Thus the spherical harmonics are defined[12] as

$$Y_l^m(\theta, \phi) = (-1)^m \left[\frac{2l + 1}{4\pi} \frac{(l - m)!}{(l + m)!} \right]^{1/2} P_l^m(\cos\theta)e^{im\phi} \qquad (8.3.1)$$

As long as the star remains spherical, the equation describing the nonradial oscillations will also be separable in spherical coordinates, and the orthogonality of the spherical harmonics will guarantee that the full solution can be represented in terms of them. Since some of the solutions to the radial wave equation were standing waves, we should not be surprised if that were also the case for the nonradial oscillations. The linearity of the small-oscillation equations guarantees that the sum of any two solutions is also a solution. So a standing wave is just the interference pattern of two oppositely directed traveling waves of similar amplitude. Just as there could be higher-order modes of the radial wave equation excited, so higher-order waves in the azimuthal and polar angles can be present. However, different modes in each of the coordinates may combine to provide a much richer multiplicity of possible oscillations for the nonradial case. Thus the solution will have the form

$$\xi(r, \theta, \phi, t) = \text{Re}\left|\mathfrak{R}(r)P_l^m(\cos \theta)e^{i(m\phi + \omega t)}\right| \tag{8.3.2}$$

The quantity $\mathfrak{R}$ is a function of the radial coordinate only and represents the eigenfunctions that are possible in the radial direction. Different orders are usually denoted by the letter n. The periodic behavior of the spherical harmonics implies that the parameters l and m will denote the different eigenfunctions in the polar and azimuthal coordinates, respectively.

Deubner and Gough[13] give a nice way of seeing this. Consider cases when the wavelength of the oscillation is much less than the stellar dimensions. Under these conditions Lamb[14] showed that the equations of motion can be reduced to the familiar form

$$\frac{d^2\psi}{dr^2} + K^2\psi = 0 \tag{8.3.3}$$

where

$$\psi = \rho^{1/2}c_s^2\nabla \cdot \vec{\xi} \tag{8.3.4}$$

and c_s is just the local speed of sound. This is simply the equation for a simple harmonic oscillator; here K is the local wave number and in this instance is related to the frequency of oscillation ω, scale height h, and local gravity g by

$$K^2 = \frac{\omega^2 - \omega_c^2}{c_s^2} + \frac{l(l + 1)}{r^2}\left(\frac{N^2}{\omega^2} - 1\right) \tag{8.3.5}$$

where

$$\omega_c^2 = \frac{c_s^2(1 - 2\,dh/dr)}{4h^2} \qquad N^2 = g(1/h) - g/c_s^2$$

Thus we may expect that the general solution for the equations of motion will consist of a complicated interplay of waves propagating in all three coordinates. Also the specific nature of these waves will depend on the structure of the star, with the low-frequency wave anchored deep in the interior and the high-frequency waves determined largely by the local structure of the star nearer the surface. To try to bring some order to the multiplicity of oscillations that may be present in stars, let us consider an idealized case.

b   Homogeneous Model and Classification of Modes

Consider a homogeneous star of uniform density. Admittedly this is an unrealistic case in the real world, but it has the virtue that the eigenfrequencies of the equations of motion can be found and have a particularly simple form. Cowling[15] found that the eigenfrequencies could be organized into several groups based on the physical mechanisms primarily responsible for their propagation. These modes all have their counterparts in the solutions of more realistic models, and so Cowling's classification scheme provides a useful basis for identifying the types of modes to be expected in real stars. Following Cox[7] (p. 235), we can define a dimensionless frequency

$$\omega_{ln}^2 \equiv \frac{\sigma_{ln}^2 R^3}{GM} \tag{8.3.6}$$

The radial oscillation modes are then given by

$$\omega_{ln}^2 = Q_{ln} \pm [Q_{ln}^2 + l(l+1)]^{1/2}$$

$$Q_{ln} \equiv -2 + \frac{\Gamma_1[n(2l+2n-5)+2l+3]}{2} \tag{8.3.7}$$

which for large n are approximately given by

$$\omega_{ln}^2 \approx \begin{cases} 2\Gamma_1 n^2 + \dfrac{l(l+1)}{2\Gamma_1 n^2} & p \text{ modes} \\[3mm] \dfrac{-l(l+1)}{2\Gamma_1 n^2} & g \text{ modes} \end{cases} \tag{8.3.8}$$

The negative root in equation (8.3.7) and its asymptotic counterpart in the g modes of equations (8.3.8) imply that $\omega_{ln}^2 < 0$. So the star is dynamically unstable, and this is a result of the homogeneous model's being unstable to convection. In real stars this is not generally the case, and the g modes can be real.

The terminology has its roots in the nature of the oscillations corresponding to each of the modes. The p modes are known as *pressure modes*; they can be viewed as pressure or acoustic waves and are characterized by relatively large radial pressure disturbances. For $n = 0$ they correspond to the radial oscillations studied in the previous two sections. Thus, as n increases, the p modes can be roughly viewed as radial standing waves having n nodes. It would be reasonable to call them longitudinal waves. On the other hand, stable oscillations characterized by the g modes can be viewed as transverse waves. They are also known as *gravity waves* (not to be confused with gravitational waves, which are a phenomenon of the general theory of relativity), because the primary force acting as a restoring force for the oscillation is the local gravity. These waves are characterized by relatively small pressure and density variations and are largely transverse in their physical displacement. The most common analogy to these waves is water waves where the restoring force is clearly that of gravity and virtually no pressure or density changes are involved. Curiously the case for $n = l$ and $n \gg 1$ leads to

$$\omega_{nn}^2 \approx -\frac{1}{4\Gamma_1} \tag{8.3.9}$$

and a characteristic frequency that is independent of the order n. Physically such a condition would correspond to small blobs of material having a typical size much less than the stellar dimension, moving radially, and exhibiting small pressure and density changes. This is a fairly good description of a convective element and is often taken as a basis for describing the expected spectrum for convective blobs in a region unstable to convection. Thus the presence of g modes in a region stable against convection may be the result of excitation by a lower-lying convective region. The actual frequencies of oscillation for the g modes are always less than those for the corresponding p modes (see Figure 8.1).

Lying between these two classes of modes is a solitary mode called the f *mode*. This mode is generally attributed to Lord Kelvin and is characterized by $\nabla \cdot \delta r = 0$ for the homogeneous model. This implies that both δP and $\delta \rho$ are zero, as for an incompressible fluid. However, this is not true for stellar models in general, and so this mode is sometimes referred to as the *pseudo-Kelvin mode*. Its dimensionless eigenfrequency for the homogeneous model is given by

$$\omega_f^2 = \frac{2l(l - 1)}{2l + 1} \tag{8.3.10}$$

This eigenfrequency is independent of Γ_1 which is to be expected since δP and $\delta \rho$ are both zero. That condition also implies a link between the radial and angular displacements, so that oscillations in the radial coordinate are uniquely linked to displacements in the radial coordinate. There are no stable modes for $l < 2$.

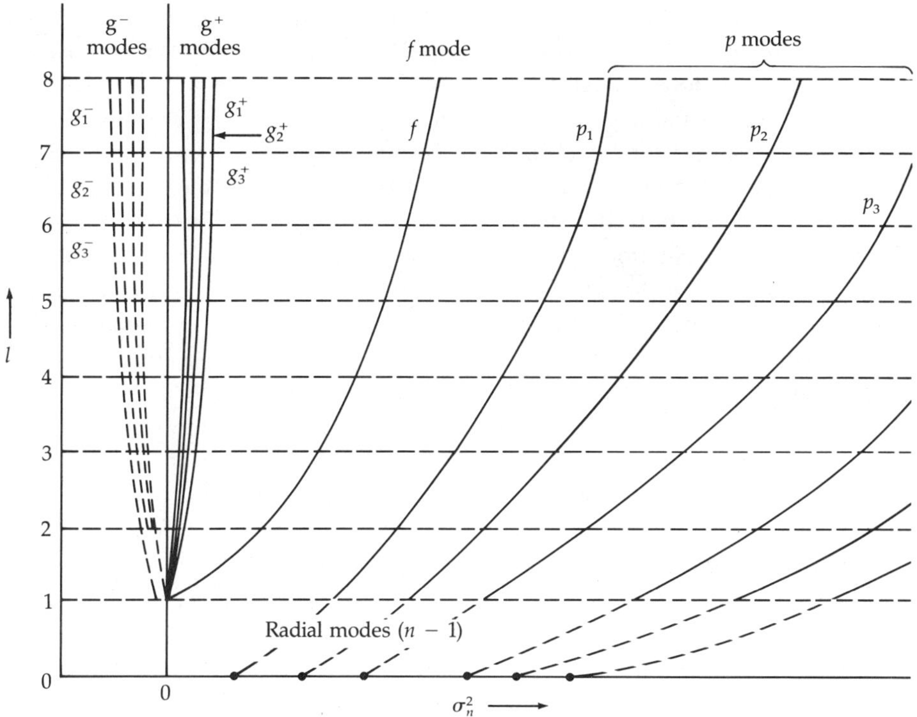

Figure 8.1 Spectral distributions of modes to be expected for nonradial oscillations. The vertical axis indicates the angular eigenvalues while the horizontal axis displays the corresponding oscillation frequency. Therefore the pure radial modes lie on the horizontal axis ($l = 0$) at the end of the spectra for p modes. The negative g modes are those for which the model is unstable, and their frequencies indicate the growth of the instability.

c Toroidal Oscillations

There remains one last class of oscillations that we have not considered. So far we have been faithful to our assumption of spherical symmetry and discussed no modes that have a dependence on the azimuthal angle ϕ so that $m = 0$. To have included cases where $m \neq 0$ would have been to admit the existence of a preferred plane and thereby violate the assumption of spherical symmetry. Thus the modes described so far will present surface phenomena that are independent of the orientation of the star. This will not be the case for $m \neq 0$. However, there is no a priori reason why azimuthal modes cannot exist. Indeed, for each value of l there are $2l + 1$ allowed values of m (that is, $m = 0, \pm 1, \pm 2, \ldots, \pm l$) which, represent waves that propagate in the $\pm \phi$ direction. Of course, for a nonrotating star there is no preferred direction of propagation, so these modes are degenerate and there are only $l + 1$ distinct possibilities.

Papaloizou and Pringle[16] have shown that, for rotation, this degeneracy is broken and the resulting modes correspond to traveling waves around the rotational axis   219 of the star similar to Rossby waves in the earth's atmosphere, so they designated them *r modes*. These waves travel with a characteristic velocity that is approximately $1/m$ times the rotational velocity of the star. Thus such a wave would be seen by an observer to be moving at a rate that is slightly faster $(+m)$ and slightly slower $(-m)$ than the rotational speed of the star. Any comprehensive analysis of the effects of rotation must deal with the effects of angular momentum conservation as well as shape distortion and is therefore quite difficult. However, there can be little doubt that rotation will influence the values for the eigenfrequencies for the p and g modes. Although the overall effects of rotation are extremely complicated, there is some evidence from nonadiabatic studies that the prograde modes are somewhat less stable, and therefore more likely to be excited, than the retrograde modes.

It is possible to have such modes in a star for which the total angular momentum is zero. In the case where $l = 1$, such modes represent uniform rotation of the object. For $l > 1$, the modes represent torsional oscillations. In the simplest case, where $l = 2$, one hemisphere would rotate, say clockwise, while the other hemisphere rotated counterclockwise. Unfortunately, to stabilize such an oscillation, some sort of shear would have to be sustained by the stellar material. Such restoring shears are simply absent in normal stars. However, in white dwarfs and neutron stars, at least part of the star is expected to be in a crystalline phase, and this matter could perhaps sustain such oscillations.

d   Nonradial Oscillations and Stellar Structure

From all that we have discussed so far, clearly the spectrum of oscillations present in a star depends critically on the structure of the star. If it were possible to observe the full spectrum of these oscillations, including their frequency and amplitude, quite sensitive tests of the internal structure of the equilibrium model would be possible. The analogy has been made to geophysicists who deduce the internal structure of the earth from the propagation of seismic waves produced by earthquakes. So strong is this analogy that the term *helioseismology* has come into fairly common use to describe oscillation analysis as applied to the sun.

Clearly, in the case of the sun, we have an opportunity to map the oscillation structure with a high degree of accuracy. While research in this area is ongoing, both p and g modes have been detected. The p modes are usually characterized by the generic term *5-minute oscillations*. Oscillations described by $1 < l < 1000$ have been detected, and the power (amplitude) of those oscillations has been determined. This is done by mapping the radial velocity field of the entire solar disk over extended periods and performing a Fourier analysis of the result for periodic structures. Clearly the velocity of an oscillatory displacement is simply related to the displacement and to the frequency itself. Thus the highly accurately determined velocity measurements provide accurate knowledge of the amplitude and frequencies of the oscillations that are present. The ability to resolve closely spaced frequencies simply requires a long, continuous series of observations.

In addition to the p modes, g modes with characteristic periods ranging from just under 3 hours to nearly 6 days have been reported. There is some disagreement among observers as to what constitutes actual periodic waves and which modes are "aliases" of other periodic phenomena and the data sampling procedure. However, the existence of the g modes is quite likely. In general, the low-order modes represent wave structures that penetrate deeply into the interior, while the higher-frequency modes are confined to the outer layers of the sun. Indeed, the highest-frequency p modes are probably confined to the solar atmosphere itself. Thus the full spectrum of the solar oscillations allows a continuous depth probe of the internal solar structure. The standard solar model reproduces the overall properties of the oscillation spectrum, but fails to fit that spectrum in detail. There is some indication that the standard solar model may have a helium abundance that is somewhat low. Ulrich and Rhodes[17] conclude that the failure of the standard solar model to fit the observations of the oscillation spectrum is real and lies outside the errors of either theory or observation. This would place it in the same category as the solar neutrino problem. Hopefully the solution of one can provide a solution for the other.

There are strong indications that nonradial modes have been detected in other stars. β Cephei stars are suspected to exhibit the effects of traveling waves on their surfaces in their spectra. Papaloizou and Pringle[16] attribute the short period oscillations seen in some cataclysmic variables to the r modes of surface traveling waves. In addition,[18,19] sharp absorption features that move through the broad absorption lines of some rapidly rotating stars have been interpreted as representing nonradial oscillations. If this proves to be the case, then the observations indicate the existence of a phenomenon for which there is no clear theoretical description. For reasons already mentioned, pulsation theory in the presence of extreme rotation is extremely difficult and far from well developed. However, should nonradial oscillations be unambiguously measured for these stars, the potential for detailed understanding of their internal structure is considerable. Given the uncertainties regarding the effects of rotation on the internal structure of these stars, every effort should be made to explore these observations as a probe of the stellar interior.

Problems

1. Show how equations (8.3.8) and (8.3.9) can be obtained from equation (8.3.7).

2. Use the virial theorem to find the fundamental radial pulsation period for a homogeneous star where the equation of state is

$$P = K\rho^{\gamma}$$

3. Compute the lowest-order mode for polytropes with indices n of 1.5, 2.5, and 3 for stellar masses $M = 1.5M_\odot$ and $30M_\odot$.

4. Assuming that the shear forces resulting from the crystalline structure of a white dwarf near the Chandrasekhar limit were sufficient to permit torsional oscillations, estimate the frequency of the lowest-order mode.

5. Show how the wave equation [equation (8.1.24)] is obtained from the equations of motion [equation (8.1.19)].

References and Supplemental Reading

1. Eddington, A. S., "On the Pulsations of a Gaseous Star and the Problem of the Cepheid Variables, Part I," *Mon. Not. R. astr. Soc.* 79, 1918, pp. 2–22.

2. Eddington, A. S., *The Internal Constitution of the Stars* [1926], Dover, New York, 1959, chap. 8.

3. Ledoux, P., "On the Radial Pulsation of Stars," *Ap. J.* 102, 1945, pp. 143–153.

4. Ledoux, P., and Th. Walraven, *Variable Stars*, vol. 51, *Handb. d. Phys.* Springer-Verlag, Berlin, 1958, pp. 353–601.

5. Chandrasekhar, S., *Hydrodynamic and Hydromagnetic Stability*, Oxford University Press, London, 1961.

6. Collins G. W., II, *The Virial Theorem in Stellar Astrophysics*, Pachart, Tucson, Ariz., 1978.

7. Cox, J. P., *Theory of Stellar Pulsation*, Princeton University Press, Princeton, N.J., 1980.

8. Baker, N., and R. Kippenhahn, "The Pulsations of Models of δ Cephei Stars," *z. f. Astrophys.* 54, 1962, pp. 114–151.

9. Cox, J. P., "On Second Helium Ionization as a Cause of Pulsational Instability in Stars," *Ap. J.* 138, 1963, pp. 487–536.

10. Castor, J. I., "On the Calculation of Linear, Nonadiabatic Pulsations of Stellar Models," *Ap. J.* 166, 1971, pp. 109–129.

11. Iben, I., Jr., "On the Specification of the Blue Edge of the RR Lyrae Instability Strip, *Ap. J.* 166, 1971, pp. 131–151.

12. Arfken, G., *Mathematical Methods for Physicists*, 2d ed., Academic, New York, 1970, pp. 569–572.

13. Deubner F.-L., and D. Gough, "Helioseismology: Oscillations as a Diagnostic of the Solar Interior," *Annual Reviews of Astronomy and Astrophysics*, vol. 22, Annual Reviews, Palo Alto, Calif., 1984, pp. 593–619.

14. Lamb, H., *Hydrodynamics*, 5th ed., Cambridge University Press, New York, 1924, pp. 467–484.

15. Cowling, T. G., "The Non-radial Oscillations of Polytropic Stars," *Mon. Not. R. astr. Soc.*, 101, 1941, pp. 367–373.

16. Papaloizou, J. C. B., and J. E. Pringle, "Non-radial Oscillations of Rotating Stars and Their Relevance to the Short-Period Oscillations of Cataclysmic Variables," *Mon. Not. R. astr. Soc.*, 182, 1978, pp. 423–442.

17. Ulrich, R. K., and E. J., Rhodes Jr., "Testing Solar Models with Global Solar Oscillations in the 5-Minute Band," *Ap. J.* 265, 1983, pp. 551–563.

18. Penrod, G. D. "Nonradial Pulsations and the Be Phenomenon," *Physics of Be Stars* (Ed.: A. Slettebak and T. P. Snow), Cambridge University Press, Cambridge, 1987, p. 463.

19. Smith, M. A., D. R. Gies, and G. D. Penrod, "Spectral Transients in the Line Profiles of λ Eridani," *Physics of Be Stars* (Ed.: A. Slettebak and T. P. Snow), Cambridge University Press, Cambridge, 1987, pp. 464–467.

For a background of the observational material that initiated much of the interest in pulsating stars, read

• Shapley, H.: "On the Nature and Cause of Cepheid Variables," *Ap. J.*, 40, 1914, pp. 448–465.

One of the most readable books on stellar pulsation remains

• Rosseland, S.: *The Pulsation Theory of Variable Stars*, Dover, New York, 1964.

An excellent review of the nonadiabatic effects in stellar pulsation theory can be found in

• Christy, R. F.: "Pulsation Theory," *Annual Reviews of Astronomy and Astrophysics*, vol. 4, Annual Reviews, Palo Alto, Calif., 1966, pp. 353–392.

▼ ▼ ▼

Epilogue to Part I: Stellar Interiors

We have presented the study of stellar interiors as a subject that builds from a minimal base of assumptions to the description of some of the most exotic objects in the universe. To describe the structure of most stars in a qualitative manner, we need only construct equilibrium models which follow a polytropic law. To improve that description and to understand the evolution of stars, it is necessary to add a substantial amount of physics and to construct steady-state models. We illustrated the nature of the physics to be added, but many more details need to be included to create models of contemporary accuracy. We did not examine in detail those instances for which steady-state models fail because of rapid evolution which requires hydrodynamic equilibrium to be invoked. However, we did sketch some consequences of this rapid evolution.

We saw that it is possible to understand all the stellar evolution that takes place on a time scale longer than the dynamical time in light of a sequence of equilibrium models linked by estimates of the dominant mode of energy transport required for the construction of steady-state models. This approach yields a good qualitative picture of stellar evolution from pre-main sequence contraction through the helium-burning phases. Finally, we investigated some problems that lie outside the normal theory of stellar structure and evolution, primarily to indicate the direction in which research in these areas has been proceeding.

The one axiom that has dominated Part I of the book is STE. This axiom enabled us to characterize the stellar gas and radiation field entirely in terms of the state variables P, T, and ρ. This was possible because the mean free path for particles and photons was much smaller than the characteristic size of the star. As the boundary is approached, systematic differences from the statistical energy distribution expected from STE will appear, but by the time this begins to happen, the structure of the star will usually be determined. However, to know what the star will look like, we have to consider this region in greater detail. Such an investigation is the aim of Part II.

Part

II

· · ·

Stellar Atmospheres

9

Flow of Radiation through the Atmosphere

· · ·

In Part I, we discussed the internal structure and history of stars with little reference to the actual appearance of the stars themselves. Yet virtually all we know of stars rests on the information that we receive from their surfaces, and so we need to understand those processes that affect the light radiated into space. At several points in the evolution of stars, their evolution was determined by the efficiency with which radiation could be lost from their atmospheres. Thus, the structure of the atmosphere may be expected to play a role in the evolution of the star itself. In addition, for some stars, the region which we call the atmosphere represents a substantial fraction of the radial extent of the star, so that the surface boundary conditions on the equations of stellar structure are set by the atmospheric structure. When this fact is important, the very meaning of a stellar radius becomes intertwined with the details of the atmospheric structure. With the absence of a clearly defined radius, the notion of an effective temperature linked to the stellar luminosity and radius becomes meaningless. Thus, the physical situation near the surface of a star must be treated differently from that of the interior. This transition zone from the relatively simple physics of the stellar interior to the emptiness of interstellar space is commonly known as the *stellar atmosphere.*

There is a tendency to think of the difference between the interior and the atmosphere of a star as distinct, as it is with earth. To be sure, the relative extent of the solar atmosphere compared to the interior is similar to that of earth, but the similarity ends there. There is no sharp interface between stellar atmospheres and interiors as commonly exists with planets. There is no material phase change at the interface. Indeed, for stars, the distinction between atmospheres and interiors is denoted by the failure of certain assumptions used in the study of stellar interiors.

The solution of the problem posed by the surface layers of a star is similar to that for the interior. We have to describe the behavior of the state variables P, T, and ρ with position in the star. However, an additional problem is posed by the atmosphere. We have to describe the energy distribution of photons as they leave the star, for this specifies the appearance of the star which is the fundamental tie with observation. Only if this description of the stellar spectrum agrees with that which is observed can we say that we have provided a successful description of the star.

The approach to finding the structure of the atmosphere can be largely divided into two parts. First, one determines the flow of radiation through the atmosphere, given the structure of the atmosphere. Second, having determined the radiation field throughout the surface layers, one corrects the atmospheric structure so that energy is conserved at all levels of the atmosphere. Since most of the energy is carried by radiation, the second condition usually amounts to the imposition of radiative equilibrium throughout the atmosphere. One then uses the improved structure to correct the radiation field and repeats the process until a self-consistent model is found. To carry out this procedure, it is necessary to make some assumptions about the conditions that prevail in this transition zone between the interior and the space surrounding the star.

9.1   Basic Assumptions for the Stellar Atmosphere

a   Breakdown of Strict Thermodynamic Equilibrium

The description of the energy distribution of the photons in the stellar interior was made particularly simple by the assumption that all constituents of the gas that made up the star were in their most probable macrostate, resulting from random or uncorrelated collisions. That is, they were in thermodynamic equilibrium. All aspects of such a gas can then be characterized by a single parameter, the temperature, which specifies the mean energy of the gas. All other aspects of the distribution function of the gas particles are described by the equilibrium distribution function for the respective kinds of particles.

The validity of this assumption relied on the fact that various components of the gas would undergo randomizing collisions within a volume where the state variables (specifically the temperature) could be considered constant. Within the deep interior of a star, these conditions are met as well as anywhere in the universe. However, any configuration must have a boundary, and it is there that we should expect this assumption to fail. Such is the case for stars. However, the manner of that failure has a peculiar characteristic in that the particles that make up a star are of two distinctly different types. The photons that make up such an important component of the gas behave quite differently from the particles that have a material rest mass. These photons follow different quantum statistics so that their equilibrium distribution functions are different—Bose-Einstein for the photons and generally maxwellian for everything else. In addition, the mean free path between collisions for material particles is very much less than that for photons. Thus, we would expect that the photons would be the first species of particles to be affected by the presence of a boundary, and this is indeed the case. As one moves outward through a star, the presence of the surface begins to affect the state of the gas when photons first begin to escape directly into space and fail to interact any longer with the material particles of the gas. Since the probability that a specific photon will escape depends on the atomic physics of the opacity corresponding to the photon's energy, we should not expect all photons to escape with equal facility. Thus, the photon distribution will depart progressively from that of Planck's law as one approaches the boundary, and our notion of STE will have broken down.

The increase in the photon mean free path brought about by the decreasing density introduces another problem not unrelated to that posed by the boundary. The variation of the state variables over a "typical" photon mean free path will become a significant fraction of the value of the variables themselves. Thus, the radiation field at any point near the boundary will be made up of photons originating in rather different physical environments. Thus, the characteristics of the radiation field will no longer be determined by the local values of the state variables, but will depend on the structure solution of the entire atmosphere. This global aspect of the properties of the local radiation field completely changes the mathematical formalism that describes the flow of radiation from that used in the interior.

b Assumption of Local Thermodynamic Equilibrium

It is a happy consequence of the difference between photons and particles with material rest mass that the mean free path for photons is generally very much greater than that for other particles. Thus, while the photons may sense the boundary, there is a substantial region where the material particles do not. The material particles continue to undergo collisions with other material particles and photons, the majority of which still represent their thermodynamic equilibrium distribution. Thus, the material particles of the gas will continue to behave as if they were in thermodynamic equilibrium as one approaches the boundary. Certainly, the point will be reached when collisions between material particles and other constituents of the gas will become sufficiently infrequent that the nonequilibrium photons of the gas will force departures from the Maxwell-Boltzmann energy distribution expected for particles in thermodynamic equilibrium. But by this point in the atmosphere (in many stars), the majority of the photons will have escaped, much of the stellar spectrum will have been established, and the atmospheric structure below this point will be determined. Thus, the notion that the distribution function for the material particles remains that obtained from the local values of the state variables in thermodynamic equilibrium, while the photon distribution does not, is a useful notion. It is called *local thermodynamic equilibrium* (LTE), and it is one of the central assumptions for much of the remainder of this book. To understand the physical situation that prevails when LTE fails, one must first understand the solution to the problems for which LTE is valid.

The effect of the boundary upon particles that lie within a mean free path of the boundary extends to convective blobs. In the stellar interior, we were able to make do with the crude mixing-length theory because the differences between the adiabatic gradient and that predicted by the mixing-length theory were so small that large errors in this difference became rather small errors in the actual gradient. This was due to the large size of the mixing length, which implied great efficiency for convective transport. This will no longer be the case in the stellar atmosphere, for it is not possible to have a mixing length greater than the local distance to the boundary, and that is the order of a photon mean free path. Thus, convection, should it even occur in the deeper sections of the atmosphere, will be nowhere as efficient as it was in the interior. The mixing-length theory, while crude, can be used to estimate the impact of convection on the atmospheric structure. Fortunately, radiation dominates, by definition, in the outer sections of the atmosphere, and so convection will not be a major concern.

c Continuum and Spectral Lines

In describing the spectral energy distribution of the photons emerging from a star, it is traditional to distinguish between the smooth distribution of photons and the dark interruptions, or lack of photons, called *spectral absorption lines*. These features arise because the opacity of atomic bound-bound transitions is so large compared to that of bound-free and free-free processes that photons with energies corresponding to those bound-bound transitions do not sense the boundary until they are relatively near it. At this point in the atmosphere, the temperature has declined to the point where the emitted radiation is less intense than that originating deeper in the atmosphere. Thus, there will be fewer photons at the frequencies corresponding to the bound-bound transitions, giving rise to the absorption lines of stellar spectra.

Remember that the distinction between *continuum* and *line* is largely artificial, and often the continuum is shot through with myriads of weak lines. The utility of the concept persists, and we are careful to explain exactly what is meant by the distinction. Since a large section of this book is devoted to the processes that give rise to spectral lines (and throughout that section we assume that the structure of the atmosphere is known), we assume that continuum processes and photons involved in those processes are the photons that determine the structure of the atmosphere.

d Additional Assumptions of Normal Stellar Atmospheres

Although some of the development of the theory of stellar atmospheres is presented in great generality, the basic focus of this book is on the theory of "normal" stars. This development is appropriate for most of the stars on the main sequence and some others. We indicate where the assumptions fail in the description of the atmospheres of other stars and what can be done about them, but for now we adopt the traditional assumptions of stellar atmospheres.

In addition to the assumption of LTE, we assume that the thickness of the atmosphere is small compared to the radius of the star. Under these conditions, the surface geometry may be assumed to be that of a plane-parallel slab of infinite thickness possessing a surface extending to infinity in all directions (see Figure 9.1). Since most of the stellar mass will reside inside the atmosphere, it is consistent with the plane-parallel atmosphere approximation to assume that the surface gravity is constant. Thus, the notion of hydrostatic equilibrium given in equation (2.1.6) simplifies to

$$\frac{dP(r)}{dr} = -\frac{GM(r)\rho}{r^2} = -g\rho \tag{9.1.1}$$

Furthermore, since no sources of energy are likely to be present in the stellar atmosphere and we need not worry about time-dependent entropy terms, the conservation of energy [equation (7.1.1)] becomes

$$\nabla \cdot \vec{F} = 0 = \frac{dF}{dx} \qquad F = \text{const} = \sigma T_e^4 \tag{9.1.2}$$

If all the energy is to be carried by radiation, equation (9.1.2) ensures that the radiant flux in the atmosphere will be constant.

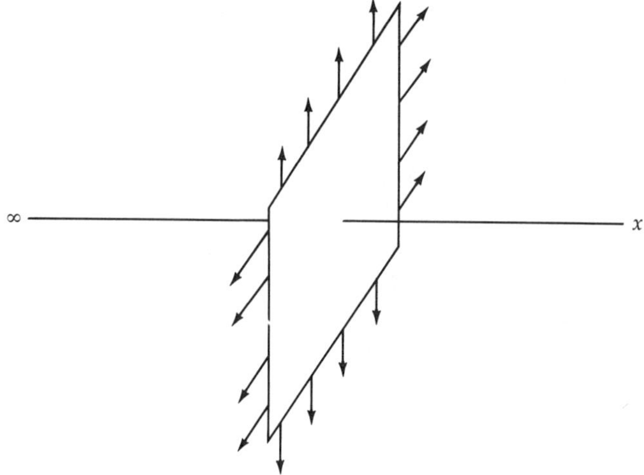

Figure 9.1 Semi-infinite plane that is appropriate for describing the local conditions for stars with thin atmospheres.

Thus these are the fundamental assumptions for the theory of normal stellar atmospheres:

1. LTE prevails. All properties of the material gas can be specified in terms of the local thermodynamic variables.

2. The atmospheric structure is affected by the continuum opacity only.

3. The local geometry is that of a plane-parallel slab.

4. The local surface gravity can be regarded as constant throughout the atmosphere.

5. All energy is carried by radiation, and there are no sources of energy within the atmosphere.

Under these conditions, in addition to the chemical composition, only two parameters are required to specify the structure of the atmosphere:

$$g = \frac{GM(r)}{R^2} \qquad T_e^4 = \frac{L}{4\pi\sigma R^2} \qquad (9.1.3)$$

Since R^{-2} appears in both the expressions for T_e and g, it is no longer an independent parameter required for specifying the atmospheric structure. This is a result of the plane-parallel approximation and does not represent a fundamental difference between the theory of stellar atmospheres and the theory of stellar interiors. If that approximation were to be relaxed, R would be required, indicating that the same parameters (M, L, and R) are necessary for the specification of the model's structure as were required for stellar interiors.

9.2 Equation of Radiative Transfer

In this section we describe, with some generality, the flow of radiation through the outer layers of the star. We developed the formalism for this in Chapter 1 in the form of the Boltzmann transport equation. This formalism basically allows us to describe the flow of any ensemble of particles from one point to another as long as we include all mechanisms for the "creation" and "destruction" of those particles in phase space. In Chapter 1, we used the Boltzmann transport equation to describe the flow of material particles and their momentum through an arbitrary medium. Now we consider the analogous flow of photons.

For material particles, three of the phase space coordinates were velocity. But such coordinates are clearly inappropriate for photons, so we replaced those coordinates with the three components of the photon momentum. This enables us to write equation (1.2.5) in momentum coordinates so that

$$\frac{\partial f}{\partial t} + \sum_{i=1}^{3} \left(\dot{x}_i \frac{\partial f}{\partial x_i} + \dot{p}_i \frac{\partial f}{\partial p_i} \right) = S \tag{9.2.1}$$

For describing the flow of photons, f represents the density in phase space of photons while S describes their creation and destruction at a local point in phase space. However, it is traditional to describe the photon phase density in terms of a quantity called the specific intensity.

a   *Specific Intensity and Its Relation to the Density of Photons in Phase Space*

The specific intensity is an energylike quantity that describes the flow of energy in a particular direction, through a differential area, into a differential solid angle, per unit frequency and time (see Figure 9.2). Remember that the momentum of a photon is just its energy divided by the speed of light:

$$p = \frac{h\nu}{c} \tag{9.2.2}$$

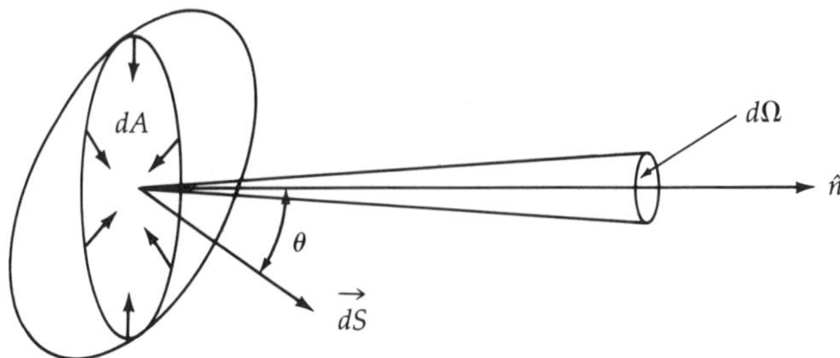

Figure 9.2 Differential parameters defining the specific intensity. Since dA is a differential area, the end of the differential solid angle $d\Omega$ covers it, and all photons passing through dA in the direction $\hat{n}$ flow into $d\Omega$.

We let the energy carried by photons with momentum p, moving in a direction $\hat{n}$, passing through a differential area dA, into a differential solid angle $d\Omega$, in a time dt and frequency interval dv be $dE_v(p, \hat{n})$. We can then define the *specific intensity* as

$$I_v(p, \hat{n}) \equiv \frac{dE_v}{dA \cos \theta \, d\Omega \, dv \, dt} \tag{9.2.3}$$

Now the number of photons traveling in a direction $\hat{n}$ and crossing dA in a time dt comes from a physical volume

$$dV = c \, dA \, dt \tag{9.2.4}$$

However, the number of photons occupying that volume is just

$$dN = f(p, x) \, dV_p \, dV \tag{9.2.5}$$

For photons in that volume, there is no preferred direction so that the differential volume of momentum space is

$$dV_p = 4\pi p^2 \, dp \tag{9.2.6}$$

[see equation (1.3.6)]. Some of these photons will flow in a direction $\hat{n}$ and into the differential solid angle $d\Omega$, each carrying energy hv. Therefore, the differential energy in our definition of specific intensity becomes

$$dE_v = hv \, dN \, \frac{d\Omega}{4\pi} \tag{9.2.7}$$

Combining equations (9.2.2) through (9.2.7), we can relate the specific intensity to the phase space density of photons

$$I_v(p, \hat{n}) = \frac{h^4 v^3}{c^2} f(p, x) \qquad f(p, x) = \frac{c^2}{h^4 v^3} I_v(p, \hat{n})$$

(9.2.8)

b General Equation of Radiative Transfer

Now let us rewrite equation (9.2.1) in vector form:

$$\frac{\partial f(p, \vec{r})}{\partial t} + \vec{r} \cdot \nabla f + \vec{p} \cdot \nabla_p f = S$$

(9.2.9)

If we assume that the photons are moving under the influence of a strong potential gradient $\nabla \Phi$, then we can write for photons that

$$\vec{r} = c\hat{n} \qquad \vec{p} = \frac{hv}{c} \hat{n} \qquad \dot{\vec{p}} = \frac{hv}{c^2} \nabla\Phi \qquad \nabla_p = \frac{\hat{n} c \partial}{h \partial v}$$

(9.2.10)

Substitution of equations (9.2.10) and (9.2.8) into equation (9.2.9) yields an extremely general form of the equation of radiative transfer:

$$\frac{1}{c} \frac{\partial I_v}{\partial t} + \hat{n} \cdot \nabla I_v + \hat{n} \cdot \nabla\Phi \left(\frac{v}{c^2} \right) \left(\frac{\partial I_v}{\partial v} - \frac{3 I_v}{v} \right) = \frac{h^4 v^3 S}{c^3}$$

(9.2.11)

This equation gives the correct description of the transfer of radiation in an arbitrary coordinate system, even if the boundary conditions are changing on a time scale comparable to the photon diffusion time. It is even correct if the photons are subject to energy loss by virtue of their moving through a strong gravitational field, although some care must be exercised in the choice of coordinates. However, if the propagation takes place in a dispersive medium, then $\vec{r}$ must be replaced by

$$\vec{r} = \frac{c}{n} \hat{n}$$

(9.2.12)

and the unit of length is changed by n, where n is the index of refraction of the medium as well.

Fortunately, in normal stellar atmospheres, the radiation field is time-independent, and the gravitational potential gradient is usually negligible so that equation (9.2.11) becomes

$$\hat{n} \cdot \nabla I_v = \frac{h^4 v^3 S}{c^3}$$

(9.2.13)

The assumption of plane parallelism will simplify this even further, but first let us turn to the "creation" rate S.

c "Creation" Rate and Source Function

The "creation" rate S is just a measure of the rate at which photons that contribute to the flow through dA into $d\Omega$ are lost to the volume $dV\,dV_p$. Any absorption process that takes place in that phase space volume will result in the loss of a photon. However, photons can be "lost" from the volume without being destroyed. Any scattering process that changes the momentum of the photon can remove the photon from the volume. Thus, we can write the number lost to the differential volume as

$$dn_l = \alpha f\,dV\,dV_p \qquad (9.2.14)$$

where α is just the fraction of particles present that are lost due to scattering and absorption. Particles may also "appear" in the volume or be "created" by thermal emission or scattering processes. We assume that the thermal emission processes are isotropic so that the number gained in the volume and radiated into a unit solid angle is

$$dn_{gt} = \frac{\epsilon\,dV\,dV_p}{4\pi} \qquad (9.2.15)$$

where ϵ is the thermal emission per unit volume of phase space.

The situation for scattering is somewhat more complicated. Photons may appear in the volume and be scattered by matter in the volume into direction $\hat{n}$ with the appropriate momentum. These photons appear to be created just as surely as the thermal photons do, but with a difference. The thermal emission rate depends only on the thermodynamic characteristics of the material gas, whereas the scattered photons have their origin directly in the radiation field. This dependence of the "creation" rate, and hence the specific intensity, on the radiation field itself is one of the hallmarks of radiative transfer in stellar atmospheres. The scattering process is the means by which the local value of the radiation field depends on the values of the radiation field throughout the medium. This coupling of the local radiation field to the global radiation field generates mathematical problems of an entirely different character from those found in stellar interiors.

Definition of Redistribution Function For scattering to act as a source of photons in the direction and solid angle of interest, the process must take a photon of a given momentum and change its direction and momentum to coincide with that of the beam (i.e., the direction and frequency of the specific intensity). The processes that can do this are characterized by a function known as the *redistribution* *function*. This function is essentially the probability that a photon with a initial momentum p' coming from an initial solid angle Ω' will be scattered into a solid angle Ω with final momentum p. We call this probability density function $R(p', p, \Omega', \Omega)$ the *redistribution function* because it describes how interacting photons will be redistributed in momentum and direction. It is normalized so that

$$\int_0^\infty \int_0^\infty \oint_{4\pi} \oint_{4\pi} R(p', p, \Omega', \Omega)\, d\Omega'\, d\Omega\, dp'\, dp = 1 \qquad (9.2.16)$$

The specific nature of the redistribution function depends on the details of the physical scattering mechanism and is discussed later. At this point, it is necessary only to know that the redistribution function exists and can be calculated for specific physical processes. Since the redistribution function has been normalized in accordance with equation (9.2.16), it represents the redistribution of a *scattered* photon. To calculate the number of particles gained from scattering, we still must include a measure of the fraction entering the volume $dV\,dV_p$ that undergo a scattering. Therefore, the number of particles gained from scattering processes is

$$dn_{gs} = \frac{\sigma'}{4\pi} \int_0^\infty \oint_{4\pi} R(p', p, \Omega', \Omega) f(p', \Omega')\, d\Omega'\, dp'\, dV\, dV_p \qquad (9.2.17)$$

where σ' is simply that fraction. The integrals run over all values of p' and Ω' so that photons entering the volume from all possible directions and with all possible values of momentum are included.

"Creation" Rate in Terms of Scattering and Absorption Processes   The net change of particles in volume $dV\,dV_p$ is obtained by combining equations (9.2.14), (9.2.15), and (9.2.17) and replacing the momentum derivatives and photon phase densities by equations (9.2.2) and (9.2.8). This process yields

$$
\begin{aligned}
dn &= dn_{gt} + dn_{gs} - dn_l \\
&= \frac{c^2}{h^4 v^3} \left[\frac{h^4 v^3}{c^2} \frac{\epsilon}{4\pi} + \frac{\sigma' h}{4\pi c} \int_0^\infty \oint_{4\pi} \left(\frac{v}{v'} \right)^3 R(v, v', \Omega', \Omega) I_{v'}(\Omega')\, d\Omega'\, dv' \right. \\
&\quad \left. - \alpha I_v(\Omega) \right] dV\, dV_p
\end{aligned}
\qquad (9.2.18)
$$

Now the "creation" rate S is the number of photons created per *unit* phase space volume per unit time. But in deriving the transformations from phase density to specific intensity given by equations (9.2.8), we did not choose an arbitrary spatial volume dV because it had a length $c\,dt$. Therefore, to relate the number of particles created

237

in an arbitrary phase space volume $dV\,dV_p$ to S, we must normalize by that length so that

$$dn = \frac{S\,dV\,dV_p}{c} \tag{9.2.19}$$

Using this and equation (9.2.18) to express the "creation" rate S in terms of the physical processes taking place in the volume, we have

$$\frac{h^4 v^3}{c^2} S = \rho j_v + \frac{\rho \sigma_v}{4\pi} \int_0^\infty \oint_{4\pi} R(v,\,v',\,\Omega,\,\Omega')I_{v'}(\Omega')\,d\Omega'\,dv'$$

$$- (\kappa_v + \sigma_v)\rho I_v(\Omega) \tag{9.2.20}$$

where we have introduced the volume emissivity j_v, the mass scattering coefficient σ_v, and the mass absorption coefficient κ_v and replaced R from

$$j_v = \frac{h^4 v^3 \epsilon/(4\pi c^2)}{\rho} \qquad \sigma_v = \frac{\sigma'}{\rho}$$

$$R(v,\,v',\,\Omega,\,\Omega') = \frac{h(v/v')^3}{c}\,R(v,\,v',\,\Omega,\,\Omega') \tag{9.2.21}$$

$$\kappa_v + \sigma_v = \frac{\alpha}{\rho}$$

Thermal Emission   For a gas that is in thermal equilibrium, the relationship between the rate of absorption and emission is not arbitrary. This is the first use of LTE. Since we are assuming that the gas is in thermal equilibrium with its surroundings (LTE), we may invoke Kirchhoff's law for the relationship between the thermal emissivity and absorptivity, namely,

$$j_v = \kappa_v B_v(T) \tag{9.2.22}$$

where $B_v(T)$ is the Planck function which depends only on the local temperature. If we use this and equation (9.2.20), the equation of radiative transfer given by equation (9.2.13) becomes

$$\hat{n} \cdot \nabla I_v = \rho \left[\kappa_v B_v(T) - (\kappa_v + \sigma_v)I_v(\Omega) \right.$$

$$\left. + \frac{\sigma_v}{4\pi} \int_0^\infty \oint_{4\pi} R(v,\,v',\,\Omega,\,\Omega')I_{v'}(\Omega')\,d\Omega'\,dv' \right] \tag{9.2.23}$$

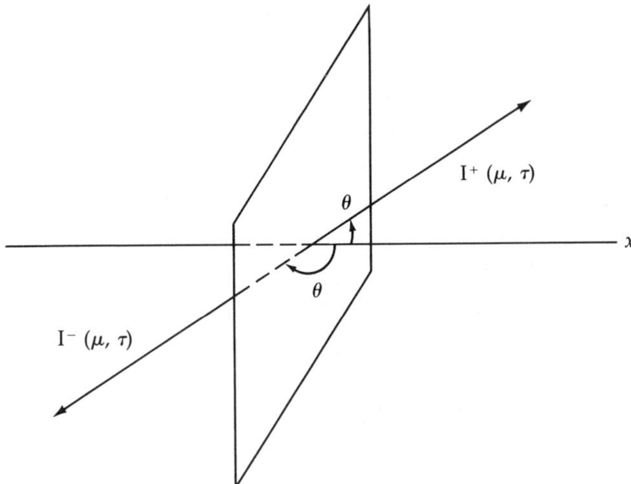

Figure 9.3 Geometry of a plane-parallel atmosphere.

We may further simplify the equation of radiative transfer by invoking the plane-parallel approximation so that ∇ becomes $\hat{x}\,d/dx$ (see Figure 9.3), yielding

$$\frac{\cos\theta}{-(\kappa_v + \sigma_v)\rho}\frac{dI_v}{dx} = I_v - \frac{\kappa_v B_v}{\kappa_v + \sigma_v}$$

$$-\frac{\sigma_v}{4\pi(\kappa_v + \sigma_v)}\int_0^\infty \oint_{4\pi} R(v,\,v',\,\Omega,\,\Omega')I_{v'}(\Omega')\,d\Omega'\,dv' \qquad (9.2.24)$$

Optical Depth The notion of a dimensionless depth parameter called *optical depth* is central to the study of stellar atmospheres. It is usually taken to increase inward as one moves into the star, and it can be viewed physically in the following manner. Optical depth of unity is that depth of material wherein $1/e$ of the photons will not be scattered or absorbed while traversing the depth. In terms of the mass absorption and scattering coefficients and the differential distance parameter, it is defined as

$$d\tau_v = -(\kappa_v + \sigma_v)\rho\,dx \qquad (9.2.25)$$

Making use of the definition of optical depth, we can write the equation of radiative transfer for a plane-parallel atmosphere as

$$\mu\frac{dI_v(\mu,\,\tau_v)}{d\tau_v} = I_v(\mu,\,\tau_v) - S_v(\mu,\,\tau_v) \qquad (9.2.26)$$

where

$$S_v \equiv \frac{\kappa_v B_v}{\kappa_v + \sigma_v} + \frac{\sigma_v}{4\pi(\kappa_v + \sigma_v)} \int_0^\infty \oint_{4\pi} R(v, v', \Omega, \Omega') I_{v'}(\Omega') \, d\Omega' \, dv' \quad (9.2.27)$$

The parameter S_v is known as the *source function* of the radiation field. Since the quantity $\kappa_v + \sigma_v$ appears so frequently, it is customary to call it the *mass extinction coefficient*. The name is reasonable as it is, indeed, a measure of the total ability of material to attenuate the flow of photons.

d   Physical Meaning of the Source Function

The source function is one of the most important concepts in the theory of radiative transfer, and it is important to have a good intuitive feeling for its meaning. As the name implies, the source function represents the local contribution to the radiation field. It is a measure of the energy contributed to the radiation field by physical processes taking place at a particular spot in the atmosphere. Consider the case where scattering is unimportant so that $\sigma_v = 0$. Under these conditions the expression for the source function [equation (9.2.27)] becomes

$$S_v = B_v \quad (9.2.28)$$

and all photons locally contributed to the radiation field can be characterized by the Planck function since they arise from thermal processes. This is a consequence of the assumption of LTE which enabled us to use Kirchhoff's law to characterize the local emissivity of the gas in terms of its absorptivity. Some authors take this as a definition of LTE, but as such, it would be unduly restrictive. The presence of scattering, say by electrons, will require a more complicated source function such as that given by equation (9.2.27), but the excitation and ionization characteristics of the gas may still be those expected for a gas in thermodynamic equilibrium. Thus, $S_v = B_v$ is normally a sufficient condition for the existence of LTE, but not a necessary one.

Now consider the case when pure absorption processes are negligible and virtually all the opacity of the material arises from scattering processes. Then

$$S_v = \frac{1}{4\pi} \int_0^\infty \oint_{4\pi} R(v, v', \Omega, \Omega') I_{v'}(\Omega') \, d\Omega' \, dv' \quad (9.2.29)$$

Here the source function depends only on the incident radiation field. Since the redistribution function is normalized to unity, the integral in equation (9.2.29) simply represents some sort of average of the local specific intensity over all frequencies and angles. The factor of $1/(4\pi)$ then represents that part of the average that is scattered into the differential solid angle appropriate for I_ν.

Thus, under the conditions of pure scattering, the source function becomes totally independent of the local physical conditions and is completely determined by the local radiation field. If this condition were to prevail throughout the atmosphere, one would have the curious result that the radiation field would be independent of the local values of the state variables (P, T, and ρ) and depend only on the ability of particles to scatter photons and the details of how the particles do it. In some real sense, the radiation field would become decoupled from the physical properties of the gas. Indeed, one can learn little about the physical conditions that prevail in a fog by observing the light transmitted through it from, say, an automobile headlight. This independence of the radiation field from the state variables of the gas enables one to solve the entire problem of radiative transfer for pure scattering without knowing anything about the gas other than the redistribution function. We use this property later to discuss methods of solving the equation of radiative transfer. However, as the case of the fog illustrates, this is a two-edged sword. The decoupling of the radiation field from the state variables of the gas, in the case of pure scattering, means that we cannot use the radiation field to determine the run of state variables with depth.

e Special Forms of the Redistribution Function

Since the redistribution function plays such an important role in specifying the nature of scattering in the source function, we examine some common physical situations and the corresponding redistribution functions.

Coherent Scattering   The term *coherent scattering* refers to the case where photons are scattered in direction but not in frequency. Thomson scattering by electrons is of this form. Such processes are generally known as *conservative processes* because no energy is exchanged between the radiation field and the particles. While this is never strictly true, in many cases it is an excellent approximation. This is certainly true for the scattering of optical photons by the electrons present in a stellar atmosphere. Under these conditions we can write the redistribution function as

$$R(\nu', \nu, \vec{\Omega}, \vec{\Omega'}) = h(\vec{\Omega}, \vec{\Omega'})\,\delta(\nu - \nu') \tag{9.2.30}$$

where $\delta(\nu - \nu')$ is the Dirac delta function. The delta function on frequency causes the frequency integral in equation (9.2.27) to collapse, simplifying the source function considerably.

Noncoherent Scattering This phrase has come to mean considerably more than the opposite of coherent scattering. For fully noncoherent scattering, the frequency of a scattered photon is completely uncorrelated with the frequency of the incident photon. In some sense, the photon "forgets" its prior frequency. Like coherent scattering, this case also represents an approximation. Clearly, if the situation were to apply to the entire frequency range from zero to infinity, the value of the redistribution function at any specific value of v would have to be arbitrarily small. Thus, the common use of the approximation is confined to a finite frequency range such as a spectral line. As we shall see later, very strong spectral lines often possess the property that an electron in the upper state is so perturbed by interactions with other particles of the gas that the specific value of the absorbed energy is irrelevant in determining the energy of the photon that will be emitted in the subsequent transition. Thus, over a finite frequency interval, the wavelength of the emitted photon will be totally uncorrelated with the wavelength of the absorbed photon. Under these conditions, the frequency simply does not appear in the redistribution function and

$$R(v, v', \vec{\Omega}, \vec{\Omega}') = h(\vec{\Omega}, \vec{\Omega}') \qquad (9.2.31)$$

Redistribution functions of this form are often called *complete redistribution functions*.

Isotropic Scattering As with complete redistribution, the photon undergoing isotropic scattering suffers from "amnesia". The direction of the scattered photon is completely uncorrelated with the direction of the incident photon. Thus, the angular dependence of the redistribution function vanishes and

$$R(v, v', \vec{\Omega}, \vec{\Omega}') = g(v, v') \qquad (9.2.32)$$

This also considerably simplifies the source function in equation (9.2.27). If the radiation field were isotropic, the integral over the solid angle merely produces a factor of 4π, which cancels the corresponding factor in front of the integral. In general, this is also an approximation. Although it is far from obvious, we shall see that it is an excellent approximation for electron scattering of optical photons in a stellar atmosphere. So great is the simplification introduced by the assumption of isotropic scattering that there is a tendency to invoke it even when it is totally inappropriate. Later, we shall see what sorts of methods can be used to incorporate the full redistribution function in the solution of the equation of radiative transfer. Such cases are often called *partially coherent anisotropic scattering*, and their solution poses one of the most difficult problems in radiative transfer. However, before we consider these formidable problems, we must understand how to approach the solution of more basic problems. The dominant form of scattering in normal stellar atmospheres is Thomson scattering by electrons, and for purposes of determining the atmospheric structure it is an excellent approximation to assume that such scattering is isotropic. Under the assumption of coherent isotropic scattering, the source function given by equation (9.2.27) becomes

$$S_v = \frac{\kappa_v B_v}{\kappa_v + \sigma_v} + \frac{\sigma_v}{4\pi(\kappa_v + \sigma_v)} \oint_{4\pi} I_v(\Omega')\, d\Omega' \qquad (9.2.33)$$

9.3 Moments of the Radiation Field

In Chapter 1 we saw that a good deal of information was gleaned and simplification achieved by taking moments of the phase density of the particles that made up the gas in question. By such methods we were able to obtain equations for the continuity of matter and momentum and eventually to develop expressions for the hydrodynamic flow of a gas and hydrostatic equilibrium. The basic approach was to throw away information contained in the phase density by averaging it over some appropriate part of the phase space volume. That part of the volume was generally taken to be described by coordinates for which we did not require specific knowledge of the phase density. Since we were to invoke STE for the gas, we knew that the details of the velocity distribution could be ignored since in thermodynamic systems the velocity distributions are specified by a single parameter (the temperature) which is related to the mean velocity. Thus, averaging the phase density over velocity or momentum space made good sense.

We may expect the same sort of benefits by taking moments of the radiation field and particularly the specific intensity, for there is a simple relation [equation (9.2.8)] between the specific intensity and the phase density of photons. However, here we must be careful because it is the momentum distribution of photons in which we are interested so that averaging over momentum space would remove the very information we seek. We must look to other coordinates of phase space to find those which can be considered unimportant.

One of our initial assumptions is that the atmosphere is well approximated by a plane-parallel slab. By symmetry, the radiation flow through such a slab will be isotropic about the normal to the slab. Hence, no important information will be contained in the azimuthal coordinate (see Figure 9.3). In addition, we might expect that information in the polar angle θ will not play a central role in the interaction of the radiation field with matter. It is this interaction that determines the emergent spectrum and the atmospheric structure. For these reasons, we can expect that the angular coordinates of phase space may prove expendable and that averages of the radiation field over these coordinates could prove useful in describing the flow of radiation through the atmosphere. Thus, we shall average over two of the three spatial coordinates, choosing the third to represent the direction of net energy flow. In the case of the plane-parallel atmosphere, this clearly is the direction of the atmosphere normal. Also, because of the simple transformation between the specific intensity and the photon phase density, the quantity to be averaged should be the specific intensity itself. In addition, the higher-order moments should involve the spatial coordinates just as the higher-order moments in Chapter 1 involved the velocity itself. Such angular moments will then describe various aspects of the net flow of energy.

a Mean Intensity

Averaging over the angular coordinates described in Figure 9.3 is equivalent to averaging over all solid angles, so with some generality we can define the lowest-order moment of the radiation field as

$$J_\nu(\tau_\nu) \equiv \frac{\oint_{4\pi} I_\nu(\tau_\nu, \theta, \phi)\, d\Omega}{\oint_{4\pi} d\Omega} = \frac{1}{4\pi} \oint_{4\pi} I_\nu(\tau_\nu, \theta, \phi)\, d\Omega \qquad (9.3.1)$$

For a plane-parallel atmosphere, where the intensity has no ϕ dependence and $\cos\theta$ is replaced by μ, equation (9.3.1) is equivalent to

$$J_\nu(\tau_\nu) = \frac{1}{2} \int_{-1}^{+1} I_\nu(\mu, \tau_\nu)\, d\mu \qquad (9.3.2)$$

This quantity, known as the *mean intensity*, is analogous to the particle density of Chapter 1 and differs from the photon energy density by a factor of $4\pi/c$.

b Flux

The next-highest-order moment is related to the net flow of energy in a specific direction $\hat{n}$, and it is defined, in a manner analogous to that for the mean intensity J_ν, as follows:

$$\vec{H}_\nu(\tau_\nu) = \frac{\oint_{4\pi} I_\nu(\tau_\nu, \theta, \phi)\hat{n}\, d\Omega}{\oint_{4\pi} d\Omega} = \frac{1}{4\pi} \oint_{4\pi} I_\nu(\tau_\nu, \theta, \phi)\hat{n}\, d\Omega \qquad (9.3.3)$$

If we break $\hat{n}$ into its components, then for the axis-symmetric case of a plane parallel atmosphere, this becomes

$$\vec{H}_\nu(\tau_\nu) = \frac{\hat{n}}{2} \int_{-1}^{+1} I_\nu(\tau_\nu, \mu)\mu\, d\mu \qquad (9.3.4)$$

where $\hat{n}$ points along the normal to the atmosphere. Indeed, it is fair to describe the flux as an intensity-weighted unit vector pointing in the direction of the flow of energy. Although the flux as defined here is a vector quantity, it is common to drop the vector properties since they are generally obvious from the geometry of the atmosphere. However, the vector nature does point to the similarity with the moments of Chapter 1 where the first moment of the phase density was the mean flow velocity. This definition of the first moment of the radiation field is sometimes known as the *Harvard flux* because it is heavily employed by the ATLAS atmosphere computer code developed at Harvard University, where the analogy to the mean intensity was deemed more important than the physical interpretation.

The actual energy crossing a differential area dA in the direction $\hat{n}$ is

$$\mathbf{F}_\nu(\tau_\nu) = \oint_{4\pi} I_\nu(\tau_\nu, \theta, \phi)\,\hat{n}\,d\Omega = 4\pi H_\nu(\tau_\nu) \tag{9.3.5}$$

The quantity $\mathbf{F}_\nu$ is often called the *physical flux* because it represents the actual flow of energy. For a plane-parallel atmosphere this reduces to

$$\mathbf{F}_\nu(\tau_\nu) = 2\pi \int_{-1}^{+1} I_\nu(\tau_\nu, \mu)\mu\,d\mu \tag{9.3.6}$$

The quantity π appears so regularly that many early authors, who were primarily concerned with plane-parallel atmospheres, defined a third form of the flux as

$$F_\nu(\tau_\nu) \equiv 2 \int_{-1}^{+1} I_\nu(\tau_\nu, \mu)\mu\,d\mu = \frac{\mathbf{F}_\nu(\tau_\nu)}{\pi} = 4H_\nu(\tau_\nu) \tag{9.3.7}$$

This has become known as the *radiative flux*, and it neither represents a physical quantity directly nor is analogous to the mean intensity. However, it is the most widely used definition of the first moment of the radiation field, so the student is to be warned to determine which definition of the flux a particular author is using, or else all sorts of confusion may result. Throughout this book, we use all three definitions, but we are quite clear as to which is which and why a specific choice is made.

c Radiation Pressure

The analogy between this moment and the pressure tensor in Chapter 1 is very close, and the formal definition has the same normalization properties as J_ν. So

$$\mathbf{K}_\nu(\tau_\nu) = \frac{\oint_{4\pi} I_\nu(\tau_\nu, \theta, \phi)(\hat{n}\hat{n})\,d\Omega}{\oint_{4\pi} d\Omega} = \frac{1}{4\pi} \oint_{4\pi} (\hat{n}\hat{n})I_\nu(\tau_\nu, \theta, \phi)\,d\Omega \tag{9.3.8}$$

In a manner similar to the physical flux $\mathbf{F}_\nu$, $\mathbf{K}_\nu$ can be regarded as an intensity-weighted unit dyadic (not to be confused with the unit tensor $\mathbf{1}$ that has components δ_{ij}). Now $\mathbf{K}_\nu$ is known as the *radiation pressure tensor* and is completely analogous to the pressure tensor $\mathbf{P}$ obtained in Chapter 1 [equation (1.2.25)]. The meaning of the unit dyadic (in this case the vector outer product of a unit vector with itself) can be seen by writing out the various cartesian components of $\mathbf{K}_\nu$ in spherical coordinates:

$$\mathbf{K}_\nu = \frac{1}{4\pi} \int_0^\pi \int_0^{2\pi} \begin{bmatrix} \hat{i}\hat{i}\sin^2\theta\cos^2\phi & \hat{i}\hat{j}\sin^2\theta\cos\phi\sin\phi & \hat{i}\hat{k}\sin\theta\cos\theta\cos\phi \\ \hat{j}\hat{i}\sin^2\theta\sin\phi\cos\phi & \hat{j}\hat{j}\sin^2\theta\sin^2\theta & \hat{j}\hat{k}\sin\theta\cos\theta\sin\phi \\ \hat{k}\hat{i}\sin\theta\cos\theta\cos\phi & \hat{k}\hat{j}\sin\theta\sin\theta\cos\phi & \hat{k}\hat{k}\cos^2\theta \end{bmatrix}$$
$$\cdot\, I_\nu \sin\theta \, d\theta \, d\phi \tag{9.3.9}$$

For the axis-symmetric case this becomes

$$\mathbf{K}_\nu(\tau_\nu) = \frac{1}{2} \int_0^\pi \begin{bmatrix} \hat{i}\hat{i}(\tfrac{1}{2}\sin^2\theta) & 0 & 0 \\ 0 & \hat{j}\hat{j}(\tfrac{1}{2}\sin^2\theta) & 0 \\ 0 & 0 & \hat{k}\hat{k}\cos^2\theta \end{bmatrix} I_\nu(\tau_\nu, \theta)\sin\theta\, d\theta \tag{9.3.10}$$

or

$$\mathbf{K}_\nu(\tau_\nu) = \frac{1}{2} \int_{-1}^{+1} \left[\frac{\hat{i}\hat{i}(1-\mu^2)}{2}, \; \frac{\hat{j}\hat{j}(1-\mu^2)}{2}, \; \hat{k}\hat{k}\mu^2 \right] I_\nu(\tau_\nu, \mu)\, d\mu \tag{9.3.11}$$

Now consider the case where the radiation field is nearly isotropic so that we may expand $I_\nu(\tau_\nu, \mu)$ in a rapidly converging series as

$$I_\nu(\tau_\nu, \mu) = \sum_{i=0}^{\infty} I_i(\tau_\nu)\mu^i \tag{9.3.12}$$

where the lead term $I_0(\tau_\nu)$ is the dominant term. The components of the radiation pressure tensor then become

$$\mathbf{K}_\nu(\tau_\nu) = \begin{bmatrix} \dfrac{\hat{i}\hat{i}\sum_{i=0}^{\infty} I_{2i}(\tau_\nu)}{(2i+1)(2i+3)} \\[2em] \dfrac{\hat{j}\hat{j}\sum_{i=0}^{\infty} I_{2i}(\tau_\nu)}{(2i+1)(2i+3)} \\[2em] \dfrac{\hat{k}\hat{k}\sum_{i=0}^{\infty} I_{2i}(\tau_\nu)}{2i+3} \end{bmatrix} \approx \frac{1}{3} I_0(\tau_\nu)\mathbf{1} \tag{9.3.13}$$

Define the scalar moment $K_\nu(\tau_\nu)$ so that

$$K_\nu(\tau_\nu) = \frac{1}{2} \int_{-1}^{+1} I_\nu(\tau_\nu, \mu)\mu^2 \, d\mu \approx \frac{I_0(\tau_\nu)}{3} \tag{9.3.14}$$

The identity of this moment to the magnitude of the radiation pressure tensor in the case of near isotropy ensures that

$$\nabla \cdot \mathbf{K}_\nu(\tau_\nu) = \nabla K_\nu(\tau_\nu) \tag{9.3.15}$$

The isotropy condition was required in Chapter 1 in order for the divergence of the pressure tensor to be replaced by the gradient of the scalar pressure. Thus, in every sense of the word, $K_\nu(\tau_\nu)$ may be considered to be related to the pressure of radiation. There remains only the problem of units. Since $\mathbf{P}$ represents the transfer of momentum across a surface, the exact relationship is

$$\mathbf{K}_\nu(\tau_\nu) = \frac{c\mathbf{P}_\nu(\tau_\nu)}{4\pi} \qquad P_r(\tau_\nu) = \frac{4\pi K_\nu(\tau_\nu)}{c} \tag{9.3.16}$$

Although these expressions give the correct formulation of the radiation pressure in terms of moments of the radiation field, it is important to remember that the radiation pressure is not identical to the force per unit area exerted by photons. That will involve the opacity, for to exert a force, the photon must interact with the matter. In the stellar interior, this was no problem because the mean free path was so short as to guarantee that all photons would interact in a short distance. However, in a stellar atmosphere, this is no longer true, for some of the photons escape. We return to this point when we consider the forces acting on the gas.

9.4 Moments of the Equation of Radiative Transfer

In Chapter 1 we saw that much useful information could be obtained about the gas by taking moments of the Boltzmann transport equation. The process always generated moments of phase density that were of one order higher than that used to generate the equation itself. Thus, to be useful, a relation between the higher-order moment and one of lower order had to be found. If this could be done, a self-consistent set of moment equations could be found and solved, yielding the values of those moments throughout the configuration. A similar set of circumstances will exist for the equation of radiative transfer.

To maintain a high level of generality, let us consider the general equation of radiative transfer given by equation (9.2.11) but with the "creation rate" replaced by the source function and the potential gradient taken to be zero. Thus

$$\frac{1}{c}\frac{\partial I_v}{\partial t} + \hat{n} \cdot \nabla I_v = -\rho(\kappa_v + \sigma_v)(I_v - S_v) \tag{9.4.1}$$

Furthermore, assume that the scattering is isotropic and coherent so that the source function in equation (9.2.27) becomes

$$S_v = \frac{\kappa_v B_v}{\kappa_v + \sigma_v} + \frac{\sigma_v J_v}{\kappa_v + \sigma_v} \tag{9.4.2}$$

Now we integrate equation (9.4.1) over all solid angles, using the form of the source function given by equation (9.4.2), and get

$$\frac{1}{c}\frac{\partial J_v}{\partial t} + \frac{1}{4}\nabla \cdot \vec{F}_v = \kappa_v \rho(B_v - J_v) \tag{9.4.3}$$

This is basically a conservation equation for the radiative flux. Note that the effects of scattering have disappeared from this equation. This is an expression of the conservative nature of scattering. Since no energy is gained or lost in each individual scattering event, the average can contribute nothing to the energy balance for the radiative flux and so all scattering terms must vanish.

a   Radiative Equilibrium and Zeroth Moment of the Equation of Radiative Transfer

Consider the right-hand side of equation (9.4.3). This is essentially the right-hand side of the Boltzmann transport equation, which denotes the creation and destruction of particles in phase space, suitably averaged over direction. Thus, if there is no net production of photons in the atmosphere, this term, integrated over frequency, must be zero. Therefore, integrating equation (9.4.3) over all frequencies, we get

$$\frac{1}{c}\frac{\partial}{\partial t}\int_0^\infty J_v\, dv + \frac{1}{4}\nabla \cdot \int_0^\infty \vec{F}_v\, dv = 0 = \int_0^\infty \kappa_v \rho(B_v - J_v)\, dv \tag{9.4.4}$$

This is a very general statement of radiative equilibrium, and either side of this equation is an equivalent statement of it. If we let $F = \int_0^\infty F_v\, dv$, then for a static plane-parallel atmosphere

$$\frac{dF}{dx} = 0 \qquad \text{or} \qquad F = \text{const} = \frac{\sigma T_e^4}{\pi} \tag{9.4.5}$$

This will serve as a definition of the local effective temperature T_e.

b First Moment of the Equation of Radiative Transfer and the Diffusion Approximation

We multiply equation (9.4.1) [with the source function of equation (9.4.2) replacing the creation rate] by a unit vector $\hat{n}$ pointing in the direction of flow of radiant energy and integrate over all solid angles to obtain

$$\frac{1}{c}\frac{\partial \vec{F}_v}{\partial t} + 4\nabla \cdot \mathbf{K}_v = -\rho(\kappa_v + \sigma_v)\vec{F}_v \tag{9.4.6}$$

Now we make the approximation of near isotropy for the radiation field that was done in equation (9.3.12) and evaluate $J_v(\tau_v)$ from its definition [equation (9.3.1)] to get

$$J_v(\tau_v) = \sum_{i=0}^{\infty} \frac{I_{2i}(\mu, \tau_v)}{2i + 1} \approx I_0 \tag{9.4.7}$$

We have already shown that under similar assumptions $K_v(\tau_v) = I_0/3$, so for conditions of near isotropy

$$K_v(\tau_v) \approx \frac{J_v(\tau_v)}{3} \tag{9.4.8}$$

This is known as the *diffusion approximation*, and it can be used to close the moment equation (9.4.6), yielding

$$\frac{1}{c}\frac{\partial \vec{F}_v}{\partial t} + \frac{4}{3}\nabla J_v = -\rho(\kappa_v + \sigma_v)\vec{F}_v \tag{9.4.9}$$

Now equations (9.4.3) and (9.4.9) can be combined, by utilizing radiative equilibrium [equation (9.4.4)], to produce a "wave equation" for the radiative flux F

$$\frac{1}{3}\nabla(\nabla \cdot \vec{F}) - \frac{1}{c}\frac{\partial}{\partial t}\left[\rho \int_0^{\infty} (\kappa_v + \sigma_v)\vec{F}_v\,dv\right] - \frac{1}{c^2}\frac{\partial^2 \vec{F}}{\partial t^2} = 0 \tag{9.4.10}$$

which has all the properties of the usual wave equation. Such an equation is useful in solving problems in radiative transfer when the boundary conditions change on a time scale comparable to the photon diffusion time through the medium. Such situations may occur in some nebulae, novae and supernovae, or possibly quasars. For stellar atmospheres, the time-independent solutions will generally be sufficient. For a plane-parallel atmosphere in which the radiation field can be viewed as static, equations (9.4.3) and (9.4.9) become, respectively,

$$\frac{dF_v}{dx} = 4\kappa_v\rho(B_v - J_v) \qquad \frac{dJ_v}{dx} = -\frac{3}{4}(\kappa_v + \sigma_v)\rho F_v \qquad (9.4.11)$$

That the static equations will be appropriate for normal stellar atmospheres becomes apparent when we consider that the diffusion time for a photon through a stellar atmosphere is only a few orders of magnitude times the light travel time. An atmosphere is a place from which photons escape after perhaps a few dozen interactions. Normal stars do not change on so short a time scale.

c Eddington Approximation

Although the diffusion approximation provides a method for closing the moment equations of the equation of radiative transfer, it does not allow the complete solution of the problem. The moment equations are, after all, differential equations and are subject to boundary conditions. Specification of these boundary conditions will provide a complete and unique solution for the radiation field. Sir Arthur Stanley Eddington suggested an additional approximation, inspired by the diffusion approximation, that allows for the sufficient specification of boundary conditions to permit the solution of equations (9.4.11).

We consider the situation at the surface, and we assume the emergent radiation field to be isotropic. Since there is generally no incident radiation at the surface of a star, and using the condition of near isotropy given by equation (9.3.12), we get

$$J_v(0) = \frac{1}{2}\int_0^1 I(\mu, 0)\,d\mu = \frac{1}{2}\sum_{i=0}^{\infty}\frac{I_i(0)}{i+1} \approx \frac{1}{2}I_0(0)$$

$$F_v(0) = 2\int_0^1 I(\mu, 0)\mu\,d\mu = 2\sum_{i=0}^{\infty}\frac{I_i(0)}{i+2} \approx I_0(0) \qquad (9.4.12)$$

Hence,

$$J_v(0) = \tfrac{1}{2}F_v(0) \qquad (9.4.13)$$

This and the condition of radiative equilibrium given by equation (9.4.4) provide the two additional constraints necessary to solve equations (9.4.11). For the case of the gray atmosphere (see Section 10.2) a particularly simple solution is given by equation (10.2.15). Although the emergent radiation field is only approximately isotropic, it is the genius of this approximation that the errors introduced by the surface approximation are somewhat offset by the errors incurred by the assumption of the diffusion approximation. Thus, as we shall see later, the Eddington approximation produces solutions for the radiation field that are usually accurate to about 10 percent. As a result, the Eddington approximation is frequently used to solve problems in radiative transfer. To do better, we shall have to do a great deal more.

We have seen that it is possible to describe the flow of radiation through a stellar atmosphere. The derivation involves the same formalisms that we developed in Chapter 1 to describe the flow of matter. The resulting description of this flow is known as the equation of radiative transfer, and it differs significantly from the simple result developed for the study of stellar interiors. The differences point up one of the central differences between stellar interiors and stellar atmospheres. Deep inside a star, the structure of the gas and radiation field is fully determined by the local values of the state variables of the gas. This is not the case in the stellar atmosphere. At any given point in the atmosphere, the local radiation field is composed of photons which originated in an environment that differed significantly from the local environment. Thus, the solution for the equation of radiative transfer locally will depend on the solution everywhere. This global nature of radiative transfer in a stellar atmosphere is one of the central differences between the interior and the outer layers of a star. We now turn our attention to solving the equation of radiative transfer.

Problems

1. Show that the specific intensity along a ray in empty space in constant.

2. Compute the specific intensity and the radiative flux at a distance r on the axis of an emitting disk having radius ρ and temperature T_e. Assume the disk to be located at $r = 0$.

3. Derive the equation of radiative transfer that is appropriate for spherical geometry. List carefully all the assumptions that you make.

4. Derive the plane-parallel equation of radiative transfer appropriate for a dispersive medium with an index of refraction n which is different from unity and which may vary with position.

5. Show that for any diagonal tensor **A**, in spherical coordinates,

$$(\nabla \cdot \mathbf{A})_r = \frac{\partial A_{rr}}{\partial r} + \frac{2A_{rr} - A_{\theta\theta} - A_{\phi\phi}}{r}$$

6. Use the above equation to show that if **K** is a diagonal tensor with all elements equal to K, then

$$(\nabla \cdot \mathbf{K})_r = \frac{\partial K}{\partial r} + \frac{3K - J}{r}$$

Here K, J, and **K** have their usual meanings for radiative transfer [see equations (9.3.1), (9.3.8), and (9.3.14)].

7. Derive equation (9.2.11) from equations (9.2.8) through (9.2.10). Show all your work.

8. Derive equation (9.2.18) from equations (9.2.2), (9.2.8), (9.2.14), (9.2.15), and (9.2.17). Show all your work.

9. Derive equation (9.4.10) from first principles and axioms. Clearly list all assumptions that you make.

Supplemental Reading

A number of books provide an excellent description of the processes taking place in a stellar atmosphere. For excellent, clear, and correct definitions of the quantities that appear in the theory of radiative transfer, see

• Mihalas, D.: *Stellar Atmospheres*, 2d ed., W. H. Freeman, San Francisco, 1978, chap. 1, pp. 1–18.

Strong insight into problems posed by scattering can be found in

• Sobolev, V. V.: *A Treatise on Radiative Transfer* (Trans.: S. I. Gaposchkin), Van Nostrand, Princeton, N.J., 1963, chap. 1, pp. 1–37.

An excellent overall statement of the problem can be found in

• Mustel, E. R.: *Theoretical Astrophysics* (Ed.: V. A. Ambartsumyan, trans.: J. B. Sykes), Pergamon, New York, 1958, pp. 1–8.

To have some feeling for just how long people have been worrying about problems like these and to sample the physical insight of one of the most insightful men of the twentieth century, read

• Eddington, A. S.: "On the Radiative Equilibrium of the Stars," *Mon. Not. R. astr. Soc.* 77, 1916, pp. 16–35.

10

Solution of the Equation of Radiative Transfer

・ ・ ・

One-half of the general problem of stellar atmospheres revolves around the solution of the equation of radiative transfer. Although equation (9.2.11) represents a very general formulation of radiative transfer, clearly the specific nature of the equation of transfer will depend on the geometry and physical environment of the medium through which the radiation flows. The nature of the physical medium will also influence the details of the source function so that the source function may depend on the radiation field itself. Thus, the mode of solution may be expected to be different for the different conditions that exist. However, the notion of plane parallelism is common to so many stars and other physical situations that we spend a significant amount of effort investigating the solution of the equation of transfer for plane-parallel atmospheres.

10.1 Classical Solution to the Equation of Radiative Transfer and Integral Equations for the Source Function

There are basically two schools of approach to the solution of the equation of transfer. One involves the solution of an integral equation for the source function, while the other deals directly with the differential equation of transfer. Both have their merits and drawbacks. Since both are widely used, we give examples of each. Both involve the classical solution, so we begin the discussion with that solution.

a Classical Solution of the Equation of Transfer for the Plane-Parallel Atmosphere

The equation of transfer is a linear differential equation, which implies that a formal solution exists for the radiation field in terms of the source function. This linear property is a marked difference from the situation in stellar interiors where the structure equations were all highly nonlinear. Although under some conditions the solution [that is, $I_\nu(\tau_\nu, \mu)$] itself is involved in the source function, this involvement is still linear. Let us consider a fairly general equation of radiative transfer for a plane-parallel atmosphere, but one where we may neglect time-dependent effects and the presence of the potential gradient on the radiation field:

$$\mu \frac{dI_\nu(\tau_\nu, \mu)}{d\tau_\nu} = I_\nu(\tau_\nu, \mu) - S_\nu(\tau_\nu, \mu) \tag{10.1.1}$$

Since this equation is linear in $I_\nu(\mu, \tau_\nu)$, we may write the complete solution as the sum of the solution to the homogeneous equation plus any particular solution. So let us choose as homogeneous and particular solutions

$$I_h(\mu, \tau_\nu) = c_1 e^{-\tau_\nu/\mu} + c_2 e^{+\tau_\nu/\mu} \qquad I_p(\mu, \tau_\nu) = f(\tau_\nu)e^{\tau_\nu/\mu} \tag{10.1.2}$$

Substitution into the equation of transfer places constraints on c_1 and $f'(\tau_\nu)$, namely,

$$c_1 = 0 \qquad f'(\tau_\nu) = \frac{-S_\nu(\tau_\nu)e^{-\tau_\nu/\mu}}{\mu} \tag{10.1.3}$$

While we have assumed that the geometry of the atmosphere is plane-parallel, we have not yet specified the extent of the atmosphere. For the moment, let us assume that the atmosphere consists of a finite slab of thickness τ_0 (see Figure 10.1).

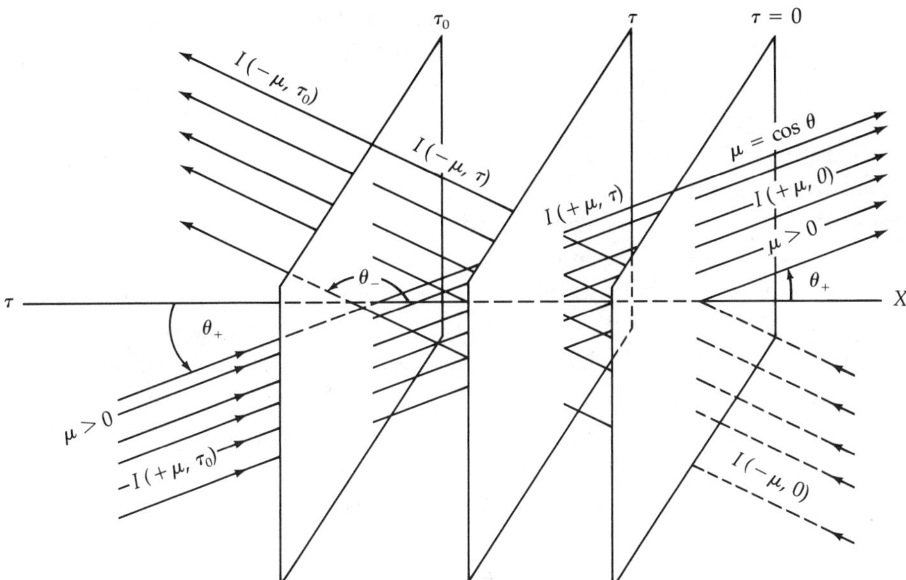

Figure 10.1 Geometry for a plane-parallel slab. Note that there are inward- ($\mu < 0$) and outward-directed ($\mu > 0$) streams of radiation. The boundary conditions necessary for the solution are specified at $\tau_v = 0$ and $\tau_v = \tau_0$.

The general classical solution for the plane-parallel slab is then

$$I_v(\mu, \tau_v) = c_2 e^{\tau_v/\mu} - \int S(t) e^{-(t-\tau_v)/\mu} \frac{dt}{\mu} \tag{10.1.4}$$

Since the equation of transfer is a first-order linear equation, only one constant must be specified by the boundary conditions. However, even though the depth variable τ_v is the only independent variable that appears in a derivative, we must always remember that $I_v(\mu, \tau_v)$ is a function of the angular variable μ. Thus, in general, the constant of integration c_2 will depend on the direction taken by the radiation. For radiation flowing outward in the atmosphere (that is, $\mu > 0$), the constant c_2 will be set equal to the radiation field at the base of the atmosphere [that is, $I_v(\mu, \tau_0)$], and the integral will include the contribution from the source function from all depths ranging from τ_0 to the point of interest τ_v. If we were concerned about radiation flowing into the atmosphere (that is, $\mu < 0$), then the integral in equation (10.1.4) would cover the interval from 0 to τ_v and c_2 would be chosen equal to the incident radiation field [$I_v(-\mu, 0)$].

At this point we encounter one of the notational problems that often leads to confusion in understanding the literature in radiative transfer. For most problems in stellar atmospheres, there is a significant difference between the radiation field represented by the inward-directed streams of radiation and that represented by those flowing outward. In modeling the normal stellar atmosphere, there is no incident radiation present so that the incident intensity $I_\nu(-\mu, 0) = 0$. However, the outward-directed streams always result from a lower boundary condition which is nonzero. Thus it is useful to distinguish between the inward- and outward-directed streams in some notational way. We have already used a standard method of indicating this difference; namely, we explicitly labeled the inward-directed streams by $-\mu$. Thus, we usually regard the angular variable μ as an intrinsically positive quantity that is bounded by $0 < \mu < 1$. The sign of μ must then be explicitly indicated, and we do this when we use this convention. Thus, to gain a physical understanding of the meaning of any solution for the radiation field, one must always keep in mind which streams of radiation are being considered.

The general classical solution for the two streams can then be written as

$$I_\nu(+\mu, \tau_\nu) = -\int_{\tau_0}^{\tau_\nu} S(t)e^{(\tau_\nu - t)/\mu}\,\frac{dt}{\mu} + I_\nu(+\mu, \tau_0)e^{(\tau_\nu - \tau_0)/\mu}$$

$$\mu \geq 0 \quad (10.1.5)$$

$$I_\nu(-\mu, \tau_\nu) = \int_0^{\tau_\nu} S(t)e^{-(\tau_\nu - t)/\mu}\,\frac{dt}{\mu} + I_\nu(-\mu, 0)e^{-\tau_\nu/\mu}$$

While τ_ν represents the vertical depth in the atmosphere increasing inward, τ_ν/μ is the actual path along the direction taken by the radiation. In general, extinction by scattering or absorption will exponentially diminish the strength of the intensity by $e^{\tau/\mu}$. Since the source function represents the local source of photons from all processes, and since it is attenuated by the optical distance along the path of the radiation, the integrand of the integral represents the local contribution of the source function to the value of the intensity at τ_ν. The remaining term simply represents the local contribution to the specific intensity of the attenuated incident radiation.

One further complication must be dealt with before we can use this description of a stellar atmosphere. In general, stellar atmospheres can be regarded as being infinitely thick. Since the influence of the lower boundary diminishes as $e^{\tau - \tau_0}$, and since this optical depth will exceed several hundred within a few thousand kilometers of the surface for main sequence stars, we can take it to be infinity. In addition, we should require the radiative flux to be finite everywhere. This will force the constant c_2 in equation (10.1.4) to vanish. Furthermore, the surface is generally unilluminated. So we can write the classical solution for the semi-infinite plane-parallel atmosphere as

$$I_\nu(+\mu, \tau_\nu) = -\int_\infty^{\tau_\nu} S(t)e^{+(\tau_\nu - t)/\mu}\,\frac{dt}{\mu}$$

$$(10.1.6)$$

$$I_\nu(-\mu, \tau_\nu) = \int_0^{\tau_\nu} S(t)e^{-(\tau_\nu - t)/\mu}\,\frac{dt}{\mu}$$

b Schwarzschild-Milne Integral Equations

One reason that the equation of transfer admits such a simple solution compared to the equations of stellar structure is that we have confined most of the difficult physics to the source function. What is left is largely geometry and hence affords a simple solution. However, the classical solution does allow for the generation of the entire radiation field, should it be possible to specify the source function. It also allows us to remove the explicit structure of the radiation field and to generate an expression for the source function itself. The result is an integral equation, that is, an equation where the unknown appears under the integral sign as well as outside it.

While much attention has been paid to the solution of differential equations, far less has been given to integral equations. However, it is very often numerically more efficient and accurate to solve an integral equation as opposed to the corresponding differential equation. Therefore, we spend some time and effort with these integral equations, for they provide a very productive path toward the solution of problems in radiative transfer.

Integral Equation for the Source Function   In Chapter 9 we showed that, for coherent isotropic scattering, we could write a quite general expression for the source function [equation (9.2.33)]. If we reexpress that result in terms of the mean intensity, we get

$$S_\nu(\tau_\nu) = \epsilon_\nu B_\nu[T(\tau_\nu)] + (1 - \epsilon_\nu)J_\nu(\tau_\nu) \qquad (10.1.7)$$

where

$$\epsilon_\nu = \frac{\kappa_\nu}{\kappa_\nu + \sigma_\nu} \qquad (10.1.8)$$

Now the role of the classical solution becomes evident. The source function contains the mean intensity $J_\nu(\tau_\nu)$, which can be generated from the classical solution that contains the source function itself. Thus, if we substitute the classical solution [equation (10.1.6)] into the definition for $J_\nu(\tau_\nu)$ [equation (9.3.2)], we get

$$J_\nu(\tau_\nu) = \frac{1}{2}\int_0^1 I(+\mu', \tau_\nu)\,d\mu' + \frac{1}{2}\int_0^1 I(-\mu', \tau_\nu)\,d(-\mu')$$

$$J_\nu(\tau_\nu) = \frac{1}{2}\int_0^1 \left[\int_\infty^{\tau_\nu} S(t)e^{+(\tau_\nu - t)/\mu'}\,\frac{dt}{\mu'}\right]d\mu' \qquad (10.1.9)$$

$$+ \frac{1}{2}\int_0^1 \left[\int_0^{\tau_\nu} S(t)e^{-(\tau_\nu - t)/\mu'}\,\frac{dt}{\mu'}\right]d\mu'$$

Now notice that the argument of the exponential is always negative and that the two integrals over t are contiguous. Thus, we can combine these integrals into a single integral that ranges from 0 to ∞. In addition, t and μ are independent variables so that we may interchange the order of integration and get

$$J_\nu(\tau_\nu) = \frac{1}{2} \int_0^\infty S(t) \left[\int_0^1 e^{-|\tau_\nu - t|/\mu'} \frac{d\mu'}{\mu'} \right] dt \qquad (10.1.10)$$

The quantity in brackets is a well-known function in mathematical physics known as the *exponential integral*. It depends only on the independent variables of the problem and therefore can be regarded as a largely geometric function. Its formal definition is

$$E_n(z) = \int_1^\infty \frac{e^{-zt}\,dt}{t^n} = \int_0^1 e^{-z/y} y^{n-2}\,dy = \int_0^1 \frac{e^{-z/y} y^{n-1}\,dy}{y} \qquad (10.1.11)$$

and when expressed by the final integral, it has the same form as the integral in brackets in equation (10.1.10). While the exponential integral may not be terribly familiar, it should be regarded with no more fear and trepidation than sines and cosines. There is an entire set of these functions where each member is denoted by n, and they have a single argument, which for our purposes will be confined to the real line. These functions (except for the first exponential integral at the origin) are well behaved and resemble $e^{-x}/(nx)$ for large x. Some useful properties of exponential integrals are

$$nE_{n+1}(x) = e^{-x} - xE_n(x) \qquad n > 1$$

$$\frac{dE_{n+1}(x)}{dx} = -E_n(x) \qquad\qquad (10.1.12)$$

$$E_n(0) = \frac{1}{n-1} \qquad E_n(\infty) = 0$$

Making use of the first exponential integral, we can rewrite our expression for the mean intensity [equation (10.1.10)] as

$$J_\nu(\tau_\nu) = \tfrac{1}{2} \int_0^\infty S(t) E_1 |\tau_\nu - t|\,dt \qquad (10.1.13)$$

Combining this with equation (10.1.7) for the source function, we arrive at the desired integral equation for the source function:

$$S_\nu(\tau_\nu) = \epsilon_\nu B_\nu[T(\tau_\nu)] + (1 - \epsilon_\nu)\tfrac{1}{2} \int_0^\infty S_\nu(t) E_1 |\tau_\nu - t|\,dt \qquad (10.1.14)$$

Any function that multiplies the unknown in the integrand of an integral equation is called the *kernel* of the integral equation. Thus, the first exponential integral is the kernel of the integral equation for the source function. The connection between the physical state of the gas and the source function is contained in the term that makes the equation inhomogeneous, namely, the one involving the Planck function $B_\nu[T(\tau_\nu)]$. A solution of this equation, when combined with the classical solution, will yield the full solution to the radiative transfer problem since $I_\nu(\mu, \tau)$ will be specified for all values of μ and τ_ν.

It is possible to understand equation (10.1.14) from a physical standpoint. Now $\epsilon(\tau_\nu)$ is the fraction of locally generated photons that arise from thermal processes, so that the first term is simply the local contribution to the source function from thermal properties of the gas. The second term represents the contribution from scattering. We have already said that a fundamental aspect of stellar atmospheres is the dependence of the local radiation field on the global solution for the radiation field. Nowhere is this more clearly demonstrated than in this term. The scattering contribution to the source function is made up of contributions from the source function throughout the atmosphere. However, these contributions decline with increasing distance from the point of interest, and they decline roughly exponentially.

One may object that this integral equation is a very specialized equation since it relies on the source function's being expressible in terms of the mean intensity and therefore is valid only for isotropic scattering. However, consider the very general expression for the source function given by equation (9.2.27). As long as the angular dependence of the redistribution function is known, it will be possible to carry out the integrals over solid angle and express the source function as a combination of the moments of the radiation field. As long as this can be done, the appropriate moments can be generated from the classical solution for the equation of transfer which will, in turn, involve only the source function. Thus, the moments can be eliminated from the moment expression for the source function, yielding an integral equation. To be sure, this will be a more complicated integral equation, but it will still be solvable by the same techniques that we shall apply to equation (10.1.14). Thus, the existence of an integral equation for the source function is a quite general result and represents the separation of the depth dependence of the radiation field from the angular dependence, which can be obtained from the classical solution.

Integral Equations for Moments of the Radiation Field Useful as the integral equation for the source function is, it is often convenient to have similar expressions for the moments of the radiation field. We should not be surprised that such expressions exist since the angular moments are free, by definition, of the angular dependence characteristic of the classical solution. Indeed, we have already supplied the required expressions to obtain an integral equation for the 259 mean intensity. We simply use equation (10.1.7) to eliminate $S_\nu(t)$ from equation (10.1.13), and we have

$$J_\nu(\tau_\nu) = \tfrac{1}{2} \int_0^\infty \epsilon_\nu(t) B_\nu[T(t)] E_1 |\tau_\nu - t| \, dt$$

$$+ \tfrac{1}{2} \int_0^\infty [1 - \epsilon_\nu(t)] J_\nu(t) E_1 |\tau_\nu - t| \, dt \qquad (10.1.15)$$

It is now clear how to develop similar expressions for the remaining moments, since equation (10.1.13) was obtained by taking moments of the classical solution to the equation of transfer. Let us define an operator which is commonly used to represent this process:

$$\Lambda_n(\tau_\nu) |G(t)| \equiv \int_0^\infty \left(\frac{(t - \tau_\nu)}{|t - \tau_\nu|} \right)^{n+1} E_n |\tau_\nu - t| G(t) \, dt \qquad (10.1.16)$$

The Λ_n operator is an integral operator which operates on a function by employing an exponential integral kernel. The term in large parentheses simply denotes the sign of the kernel throughout the region. With this integral operator, we can express the first three moments of the radiation field in terms of the source function as follows:

$$J_\nu(\tau_\nu) = \tfrac{1}{2}\Lambda_1(\tau_\nu) |S_\nu(t)|$$

$$H_\nu(\tau_\nu) = \tfrac{1}{2}\Lambda_2(\tau_\nu) |S_\nu(t)| \qquad (10.1.17)$$

$$K_\nu(\tau_\nu) = \tfrac{1}{2}\Lambda_3(\tau_\nu) |S_\nu(t)|$$

Such equations are known as *Schwarzschild-Milne type of equations* and are extremely useful for the construction of model stellar atmospheres. For example, consider the condition of radiative equilibrium where it is necessary to know the radiative flux throughout the atmosphere, but not the complete radiation field. This information can be obtained directly with the aid of the flux equation of equations (10.1.17) and the source function. Thus, determination of the source function provides a complete solution of the radiative transfer problem.

c Limb-Darkening in a Stellar Atmosphere

There is one property of the classical solution of the equation of transfer that we should address before moving on. If we consider the classical solution for the emergent intensity, we see that it basically represents the Laplace transform of the source function, namely

$$I(\mu, 0) = \int_0^\infty S(t) e^{-t/\mu} \frac{dt}{\mu} = \frac{\mathscr{L}[S(t)]}{\mu} \tag{10.1.18}$$

where $\mathscr{L}[S(t)]$ is the Laplace transform of the source function. Thus determination of the angular distribution of the emergent intensity is equivalent to determining the behavior of the source function with depth. Since the source function is determined by the temperature, determination of the depth dependence of the source function is equivalent to determining the depth dependence of the temperature. This is of considerable significance for stars where this dependence can be measured directly for it provides a direct observational check on the models of those stellar atmospheres.

If we anticipate some later results and assume that the source function can be approximated by

$$S(t) = at + b \tag{10.1.19}$$

then

$$I(\mu, 0) = a\mu + b \tag{10.1.20}$$

Thus, the coefficient a that multiplies the angular parameter μ in the emergent intensity is a direct measure of the source function gradient, while the constant term b denotes the value of the source function at the boundary. The decrease in brightness as one approaches the limb of the apparent stellar disk implied by equation (10.1.20) is called *limb-darkening*. Since for spherical stars the variation across the apparent disk is the same as the local angular dependence of the emergent intensity, measurement of the limb-darkening coefficient a yields a measurement of the source function gradient. This is of particular interest for the sun where such measurements are possible. Unfortunately, the poorest theoretical representation of the model atmosphere occurs near the surface, and this corresponds to just that region of the stellar disk (i.e., near the limb where $\mu \to 0$) where confirmatory measurements are most difficult to make. Although we have made an approximation to the depth dependence of the source function in equation (10.1.19), the approximation is unnecessary and more rigorous studies of this depth dependence would deal directly with the Laplace transform itself as given by equation (10.1.18). We have now compiled methods by which we can theoretically relate the emergent intensity to the source function and provided a potential observational method to verify our result. However, before discussing methods for the solution for the integral equation for the source function [equation (10.1.14)], we consider the solutions to a somewhat simpler problem, in order to gain an appreciation for the behavior of these solutions.

Empirical Determination of $T(\tau_\nu)$ for the Sun   In the sun and some eclipsing binary stars, it is possible to determine the variation of the specific intensity across the apparent disk. If we approximate that variation by

$$\frac{I_\nu(0, \mu)}{I_\nu(0, 1)} \approx \sum_{i=0}^{n} a_i \mu^i \tag{10.1.21}$$

we can use equation (10.1.18) to obtain a power series representation of the source function with optical depth. Let us further assume that the source function can be represented by the Planck function, which in turn can be expanded in a power series in the optical depth so that

$$\frac{S_\nu(\tau_\nu)}{I_\nu(0, 1)} \approx \frac{B_\nu[T(\tau_\nu)]}{I_\nu(0, 1)} = \sum_{i=0}^{n} \beta_i \tau_\nu^i = \sum_{i=0}^{n} \beta_i \left(\frac{\tau_\nu}{\mu}\right)^i \mu^i \tag{10.1.22}$$

Then the substitution of this power series representation into equation (10.1.18) yields

$$\sum_{i=0}^{n} \beta_i i! \, \mu^i = \sum_{i=0}^{n} a_i \mu^i \tag{10.1.23}$$

Since for the sun the a_i's and $I_\nu(0, 1)$ may be determined from observation, the β_i's may be regarded as known. Thus, the temperature variation with monochromatic optical depth may be recovered from

$$\frac{2h\nu^3/c^2}{e^{h\nu/[kT(\tau_\nu)]} - 1} = I_\nu(0, 1) \sum_{i=0}^{n} \beta_i \tau_\nu^i \tag{10.1.24}$$

In the sun, the assumption that $S_\nu(\tau_\nu) = B_\nu(\tau_\nu)$ is a particularly good one, so that for the sun the optical depth variation of the temperature can be determined with the same sort of accuracy that attends the determination of the limb-darkening.

Empirical Determination of $\kappa(\tau_1)/\kappa(\tau_2)$ for the Sun   This type of analysis can be continued under the above assumptions to obtain the variation with optical depth of the ratio of two monochromatic absorption coefficients. Since by definition

$$d\tau_\nu = -\kappa_\nu \rho \, dx \tag{10.1.25}$$

the ratio of two monochromatic optical depths is

$$\frac{d\tau_1}{d\tau_2} = \frac{\kappa_1(\tau_\nu)}{\kappa_2(\tau_\nu)} \tag{10.1.26}$$

Differentiating equation (10.1.22) with respect to temperature and substituting the result into equation (10.1.26), we get

$$\frac{\kappa_1(\tau_\nu)}{\kappa_2(\tau_\nu)} = \frac{I_2(0, 1)}{I_1(0, 1)} \frac{dB(\nu_1)/dT}{dB(\nu_2)/dT} \frac{\sum_i i\beta_i(\tau_2)\tau_2^{(i-1)}}{\sum_i i\beta_i(\tau_1)\tau_1^{(i-1)}} \tag{10.1.27}$$

Thus it is possible to determine the approximate wavelength dependence of the opacity for stars like the sun from the observed limb-darkening. Such observations provide a valuable check on the theory of stellar atmospheres.

10.2 Gray Atmosphere

For the better part of this century, theoretical astrophysicists have been concerned with the solution to an idealized radiative transfer problem known as the *gray atmosphere*. Although it is an idealized situation, it has some counterparts in nature. In addition, this problem possesses the virtue that a complete solution can be obtained for the radiation field without recourse to the physical details of the atmosphere. In this regard, the gray atmosphere model is rather like polytropic models for stellar interiors. As was the case for polytropes and stellar interiors, we may expect to gain significant insight into the properties of stellar atmospheres by understanding the solution to the gray atmosphere problem. The additional assumption required to turn our study of radiative transfer into that of a gray atmosphere is simple. Assume that the opacity, whether it be absorption or scattering, is independent of frequency. Thus, any frequency can be treated as any other frequency, as far as the radiative transfer is concerned. This independence of the radiative transfer from frequency has the interesting consequence that the mathematical solution to the equation of transfer for any frequency will be the solution for all frequencies, and thus must be the solution for the sum of all frequencies. Hence, the aspect of the solution that specifies the radiative flux also refers to the total flux, making the condition of radiative equilibrium relatively simple to apply. Since all aspects of the mathematical description are independent of frequency, we drop the subscript ν for the balance of this discussion.

Knowing what we do about the physical processes of absorption, it is reasonable to ask if the gray atmosphere is anything more than an interesting mathematical exercise. Certainly bound-bound transitions are anything but gray. However, there are some bound-free transitions that exhibit only weak frequency dependence over substantial regions of the spectrum. If those regions of the spectrum correspond to that part of the spectrum containing most of the radiant flux, then the atmosphere will be very similar to a gray atmosphere. Absorption due to the H^- ion is relatively frequency-independent throughout the visible part of the spectrum and in some stars is the dominant source of opacity. However, the premier example of a gray opacity source is electron scattering. Thomson scattering by free electrons is frequency-independent by definition, and for stars hotter than about 25,000 K, it is the dominant source of opacity throughout the range of wavelengths encompassing the maximum flow of energy. Thus, the early O and B stars have atmospheres that, to a very high degree, may be regarded as gray.

Since frequency dependence has been removed from the problem, we may write the equation of radiative transfer for a plane-parallel static atmosphere as

$$\mu \frac{dI(\tau, \mu)}{d\tau} = I(\tau, \mu) - S(\tau) \qquad (10.2.1)$$

where, for isotropic coherent scattering, the source function is

$$S(\tau) = \frac{\kappa B + \sigma J}{\kappa + \sigma} \qquad (10.2.2)$$

Now the independence of the opacity on frequency makes the condition of radiative equilibrium given by equation (9.4.4) particularly simple:

$$\int_0^\infty \kappa_v \rho (B_v - J_v) \, dv = 0 = B - J \qquad (10.2.3)$$

or simply

$$B = J \qquad (10.2.4)$$

Substitution of this result into equation (10.2.2) yields

$$S(\tau) = B[T(\tau)] = J(\tau) \qquad (10.2.5)$$

The fact that the mean intensity is equal to the Planck function and that either can be taken to be the source function has the interesting result that the solution to the gray atmosphere is independent of the relative roles of scattering and absorption. Thus, the radiation field for a pure absorbing gray atmosphere, where the source function is clearly the Planck function, will be indistinguishable from the radiation field of a pure scattering gray atmosphere. In addition, since there is a general independence on frequency, the spectral energy distribution will be that resulting from a gray atmosphere where the source function is the Planck function.

The gray atmosphere implies that all the development of Chapters 9 and 10 will apply at each frequency. This is indeed the easiest way to obtain equations (10.2.1) through (10.2.4). But there is much more. The integral equation for the source function [equation (10.1.14)] and that for the moments of the radiation field [equations (10.1.17)] become

$$B(\tau) = \tfrac{1}{2} \int_0^\infty B(t) E_1 |t - \tau| \, dt \qquad J(\tau) = \tfrac{1}{2} \int_0^\infty J(t) E_1 |t - \tau| \, dt$$

$$H(\tau) = \tfrac{1}{2} \int_\tau^\infty B(t) E_2(t - \tau) \, dt - \tfrac{1}{2} \int_0^\tau B(t) E_2(\tau - t) \, dt \qquad (10.2.6)$$

$$K(\tau) = \tfrac{1}{2} \int_0^\infty B(t) E_3 |t - \tau| \, dt$$

Solution of these equations, combined with the classical solution to the equation of transfer, yields a complete description for the radiation field at all depths in the atmosphere. The method of solution for the gray atmosphere equation of transfer is also illustrative of the methods of solution for the more general nongray problem.

a   Solution of Schwarzschild-Milne Equations for the Gray Atmosphere

In general, an accurate solution of these equations must be accomplished numerically because the solution, even for the gray atmosphere, is not analytic everywhere. Particular care must be taken with these equations because the first exponential integral behaves badly as its argument approaches zero. Specifically

$$\lim_{x \to 0} E_1(x) = -\ln x \to \infty \qquad (10.2.7)$$

Thus, the kernel of first two of equations (10.2.6) has a singularity when $t = \tau$. However, this singularity is integrated over, and the integral is finite and well behaved. For years this singularity was regarded as an insurmountable barrier, and interest in the solution of the integral equations of radiative transfer languished in favor of more direct methods applicable to the differential equation of transfer itself. However, the singularity of the kernel is not an essential one and may be easily removed. Simply adding and subtracting the solution $B(\tau)$ from the right-hand side of the first of equations (10.2.6) yields

$$B(\tau) = \tfrac{1}{2} \int_0^\infty [B(t) - B(\tau)] E_1 |t - \tau| \, dt + \tfrac{1}{2} B(\tau) \int_0^\infty E_1 |t - \tau| \, dt \qquad (10.2.8)$$

The integrand of the first of these integrals is now well behaved for all values of t since $[B(t) - B(\tau)]$ will go to zero faster than the exponential integral diverges as $t \to \tau$. The only condition placed on the solution is that $B(\tau)$ satisfy a Lipschitz condition which is a weaker condition than requiring the solution to be continuous. The second integral is analytic and can be evaluated by using the properties of exponential integrals given in equations (10.1.12). This approach yields a slightly different integral equation, but one that has a well behaved integrand:

$$B(\tau) = \frac{\int_0^\infty [B(t) - B(\tau)]E_1|t - \tau|\,dt}{E_2(\tau)} \qquad (10.2.9)$$

A simple way to deal with this type of integral equation is to replace the integral with some standard numerical quadrature formula. While Simpson's rule enjoys a great popularity, a gaussian-type quadrature scheme offers much greater accuracy for the same number of points of evaluation of the integrand. When the integral is so replaced, we obtain

$$B(\tau) = \frac{\sum_{i=0}^{n} [B(t_i) - B(\tau)]E_1|t_i - \tau|W_i}{E_2(\tau)} \qquad (10.2.10)$$

which is a functional equation for $B(\tau)$ in terms of the solution at a discrete set of points t_i. The quantities W_i are just the weights of the quadrature scheme appropriate for the various points t_i. Evaluating the functional equation for $B(\tau)$ with τ equal to each value of t_j and rearranging terms, we can obtain a system of linear algebraic equations for the solution at the specific points t_i:

$$\sum_{k=1}^{n} B(t_k)\left[\sum_{i=1}^{n} \frac{(\delta_{ik} - \delta_{jk})E_1|t_i - t_j|W_i}{E_2(t_j)} - \delta_{kj}\right] = 0 \qquad j = 1, \ldots, n \quad (10.2.11)$$

The term governed by the summation over i depends only on the type of quadrature scheme chosen, and so equation (10.2.11) represents n linear homogeneous algebraic equations that have the standard form

$$\sum_{k=1}^{n} B(t_k)A_{kj} = 0 \qquad j = 1, \ldots, n \qquad (10.2.12)$$

The fact that these equations are homogeneous points out an observation made earlier. For the gray atmosphere, the radiation field is decoupled from the values of the physical state variables. Thus, the homogeneous equations constitute an eigenvalue problem, and, as we shall see later, the eigenvalue is the value of the total radiative flux or alternately the effective temperature. One approach to the solution of equation (10.2.12) would be to define a new set of variables $B(t_i)/B(t_1)$, say, and to generate a system of inhomogeneous equations that can then be solved for the ratio of the source function to its value at one of the given points. Once the source function (or its ratio) has been found at the discrete points t_i, the solution can be obtained everywhere by substitution into equation (10.2.10). Since this is a functional equation, the results will have the same level of accuracy as that obtained for the values of $B(t_i)$. To achieve a level of accuracy significantly greater than that offered by the Eddington approximation, we will have to use a particularly accurate quadrature formula. Also the exponential nature of the exponential integral implies that the quadrature scheme should be chosen with great care.

b Solutions for the Gray Atmosphere Utilizing the Eddington Approximation

We have already seen that the diffusion approximation yields moment equations from the equation of transfer given by equation (9.4.11). For the gray atmosphere, these take the particularly simple form

$$\frac{dF}{d\tau} = 0 \qquad \frac{dJ}{d\tau} = \frac{3}{4} F \tag{10.2.13}$$

The first is a statement of radiative equilibrium which says that for a gray atmosphere F_ν is constant, and its integrated value can be related to the effective temperature. The second equation is immediately integrable, yielding a constant of integration. Thus,

$$F(\tau) = \text{const} = \frac{\sigma T_e^4}{\pi} \qquad J(\tau) = \tfrac{3}{4} F\tau + \text{const} \tag{10.2.14}$$

Using the Eddington approximation as given by equation (9.4.13), we can evaluate the constant and arrive at the dependence of the mean intensity with depth in the atmosphere:

$$J(0) = \tfrac{1}{2} F = \text{const} \qquad J(\tau) = \tfrac{3}{4} F(\tau + \tfrac{2}{3}) \tag{10.2.15}$$

Remembering that $J = S = B$ for a gray atmosphere in radiative equilibrium, we find that the temperature of the atmosphere should vary as

$$\left[\frac{T(\tau)}{T_e}\right]^4 = \tfrac{3}{4}(\tau + \tfrac{2}{3}) \tag{10.2.16}$$

Thus, we see that at large depths, where we should expect the diffusion approximation to yield accurate results, the source function becomes linear with depth. Also, when $\tau = \frac{2}{3}$, the local temperature equals the effective temperature. So, in some real sense, we can consider the optical "surface" to be located at $\tau = \frac{2}{3}$. This is the depth from which the typical photon emerges from the atmosphere into the surrounding space. Only at depths less than $\frac{2}{3}$ does the source function begin to depart significantly from linearity with depth. Unfortunately, this is the region in which most of the spectral lines that we see in stellar spectra are formed. Thus, we will have to pay special attention to that part of the atmosphere lying above optical depth $\frac{2}{3}$.

We may check on the accuracy of the Eddington approximation by seeing how well this solution reproduces the surface boundary condition that it assumes. Using the definition for the mean intensity, the classical solution for the equation of transfer [equation (10.1.5)], and the fact that the source function is J itself, we obtain

$$J(0) = \frac{1}{2} \int_0^1 I(\mu, 0)\, d\mu = \frac{1}{2} \int_0^1 \int_0^\infty J(t) e^{-t/\mu} \frac{dt}{\mu}\, d\mu = \frac{7F}{16} \qquad (10.2.17)$$

So the Eddington approximation fails to be self-consistent by about 1 part in 8 (or 12.5 percent) in reproducing the surface value for the flux. To improve on this result, we will have to take a rather more complicated approach to the radiative problem.

c Solution by Discrete Ordinates: Wick-Chandrasekhar Method

The following method for the solution of radiative transfer problems has been extensively developed by Chandrasekhar,[1] and we only briefly sketch it and its implications here. The method begins by noting that if one takes the source function to be the mean intensity J, then the equation of transfer can be written in terms of the specific intensity alone. However, the resulting equation is an integrodifferential equation. That is, the intensity, which is a function of the two variables μ and τ, appears differentiated with respect to one of them and is integrated over the other. Thus,

$$\mu \frac{dI(\mu, \tau)}{d\tau} = I(\mu, \tau) - \frac{1}{2} \int_{-1}^{+1} I(\mu', \tau)\, d\mu' \qquad (10.2.18)$$

Now, as we did in the integral equation for the source function, we can replace the integral by a quadrature summation so that

$$\mu \frac{dI(\mu, \tau)}{d\tau} = I(\mu, \tau) - \frac{1}{2} \sum_{j=1}^{n} I(\tau, \mu_j) a_j \qquad (10.2.19)$$

Here the a_j values are the weights of the quadrature scheme. This is a functional differential equation for $I(\tau, \mu)$ in terms of the solution at certain discrete values of μ_i. Chandrasekhar[1] is very explicit about using a gaussian quadrature scheme; a scheme that yields exact answers for polynomials of degree $2n - 1$ or less utilizes the zeros of the Legendre polynomials of degree n as defined in the interval -1 to $+1$. A more accurate procedure is to divide the integral in equation (10.2.18) into two integrals, one from -1 to 0 and the other from 0 to $+1$, and to approximate these integrals separately. The reason is that since there is no incident radiation, the intensity develops a discontinuity in μ at $\tau = 0$. Numerical quadrature schemes rely on the function to be integrated, in this case $I(\mu, \tau)$, being well approximated by a polynomial throughout the range of the integral. Splitting the integral at the discontinuity allows the resulting integrals to be well approximated where the single integral could not be. This procedure is sometimes called the *double-gauss quadrature scheme*. However, this "engineering detail" in no way affects the validity of the basic approach.

As we did with equation (10.2.10), we evaluate the functional equation of transfer [equation (10.2.19)] at the same values of μ as are used in the summation so that

$$\mu_i \frac{dI(\tau, \mu_i)}{d\tau} = I(\tau, \mu_i) - \frac{1}{2} \sum_{j=1}^{n} I(\tau, \mu_j)a_j \qquad (10.2.20)$$

We now have a system of n homogeneous linear differential equations for the functions $I(\tau, \mu_i)$. Each of these functions represents the specific intensity along a particular direction specified by the value of μ_i. Since the value $\mu_i = 0$ represents the point of discontinuity in $I(\mu, \tau)$ at the surface, this value should be avoided. Thus, there will normally be as many negative values of μ_i as positive ones. To solve the problem, we must find n constants of integration for the n first-order differential equations.

Inspired by the general exponential attenuation of a beam of photons passing through a medium, let us assume a solution of the form

$$I(\tau, \mu_i) = g_i e^{-k\tau} \qquad (10.2.21)$$

Substitution of this form into this set of linear differential equations (10.2.20), will satisfy the equations if

$$g_i = \frac{\text{const}}{1 + k\mu_i} \qquad (10.2.22)$$

and k satisfies the eigenvalue equation

$$1 = \sum_{j=1}^{n/2} \frac{a_j}{1 - \mu_j^2 k^2} \qquad \mu_j > 0 \qquad (10.2.23)$$

Thus equation (10.2.22) provides a constant of integration for every distinct value of k. Since in all quadrature schemes the sum of the weights must equal the interval, $k^2 = 0$ will satisfy equation (10.2.23). Thus, since equation (10.2.23) is essentially polynomic in form, there will be $n/2 - 1$ distinct nonzero values of k^2 and thus $n - 2$ distinct nonzero values of k, which we denote as $\pm k_\alpha$. When these are combined with the value $k = 0$, we are still missing one constant of integration. Wick, inspired by the Eddington approximation, suggested a solution of the form

$$I(\tau, \mu_i) = b(\tau + q_i) \tag{10.2.24}$$

Substitution of this form into equations (10.2.20) also satisfies the equation of transfer provided that

$$q_i = \mu_i + Q \tag{10.2.25}$$

The product constant bQ can be identified with the constant obtained from $k^2 = 0$, so it cannot be regarded as a new constant of integration; but the term $b\tau$ can be regarded as such and therefore completes the solution, so that

$$I(\tau, \mu_i) = b\left[\sum_{\alpha=1}^{m} \left(\frac{L_{+\alpha}e^{-k_\alpha\tau}}{1 + \mu_i k_\alpha} + \frac{L_{-\alpha}e^{+k_\alpha\tau}}{1 - \mu_i k_\alpha} \right) + \mu_i + \tau + Q \right] \tag{10.2.26}$$

where

$$m = \frac{n}{2} - 1 \tag{10.2.27}$$

and the values of μ_i range from -1 to $+1$. The constants $L_{\pm\alpha}$ are the constants that result from equation (10.2.22) and the distinct values of k_α.

Moments of the Radiation Field from Discrete Ordinates   We can generate the moments of the radiation field at a level of approximation which is consistent with the solution given by equation (10.2.26) by using the same quadrature scheme for the evaluation of the integrals over μ that was used to replace the integral in the integrodifferential equation of radiative transfer. Thus,

$$J(\tau) = \tfrac{1}{2} \int_{-1}^{+1} I(\tau, \mu)\, d\mu = \tfrac{1}{2} \sum_{i=1}^{n} I(\tau, \mu_i) a_i \tag{10.2.28}$$

We already have the values $I(\tau, \mu_i)$ required to evaluate the resulting sums. For the gaussian quadrature schemes suggested, the a_i's are symmetrically distributed in the interval -1 to $+1$, while the μ_i's are antisymmetrically distributed. Making use of these facts, substituting the solution [equation (10.2.26)] into equation (10.2.28), and manipulating, we get

$$J(\tau) = b\left[\sum_{\alpha=1}^{m}\left(L_{+\alpha}e^{-k_{\alpha}\tau} + L_{-\alpha}e^{+k_{\alpha}\tau}\right) + \tau + Q\right] \tag{10.2.29}$$

Following the same procedure for the flux, we get

$$F(\tau) = \frac{4b}{3} = \text{const} \tag{10.2.30}$$

so that the constant b of the Wick solution is related to the constant flux. All that remains to complete the solution is to determine the constants $L_{\pm\alpha}$ from the boundary conditions.

Application of Boundary Values to the Discrete Solution   At no point in the derivation have we used the fact that the atmosphere is assumed semi-infinite. So, in principle, the solution given by equation (10.2.26) is correct for finite slabs. Some applications of the approach have been used in the study of planetary atmospheres, and so for generality let us consider the application to an atmosphere which has a finite thickness τ_0. For such an atmosphere, we must know the distribution of the intensity entering the atmosphere at the base τ_0 as well as that which is incident on the surface. Given that, it is a simple matter to equate the solution [equation (10.2.26)] to the boundary values, and we get

$$I(-\mu_i, 0) = \frac{3}{4}F\left[\sum_{\alpha=1}^{m}\left(\frac{L_{+\alpha}}{1 - \mu_i k_{\alpha}} + \frac{L_{-\alpha}}{1 + \mu_i k_{\alpha}}\right) - \mu_i + Q\right]$$

$$I(+\mu_i, \tau_0) = \frac{3}{4}F\left[\sum_{\alpha=1}^{m}\left(\frac{L_{+\alpha}e^{-k_{\alpha}\tau_0}}{1 + \mu_i k_{\alpha}} + \frac{L_{-\alpha}e^{+k_{\alpha}\tau_0}}{1 - \mu_i k_{\alpha}}\right) + \mu_i + \tau_0 + Q\right]$$

$$i = 1, \ldots, \frac{n}{2} \tag{10.2.31}$$

These equations represent n equations in n unknowns. There are $2n - 2$ values of $L_{\pm\alpha}$, F, and Q, all specified by the n values of the boundary intensity. Here we explicitly incorporated the sign of μ_i into the equation so that all values of μ_i should be taken to be positive. Although the equations are effectively linear in the unknowns, note that the coefficients of those equations grow exponentially with optical depth. Indeed, since the nonzero values of k_{α} are all greater than unity, that growth is quite rapid. In practice, it is virtually impossible to solve these equations for any value of $\tau_0 > 100$. Indeed, if the order of approximation is large, the practical upper limit is near 10. This instability is inherent in all discrete ordinate methods used for finite atmospheres.

The reason is fairly straightforward. Each of the k_α's corresponds to a stream of radiation with a particular value of μ_i. The total optical path for this radiation stream is τ_0/μ_i. Since the solution of equation (10.2.26) is essentially a linear two-point boundary-value problem, the solution at one boundary is determined by the solution at the other boundary. If part of the solution at one boundary is optically remote from the other boundary, it will decouple from the solution, causing the solution to become singular or poorly determined. Physically, the photons from the remote boundary have been so randomized by scatterings or absorptions that all information pertaining to their direction of entrance into the atmosphere has been lost. In the case of the semi-infinite atmosphere, this has explicitly been taken into account, and the information from the lower boundary is contained in the finite and constant radiative flux.

We can see the effect of this constraint on the discrete solution by examining the behavior of the solution [equation (10.2.31)] as $\tau_0 \to \infty$. Since we require the radiation field to remain finite as $\tau_0 \to \infty$, the $L_{-\alpha}$'s must go to zero. Thus, the influence of the deep radiation field explicitly disappears from the solution, and the radiative flux becomes the eigenvalue of the problem. So the complete solution for the semi-infinite gray atmosphere for the method of discrete ordinates is

$$B(\tau) = J(\tau) = \frac{3}{4} F\left(\sum_{\alpha=1}^{m} L_{+\alpha} e^{-k_\alpha \tau} + \tau + Q \right)$$

$$I(-\mu_i, 0) = 0 = \frac{3}{4} F\left(\sum_{\alpha=1}^{m} \frac{L_{+\alpha}}{1 - \mu_i k_\alpha} - \mu_i + Q \right)$$

$$F = \text{const} = \frac{\sigma T_e^4}{\pi} \tag{10.2.32}$$

$$1 = \sum_{j=1}^{m+1} \frac{a_j}{1 - \mu_j^2 k_\alpha^2}$$

Table 10.1 contains some values of $L_{+\alpha}$, k_α, and Q for various orders of approximation for the semi-infinite gray atmosphere for the single-gauss quadrature scheme. By analogy to the Eddington approximation, the source function is sometimes written as

$$J(\tau) = \tfrac{3}{4} F[\tau + q(\tau)] \tag{10.2.33}$$

where

$$q(\tau) = Q + \sum_{\alpha=1}^{m} L_{+\alpha} e^{-k_\alpha \tau} \tag{10.2.34}$$

is known as the *Hopf function*. It is clear that for the Eddington approximation the appropriate Hopf function would be $q(\tau) = \tfrac{2}{3}$. The Eddington approximation also

Table 10.1 Values of the Eigenvalues k_α and Integration Constants L_α and Q for the Semi-infinite Plane-Parallel Gray Atmosphere

			k_α		
$n/2$	2	4	10	20	
Q	0.69402480	0.70691789	0.70991539	0.71031562	$/\alpha$
	1.972027	1.103185	1.012222	1.002743	1
	xxxxxxxx	1.591779	1.054805	1.012774	2
19	−2.0932	4.458086	1.133904	1.028846	3
18	−1.7809		1.263385	1.052985	4
17	−1.5440		1.471808	1.085545	5
16	−1.3454		1.822337	1.127770	6
15	−1.1717		2.481124	1.181417	7
14	−1.0164		4.059775	1.248953	8
13	−0.87584	xxxxxxxx	12.068353	1.333858	9
12	−0.74792			1.441246	10
11	−0.63132			1.578567	11
10	−0.52526			1.757429	12
9	−0.42926	−3.9483	xxxxxxxx	1.996629	13
8	−0.34305	−2.9333		2.328706	14
7	−0.26646	−2.2033		2.815209	15
6	−0.19945	−1.6167		3.588119	16
5	−0.14198	−1.1331		4.990712	17
4	−0.094949	−0.73841		8.281846	18
3	−0.055675	−0.42775	−8.3920	24.791774	19
2	−0.026843	−0.20005	−3.6186	xxxxxxxx	
1	−0.0078584	−0.055516	−0.94609	−11.667	
α	20	10	4	2	$n/2$

$$L_{+\alpha} \times 10^2$$

avoids the problem of the solution's becoming unstable with increasing depth, by the use of the diffusion approximation, which basically assumes that the radiation field has been directionally randomized.

Nonconservative Gray Atmosphere   The notion of a nonconservative gray atmosphere may sound like a contradiction in terms, and if it were meant to apply to all frequencies, it would be. However, consider the case where the opacity is essentially gray over the part of the spectrum containing most of the emergent radiation, but radiative equilibrium does not apply because some energy is lost from the radiation field to perhaps convection. Or consider an atmosphere where the dominant opacity source is the scattering of light from a hot external source, but the atmosphere itself is so cold that the thermal emission can be neglected. Planetary atmospheres often fit into this category.

Under these conditions, the equation of transfer becomes

$$\mu \frac{dI(\tau, \mu)}{d\tau} = I(\tau, \mu) - \frac{p}{2} \int_{-1}^{+1} I(\tau, \mu')\, d\mu' \qquad (10.2.35)$$

which, by the same methods used to generate equation (10.2.23), yields the eigenvalue equation

$$1 = p \sum_{j=1}^{n/2} \frac{a_j}{1 - \mu_j^2 k_\alpha^2} \qquad \mu_j > 0 \qquad (10.2.36)$$

Here p is the scattering albedo, or the fraction of interacting photons that are scattered. Since $p < 1$ for a nonconservative atmosphere, there will now be n distinct k_α's and n distinct $L_{\pm\alpha}$'s, so that the n values of the boundary radiation field completely specify the solution. The $L_{\pm\alpha}$'s are specified by the boundary equations

$$I(-\mu_i, 0) = \sum_{\alpha=1}^{n/2} \left(\frac{L_{+\alpha}}{1 - \mu_i k_\alpha} + \frac{L_{-\alpha}}{1 + \mu_i k_\alpha} \right)$$

$$I(+\mu_i, 0) = \sum_{\alpha=1}^{n/2} \left(\frac{L_{+\alpha}}{1 + \mu_i k_\alpha} + \frac{L_{-\alpha}}{1 - \mu_i k_\alpha} \right) \qquad (10.2.37)$$

and the source function for the atmosphere is given by

$$S(\tau) = J(\tau) = \sum_{\alpha=1}^{n/2} L_{+\alpha} e^{-k_\alpha \tau} + L_{-\alpha} e^{+k_\alpha \tau} \qquad (10.2.38)$$

We need not consider the unilluminated semi-infinite atmosphere since all radiation moving up through a nonconservative semi-infinite atmosphere will eventually be lost before it emerges. Thus, only the finite slab or an illuminated semi-infinite nonconservative atmosphere will yield anything other than the trivial solution.

10.3 Nongray Radiative Transfer

While the elimination of the assumption of a gray opacity removes the easy incorporation of radiative equilibrium into the solution of the equation of radiative transfer, most methods described in Section 10.2 can be used for the nongray case.

In spite of the diversity of methods available to the researcher for the solution of radiative transfer problems (there are more than are described here), most practical approaches can be divided into two categories: the solution of the integral equation for the source function and methods based on the solution of the differential equations for the radiation field. The solution of the integral equation for the source function is highly efficient, since no more information is generated than is necessary for the solution of the problem, and has also proved effective in dealing with problems of polarization, where complex redistribution functions are required (see Chapter 16). The differential equation approach is perhaps more widely used because a highly efficient algorithm has been developed which enables the investigator to utilize existing and proven mathematical packages for much of the numerical work. In addition, the differential equation approach has proved effective where geometries other than plane-parallel ones are required, and lends itself naturally to the incorporation of time-dependent and hydrodynamic terms when they may be needed. Of the myriads of specific applications, we will be concerned with only two.

a Solutions of the Nongray Integral Equation for the Source Function

We derived the integral equation for the nongray source function in Section 10.1 [equation (10.1.14)]. The approach we take is basically that described for the solution of the Schwarzschild-Milne equations in Section 10.2. Replacing the integral in equation (10.1.14) with a suitable quadrature scheme, after removing the singularity of the first exponential integral as described in equation (10.2.8), we get

$$S_\nu(\tau_\nu) = \epsilon_\nu B(\tau_\nu) + \tfrac{1}{2}(1 - \epsilon_\nu)\left\{ \sum_{j=1}^{n} [S_\nu(t_j) - S_\nu(\tau_\nu)]E_1|t_j - \tau_\nu|W_j \right.$$

$$\left. + S_\nu(\tau_\nu)[2 - E_2(\tau_\nu)] \right\} \tag{10.3.1}$$

This functional equation, evaluated at the points of the quadrature, yields a set of linear algebraic equations for the source function at the quadrature points. These, in turn, can be put into standard form so that

$$\sum_{k=1}^{n} S_\nu(t_k)\left\{ \tfrac{1}{2} \sum_{j=1}^{n} [1 - \epsilon(t_i)][(\delta_{kj} - \delta_{ki})E_1|t_i - t_k|W_i + \delta_{ik}E_2(t_i)] + \epsilon(t_i)\delta_{ik} \right\}$$

$$= \epsilon_\nu(t_i)B_\nu(t_i) \qquad i = 1, \ldots, n \tag{10.3.2}$$

These equations are strongly diagonal since the dominant contribution to the source function is always the local one. That contribution is measured by the last

275

term in equation (10.3.1), and it represents the addition made to the equation to compensate for the removal of the local contribution within the integral. The strongly diagonal nature of the equations ensures that the solution is numerically stable. Indeed, when $S_v = B_v$ and $\epsilon_v = 1$, the equations are formally diagonal. Thus, in practice they may be solved rapidly by means of the Gauss-Seidel iteration with $S_v(t_i) = B_v(t_i)$ as the initial guess. We remarked earlier that some care should be taken in choosing the quadrature scheme. It is a good practice to split the integral into two parts, with the first ranging from 0 to 1 and the second from 1 to ∞. A 10-point Gauss-Legendre quadrature provides sufficient accuracy for the rapid change of the source function near the surface, while a 4-point Gauss-Laguerre quadrature scheme is adequate for the second as the source function approaches linearity.

A slightly different approach is taken by the Harvard group[2] in the widely used atmosphere program called ATLAS. They also solve the integral equation for the source function, but they deal with the singularity of the exponential integral in a somewhat different fashion. Instead of formally removing the singularity, they approximate the source function with cubic splines over a small interval. With an analytic form for the source function, it is possible to evaluate the integral, resulting in a multiplicative weight for the coefficients of the splines. This results in a series of weights which are numerically very similar to those present in equation (10.3.2). Again, a set of linear algebraic equations is produced for the source function at a discrete set of optical depths. The results of the two methods are nearly identical, with the gaussian quadrature scheme being somewhat more efficient.

b *Differential Equation Approach: The Feautrier Method*

This method replaces the differential equations of radiative transfer with a set of finite difference equations for parameters related to the specific intensity at a discrete set of values for the angular variable μ_i. However, the choice of values of μ_i is irrelevant to understanding the method, so we leave that choice arbitrary for the moment. Instead of solving the equation of transfer for the specific intensity, we write equations of transfer for combinations of inward- and outward-directed streams.

Feautrier Equations   Consider the variables

$$u \equiv \tfrac{1}{2}[I(+\mu, \tau_v) + I(-\mu, \tau_v)] \qquad v \equiv \tfrac{1}{2}[I(+\mu, \tau_v) - I(-\mu, \tau_v)] \quad (10.3.3)$$

Here we have paired the outward-directed stream $I(+\mu, \tau)$ with its inward-directed counterpart $I(-\mu, \tau)$ into quantities that resemble a "mean" intensity u and a "flux" v. One of the benefits of the linearity of the equation of transfer is that we can add or subtract such equations and still get a linear equation. Thus, by adding an equation for a $+\mu$ stream to one for a $-\mu$ stream, we get

$$\mu \frac{dv}{d\tau_v} = u - S \qquad \mu > 0 \tag{10.3.4}$$

Similarly, by subtracting one from the other, we get

$$\mu \frac{du}{d\tau_v} = v \qquad \mu > 0 \tag{10.3.5}$$

Using this result to eliminate v from equation (10.3.4), we have

$$\mu^2 \frac{d^2u}{d\tau_v^2} = u - S \qquad \mu > 0 \tag{10.3.6}$$

This is a second-order linear differential equation, so we will need two constraints or constants of integration. At the surface $I(-\mu, 0) = 0$, so $v(0) = u(0)$, and from equation (10.3.5) we have

$$\mu \frac{du}{d\tau_v}\bigg|_{\tau_v = 0} = u(0) \tag{10.3.7}$$

The other constraint on the differential equation comes from invoking the diffusion approximation at large depths. Under this assumption

$$I_v \simeq J_v \simeq B_v \simeq S_v \tag{10.3.8}$$

We may now use the equation of transfer itself to generate a perturbation expression for $I(\mu, \tau_v)$ at large depths:

$$I_v(\mu, \tau_v) = \mu \frac{dI_v(\mu, \tau_v)}{d\tau_v} + S_v(\tau_v)$$

$$\approx \mu \frac{dB_v(\tau_v)}{d\tau_v} + B_v(\tau_v) \qquad \tau_v \gg 1 \tag{10.3.9}$$

Substituting the left-hand side into the definition for v, we get

$$v = \mu \frac{dB_v(\tau_v)}{d\tau_v} = \mu \frac{du}{d\tau_v} \qquad \tau \gg 1 \tag{10.3.10}$$

Equations (10.3.7) and (10.3.10) are the two constraints needed to specify the solution. Now consider the finite difference approximations required to solve equation (10.3.6) subject to these constraints.

Solution of the Feautrier Equations We saw earlier, in Chapter 4, how the method of solution used to solve the Schwarzschild equations of stellar structure was supplanted by the Henyey method utilizing finite differences. Many of the reasons that lead to the superiority of the Henyey method are applicable to the Feautrier method for solution of the equations of radiative transfer. It is for that reason that we describe the numerical method in some detail.

First, we must pick a set of τ_k's for which we desire the solution. We must be certain that the largest τ_N is deep enough in the atmosphere to ensure that the assumptions resulting in the boundary condition given in equation (10.3.9) are met. In addition, it is useful if the density of points near the surface is large enough that the solution will be accurately described. This is particularly important when we are dealing with the transport of radiation within a spectral line. Now we define the following finite difference operators:

$$\Delta f_{k+1/2} \equiv f(\tau_{k+1}) - f(\tau_k)$$

$$\Delta \tau_{k+1/2} \equiv \tau_{k+1} - \tau_k \qquad (10.3.11)$$

$$\Delta \tau_k \equiv \tfrac{1}{2}(\Delta \tau_{k+1/2} + \Delta \tau_{k-1/2})$$

The subscript $k + \frac{1}{2}$ simply means that this is an estimate of the parameter appropriate for the value of τ midway between k and $k + 1$. Unlike the Henyey scheme, where this information was obtained from an earlier model structure, the Feautrier method obtains the information by linear interpolation from the existing solution. Now we replace the derivatives with the following finite difference operators:

$$\left. \frac{df(\tau)}{d\tau} \right|_{\tau=\tau_k} \simeq \frac{\Delta f_{k+1/2}}{\Delta \tau_{k+1/2}}$$

$$\left. \frac{d^2 f(\tau)}{d\tau^2} \right|_{\tau=\tau_k} \simeq \frac{\Delta f_{k+1/2}/\Delta \tau_{k+1/2} - \Delta f_{k-1/2}/\Delta \tau_{k-1/2}}{\Delta \tau_k} \qquad (10.3.12)$$

The second derivative in equation (10.3.6) can now be replaced by these operators operating on $u(\mu)$ to yield the following linear algebraic equations for $u(\mu)$ at the chosen optical depth points τ_k:

$$\mu^2 \frac{u_{k-1}(\mu)}{\Delta \tau_k \Delta \tau_{k-1/2}} - \frac{\mu^2}{\Delta \tau_k}\left(\frac{1}{\Delta \tau_{k-1/2}} + \frac{1}{\Delta \tau_{k+1/2}}\right) u_k(\mu) + \frac{\mu^2 u_{k+1}(\mu)}{\Delta \tau_k \Delta \tau_{k+1/2}}$$

$$= \mu_k(\mu) - S_k \qquad (10.3.13)$$

Now it is time to pick those values of μ for which we desire the solution. Let u be considered a vector whose elements are the values of u at the particular values of μ_i, so that

$$\vec{u} = [u(\mu_1),\, u(\mu_2),\, u(\mu_3),\, \ldots,\, u(\mu_n)] \qquad (10.3.14)$$

The linear equations (10.3.13) can now be written as a system of matrix-vector equations of the form

$$\mathbf{A}_k \cdot \vec{u}_{k-1} - \mathbf{B}_k \cdot \vec{u}_k + \mathbf{C}_k \cdot \vec{u}_{k+1} = -\vec{S}_k \qquad (10.3.15)$$

The elements of matrices $\mathbf{A}$, $\mathbf{B}$, and $\mathbf{C}$ involve only the values of μ_i and τ_k that were chosen to describe the solution. The constraints given by equations (10.3.7) and (10.3.9) require that

$$\mathbf{A}_1 = \mathbf{0} \qquad \mathbf{C}_N = \mathbf{0} \qquad (10.3.16)$$

Thus we have set the conditions required to solve the equations for $u_k(\mu_i)$ from which the specific intensity can be recovered and all the moments that depend on it. Equations (10.3.16) happen to be tridiagonal, which ensures that they can be solved efficiently and accurately. We have glossed over the source function S_k in our discussion by assuming that it is known everywhere and depends only on τ_k. However, the property of the source function that caused so many problems for earlier methods (and, indeed, resulted in the integral equations in the first place) is that the source function usually depends on the intensity itself. However, for scattering, the source function does depend on the intensity in a linear manner. Therefore, it is possible to represent the source function in terms of the unknowns u_k and v_k and to include them in the equations, still preserving their tridiagonal form.

There is one caveat to this. The Feautrier method imposes a certain symmetry on the solution to the radiative transfer problem by combining inward- and outward-directed streams. If the redistribution function does not share this symmetry, it will not be possible to represent the scattering in terms of the functions $u(\mu)$ and $v(\mu)$. Thus, for some problems involving anisotropic scattering, the Feautrier method may not be applicable. In addition, when the redistribution function involves redistribution in frequency, the optical depth points must be chosen so that the deepest point will satisfy the assumptions required for the approximation given in equation (10.3.9) for all frequencies. If this is not done, errors incurred at those frequencies for which the assumptions fail can propagate in an insidious manner throughout the entire solution.

The Feautrier method does not suffer from the exponential instabilities described for the discrete ordinate method, because it invokes the diffusion approximation at large depths (specifically the inner boundary). This basically contains the information that the radiation field has been randomized in direction and thereby stabilizes the solution in the same manner as it stabilizes the Eddington solution. As we shall see in Chapter 11, knowledge of the mean intensity, the radiative flux, and occasionally the radiation pressure is usually sufficient to calculate the structure of the atmosphere. The Feautrier method finds more information than that and therefore is not as efficient as it might be. However, the numerical methods for solving the resulting linear equations are so fast that the overall efficiency of the method is quite good, and it provides an excellent method of solution for most problems of radiative transfer in stellar atmospheres. Remember that, like any numerical method, the Feautrier method should be used with great care and only on those problems for which it suited.

10.4 Radiative Transport in a Spherical Atmosphere

Any discussion of the solution of radiative transfer problems would be incomplete without some mention of the problem introduced by a departure from the simplifying assumption of plane-parallel geometry. In addition, there are stars for which the plane-parallel approximation is inappropriate, and we would like to model these stars as well as the main sequence stars for which the plane-parallel approximation is generally adequate. The density in the outer regions of red supergiants is so low that the atmosphere will occupy the outer 30 to 40 percent of what we would like to call the radius of the star. Here, the plane-parallel assumption is clearly inappropriate for describing the star. We must include the curvature of the star in any description of its atmosphere. In doing so, we will require a parameter that was removed by the plane-parallel assumption—the *stellar radius*. This parameter can be operationally defined as the distance from the center to some point where the radial optical depth to the surface is some specified number (say, unity). In doing so, we must remember that the radius may now become a wavelength-dependent number, and so some mean value from which the majority of the energy escapes to the surrounding space may be appropriate for describing the star when a single value for the radius is required. However, for the calculation of the stellar interior, we need to know only the surface structure at a given distance from the center in order to specify the interior structure. Whether the distance corresponds to our idea of a stellar radius is irrelevant. In addition, we assume that the star is spherically symmetric.

a Equation of Radiative Transport in Spherical Coordinates

In Chapter 9 we developed a very general equation of radiative transfer which was coordinate-independent [equation (9.2.11)]. Writing the time-independent form for which the gravity gradient does not significantly affect the photon energy, we get

$$\hat{n} \cdot \nabla I_\nu = \rho(\kappa_\nu + \sigma_\nu)(S_\nu - I_\nu) \tag{10.4.1}$$

Writing the ∇ operator in spherical coordinates and making the usual definition for μ (see Figure 10.2), we get

$$\mu \frac{\partial I_v(r,\mu)}{\partial r} + \frac{1-\mu^2}{r}\frac{\partial I_v(r,\mu)}{\partial \mu} = \rho(\kappa_v + \sigma_v)[S_v(r,\mu) - I_v(r,\mu)] \quad (10.4.2)$$

where we take the source function to be that of a nongray atmosphere with coherent isotropic scattering, so that

$$S_v = \epsilon_v B_v[T(r)] + (1 - \epsilon_v)J_v(r) \quad (10.4.3)$$

Our approach to the solution of the equation of transfer will be to obtain and solve some equations for the important moments of the radiation field.

Radiative Equilibrium and Moments of the Radiation Field For a steady state atmosphere, our condition for radiative equilibrium [equation (9.4.4)] becomes

$$\nabla \cdot \int_0^\infty \vec{F}_v \, dv = 4 \int_0^\infty \kappa_v \rho (B_v - J_v) \, dv = 0 \quad (10.4.4)$$

However, in spherical coordinates, the divergence of the total flux yields the same condition that we obtained for stellar interiors [equation (4.2.1)]:

$$\pi F \equiv \pi \int_0^\infty F_v \, dv = \frac{L}{4\pi r^2} \quad (10.4.5)$$

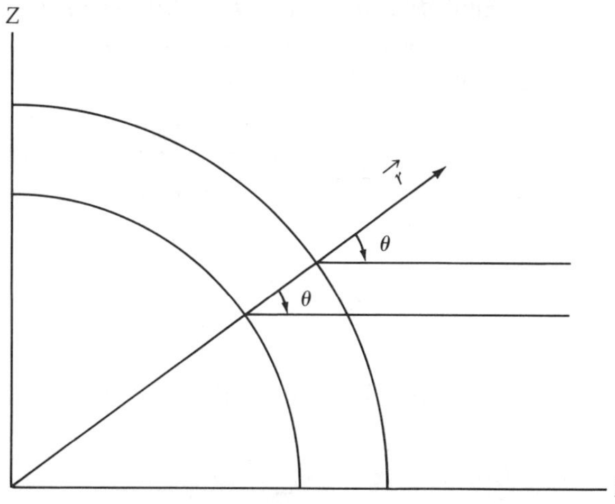

Figure 10.2 Geometry assumed for the spherical equations of radiative transfer. The angle θ for which $\mu = \cos\theta$ is defined with respect to the radius vector. Unlike in the plane-parallel approximation, the depth variable is the radius and increases outward.

Now the condition of radiative equilibrium is obtained from the zeroth moment of the equation of transfer [equation (9.4.3)], while the first moment of the equation of transfer [equation (9.4.6)] yields an expression for the radiation pressure tensor. For an atmosphere with no time-dependent processes, these moment equations become

$$\nabla \cdot \vec{F}_\nu = 4\kappa_\nu \rho (B_\nu - J_\nu) \qquad \nabla \cdot \mathbf{K}_\nu = -\frac{\rho(\kappa_\nu + \sigma_\nu)\vec{F}_\nu}{4} \qquad (10.4.6)$$

Noting that there is no net flow of radiation in either the θ or ϕ coordinates for a spherically symmetric atmosphere, we see that the divergence of the flux in spherical coordinates becomes

$$\frac{\partial(r^2 F_\nu)}{\partial r} = 4r^2 \kappa_\nu \rho (B_\nu - J_\nu) \qquad (10.4.7)$$

If we make the assumption that the radiation field is nearly isotropic, then $\nabla \cdot \mathbf{K}_\nu$ becomes ∇K_ν, where K_ν is the scalar moment that we have identified with the radiation pressure [see equations (9.3.14) through (9.3.16)]. Perhaps the easiest way to find the representation of equation (10.4.6) in spherical coordiates is to multiply (10.4.2) by μ and integrate over all μ. This yields the second of the required moment equations,

$$\frac{\partial K_\nu}{\partial r} + \frac{3K_\nu - J_\nu}{r} = -\frac{\rho(\kappa_\nu + \sigma_\nu)F_\nu}{4} \qquad (10.4.8)$$

Closing the Moment Equations and the Eddington Factor In Chapter 9 we observed [equation (9.4.8)] that under conditions of near isotropy $K_\nu = J_\nu/3$. This was the moment approximation needed to close the moment equations, and it is known as the *diffusion approximation*. However, such conditions do not prevail throughout the atmosphere, so it is common to assume that the two moments can be related by a scale factor, which has come to be known as the *Eddington factor*, defined as

$$f_\nu(r) = \frac{K_\nu(r)}{J_\nu(r)} \qquad (10.4.9)$$

We can replace the radiation pressure by the Eddington factor and obtain

$$\frac{\partial(f_\nu J_\nu)}{\partial r} + \frac{(3f_\nu - 1)J_\nu}{r} = \frac{-\rho(\kappa_\nu + \sigma_\nu)F_\nu}{4} \qquad (10.4.10)$$

　　for the second moment equation.

Equation (10.4.10) combined with equation (10.4.7) forms a complete system for F_v and J_v, subject to the appropriate boundary conditions. Of course, we have not fundamentally changed the problem since the Eddington factor is unknown and presumably a function of depth. It must be found so that any atmosphere produced is self-consistent under the constraint of radiative equilibrium. The Eddington factor basically measures the isotropy of the radiation field, since for isotropic radiation it is $\frac{1}{3}$. Imagine a radiation field entirely directed along $\mu = \pm 1$. For such a field $f_v = 1$, while for a radiation field confined to a plane normal to this direction, $f_v = 0$. If we consider the normal radiation field emerging from a star, the temperature gradient normally produces limb-darkening, implying that the radiation field near the surface becomes more strongly directed along the normal to the atmosphere. Thus, we should expect the Eddington factor to increase as the surface approaches. This effect should be enhanced for stars with large spherical atmospheres. Thus, for normal stellar atmospheres

$$\tfrac{1}{3} \le f_v(r) < 1 \tag{10.4.11}$$

b   An Approach to Solution of the Spherical Radiative Transfer Problem

Sphericality Factor   This factor is introduced purely for mathematical convenience and as such has no major physical importance. However, it does tend to make the spherical moment equations resemble their plane-parallel counterparts. We define

$$\ln (r^2 q_v) \equiv \int_{r_c}^{r} \frac{3f_v - 1}{x f_v} \, dx + \ln r_c^2 \tag{10.4.12}$$

so that

$$\frac{1}{r^2 q_v} \frac{\partial (r^2 q_v)}{\partial r} = \frac{3f_v - 1}{r f_v} \tag{10.4.13}$$

The parameter r_c is the deepest radius for which the problem is to be solved. Given f_v, we can find the sphericality factor q_v by numerically integrating equation (10.4.12). Using this definition of q_v, we may rewrite the second moment equation (10.4.10), as

$$\frac{-4}{\rho(\kappa_v + \sigma_v) q_v} \frac{\partial (r^2 q_v f_v J_v)}{\partial r} = r^2 F_v \tag{10.4.14}$$

This form is suitable for combining with the first moment equation, (10.4.7), to eliminate F_v and get

$$\frac{\partial^2 (r^2 q_v f_v J_v)}{\partial \tau_v^2} = \frac{r^2 \epsilon_v (J_v - B_v)}{q_v} \tag{10.4.15}$$

where

$$\partial\tau_\nu = -q_\nu\rho(\kappa_\nu + \sigma_\nu)\,\partial r \qquad (10.4.16)$$

and ϵ_ν has the same meaning as before [see equation (10.1.8)].

We have now generated a second-order differential equation for J_ν that is similar to the one obtained for the Feautrier method, and we solve it in a similar manner.

Boundary Conditions The boundary conditions are determined in much the same manner as for the Feautrier method. For the lower boundary we make the same assumptions of isotropy as were made for equation (10.3.9). Indeed, we multiply equation (10.3.9) by μ and integrate over all μ, to get

$$F_\nu \simeq 2\int_{-1}^{+1}\mu^2\frac{dB_\nu}{d\tau_\nu}\,d\mu + 2\int_{-1}^{+1}\mu B_\nu\,d\mu = \frac{4}{3}\frac{dB_\nu}{d\tau_\nu} \qquad (10.4.17)$$

This and equation (10.4.14) allow us to specify the derivative of J_ν at the lower boundary as

$$\frac{\partial(r^2 q_\nu f_\nu J_\nu)}{\partial\tau_\nu}\bigg|_{r=r_c} = \frac{r_c^2}{3}\frac{dB_\nu}{d\tau_\nu}\bigg|_{r=r_c} \qquad (10.4.18)$$

Again r_c is the deepest point for which the solution is desired. Equation (10.4.14) also sets the upper boundary condition at R as

$$4\frac{\partial(r^2 q_\nu f_\nu J_\nu)}{\partial\tau_\nu}\bigg|_{r=R} = R^2 F_\nu = R^2\frac{\int_0^1 \mu I(R,\mu)\,d\mu}{\int_0^1 I(R,\mu)\,d\mu}\,J_\nu(R) \qquad (10.4.19)$$

so that we again have a two-point boundary-value problem and a second-order differential equation for J_ν which we can solve by the same finite difference techniques that were used for the Feautrier method [see equations (10.3.11) through (10.3.16)].

The problem can now be solved, assuming we know the behavior of the Eddington factor with depth in the atmosphere. Unfortunately, to find this, we must know the angular distribution of the radiation field at all depths. Normally, we could appeal to the classical solution, for knowledge of J_ν would provide all the information needed to calculate the source function. But the classical solution was appropriate for only the plane-parallel approximation. To find the analog for spherical coordinates, we have to use the symmetry of a spherical atmosphere and perform still another coordinate transformation.

Impact Space and Formal Solution for the Spherical Equation of Radiative Transfer Consider a coordinate frame attached to the star so that the z axis points in the direction of the observer and passes through the center of the star (see Figure 10.3). Coordinates z and p designate all places within the star with p playing the role of an impact parameter for photons directed toward the observer parallel to z. The entire solution set $I(\mu, r)$ can be represented by the radiation streams $I_+(p, z)$ and $I_-(p, z)$ by moving along surfaces of constant r. Thus any solution that gives us a complete representation of the specific intensity in the pz plane will give a complete description of the radiation field. We can immediately write the equation of transfer for the special beams directed toward or away from the observer as

$$\pm \frac{\partial I_\pm(p, z)}{\partial z} = \rho(\kappa_v + \sigma_v)[S(p, z) - I_\pm(p, z)] \qquad (10.4.20)$$

where the coordinate transformation from pz coordinates to μr coordinates is

$$r = (p^2 + z^2)^{1/2} \qquad |\mu| = \frac{z}{(p^2 + z^2)^{1/2}} \qquad (10.4.21)$$

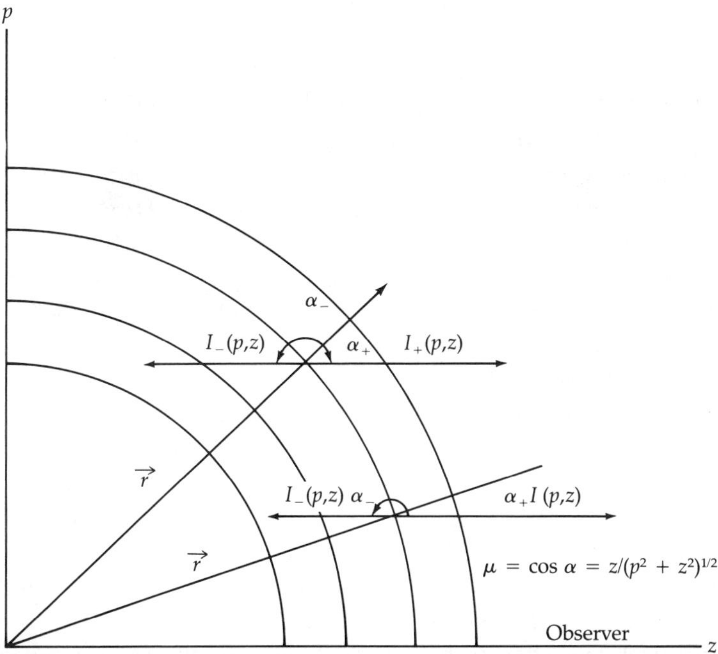

Figure 10.3 Impact space for spherical transport. The z axis points at the observer, while the p coordinate is perpendicular to z and plays the role of an impact parameter for the photons directed toward the observer. The angle α denotes the angle between a line parallel to z, directed toward the observer, and a radius vector.

For simplicity we denote

$$k_\nu = \rho(\kappa_\nu + \sigma_\nu) \tag{10.4.22}$$

Equation (10.4.20) is a linear first-order equation that has a classical solution of

$$I_-(p, z) = \int_z^{(R^2 - p^2)^{1/2}} k_\nu(\xi) S_\nu(\xi) e^{-\tau(p,\zeta,z)} \, d\zeta$$

$$I_+(p, z) = \int_0^z k_\nu(\xi) S_\nu(\xi) e^{-\tau(p,z,\zeta)} \, d\zeta + I_-(p, 0) e^{-\tau(p,z,0)} \tag{10.4.23}$$

$$\xi^2 \equiv p^2 + \zeta^2 \qquad \tau(p, a, b) = \int_a^b k(\xi) \, d\zeta$$

While this is a complicated expression, it can be evaluated numerically as long as one has a representation of the source function. Thus, it is now possible to solve for the entire radiation field and recalculate the variable Eddington factor f_ν. Equation (10.4.15) is then solved again for a new value of J_ν and hence S_ν. The entire procedure is repeated until a self-consistent solution is found. Rather than carry out the admittedly messy numerical integration, Mihalas[3] describes a Feautrier-like method to calculate the intensities directly.

A method proposed by Schmid-Burgk[4] assumes that the source function can be locally represented by a polynomial in the optical depth. This analytic function is then substituted into the formal solution in impact space so that the radiation field can be represented in terms of the undetermined coefficients of the source function's approximating polynomials. The moments of the radiation field can then be generated which depend only on these same coefficients. Thus, if one starts with an initial atmospheric structure and a guess for the source function, one can fit that source function to the local polynomial and thereby determine the approximating coefficients. These, in turn, can be used to generate the moments of the radiation field upon which an improved version of the source function rests. An excellent initial guess for the source function is $S_\nu = B_\nu$, and unless scattering completely dominates the opacity, the iteration process converges very rapidly.

It is clear that the spherical atmosphere poses significant difficulties over and above those found in the plane-parallel atmosphere. However, there are very few differences that are fundamental in nature. All present methods rely on the global symmetry of spherical stars, and it seems likely that those stars with atmospheres sufficiently extended to require the spherical treatment will also be subject to other forces, such as rotation, that further distort the atmospheres so that even this global symmetry is lost. However, such studies can offer insight into the severity of the effects that we can expect from the geometry.

We have only skimmed the surface of the methods and techniques devised to solve the equation of radiative transfer. The methods discussed merely comprise some of the more popular and successful methods currently in use. We have left to the studious reader the entire area of the "exact approximation" and the H functions of Chandrasekhar[1] (pp. 105 to 126). No mention has been made of invariant embedding and the voluminous literature written for linear two-point boundary-value problems. Many of these techniques have proved useful in solving specific radiative transfer problems, and those who would count themselves experts in this area should avail themselves of that literature. There is an entire field of study surrounding the transfer of radiation within spectral lines, some of which will be discussed later, but much of which will not be. This material is important for anyone interested in problems requiring line-transfer solutions. However, the methods presented here suffice for providing the solution to half of the task of constructing a normal stellar atmosphere, and next we turn to the solution of the other half of the problem.

Problems

1. Find the general expression for

$$\int_0^\infty \frac{(t - \tau)^n E_1 |t - \tau|}{n!}\, dt$$

2. Find the eigenvalues k_α and $L_{+\alpha}$ for the discrete ordinate solution to the semi-infinite plane-parallel gray atmosphere for $n = 8$.

3. Repeat Problem 2 for the double-gauss quadrature scheme for $n = 8$.

4. If there is an arbitrary iterative function $\Phi(x)$ such that

$$x_{k+1} = \Phi(x_k)$$

then an iterative sequence defined by $\Phi(x_k)$ will converge to a fixed point x_0 if and only if

$$\left| \frac{\partial \Phi(x)}{\partial x} \right| < 1 \qquad |x_k| \leq |x| \leq |x_0|$$

Use this theorem to prove that any fixed-point iteration scheme will provide a solution for

$$B(\tau) = \tfrac{1}{2} \int_0^\infty B(t) E_1 |t - \tau|\, dt$$

5. Find a general interpolative scheme for $I(\tau, \mu)$ when $\mu < 0$ for the discrete ordinate approximation. The interpolative formula should have the same degree of precision as the quadrature scheme used in the discrete ordinate solution.

6. Consider a pure scattering plane-parallel gray atmosphere of optical depth τ_0, illuminated from below by $I(+\mu, \tau_0) = I_0$. Further assume that the surface is not illuminated [that is, $I(-\mu, 0) = 0$]. Use the Eddington approximation to find $F(\tau)$, $J(\tau)$, and $I(+\mu, 0)$ in terms of I_0 and τ_0.

7. Show that in a gray atmosphere

$$\frac{dP_r}{dT} = \frac{16\sigma}{3c} \frac{T^3}{1 + dq(\tau)/d\tau}$$

8. Use the first of the Schwarzschild-Milne integral equations for the source function in a gray atmosphere [equation (10.2.6)] to derive an integral equation for the Hopf function $q(\tau)$.

9. Show that no self-consistent solution to the equation of radiative transfer exists for a pure absorbing plane-parallel gray atmosphere in radiative equilibrium where the source function has the form

$$S(\tau) = a + b\tau$$

10. Show that the equation of transfer in spherical coordinates

$$\cos\theta \, \frac{\partial I_\nu(r, \theta)}{\partial r} - \frac{\sin\theta}{r} \frac{\partial I_\nu(r, \theta)}{\partial \theta} = \rho(\kappa_\nu + \sigma_\nu)[S_\nu(r) - I_\nu(r, \theta)]$$

transforms to

$$\pm \frac{\partial I_\pm(p, z)}{\partial z} = \rho(\kappa_\nu + \sigma_\nu)[S_\nu(p, z) - I_\pm(p, z)]$$

in impact space where $r^2 = p^2 + z^2$ and $|\mu| = z/r$.

11. Derive an integral equation for the mean intensity $J_\nu(\tau_\nu)$ when the source function is given by

$$S_\nu(\tau_\nu) = \kappa_\nu B_\nu(\tau_\nu) + \frac{\sigma_\nu}{2} \int_{-1}^{+1} [1 - (\mu')^2] I_\nu(\mu', \tau_\nu) \, d\mu'$$

12. Numerically obtain a solution for the Schwarzschild-Milne integral equation for the source function in a gray atmosphere by solving equation (10.2.11) for the ratio of the source function at eight points in the atmosphere to its value at one point. Describe why you picked the points as you did, and compare your result with that obtained from the Eddington approximation.

13. Using equation (10.2.21), show that equations (10.2.22) and (10.2.23) follow from the discrete ordinate equation of transfer [equation (10.2.20)].

14. Show that equations (10.2.29) and (10.2.30) follow from the substitution of the solution for the discrete ordinate method [equation (10.2.26)] into the definition for the moments of the radiation field, $J(\tau)$, and F.

15. Show that equation (10.2.36) is indeed the eigenequation for the nonconservative gray atmosphere.

16. Use the Feautrier method to solve the problem of radiative transfer in a gray atmosphere.

References and Supplemental Reading

1. Chandrasekhar, S., *Radiative Transfer* [1949], Dover, New York, 1960, pp. 54–68.

2. Kurucz, R. L., *ATLAS: A Computer Program for Calculating Model Stellar Atmospheres*, Smithsonian Astrophysical Observatory Special Report 309, 1970.

3. Mihalas, D., *Stellar Atmospheres*, 2d ed., W. H. Freeman, San Francisco, 1978, pp. 250–255.

4. Schmid-Burgk, J., "Radiative Transfer through Spherical-Symmetric Atmospheres and Shells," *Astron. & Astrophy.* 40, 1975, pp. 249–255.

Virtually every book about stellar atmospheres provides an introduction to the subject that is worth perusing. Some are more valuable than others in providing insight into the physics of the atmosphere. In the area of radiative transfer, the definitive mathematical treatise is still

• Chandrasekhar, S.: *Radiative Transfer* [1949], Dover, New York, 1960.

However, students should not try to read this work until they have gained considerable familiarity with the problem. One of the clearest and most comprehensive descriptions of the gray atmosphere and various methods of solution of the radiative transfer problem is found in

• Kourganoff, V.: *Basic Methods in Transfer Problems—Radiative Equilibrium and Neutron Diffusion*, Dover, New York, 1963, pp. 86–125.

An extremely complete discussion of Λ operators is given in this same reference (pp. 40–85). Dimitri Mihalas provides a good description of the gray atmosphere in both editions of his book on stellar atmospheres, but of the two, I prefer the first edition:

• Mihalas, D.: *Stellar Atmospheres*, 1st ed., W. H. Freeman, San Francisco, 1970, pp. 34–66.

For a lucid discussion of the relative merits of solutions to the integral equations of radiative transfer, see

• Kalkofen, W.: "A Comparison of Differential and Integral Equations of Radiative Transfer," *J. Quant. Spectrosc. & Rad. Trans.* 14, 1974, pp. 309–316.

For a general background of the subject as considered by some of the finest minds of the twentieth century, everyone should spend some time reading *Selected Papers on the Transfer of Radiation*, edited by D. H. Menzel (Dover, New York, 1966). All these papers are of landmark quality, but I found this one to be most rewarding and somewhat humbling:

• Schuster, A.: "Radiation through a Foggy Atmosphere," *Ap. J.* 21, 1905, pp. 1–22.

It is clear that Arthur Schuster identified and understood most of the important aspects of scattering theory in radiative transfer without the benefit of the work of the rest of the twentieth century that is available to the contemporary student of physics. Much of the work on neutron diffusion theory deals with the same mathematical formalisms that serve radiative transfer theory, and we should be ever mindful of the physics literature on that subject if we are to appreciate the full breadth of the nature of the problems posed by the flow of radiation through the outer layers of stars. Finally, it would be a mistake to ignore the substantial contribution from the Russian school of radiative transfer theory. Perhaps the finest example of their efforts can be found in

• Sobolev, V. V.: *A Treatise on Radiative Transfer* (Trans.; S. I. Gaposchkin), Van Nostrand, Princeton, N.J., 1963.

The approaches described in this book are insightful, novel, and particularly useful in dealing with some of the more advanced problems of radiative transfer.

11

Environment of the Radiation Field

. . .

Thus far, we have said little or nothing about the gas through which the radiation is flowing. This constitutes the second half of the problem of constructing a model for the atmosphere of a star. To get the solutions for the radiation field described in Chapter 10, we must know the opacity—and not just the mean opacity that was required for stellar interiors, but the frequency-dependent opacity that determines which photons will escape from the star and from what location. We must understand enough of the physics of the gas for that opacity to be determined. Given that, we can calculate the emergent stellar spectrum by solving the equation of radiative transfer. This seems like a small requirement, but as we look more closely at the details of the opacity, the more specific knowledge of the state of the gas is required. For the construction of a model of the atmospheric structure, a little more is required. We must know how the particles that make up the gas interact with each other as well as with photons. Fortunately, with the aid of the assumption of local thermodynamic equilibrium (LTE), much of our task has already been accomplished.

11.1 Statistics of the Gas and the Equation of State

For normal stellar atmospheres, the effective temperatures range from a few thousand degrees to perhaps 50,000 K. The pressures are such as to permit the existence of spectral lines, and so the densities cannot be great enough to cause departures from the ideal-gas law. The assumption of LTE implies that the material components of the gas making up the stellar atmosphere behave as if they were in thermodynamic equilibrium characterized by the local value of the kinetic temperature. In Chapter 1 [equations (1.1.16) and (1.1.17)] we found that as long as the density of available cells in phase space is much greater than the particle phase density, such a gas should obey Maxwell-Boltzmann statistics. If this is true, the fraction of particles that have a certain kinetic energy w_i is

$$\frac{N_i}{N} = \frac{g_i e^{-w_i/(kT)}}{U(T)} \qquad (11.1.1)$$

Since the pressure is the second velocity moment of the density, the Maxwell-Boltzmann distribution formula [equation (11.1.1)] leads to the ideal-gas law [equation (1.3.4)], namely,

$$P_g = nkT \qquad (11.1.2)$$

Later [equation (4.1.6)] we found it convenient to represent the total number of particles in terms of the density and the corresponding mass of hydrogen atoms μm_h that would yield the required number of particles, so that

$$P_g = \frac{\rho kT}{\mu m_h} \qquad (11.1.3)$$

The parameter μ is called the *mean molecular weight*.

a Boltzmann Excitation Formula

Under the assumption of LTE, the energy distribution of all particles represents the most probable macrostate for the system. This state is arrived at through random collisions between the gas particles themselves. Such a gas is said to be *collisionally relaxed* and is in stationary equilibrium. This means that all aspects of the gas will exhibit the same energy distribution, including those energy aspects of the gas which do not allow for a continuum distribution of energy states—specifically those states described by the orbital electrons. Thus, an ensemble of atoms will exhibit a distribution of states of electronic excitation that follows the Maxwell-Boltzmann distribution law. Were the energy not shared between the excitation energy and the kinetic energy of the particles, collisions would ensure that energy differences were made up in the deficit population at the expense of the population that had the relative surplus.

Such a situation would not represent a time-independent distribution until equilibrium between the two populations was established and therefore would not be the most probable macrostate. Since the various states of atomic excitation will be distributed according to the Maxwell-Boltzmann distribution law, the number of particles in any particular state of excitation will be

$$\frac{N_j}{N} = \frac{g_j e^{-\epsilon_j/(kT)}}{U(T)} \tag{11.1.4}$$

Here, g_j is the statistical weight, and it has the same meaning as it did in Chapter 1 [equation (1.1.17)]. The parameter ϵ_j is the excitation energy above the ground state, and $U(T)$ is the partition function. $U(T)$ is nothing more than a normalization parameter that reflects the total number of particles available for distribution among the various energy states. It also has the same meaning as it did for the continuum distribution of energies discussed in Chapter 1, but now will be determined from the sum over the discrete states of excitation. Its role as a normalization parameter of the distribution is most clearly demonstrated by summing equation (11.1.4) over all particles and their energy states so that the left-hand side is unity. This then confirms the form of the partition function as

$$U(T) = \sum_j g_j e^{-\epsilon_j/(kT)} \tag{11.1.5}$$

b. Saha Ionization Equilibrium Equation

A completely rigorous derivation of the Saha equation from first principles is long and not particularly illuminating. So instead of performing such a derivation, we appeal to arguments similar to those for the Boltzmann excitation formula. The object now is to find the equilibrium distribution formula for the distribution of the various states of ionization for a collection of atoms. Again, we assume that a time-independent equilibrium exists between the electrons and the ions. However, now the electron population will depend on the equilibrium established for all elements, which appears to make this case quite different from the Boltzmann excitation formula. However, we proceed in a manner similar to that for the Boltzmann excitation formula. Let us try to find the probability of excitation, not of a bound state, but of a state in the continuum where the electron can be regarded as a free particle.

Consider an atom in a particular state of ionization, and denote the number of such atoms in a particular state of excitation, say the ground state, by the quantity n_{00}. The number of these atoms excited to the ionized state where the electron can be regarded as a free particle is then given by the Boltzmann excitation formula as

$$\frac{n_{01}}{n_{00}} = \frac{g_f}{g_{00}} e^{-(\chi_0 + \frac{1}{2}mv^2)/(kT)} \tag{11.1.6}$$

Here g_f is the statistical weight of the final state of the ionized atom, and χ_0 is the ionization potential of the atom in question. The second term in the exponential is simply the energy of the electron that has been elevated to the continuum. Now we can relate the total number of ionized atoms in the ground state to the total number of ionized atoms through the repeated use of the Boltzmann excitation formula and the partition function as

$$n_{0i} = \frac{N_i g_{0i} e^{-\epsilon_0/(kT)}}{U(T)} \tag{11.1.7}$$

where N_i is the total number of atoms in the ith state of ionization. However, since the resultant ionized atoms we are considering are in their ground state, ϵ_0 is zero by definition. In addition, the statistical weight of the final state g_f can be written as the product of the statistical weight of the ground state of the ionized atom and that of a free electron. We may now use this result and equation (11.1.6) to write

$$\frac{n_{01}}{n_{00}} = \frac{N_1 g_{01}/U_1(T)}{N_0 g_{00}/U_0(T)} = \frac{g_{01} g_e}{g_{00}} e^{-(\chi_0 + \frac{1}{2}mv^2)/(kT)} \tag{11.1.8}$$

which simplifies to

$$\frac{N_1}{N_0} = \frac{U_1(T)}{U_0(T)} e^{-\chi_0/(kT)} g_e e^{-\frac{1}{2}mv^2/(kT)} \tag{11.1.9}$$

Although we picked a specific state of excitation—the ground state—to arrive at equation (11.1.9), that choice was in no way required. It only provided us with a way to use the Boltzmann excitation formula for atoms in two differing states of ionization.

   The statistical weight of a free electron is really nothing more than the probability of finding a given electron in a specific cell of phase space, so that

$$g_e = \frac{2\,dx\,dy\,dz\,dp_x\,dp_y\,dp_z}{h^3} \tag{11.1.10}$$

Again, for electrons, the familiar factor of 2 arises because the spin of an electron can be either "up" or "down." Assuming that the microscopic velocity field is isotropic, we can replace the "momentum volume" by its spherical counterpart and express it in terms of the velocity:

$$dp_x\,dp_y\,dp_z = 4\pi p^2\,dp = 4\pi m^3 v^2\,dv \tag{11.1.11}$$

The "space volume" of phase space occupied by the electron can be expressed in terms of the inverse of the electron number density, so that the statistical weight of an electron becomes

$$g_e = \frac{8\pi m^3 v^2\,dv}{N_e h^3} \tag{11.1.12}$$

and equation (11.1.9) can be written as

$$\frac{N_e N_1}{N_0} = \frac{U_1(T)}{U_0(T)}\,e^{-\chi_0/(kT)}\,\frac{8\pi m_e^3 v^2}{h^3}\,e^{-\frac{1}{2}m_e v^2/(kT)}\,dv \tag{11.1.13}$$

   So far we have assumed that the ionization produced an electron in a specific free state, that is, with a specific velocity in the energy continuum. However, we are interested in only the total number of ionizations, so we must integrate equation (11.1.13) over all allowed velocities for the electrons that result from the ionization process. Thus,

$$\begin{aligned}
\frac{N_e N_1}{N_0} &= \frac{U_1(T)}{U_0(T)}\,e^{-\chi_0/(kT)}\,\frac{8\pi m_e^3}{h^3}\int_0^\infty v^2 e^{-\frac{1}{2}m_e v^2/(kT)}\,dv \\[2mm]
&= \frac{U_1(T)}{U_0(T)}\,\frac{2(2\pi kTm_e)^{3/2}}{h^3}\,e^{-\chi_0/(kT)}
\end{aligned} \tag{11.1.14}$$

For convenience, we also assumed that the states of ionization of interest were the neutral and first states of ionization. However, the argument is correct for any two adjacent states of ionization, so we can write with some generality

$$\frac{N_{i+1}N_e}{N_i} = \frac{U_{i+1}(T)}{U_i(T)} \frac{2(2\pi kTm_e)^{3/2}}{h^3} e^{-\chi_i/(kT)} \qquad (11.1.15)$$

This expression is often written in terms of the electron pressure as

$$\frac{N_{i+1}}{N_i} P_e = \frac{2(2\pi m_e)^{3/2}(kT)^{5/2}}{h^3} \frac{U_{i+1}(T)}{U_i(T)} e^{-\chi_i/(kT)} \qquad (11.1.16)$$

Both expressions [equations (11.1.15) and (11.1.16)] are known as the *Saha ionization equation*. Its validity rests on the Boltzmann excitation formula and the velocity distribution of the electrons produced by the ionization being described by the Boltzmann distribution formula. Both these conditions are met under the conditions of LTE. Indeed, many authors take the validity of the Saha and Boltzmann formulas as a definition for LTE.

The Boltzmann excitation equation and the Saha ionization equation can be combined to yield the fraction of atoms in a particular state of ionization and excitation. Knowing that fraction, we are in a position to describe the extent to which those atoms will impede the flow of photons through the gas.

11.2 Continuous Opacity

In Section 4.1b we discussed the way in which a gas can absorb photons, and we calculated the continuous opacity due to an atom of hydrogen. The calculation of the opacity of other individual species of atoms follows the same type of argument, and so we do not deal with them in detail here since these details can be found elsewhere.[1] However, in the stellar interior, we were able to characterize the opacity of the stellar material by a single parameter known as the *Rosseland mean opacity*. This resulted from the fact that the radiation field was itself in thermodynamic equilibrium and therefore depended on the temperature alone. Therefore, all parameters that arise from the interaction of the radiation field and the gas, which is also in thermodynamic equilibrium, must be described in terms of the state variables alone. Unfortunately, in the stellar atmosphere, although the gas can still be considered to be in thermodynamic equilibrium (LTE), the radiation field is not. Thus, the opacity must be determined for each frequency for which a significant amount of radiant energy is flowing through the gas.

Traditionally, the dominant source of opacity has been considered to be that arising from bound-free atomic transitions which are called "continuous" opacity sources. However, contributions to the total opacity that result from bound-bound atomic transitions have been found to play an important role in forming the structure of the atmosphere in a wide variety of stars. Since we deal with the formation of spectral lines arising from bound-bound transitions in considerable detail later, we defer the discussion of bound-bound opacity until then. It is sufficient to know that the total opacity can be calculated for each frequency of importance and to explicitly consider some of the important sources.

a   Hydrogenlike Opacity

Any atom which has a single electron in its outer shell will absorb photons in a manner similar to that for hydrogen, so we should expect the opacity to have a form similar to that of equation (4.1.17). Indeed, the expression differs from that of hydrogen by only a factor involving the atomic weight and the atomic constants appropriate for the particular atom. Thus, the opacity per gram of the ionized species is

$$\kappa_\nu(\text{H-like}) = \frac{32\pi e^6 R Z^4 e^{-\chi_{H1}/(kT)}}{3\sqrt{3}h^3 c m_h \nu^3}\left(\sum_{n,\nu_n<\nu}^{\infty}\frac{g_n}{n^3}e^{-\epsilon_n/(kT)} + \frac{g_{ff}kT}{2\chi_{H1}}\right) \quad (11.2.1)$$

where R is the Rydberg constant. For ionized helium, $Z = 2$. Since hydrogen and helium are by far the most abundant elements in stars, neutral hydrogen and ionized helium must be considered major sources of opacity.

b   Neutral Helium

Again, because of its large abundance, the opacity of neutral helium has long been regarded as an important source of stellar opacity. However, since neutral helium has two electrons in the outer shell, the atomic absorption coefficient is much more difficult to calculate. Approximate values for the opacity of neutral helium were first given by Ueno et al.[2] in 1954. Later Stewart and Webb[3] calculated the opacity arising from the ground state, and in general the contributions from the first two excited states must be treated separately. For states of excitation greater than 2, the approximate solutions will generally suffice.

c   Quasi-atomic and Molecular States

Although molecular opacity plays an important role in the later-type stars (and will be dealt with later), one might think that molecular opacity is unimportant at temperatures greater than those corresponding to the disassociation energies of the molecules. However, some molecules may form for a short time and absorb photons before they disassociate. If the abundance of the atomic species that give rise to these quasi-states is great, they may provide a significant source of opacity. The prototypical example of a short lived or quasi-state is the H-minus ion.

Classically, the existence of the H-minus ion can be inferred from the incomplete screening of the proton by the orbiting electron of hydrogen (see Figure 11.1). Although the orbiting electron is "on one side" of the proton, it is possible to bind an additional electron to the atom for a short time. During this time, the additional electron can undergo bound-free transitions. Quantum mechanically, there exists a single weakly bound state for an electron near a neutral hydrogen atom. The negatively charged configuration is called the H-minus ion is important in stars only because of the great abundance of hydrogen and electrons at certain temperatures and densities. The Saha equation for the abundance of such ions will be somewhat different from that for normal atoms because the existence of the ion will depend on the availability of electrons as well as hydrogen atoms. Thus the Saha equation for the H-minus ion would have the form

$$\frac{N(\mathrm{H}^-)}{N(\mathrm{H})} = \Phi(T)P_e \qquad (11.2.2)$$

The parameter $\Phi(T)$ involves partition functions and the like, but depends on only the temperature. Since the Saha equation for neutral hydrogen implies that neutral hydrogen will also be proportional to the electron pressure, the abundance of the H-minus ion will depend quadratically on the electron pressure. Thus, the relative important of H-minus to hydrogen opacity will decrease with decreasing

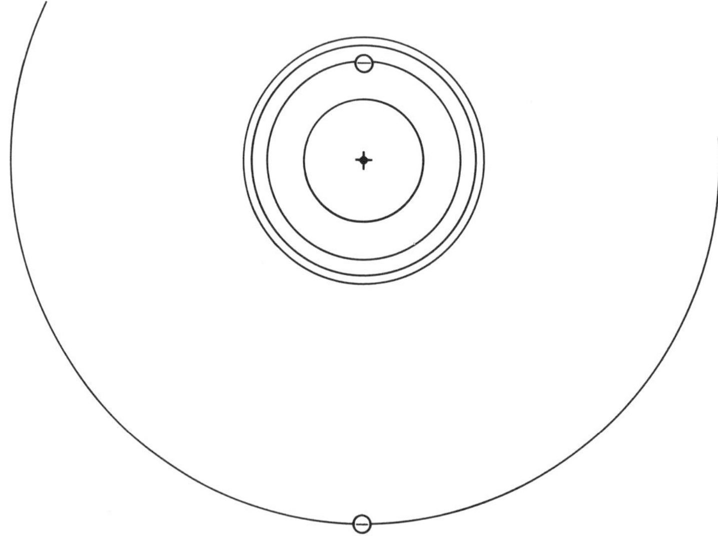

Figure 11.1 Classical representation of a Bohr hydrogen atom with an additional electron bound temporarily so as to form an H$^-$ ion.

pressure. So H-minus is less important than hydrogen as an opacity source for giants than for main sequence dwarfs of the same spectral type.

The opacity of a single H-minus ion has been calculated by a number of people over the past 40 years and is not a simple quantum mechanical calculation. The weak binding of the additional electron allows for the existence of a single bound state from which bound-free transitions make contributions to the continuous opacity. The contribution from the bound-free transitions has a peak at about 0.8 micrometer (μm) and is roughly bell-shaped in frequency with a half width of about 1 μm. The free-free contribution rises steadily into the infrared zone, becoming equal to the bound-free absorption at about 1.45 μm.

d   Important Sources of Continuous Opacity for Main Sequence Stars

OB Stars   For the hottest stars on the main sequence (the O stars), most of the hydrogen is ionized along with a substantial amount of the helium. The largest source of continuous opacity is therefore due to electron scattering. From spectral type B2 to O, neutral helium joins electron scattering as an important opacity source. In the early-type stars later than B2, hydrogen becomes a significant opacity source which increases in importance as one considers later B-type stars. In these stars, the opacity due to bound-bound (see Section 15.4) and bound-free transitions in the metals becomes increasingly important for the ultraviolet part of the spectrum.

Stars of Spectral Type A0 to F5   At spectral type A0, neutral hydrogen is the dominant source of continuous opacity. As one moves to later spectral types, the H-minus opacity increases in importance and dominates the opacity in the late A-type and F stars. For stars of spectral type F, the H_2 molecule emerges as an important opacity source. Although the H_2 molecule cannot exist for long at these temperatures, enough does exist at any instant that the once ionized form H_2^+ provides up to 10 percent of the continuous opacity. The H-minus opacity continues to grow in importance throughout this range of spectral type.

Stars of Spectral Type F5 to G   In this range H-minus continues to provide more than 60 percent of the continuous opacity. However, molecules are not of particular importance until one gets to spectral types later than G. The continuous opacity of metals is particularly important in the ultraviolet range below 3000 Å. The opacity due to atomic hydrogen diminishes steadily into the G stars, but is still of major importance.

Late Spectral Type K to M Stars   Very little of the spectral energy distribution of these stars can be considered to result from continuum processes—particularly in the later spectral types. The absorption from the myriads of discrete transitions of molecules so dominates the spectrum of the late-type stars that little is apparent except the large molecular absorption bands. However, insofar as continuum processes still take place, Rayleigh scattering from H_2 molecules is the most important source. Absorption arising from the disassociation of molecules also provides an important source of continuous opacity, particularly in the infrared region of the spectrum.

11.3 Einstein Coefficients and Stimulated Emission

An extremely useful way to view the processes of absorption is to consider the specific types of interactions between a photon and an atom. This view will become the preferred one when we consider problems where we cannot assume LTE. In quantum mechanics, it is customary to think of the probability of the occurrence of a specific event rather than the classical picture of the frequency with which the event occurs. These are really opposite sides of the same coin, but they yield somewhat philosophically different pictures of the processes in question. In considering the transition of an atom from one excited state to another, Einstein defined a set of coefficients that denote the probability of specific transition's taking place. These are known as the *Einstein coefficients*. All radiative processes and the equation of radiative transfer itself may be formulated in terms of these coefficients. The coefficients are determined by the wave functions of the atom alone and thus do not depend in any way on the environment of the atom. (As with most rules, there is an exception. If the density is sufficiently high that the presence of other atoms distorts the wave functions of the atom of interest, then the Einstein coefficients of that atom can be modified.)

Consider an atomic transition which takes an atom from a lower-energy state of excitation n' to a higher-energy state n (or vice versa). The upper state n may be a bound or continuum state, and the transition will involve the absorption or emission of a photon. The number of transitions that will occur in a given time interval for an ensemble of atoms will then depend on the probability that one transition will occur times the number of atoms available to make the transition. Three types of transitions can occur:

1. *Spontaneous emission,* where the electron spontaneously makes a downward transition with the accompanying emission of a photon.

2. *Stimulated absorption,* where a passing photon is absorbed, producing the resulting transition.

3. *Stimulated emission,* where the electron makes a downward transition with the accompanying emission of a photon. If this occurs in the presence of a photon of the same type as that emitted by the transition, the probability of the event is greatly enhanced.

The symmetric process of spontaneous absorption simply cannot occur because one cannot absorb what is not there.

We can describe the number of atoms undergoing these processes in terms of the probabilities of the occurrence of a single event:

$$N_{n \to n'} = N_n A_{nn'} \, dt$$

$$N_{n' \to n} = N_{n'} B_{n'n} I(v_{nn'}) \, dt \qquad (11.3.1)$$

$$N_{n \to n'} = N_n B_{nn'} I(v_{n'n}) \, dt$$

The dependence of the stimulated processes on the presence of photons is made clear by the inclusion of the specific intensity, corresponding to the energy of the transition, in the last two equations. The coefficients that appear in these equations are known as the Einstein coefficient of spontaneous emission $A_{nn'}$, the Einstein coefficient of absorption (stimulated) $B_{n'n}$, and the Einstein coefficient of stimulated emission $B_{nn'}$. Since there is only one kind of absorption processes, the adjective *stimulated* is usually dropped from the coefficient $B_{n'n}$.

a Relations among Einstein Coefficients

The three Einstein coefficients are not linearly independent. Indeed, the specification of any one of them allows the determination of the other two. To show this, we construct an environment where we know something about the rates at which processes should take place. Since the Einstein coefficients are atomic constants, they are independent of the environment, and thus we are free to create any environment that we choose as long as it is physically self-consistent. With this in mind, consider a gas that is in STE. Under these conditions, the atomic transition rates in and out of each level are equal (detailed balancing). If this were not the case, cyclical processes could exist that would provide for a flow of energy from one frequency to another. But the assumption of STE requires that the photon energy distribution be given by the Planck function, and that situation would not be preserved by an energy flow in frequency space. Therefore, it cannot happen in STE. In addition, in STE the Boltzmann excitation formula holds for the distribution of atoms among the various states of excitation. Thus, the second of equations (11.3.1) must be equal to the sum of the other two.

$$N_{n'} B_{n'n} B_v(T) = N_n [A_{nn'} + B_{nn'} B_v(T)] \qquad (11.3.2)$$

But since the Boltzmann formula must hold,

$$\frac{N_n}{N_{n'}} = \frac{g_n}{g_{n'}} e^{-h v_{nn'}/(kT)} \qquad (11.3.3)$$

Substituting in the correct form for the Planck function [see equation (1.1.24)] and noting that the frequency v is the same as the frequency that appears in the excitation energy of the Boltzmann formula $v_{nn'}$, we get

$$A_{nn'}\frac{g_n}{g_{n'}} = \frac{2hv^3}{c^2}B_{n'n}\frac{e^{hv/(kT)} - B_{nn'}g_n/(B_{n'n}g_{n'})}{e^{hv/(kT)} - 1} \qquad (11.3.4)$$

Now the Einstein coefficients are atomic constants and therefore must be independent of the temperature. This can happen only if the numerator of the rightmost fraction in equation (11.3.4) is identically 1. Thus,

$$\frac{B_{nn'}g_n}{B_{n'n}g_{n'}} = 1 \qquad A_{nn'} = \left(\frac{2hv^3}{c^2}\frac{g_{n'}}{g_n}\right)B_{n'n} \qquad (11.3.5)$$

These relationships must be completely general since the Einstein coefficients must be independent of the environment.

b   Correction of the Mass Absorption Coefficient for Stimulated Emission

In deriving the equation of radiative transfer, we took no notice of the concept of stimulated emission. The mass emission coefficient j_v implicitly contains the notion since it represents the total energy emitted per gram of stellar material. However, the mass absorption coefficient κ_v was calculated as the effective cross section per gram of stellar material and thus counts only those photons absorbed. Should a passing photon stimulate the production of an additional photon, that processes should be counted as a "negative" absorption. Indeed, some authors call the coefficient of stimulated emission the *coefficient of negative absorption*. Now it is a property of the stimulated emission process that the photon produced by the passage of another photon has exactly the same direction, energy, and phase. Indeed, this is the mechanism by which lasers work and which we discuss in greater depth in the chapters dealing with line formation. To correct the absorption coefficient for the phenomenon of stimulated emission, we need only conserve energy.

Consider the total energy produced within a cubic centimeter of a star and flowing into a solid angle $d\Omega$:

$$j_v\rho\,dv\,d\Omega = hvN_n(A_{nn'} + B_{nn'}I_v) = N_nA_{nn'}hv\left(1 + \frac{c^2I_v}{2hv^3}\right) \qquad (11.3.6)$$

This must be equal to the total energy absorbed in that same cubic centimeter from the same solid angle:

$$I_v\kappa_v\rho\,dv\,d\Omega = N_{n'}B_{n'n}I_vhv \qquad (11.3.7)$$

Now consider an environment that is in LTE and in which scattering is unimportant. The source function for such an atmosphere is then

$$S_\nu(\tau_\nu) = \frac{j_\nu}{\kappa_\nu} = \frac{N_n}{N_{n'}} \frac{2h\nu^3}{c^2} \frac{g_{n'}}{g_n} \left(\frac{1 + c^2 I_\nu}{2h\nu^3} \right)$$

$$S_\nu = B_\nu(1 - e^{-h\nu/(kT)}) + I_\nu e^{-h\nu/(kT)}$$

(11.3.8)

Now, if we insert this form for the source function into the plane-parallel equation of radiative transport, we get

$$\mu \frac{dI_\nu}{d\tau_\nu} = I_\nu - S_\nu = (I_\nu - B_\nu)(1 - e^{-h\nu/(kT)})$$

(11.3.9)

However, our original equation for this problem had the form

$$\mu \frac{dI_\nu}{d\tau'_\nu} = I_\nu - B_\nu$$

(11.3.10)

We can bring equation (11.3.9) into the same form as equation (11.3.10) by simply redefining the mass absorption coefficient κ_ν as

$$\kappa'_\nu = \kappa_\nu(1 - e^{-h\nu/(kT)}) \qquad d\tau'_\nu = -\kappa'_\nu \rho \, dx$$

(11.3.11)

Modifying the mass absorption coefficient by the factor $1 - e^{h\nu/(kT)}$ simply corrects it for the effects of stimulated emission. Once again we are correcting atomic parameters that are independent of their environment, and so the result is independent of the details of the derivation. That is, the correction term is a general one and applies to all problems of radiative transfer. Thus, one should always be sure that the absorption coefficients are corrected for stimulated emission. This is particularly true when one is using tabular opacities that are generated by someone else.

11.4 Definitions and Origins of Mean Opacities

In Chapter 4, we averaged the frequency-dependent opacity over wavelength to obtain the Rosseland mean opacity. We claimed that this was indeed the correct opacity to use in the case of STE. That such a mean should exist in the case of STE seemed reasonable since the fundamental parameters governing the structure of the gas should depend on only the temperature. That the appropriate average should be the Rosseland mean is less obvious. In the early days of the study of stellar atmospheres a considerable effort was devoted to reducing the nongray problem of radiative transfer to the gray problem, because the gray atmosphere had been well studied and there were numerous methods for its description. The general idea was that there should exist some "mean" opacity that would reduce the problem to the gray problem or perhaps one that was nearly gray. We know now that such a mean does not exist, but the arguments used in the search are useful to review if for no other reason than many of the mean opacities that were proposed can still be found in the literature and often have some utility in describing the properties of an atmosphere. They are often used for calculating optical depths which label various depth points in the atmosphere. For a particularly good discussion of mean opacities, see Mihalas.[4]

Consider the equation of radiative transfer for a plane-parallel atmosphere and its first two spatial moments. Table 11.1 contains these expressions for both the gray and nongray case. Ideally, we would like to define a mean opacity so that all the moment equations for the nongray case look mathematically like those for the gray case.

a *Flux-Weighted (Chandrasekhar) Mean Opacity*

Suppose we choose a mean so that the last of the moment equations takes on the gray form. We can obtain such a mean by

$$-\frac{1}{\rho}\frac{d}{dx}\int_0^\infty K_v\,dv = -\frac{1}{\rho}\frac{dK}{dx} = \frac{1}{4}\int_0^\infty \kappa_v F_v\,dv \equiv \frac{1}{4}\langle\kappa_v\rangle_F F \tag{11.4.1}$$

so that the mean is defined by

$$\langle\kappa_v\rangle_F = \int_0^\infty \frac{\kappa_v F_v\,dv}{F} \tag{11.4.2}$$

Table 11.1 Equations of Radiative Transfer for a Plane-Parallel Atmosphere

	Gray Atmosphere	Non-gray Atmosphere
(1)	$\mu\dfrac{dI}{d\tau} = I - J$	$\mu\dfrac{dI_v}{d\tau_v} = I_v - J_v$
(2)	$\dfrac{dF}{d\tau} = 0$	$\dfrac{dF_v}{d\tau_v} = 4(J_v - S_v)$
(3)	$\dfrac{dK}{d\tau} = \dfrac{F}{4}$	$\dfrac{dK_v}{d\tau_v} = \dfrac{F_v}{4}$

The use of such a mean will indeed reduce the nongray equation for the radiation pressure gradient (3) to the gray form. Such a mean opacity is often referred to as a *flux-weighted mean*, or the *Chandrasekhar mean opacity*. Unfortunately, such a mean will not reduce either of the other two moment equations from the nongray to the gray case. However, it does yield a simple expression for the radiation pressure gradient which is useful for high-temperature atmospheres where the radiation pressure contributes significantly to the support of the atmosphere, namely,

$$\frac{dP_r}{d\tau_F} = \frac{\sigma T_{\text{eff}}^4}{c} \qquad d\tau_F = -\langle \kappa_v \rangle_F \rho \, dx \qquad (11.4.3)$$

b   Rosseland Mean Opacity

Suppose we require of the third equation in Table 11.1 that

$$\int_0^\infty F_v \, dv = F \qquad (11.4.4)$$

This quite reasonable requirement means that the nongray moment equation will become

$$\int_0^\infty \frac{1}{\rho \kappa_v} \frac{dK_v}{dx} \, dv = -\frac{1}{4} \int_0^\infty F_v \, dv = -\frac{F}{4} \equiv \frac{1}{\rho \langle \kappa_v \rangle_R} \int_0^\infty \frac{dK_v}{dx} \, dv \qquad (11.4.5)$$

which implies that the mean opacity has the following form:

$$\frac{1}{\langle \kappa_v \rangle} = \frac{\int_0^\infty (1/\kappa_v)(dK_v/dx) \, dv}{\int_0^\infty (dK_v/dx) \, dv} \qquad (11.4.6)$$

As one moves more deeply into the star, the near isotropy of the radiation field and approach to STE will require that

$$K_v \rightarrow \frac{J_v}{3} \rightarrow \frac{B_v(t)}{3} \qquad (11.4.7)$$

Since the Planck function depends on the temperature alone,

$$\frac{dB_v(T)}{dx} = \frac{\partial B_v}{\partial T} \frac{dT}{dx} \qquad (11.4.8)$$

Substitution of equations (11.4.7) and (11.4.8) into equation (11.4.6) yields

$$\frac{1}{\langle \kappa_v \rangle_R} \equiv \frac{\int_0^\infty (1/\kappa_v)\partial B_v(T)/\partial T \, dv}{\int_0^\infty [\partial B_v(T)/\partial T] \, dv} \tag{11.4.9}$$

which is identical to equation (4.1.18) for the appropriate mean for stellar interiors. Thus, under the conditions of STE, we find that the Rosseland mean opacity does indeed remove the frequency dependence from the radiative transfer problem. However, in the stellar atmosphere the conditions required by equation (11.4.7) do not apply, so that the Rosseland mean will not fulfill the same function for the theory of stellar atmospheres as it does in the theory of stellar interiors.

c Planck Mean Opacity

Finally, let us consider a mean opacity that will yield a correct value for the thermal emission. Thus,

$$\langle \kappa_v \rangle_P \equiv \frac{\int_0^\infty \kappa_v B_v(T) \, dv}{\int_0^\infty Bv(T) \, dv} \tag{11.4.10}$$

To appreciate the utility of this mean, we develop the condition of radiative equilibrium for the nongray case and see how that approaches the gray result. Let us see what condition would be placed on a mean opacity in order to bring radiative equilibrium in line with the gray case. Consider a mean opacity such that

$$\int_0^\infty (\kappa_v - \bar{\kappa}_v)[J_v(\tau_v) - B_v(\tau_v)] \, dv = 0 \tag{11.4.11}$$

This is just a condition that expresses the difference between the nongray condition for radiative equilibrium and the gray condition, both of which should be zero. Now from equation (10.1.13), we can write J_v as

$$J_v(\tau_v) = \tfrac{1}{2} \int_0^\infty S_v(t)E_1 |t - \tau_v| \, dt \tag{11.4.12}$$

Under the condition that $S_v(t) = B_v[T(\tau)]$, we can expand the source function in equation (11.4.12) in a Taylor series about τ_v and integrate term by term, to obtain

$$J_v(\tau_v) \approx \tfrac{1}{2} B_v(\tau_v)[2 - E_2(\tau_v)] + \cdots + \tag{11.4.13}$$

Thus, near the surface of the atmosphere

$$J_\nu(\tau_\nu) - B_\nu(\tau_\nu) \approx -\tfrac{1}{2}B_\nu(\tau_\nu) \qquad \tau_\nu \ll 1 \qquad (11.4.14)$$

Substitution of equation (11.4.14) into equation (11.4.11) yields

$$\bar{\kappa}_\nu = \frac{\displaystyle\int_0^\infty \kappa_\nu B_\nu(T)\,d\nu}{\displaystyle\int_0^\infty B_\nu(T)\,d\nu} \equiv \langle \kappa_\nu \rangle_P \qquad (11.4.15)$$

Thus, the Planck mean opacity is the most physically relevant mean for satisfying radiative equilibrium near the surface of the atmosphere.

These various means all have their regions of validity and utility, but none can fulfill the promise of reducing the nongray atmosphere problem to that of a gray atmosphere. Indeed, clearly such a mean does not exist. Three separate moment equations are listed in Table 11.1, and a mean opacity represents only one parameter available to bring them all into conformity with their gray analogs. For an arbitrary behavior for κ_ν, this is impossible. Thus, we have to be content with solving the nongray radiative transfer problem at all frequencies for which a significant amount of flux flows through the atmosphere.

11.5 Hydrostatic Equilibrium and the Stellar Atmosphere

Any discussion of the environment of the radiation field would be incomplete without some mention of the motions to be expected within the atmosphere. Recently, considerable effort has been expended in describing radiation-driven winds that originate in the outer reaches of the atmospheres of hot stars. This more advanced topic is well beyond the scope of our interest at this point. Instead, we take advantage of the fact that radiation is the primary mode of energy transport through the stellar atmosphere, and we assume that the atmosphere is in hydrostatic equilibrium. While turbulent convection is present in many stars, the motions implied by the observed velocities are likely to cause problems only for the stars with the lowest surface gravities. For stars on the main sequence, hydrostatic equilibrium is an excellent assumption.

The notion of hydrostatic equilibrium has been mentioned repeatedly throughout this book from the general concept obtained from the Boltzmann transport equation in Chapter 1, through the explicit formulation for spherical stars given by equation (2.1.6), to the representation for plane-parallel atmospheres found in equation (9.1.1). However, the introduction of the concept of optical depth allows for a further refinement. Since the logical depth variable for radiative transfer is the optical depth variable τ_ν, it makes some sense to reformulate the other structure equations that

307

explicitly involve the depth coordinate to reflect the use of the optical depth and the independent variable of the structure. Some atmosphere modeling codes use a variant of equation (9.1.1)

$$\frac{1}{\rho}\frac{dP}{dx} = -g \qquad (11.5.1)$$

so that the quantity $\rho\, dx$ is the depth variable. A more common formulation takes advantage of the definition of the optical depth

$$d\tau_\nu = -(\kappa_\nu + \sigma_\nu)\rho\, dx \qquad (11.5.2)$$

which yields the condition for hydrostatic equilibrium in terms of the optical depth. Thus,

$$\frac{dP}{d\tau_\nu} = \frac{g}{\kappa_\nu + \sigma_\nu} \qquad (11.5.3)$$

Having described the environment of the radiation field in terms of the state variables P, T, and ρ, we can now proceed to the actual construction of a model atmosphere.

Problems

1. Find the ratio of ionized to neutral hydrogen at τ_λ ($\lambda = 5000$ Å) $= 1$ in (*a*) the sun, (*b*) Sirius, and (*c*) Rigel.

2. Find the ratio of SiI to SiII to SiIII in η Ursa Majoris.

3. Consider an atmosphere made up of pure hydrogen that is almost all in a neutral state. The radiation pressure is negligible, and the opacity is due to the H-minus ion.

(*a*) Determine how the pressure at a given optical depth depends on the surface gravity (i.e., how it would change if g were changed).

(*b*) Express the pressure as a function of the temperature. Leave the answer in integral form.

(*c*) How can one find the pressure as a function of physical depth. Display explicitly all equations.

(*d*) Suppose that the effective temperature is T_e and that one wants to find the pressure at the point in the atmosphere corresponding to that temperature. Further, suppose that the surface temperature is in error by 10 percent. Estimate the resultant error in $P(T_e)$.

4. Use a model atmosphere code to construct a model as described in problem 3 (try $T_e = 6000$ K and $\log g = 3$ and $\log g = 4$). Compare your results with those of problem 3.

5. Consider a gas composed of 60 percent hydrogen and 40 percent helium at 10,000 K in a gravity field where $\log g = 4.5$.

(*a*) Find the ratio of neutral to ionized hydrogen and the ratio of the three states of ionization for helium.

(*b*) Find the ratio of the level populations for the first four levels of excitation in neutral hydrogen.

6. Use a model atmosphere code to find how the state of ionization varies with physical depth in a star with $T_e = 10,000$ K and $\text{Log} \, g = 4.0$. Repeat the calculation for a star with $T_e = 7000$ K and $\text{Log} \, g = 1.0$. Compare the two cases.

References and Supplemental Reading

1. For additional reading on the subject of stellar atmospheric opacities, consult:

• Aller, L. H.: *The Atmospheres of the Sun and Stars*, 2d ed., Ronald Press, New York, 1963, chap. 4, pp. 141–199.

• Mihalas, D.: *Stellar Atmospheres*, W. H. Freeman, San Francisco, 1970, chap. 4, pp. 81–128.

• Mihalas, D.: *Stellar Atmospheres*, 2d ed., W. H. Freeman, San Francisco, 1978, chap. 4, pp. 77–107.

• Mihalas, D.: *Methods in Computational Physics* (Eds.: B. Alder, S. Fernbach, M. Rotenberg), Academic, New York, 1967 pp. 19–23.

2. Ueno, S., S. Saito, and K. Jugaku, "Continuous Absorption Coefficients of the Model Stellar Atmosphere," *Cont. Inst. Ap. Univ. of Kyoto*, no. 43, 1954, pp. 1–28.

3. Stewart, A. and T. Webb, "Photo-ionization of Helium and Ionized Lithium," *Proc. Phys. Soc. (London)* 82, 1963, pp. 532–536.

4. Mihalas, D. *Stellar Atmospheres*, W. H. Freeman, San Francisco, 1970, chap. 2, pp. 37–41.

12

Construction of a Model
Stellar Atmosphere

$\cdot \quad \cdot \quad \cdot$

12.1 A Statement of the Basic Problem

We have now acquired all the material necessary to construct a model of the atmosphere of a star. This material includes not only the dependence of the state variables P, T, and ρ with depth in the atmosphere but also an approximation to the emergent spectrum. That predicted spectrum will not contain the details of the stellar absorption lines, but will show the departures from the Planck function of a radiation field in thermodynamic equilibrium. The major departures are caused by the absorption edges corresponding to the ionization limits for the elements included in the calculation. Even if there were no such discontinuities in the frequency dependence of the absorption coefficient, the emergent spectra would still differ from those of a blackbody. Since the photons emerge from different depths in the atmosphere, having different temperatures, even a gray atmosphere spectrum will depart from the Planck function. The more sophisticated spectrum results from the solution of the equation of radiative transfer, the calculation of which represents a major part of the construction of a model atmosphere.

In developing this material, we have used the same conservation laws that yielded the equations of structure for the stellar interior. However, the resulting formulation is somewhat different. The conservation of momentum yielded the expression for hydrostatic equilibrium, as it did in the stellar interior. However, the assumption of a plane-parallel structure for the atmosphere and the use of a different depth coordinate have caused the expression of hydrostatic equilibrium to take a somewhat different form from that for the stellar interior. The conservation of energy is at the root of radiative equilibrium. This condition is imposed on the Boltzmann transport equation itself which was used to produce the equation of radiative transfer. However, because of the departure of the radiation field from STE, the flow of radiation was described by an integral equation, implying that the solution at any point depends on the solution at all points. As a result, we no longer have the simple differential equation for the radiative gradient that was appropriate for the stellar interior. Even the equation of state, which results from saying that the local velocity field of the particles is largely isotropic and dominates any macroscopic flow velocity, is present in basically the form used in the stellar interior. Although the calculation of the mass absorption coefficient appears to present a greater problem for stellar atmospheres, this is largely an illusion. The construction of an accurate model interior requires careful calculation of the frequency-dependent absorption coefficient, and the range of atomic phenomena that must be included is actually greater than that of an atmosphere because of the greater range of possible ionization states. However, in the stellar atmosphere, the frequency dependence of the absorption coefficient enters far more directly into the solution and plays a greater role. The use of the Rosseland mean opacity for stellar interiors tends to average out the "mistakes" in the opacity calculations whereas those mistakes in a stellar atmosphere are directly visible in the emergent spectrum. The presence of molecules is an added complication for the theory of stellar atmospheres that does not plague the theory of stellar interiors.

We can summarize the equations of atmospheric structure, obtained from these conservation laws and assumptions:

$$\frac{dP}{d\tau_\nu} = \frac{g}{\kappa_\nu + \sigma_\nu} \qquad \text{hydrostatic equilibrium}$$

$$\tag{12.1.1}$$

$$d\tau_\nu = -(\kappa_\nu + \sigma_\nu)\rho\, dx \qquad \text{definition of } \tau_\nu$$

311

$$P = \frac{\rho k T}{\mu m_h} + P_r \qquad \text{equation of state}$$

$$\mathbf{F} = \frac{\sigma T_e^4}{\pi} \qquad \text{definition of } T_e$$

$$\int_0^\infty \kappa_\nu \{ J_\nu(\tau_\nu) - B_\nu[T(\tau_\nu)] \}\, d\nu = 0 \qquad \text{radiative equilibrium}$$

$$S_\nu(\tau_\nu) = \epsilon_\nu(\tau_\nu) B_\nu[T(\tau_\nu)] + \tfrac{1}{2}[1 - \epsilon_\nu(\tau_\nu)] \int_0^\infty S_\nu(t) E_1 |\tau_\nu - t|\, dt$$

$$\text{integral equation of radiative transfer}$$

$$(12.1.1)$$

Much has been concealed by writing the opacity as a function of the state variables. But while the details are messy and LTE has been assumed, the process is straightforward. The major difference between the calculation of a stellar interior and the construction of a model stellar atmosphere can be seen in the last of equations (12.1.1). No longer do we have a situation that can be mathematically described as a linear two-point boundary-value problem. Because of the assumption of plane parallelism, the "eigenvalues" of the problem have been reduced to two, T_e and g. In addition, the four nonlinear differential equations of the interior structure have been replaced by one first-order differential equation and an integral equation for the source function from which all physically relevant moments of the radiation field can be calculated. The global nature of this integral equation forces a rather different approach to the construction of a model stellar atmosphere from that adopted for the stellar interior.

In general, we split the problem into two parts, each of which can be solved with the knowledge of the other. After making as shrewd a guess as possible for the approximate solution of one, we solve the other and use those results to improve the initial guess for the first. We can then proceed to solve these two halves of the problem alternately until we obtain an answer that is self-consistent and satisfies the conditions of radiative equilibrium throughout the atmosphere. The basic division of the problem is to calculate the depth dependence of the state variables under the assumption of the radiation field and then to use this atmospheric structure to improve the radiation field. Since the initial guess for the radiation field is not likely to be correct, we cannot expect that radiative equilibrium will be satisfied throughout the atmosphere. Thus we will try to use the calculated departures from radiative equilibrium to modify the physical structure so as to produce a radiation field that more nearly satisfies radiative equilibrium. Since we have already dealt with the solution of the equation of radiative transfer, most of this chapter involves the iterative aspect of the problem. Proper formulation of such a correction scheme will provide the basis for forming a rigorous iterative algorithm that will converge to a fully self-consistent model atmosphere with a structure that yields a radiation field satisfying radiative equilibrium throughout the entire atmosphere. However, we must begin with some comments on how to find the dependence of state variables on depth in the atmosphere, given the radiation field. This involves the solution of the differential equation for hydrostatic equilibrium.

12.2 Structure of the Atmosphere, given the Radiation Field

At the outset of any atmosphere calculation, we must decide on the particular atmosphere to be modeled. Choosing the parameters T_e, g, and μ is analogous to choosing M, L, R, and μ for the construction of a model stellar interior. Indeed, the relationship between them is straightforward:

$$T_e = \left(\frac{L}{4\pi\sigma R^2} \right)^{1/4} \qquad g = \frac{GM}{R^2} \qquad (12.2.1)$$

a Choice of the Independent Variable of Atmospheric Depth

Before beginning the model calculation itself, we must choose a depth parameter to serve as an independent variable. While the traditional choice has always been the optical depth, some more modern atmosphere codes use a "density depth" like

$$h(x_0) = \int_0^{x_0} \rho(x)\, dx \qquad (12.2.2)$$

as the independent depth variable. As we shall see, this greatly simplifies the calculation of hydrostatic equilibrium, but introduces some difficulties in the solution of the equation of transfer. For the most part, we use the traditional optical depth as our independent depth variable.

Which optical depth should we use? Early investigators would pick one of the mean opacities to generate a mean optical depth, and this dimensionless, frequency-independent variable would provide an excellent parameter for describing the atmospheric structure. Unfortunately, the calculation of the mean opacity at numerous depths in the atmosphere is a nontrivial undertaking and is completely avoidable. Since there is no particular significance for any of the mean opacities, no optical depth scale is to be preferred over any other on the basis of the physical information contained there. Therefore, it makes good sense to pick a monochromatic optical depth at some frequency $\tau(\nu_0)$ (or simply τ_0) as the depth parameter, thereby avoiding the tedious calculation of the mean opacity and the associated mean optical depths. However, since the radiative transfer equation must be solved at each frequency, it will be necessary to interpolate the solution to the reference optical

313

depth τ_0. So it would be wise to choose a frequency in the general vicinity of the maximum energy flow through the atmosphere and in a part of the spectrum where the opacity does not vary rapidly with frequency. This will tend to minimize interpolation errors when the solutions are transferred to the reference optical depth. For the majority of the development in this chapter, this is the choice that we make. The relevant optical depths are then given by

$$d\tau_0 = (\kappa_0 + \sigma_0)\,\rho dx$$

$$d\tau_\nu = \frac{\kappa_\nu + \sigma_\nu}{\kappa_0 + \sigma_0}\,d\tau_0 \equiv k_\nu\,d\tau_0 \qquad (12.2.3)$$

The parameter k_ν is just a normalized opacity which relates the differential monochromatic optical depth to its counterpart on the reference depth scale.

b Assumption of Temperature Dependence with Depth

Having specified the nature of the atmosphere and chosen the depth parameter, we can begin the calculation of the atmospheric structure with depth. We have indicated that we will split the process of making the model into two parts by assuming the results of the radiative transfer solution in order to calculate the atmospheric structure. The form that this assumption takes is the dependence of the temperature with depth. That is, the end result for the solution of the radiative transfer calculation will be to generate the dependence of the temperature with depth in a manner that is consistent with radiative equilibrium. Thus, to begin our calculation, we must assume the existence of this temperature distribution. We may obtain this information either as a result of an earlier model calculation or from an initial approximation.

In Section 10.2, we spent considerable effort in solving the equation of transfer for the gray atmosphere. One of the results of this effort was the temperature distribution in the Eddington approximation [equation (10.2.16)]. Remembering that for a gray atmosphere in radiative equilibrium

$$S(\tau) = J(\tau) = B(\tau) \qquad (12.2.4)$$

we can write the more general result

$$T^4(\tau) = \tfrac{3}{4}T_e^4[q(\tau) + \tau] \qquad (12.2.5)$$

where $q(\tau)$ is the Hopf function specified in equation (10.2.34). Although the gray atmosphere does not specify a unique physical atmospheric structure, it does provide a temperature distribution that scales with the effective temperature and is consistent with radiative equilibrium. In addition, the opacity in a wide variety of stars is relatively independent of frequency over a large part of the spectrum, so that the gray atmosphere temperature distribution provides a good first approximation to the actual temperature distribution. However, the accuracy of this assumption does depend on the choice of reference optical depth being representative of the atmosphere as a whole, so we can only expect it to form an approximate first guess.

c Solution of the Equation of Hydrostatic Equilibrium

The equation of hydrostatic equilibrium is a deceptively simple-looking, first-order differential equation. There are many sophisticated methods for obtaining the numerical solution to such an equation, but all such methods involve knowing at least one value (and usually several values) for the solution at and near the boundary. This poses both philosophical and practical problems. The boundary of a plane-parallel atmosphere, while well located in optical depth, is poorly placed in physical depth since it formally occurs where the density vanishes. In principle, this occurs at an infinite distance from the star where the plane-parallel approximation itself would no longer be valid. However, in practice, a small but finite optical depth is reasonably located with respect to the photosphere (i.e., near optical depth $\tau \approx \frac{2}{3}$) so that boundary conditions can be specified there without jeopardizing the accuracy of the solution. Since we have assumed a distribution of temperature with optical depth, there is no problem in determining the boundary temperature.

In Section 4.1a we discussed how to relate the mass fractions of hydrogen, helium, and "metals" to the mean molecular weight and thereby provide a connection between the number and mass density. The Saha equations for each element [equation (11.1.16)] provide a relationship between the relative ionization fraction, the temperature, and the electron pressure. By remembering that the sum of all the various states of ionization for a particular element is simply equal to the number abundance for the element, it is possible to parameterize the opacity in terms of the gas pressure, temperature, and electron pressure. Thus, we may write the total pressure at any optical depth as

$$P(\tau_i) = P_g[T(\tau_i), P_e(\tau_i)] + P_r[T(\tau_i)] \tag{12.2.6}$$

In a similar manner it is possible to integrate the equation of hydrostatic equilibrium [as stated in equation (11.5.3)] so that

$$P(\tau_i) = \int_0^{\tau_i} g\{\kappa_0[T(\tau), P_e(\tau)] + \sigma_0[T(\tau), P_e(\tau)]\}\, d\tau$$

$$\equiv P[T(\tau_i), P_e(\tau_i)] \tag{12.2.7}$$

We may look for a value of P_e that makes equations (12.2.6) and (12.2.7) self-consistent. Numerically this can be accomplished by creating tables of κ_0 and σ_0 as functions of P_e so that the integral in equation (12.2.7) can be done directly by any efficient quadrature scheme and then a solution found by iteration with equation (12.2.6). Details of this procedure are given by D. Mihalas.[1] In carrying out this procedure, one keeps the value of the optical depth τ_i sufficiently small that $T(0) \approx T(\tau_i)$. When one has found a self-consistent value of $P_e(\tau_i)$ (and hence all the state variables), values of the state variables may be interpolated for all intermediate optical depths between 0 and τ_i. This technique will provide all the required values of the pressure to initiate a general numerical integration procedure for the differential equation for hydrostatic equilibrium. Since all the other state variables are given in terms of algebraic expressions, the entire atmospheric structure may be obtained as a function of τ_0.

A note of caution should be interjected at this point concerning the numerical solution described by this procedure. The range in pressures to be expected from the solution is several powers of 10. For this reason, logarithmic variables are often used to improve the stability of the numerical solution. In any case, the method used to solve the equation of hydrostatic equilibrium should be reasonably sophisticated since the rapid initial growth of the pressure, if not carefully dealt with, can produce systematic errors that destroy the accuracy of the entire atmosphere as the integration proceeds. Several investigators have found it necessary to employ up to a seven-point predictor-corrector integration scheme to achieve the accuracy required. In addition, although the remaining equations are indeed algebraic, the Saha equations for the various elements and states of ionization represent a system of coupled nonlinear algebraic equations and must be solved by iteration. Furthermore, the equations for the opacity due to the different elements in their various states of ionization and excitation represent a significant amount of calculation. Thus, the calculation of $\kappa_\nu(\tau_0)$ can be quite time-consuming and represents a significant time burden for the calculation of the model structure. This is particularly evident when one remembers that a multipoint numerical integration scheme requires multiple evaluations of the function g/κ_ν to carry out one step in the integration. The situation is further exacerbated by the realization that a rapidly varying numerical solution to a differential equation usually requires that the solution proceed with very small steps, and the range required for the independent variable will be of the order of 2 powers of 10. This is the reason that some modern atmosphere codes utilize a density depth as given in equation (12.2.2) as the independent depth variable. With this choice, the calculation of the opacity is entirely avoided. However, as we shall see, this choice of an independent variable is not without its own set of problems.

The solution of the equation of hydrostatic equilibrium also provides us with the dependence on depth of all the state variables and the various states of ionization and excitation of the elements. With this information, it is possible to calculate the opacity and hence the radiation field at all points in the atmosphere.

12.3   Calculation of the Radiation Field of the Atmosphere

All Chapter 10 was devoted to solving the equation of radiative transfer, so there is no need to repeat the specifics here. However, some numerical aspects of those solutions require comment. As even the casual reader of Chapter 10 will notice, the general solution of the equation of radiative transfer is fraught with some formidable numerical difficulties. Not the least of these is ensuring the numerical accuracy of the results. Whether one chooses to solve the integrodifferential equation for the specific intensity or the integral equation for the source function of the radiation field, the spacing of the optical depth points at which the solution is to be obtained is crucial for determining the accuracy of that solution.

However, to obtain the radiative flux and source function at a sufficient number of frequencies to accurately evaluate the condition of radiative equilibrium, we will have to solve the equation of transfer repeatedly over optical depths whose range varies widely. For example, the opacity of a normal stellar atmosphere at frequencies greater than the Lyman limit of hydrogen will be enormously greater than for frequencies in the Balmer continuum for any given physical depth in the atmosphere. Hence, the corresponding optical depth will be very much larger. Since practical realities require that any radiative transfer solution for a semi-infinite atmosphere be truncated at a finite physical depth, we can expect that the monochromatic optical depth corresponding to that physical depth will vary greatly with frequency. However, we must ensure that the radiative transfer solutions, which are calculated at a finite set of depth points that yet span a wide range of optical depths, have sufficient accuracy to enable the accurate calculation of radiative equilibrium.

The normal method of accomplishing this is to carry out the numerical solution of the equation of radiative transfer at a predetermined set of optical depths τ_{ri}, chosen to ensure the accuracy of the solution. All physical parameters required for that solution are interpolated from the monochromatic optical depths corresponding to the reference depths onto the set of optical depths to be used for the radiative transfer solution. Mathematically, this amounts to mapping these parameters from the τ_v space on which they are defined onto the τ_r space in which the equation of radiative transfer will be solved. In many cases, the points in the τ_r space can be chosen to be the same points as those used for the reference depth scale τ_0, but occasionally they may be a separate set of points. In this case, a further mapping of the radiative transfer solution from the τ_r space to the τ_0 space must be carried out. The primary reason for this convoluted procedure is to separate the numerical errors into two well-defined categories—those arising from the interpolation and those arising from the solution of the equation of transfer. The latter are generally more difficult to estimate and so are controlled by carrying out the solution over a set of optical depths for which the numerical stability of the radiative transfer solution is well understood. The errors introduced by the interpolation from one optical depth scale to another are generally easier to control. However, a rapid and accurate mapping algorithm must be available. Such an algorithm is contained in the current version of the atmosphere code known as ATLAS.[2]

For the general overall accuracy of the calculation, we require that the most accurate solutions be obtained at those frequencies for which the majority of the radiative flux flows through the atmosphere. If the reference set of optical depth points is chosen to correspond closely to the monochromatic optical depths at those frequencies, then the interpolation errors incurred from the mapping procedures will be minimal. Frequencies at which the atmosphere is very much optically thicker than the reference optical depths will tend to carry less flux simply because the radiation can escape more easily at the frequencies for which the atmosphere is more transparent. Hence, the frequencies at which the atmosphere is relatively optically thick, and for which the interpolation errors of the mapping can be expected to be the greatest, will make a correspondingly smaller contribution to the total flux and to the conditions of radiative equilibrium. However, one must be careful that, at the frequencies for which the atmosphere is most transparent, a sufficient number of optical depth points are chosen to ensure that the maximum optical depth is optically remote from the surface. In practice, this generally means that $\tau_v \gg 10$.

12.4 Correction of the Temperature Distribution and Radiative Equilibrium

Having created an accurate representation of the radiation field from the previously obtained physical structure, we must see how well the solution conforms to the condition of radiative equilibrium. Departures of the radiation field from that required to satisfy radiative equilibrium will form the basis for correcting the temperature distribution throughout the atmosphere. We have developed the concept of radiative equilibrium several times in this book and most recently in Chapter 10 [equations (10.4.4) and (10.4.5)] as

$$\int_0^\infty \kappa_v \rho \{ J_v(\tau_0) - B_v[T_0(\tau_0)] \}\, dv = 0$$

$$\mathbf{F} = \int_0^\infty F_v\, dv = \frac{\sigma T_e^4}{\pi}$$

(12.4.1)

   Even though these two conditions are logically equivalent, their utilizations for generating a temperature correction scheme will be quite different. Although a substantial number of temperature correction schemes have been developed during the last 40 years, we describe only two. The first is chosen for its simplicity and historical interest while the second represents the most widely used method in contemporary use.

a   Lambda Iteration Scheme

The first of equations (12.4.1) is obtained by setting the total flux derivative to zero. In general, the radiation field obtained from our approximate structure will not satisfy this expression. If we assume that the reason for this is that the temperature used to evaluate the local Planck function is incorrect, we can replace the temperature with a first-order Taylor series expansion about the current temperature. Thus,

$$\int_0^\infty \kappa_\nu J_\nu(\tau_0)\, d\nu = \int_0^\infty \kappa_\nu \{B_\nu[T_0(\tau_0) + \delta T]\, d\nu$$

$$= \int_0^\infty \kappa_\nu B_\nu[T_0(\tau_0)]\, d\nu + \delta T \int_0^\infty \kappa_\nu \frac{\partial B_\nu}{\partial T}\, d\nu \qquad (12.4.2)$$

or solving for the temperature correction, we have

$$\delta T(\tau_0) = \frac{\displaystyle\int_0^\infty \kappa_\nu [J_\nu(\tau_0) - B_\nu(\tau_0)]\, d\nu}{\displaystyle\int_0^\infty \kappa_\nu [\partial B_\nu(\tau_0)/\partial T]\, d\nu} \qquad (12.4.3)$$

This is known as the Λ *iteration scheme* since $J_\nu(\tau_0) = \Lambda[B_\nu(\tau_0)]$ [see equation (10.1.16)]. The method yields suitable corrections to the temperature distribution as long as the opacity κ_ν is decidedly nongray. However, as one moves deeper and deeper in the atmosphere, $J_\nu \to B_\nu$ and the integrand vanishes for all frequencies. Thus, this method relies on the departure of the source function from the value it would have in statistical equilibrium to provide corrections to the local temperature. So while the method may produce a useful temperature correction near the surface, the correction will become smaller and smaller as one descends into the atmosphere. This fact will be reflected in the rate at which the atmosphere converges to a self-consistent value. Indeed, it may become difficult to even know when meaningful convergence has been achieved. To make matters worse, equation (12.4.3) guarantees—in principle—that a self-consistent atmosphere with zero total flux derivative can be calculated. However, it may not have the desired flux, $\sigma T_e^4/\pi$. Thus, we should look for a method for correcting the temperature that employs the second of equations (12.4.1) as well as the first. Such a scheme is due

319

to E. Avrett and M. Krook,[3] although it is more lucidly described by D. Mihalas[1] (pp. 35–39).

b Avrett-Krook Temperature Correction Scheme

Since the temperature correction scheme is to form the basis for an iteration algorithm, it is not essential that it produce the correct temperature the first time it is applied. However, repeated application should produce a series of temperature distributions which approach the one that is correct for radiative equilibrium. Thus, all temperature corrections must vanish asymptotically as the sequence approaches radiative equilibrium. This is the only essential criterion for an iteration scheme. Therefore, it is not necessary to justify all assumptions made in establishing the iterative equations as long as the final result converges to a temperature distribution that is consistent with radiative equilibrium.

The beauty of the Avrett-Krook scheme is that it simultaneously uses both expressions of radiative equilibrium as given in equations (12.4.1). There are two ways of correcting the temperature distribution. The first is the obvious one of simply changing the value of the temperature at some given value of the independent variable τ_0. This is the approach taken by the Λ iteration scheme. A second way to find an improved temperature distribution is to find the value of the independent variable, in this case the reference optical depth, for which the given temperature is the correct temperature. This approach amounts to inverting the problem and treating the temperature as the independent variable and perturbing τ_0.

The Avrett-Krook scheme does both, using one statement of radiative equilibrium to calculate a temperature correction and the other condition of radiative equilibrium to find a new value of the optical depth at which the *corrected* temperature is to be applied. Thus, both temperature and optical depth become independent variables in the perturbation calculation. The perturbation equations for temperature are very similar to the Λ iteration equations and therefore provide good corrections near the surface. The perturbation equations for the optical depth yield small corrections near the surface, but become significant at larger optical depths where the Λ iteration scheme is ineffective. Thus, the combination yields a temperature correction scheme which converges fairly quickly throughout the entire atmosphere. Unfortunately, the resulting temperature distribution will not directly give the corrected temperatures at the reference optical depths. However, the appropriate temperatures at the reference optical depths can be obtained from the new temperature distribution by interpolation.

The basic approach is to express both the correct temperature and the optical depth in terms of the given values and a first-order correction to them, namely,

$$\tau_0 = t + \lambda\tau^{(1)}(t) \qquad T(\tau_0) = T^{(0)}(t) + \lambda T^{(1)}(t) \qquad (12.4.4)$$

The parameter λ simply measures the order of significance for the particular term and will eventually be set to unity. Substitution of these expressions into the equation of transfer will produce similar corrections in the parameters that describe the radiation field so that

$$I_\nu(\mu, \tau_0) = I_\nu^{(0)}(t, \mu) + \lambda I_\nu^{(1)}(t, \mu)$$

$$J_\nu(\tau_0) = J_\nu^{(0)}(t) + \lambda J_\nu^{(1)}(t) \tag{12.4.5}$$

$$F_\nu(\tau_0) = F_\nu^{(0)}(t) + \lambda F_\nu^{(1)}(t)$$

We can expand the normalized opacity [equation (12.2.3)] and the Planck function in a Taylor series in t and T, respectively, and get

$$\kappa_\nu(\tau_0) = \kappa_\nu^{(0)}(t) + \lambda \tau^{(1)}(t) \frac{dk_\nu(t)}{dt}$$

$$\tag{12.4.6}$$

$$B_\nu[T(\tau_0)] = B_\nu[T^{(0)}(t)] + \lambda T^{(1)}(t) \frac{dB_\nu[T^{(0)}(t)]}{dT}$$

For simplicity, from now on we denote differentiation with respect to optical depth and temperature by

$$\frac{d}{dt} = ' , \qquad \frac{d}{dT} = \cdot \tag{12.4.7}$$

In addition, for clarity we ignore scattering and treat the problem of pure absorption only. Later we give the perturbation equations appropriate for a source function that includes scattering, justifying the results on physical grounds alone.

Perturbed Equation of Radiative Transfer The general nongray equation of transfer for a plane-parallel atmosphere for the case of pure absorption is

$$\mu \frac{dI(\mu, \tau_0)}{d\tau_0} = k_\nu \{ I_\nu(\mu, \tau_0) - B_\nu[T(\tau_0)] \} \tag{12.4.8}$$

If we insert the expansions given by equations (12.4.4) and (12.4.5) into this equation and ignore all second-order terms (i.e., terms involving λ^2), we get

$$\mu \frac{dI_\nu^{(0)}}{dt} - \lambda \mu \tau'^{(1)} \frac{dI_\nu^{(0)}}{dt} + \mu \lambda \frac{dI_\nu^{(1)}}{dt} = (k_\nu^{(0)} + \lambda \tau^{(1)} k_\nu'^{(0)})(I_\nu^{(0)} - B_\nu^{(0)})$$

$$+ \lambda k_\nu^{(0)}(I_\nu^{(1)} + T^{(1)}\dot{B}_\nu^{(0)}) \tag{12.4.9}$$

Since this equation must hold for any value of λ, we can separate the zeroth- and first-order terms. The zeroth-order equation is then

$$\mu \frac{dI_\nu^{(0)}}{dt} = k_\nu^{(0)}(I_\nu^{(0)} - B_\nu^{(0)}) \tag{12.4.10}$$

We can use this result to eliminate $dI_\nu^{(0)}/dt$ from the first-order equation so that it becomes

$$\mu \frac{dI_\nu^{(1)}}{dt} = k_\nu^{(0)}(I_\nu^{(1)} + T^{(1)}\dot{B}_\nu^{(0)}) + (\tau'^{(1)}k_\nu^{(0)} + \tau^{(1)}k_\nu'^{(0)})(I_\nu^{(0)} - B_\nu^{(0)}) \tag{12.4.11}$$

This equation can be solved by using the Eddington approximation to moments of the equation in a manner that should be familiar by now.

Forming the first two moments of equation (12.4.11) (i.e., just integrating over all μ to obtain the first and multiplying by μ and integrating to get the second), we obtain

$$F'^{(1)} = k_\nu^{(0)}(J_\nu^{(1)} + T^{(1)}\dot{B}_\nu^{(0)}) + (\tau'^{(1)}k_\nu^{(0)} + \tau^{(1)}k_\nu'^{(0)})(J_\nu^{(0)} - B_\nu^{(0)})$$

$$K_\nu'^{(0)} = \frac{J_\nu'^{(1)}}{3} = k_\nu^{(0)}F_\nu^{(1)} + (\tau'^{(1)}k_\nu^{(0)} + \tau^{(1)}k_\nu'^{(0)})F_\nu'^{(0)} \tag{12.4.12}$$

In the second equation, we have already assumed that the Eddington approximation can be applied to the first-order perturbations as it is to the entire radiation field.

Tau Perturbation Equation Now we integrate the second of equations (12.4.12) over all frequencies and get

$$\frac{1}{3} \int_0^\infty \frac{J_\nu'^{(1)}}{k_\nu^{(0)}} d\nu = \int_0^\infty F_\nu^{(1)} d\nu + \tau'^{(1)} \int_0^\infty F_\nu^{(0)} d\nu + \tau^{(1)} \int_0^\infty \frac{k_\nu'^{(0)}}{k_\nu^{(0)}} F_\nu^{(0)} d\nu \tag{12.4.13}$$

Requiring that

$$J_\nu'^{(1)}(t) = 0 \tag{12.4.14}$$

guarantees that the left-hand side of equation (12.4.13) will vanish. The assumption stated by equation (12.4.14) is justified by expediency alone. However, it is an assumption concerning the perturbation only and therefore can affect only the rate of convergence. There may be some instances where this approximation should be replaced. However, to do so, we must know something additional about the

problem.

The first term on the right-hand side of equation (12.4.13) is just the integrated flux error so that

$$\int_0^\infty F_\nu^{(1)}(t)\,d\nu = \mathbf{F} - F^{(0)} \tag{12.4.15}$$

where

$$\mathbf{F} = \frac{\sigma T_e^4}{\pi} \tag{12.4.16}$$

With this, we can rewrite equation (12.4.12) as a first-order linear differential equation for the perturbed optical depth

$$\tau'^{(1)} + \tau^{(1)}\int_0^\infty \frac{k_\nu'^{(0)}}{k_\nu^{(0)}}\frac{F_\nu^{(0)}}{F^{(0)}}\,d\nu = 1 - \frac{\mathbf{F}}{F^{(0)}} \tag{12.4.17}$$

All that remains is to specify a boundary condition for the solution of the equation. An appropriate condition is

$$\tau^{(1)}(0) = 0 \tag{12.4.18}$$

While this condition appears to be arbitrary, it anticipates the result for the T perturbation which will provide the majority of the correction at the surface. The boundary condition given in equation (12.4.18) will ensure that the tau corrections are small near the surface and thus will not compete heavily with the T corrections.

Temperature Perturbation Equation   To obtain the T perturbation equation, we begin with the first of equations (12.4.12). Since we required that the derivative of the perturbed mean intensity $J_\nu^{(1)}$ be zero at all frequencies and depths [equation (12.4.14)], we may get the last term on the right-hand side of the first of equations (12.4.12) from the second equation, so that

$$\tau'^{(1)}k_\nu^{(0)} + \tau^{(1)}k_\nu'^{(0)} = -\frac{k_\nu^{(0)}F_\nu^{(1)}}{F_\nu^{(0)}} \tag{12.4.19}$$

That same assumption on the derivative of the perturbed mean intensity will require that

$$J_\nu^{(1)}(t) = \text{const} = J_\nu^{(1)}(0) = aF_\nu^{(1)}(0) \tag{12.4.20}$$

The last term implies the Eddington approximation; so that a is usually taken to be $\frac{1}{2}$. However, some authors use somewhat different values for a based on empirical work. As with any iteration scheme, one that works is a good one. Remembering that we have assumed a boundary condition on the $\tau^{(1)}$ equation of $\tau^{(1)} = 0$, we see that equations (12.4.17) and (12.4.19) give

$$\tau'^{(1)}(0) = -\frac{F_v^{(1)}(0)}{F_v^{(0)}(0)} = 1 - \frac{\mathbf{F}}{F^{(0)}(0)} \tag{12.4.21}$$

Thus, we may obtain the perturbed value for J as

$$J_v^{(1)}(t) = -aF_v^{(0)}(0)\left[1 - \frac{\mathbf{F}}{F^{(0)}(0)}\right] = \text{const} \tag{12.4.22}$$

Inserting this result and equation (12.4.19) into the first of equations (12.4.12), we get

$$F_v'^{(1)}(t) = k_v^{(0)}(t)\left\{aF_v^{(0)}(0)\left[1 - \frac{\mathbf{F}}{F^{(0)}(0)}\right] + T^{(1)}(t)\dot{B}_v^{(0)}\right\}$$

$$= -\frac{k_v^{(0)}F_v^{(1)}}{F_v^{(0)}}[J_v^{(0)}(t) - B_v^{(0)}] \tag{12.4.23}$$

From the definition of $F_v'^{(1)}$ we know that

$$F_v'^{(1)}(t) = \frac{d}{dt}\left[\mathbf{F} - F_v^{(0)}(t)\right] = -F_v'^{(0)} \tag{12.4.24}$$

Incorporating this into equation (12.4.23), integrating over all frequencies, and remembering that the condition of radiative equilibrium applies to the zeroth-order equations, we finally get the perturbation equation for the temperature as

$$T^{(1)}(t) = \frac{\mathbf{F}\displaystyle\int_0^\infty k_v^{(0)}(t)\left\{a[1 - F_v^{(0)}(0)/\mathbf{F}] + [J_v^{(0)}(t) - B_v^{(0)}(t)]/F_v^{(0)}(t)\right\}dv}{\displaystyle\int_0^\infty \dot{B}_v^{(0)}(t)k_v^{(0)}(t)\,dv} \tag{12.4.25}$$

We now have expressions for the temperature corrections $T^{(1)}(t)$ and the corrected values of the optical depth $t + \tau^{(1)}$ to which they are to apply. Interpolation of this temperature distribution back onto the original optical depth scale completes the temperature correction procedure. A comparison of equations (12.4.25) and

(12.4.3) shows that the Avrett-Krook temperature correction equation is indeed very close to the Λ iteration equation. However, an additive constant appears in the Avrett-Krook equation which ensures that the corrections will converge to the correct flux $\mathbf{F}$.

Perturbation Equations Including Scattering The inclusion of scattering complicates the algebra of deriving the perturbation equations, but not the concept. However, the essence of the problem can be seen without suffering through the algebra of the derivation. Consider a very general source function such as that given in equation (10.1.7). The parameter ϵ_ν is a measure of the fraction of photon interactions that can be viewed as pure absorptions. Thus, $1 - \epsilon_\nu$ is the relative fraction of scatterings. Since at the microscopic level scattering is a fully conservative process, we should expect it to have no influence on the physical structure of the atmosphere. Scattering decouples the radiation field from the physical domain of the gas. Thus, any temperature correction procedure will become less well defined for an atmosphere where the opacity becomes more nearly gray.

To carry out the perturbation analysis, we must add a perturbation equation for $\epsilon_\nu(\tau_0)$ similar to equations (12.4.6). It could take the form

$$\epsilon_\nu(\tau_0) = \epsilon_\nu^{(0)}(t) + \lambda\tau^{(1)}(t)\epsilon_\nu'^{(0)}(t) \tag{12.4.26}$$

As with the opacity, an assumption is made that the derivatives with respect to optical depth are more important than the derivatives with respect to temperature. The appropriate equation of radiative transfer analogous to equation (12.4.8) is then

$$\mu\frac{dI(\mu, \tau_0)}{d\tau_0} = k_\nu\{I_\nu(\mu, \tau_0) - \epsilon_\nu(\tau_0)B_\nu[T(\tau_0)] - [1 - \epsilon_\nu(\tau_0)]J_\nu(\tau_0)\} \tag{12.4.27}$$

where k_ν is now defined by

$$k_\nu(\tau_0) \equiv \frac{\kappa_\nu(\tau_0) + \sigma_\nu(\tau_0)}{\kappa_0(\tau_0) + \sigma_0(\tau_0)} \tag{12.4.28}$$

Development of the two moment equations analogous to equations (12.4.12) will show that the second is unchanged by the presence of scattering. This leads to the happy result that the tau perturbation equation is also unchanged, so that equation (12.4.17) and its solution are correct for the more general case including scattering.

The presence of scattering does modify the first-moment equation of equations (12.4.12). This yields a somewhat different temperature perturbation equation from equation (12.4.25). With scattering, it takes the form

$$T^{(1)}(t) = \frac{\mathbf{F} \displaystyle\int_0^\infty k_\nu^{(0)}(t) \left\{ a\epsilon_\nu^{(0)}(t)[1 - F_\nu^{(0)}(0)/\mathbf{F}] + [J_\nu^{(0)}(t) - B_\nu^{(0)}(t)] \times [\tau^{(1)}(t)\epsilon_\nu'^{(0)}(t) + \epsilon_\nu^{(0)}(t)\mathbf{F}/F_\nu^{(0)}(t)]/\mathbf{F} \right\} d\nu}{\displaystyle\int_0^\infty \dot{B}_\nu^{(0)}(t) k_\nu^{(0)}(t)\epsilon_\nu^{(0)}(t)\, d\nu} \qquad (12.4.29)$$

In the limit of pure absorption where $\epsilon_\nu \rightarrow 1$, we recover immediately equation (12.4.25). As we approach the limit of a pure scattering atmosphere, $\epsilon_\nu \rightarrow 0$. All terms in the numerator of equation (12.4.29) clearly vanish. Unfortunately so does the denominator, leaving the asymptotic behavior of $T^{(1)}$ in doubt. An application of L'Hospital's rule shows that the temperature correction terms indeed formally go to zero for the case of pure scattering. However, many of the terms of equation (12.4.29) are difficult to calculate numerically so that the practical result of increased scattering will be to at first slow the rate of convergence of the iteration procedure. The iteration procedure will become unstable as the amount of scattering becomes very large. This is not surprising since the instability merely reflects the decoupling of the radiation field from the physical structure of the atmosphere.

Equations (12.4.17) and (12.4.29) provide the mechanism by which departures from radiative equilibrium can be translated to an improved temperature distribution. With this temperature distribution, we may return to the beginning of this chapter and recompute the structure and improved radiation field of the atmosphere. The entire process can be iterated until radiative equilibrium is satisfied at the appropriate level.

12.5   Recapitulation

Building on the results of the previous three chapters, we present in this chapter the basic approach to the construction of a model stellar atmosphere. The process is essentially an iterative one where an initial guess of the temperature distribution throughout the atmosphere yields the atmosphere's physical structure. To obtain this structure, one needs a lot of information about the dependence of the opacity on the state variables of the gas. One generally assumes that the Saha ionization and Boltzmann excitation formulas hold so that one can relatively easily calculate the abundance of each type of absorber in the atmosphere. This, then, allows for the solution of the equation of hydrostatic equilibrium and the calculation of the radiation field at all points in the atmosphere. Application of the condition of radiative equilibrium and a temperature correction procedure allow for the entire process to be iterated until a self-consistent model of the atmosphere is obtained.

What constitutes a converged, self-consistent atmosphere is not at all obvious and may depend on which properties of the model are of particular interest to the investigator. For example, it makes little sense to require 0.1 percent constancy in the radiative flux at optical depth 100 if one is interested in only the emergent flux. Conversely, if the atmosphere is to form the boundary layer for the calculation of a model stellar interior, then some care should be taken with the deep solution. If one is concerned about strong spectral lines, then considerable care should be taken with the surface solution. Since it is still relatively difficult to construct a model stellar atmosphere that exhibits both a constant flux and a zero flux derivative throughout the atmosphere at an arbitrary level of accuracy, such considerations regarding the use of the model should be weighed.

We have now completed the fundamental physics concerning the construction of models for both the inside and surface layers of a star. For normal stars, these models would give a reasonably accurate picture of the structure of these stars and the processes that take place within them. We have even included the departure of the radiation field from strict thermodynamic equilibrium that results from the escape of photons that are near the surface into free space. To the extent that the continuous opacity dominates the total stellar opacity, this will even yield a reasonably correct picture of the grosser aspects of the star's spectrum. However, as anyone who has looked at a stellar spectrum knows, the most salient feature of such a spectrum is the dark lines that cover it. These lines provide most of the information that we have about stars, from their composition to their motions. No description of stellar structure can hope to be taken seriously unless it provides some explanation of the occurrence of these lines. Therefore, for the majority of the rest of this book, we discuss the fundamentals of the formation of spectral absorption lines and the physics that yields their characteristic shape.

Problems

1. Assume that all particles in a normal GV star at optical depth $\tau(\lambda 5000) = 1$ have the same speed. Estimate the time required for LTE to be established.

2. Starting with the gray atmosphere temperature distribution, find the rate of convergence toward radiative equilibrium as a function of the Rosseland optical depth for a standard atmosphere (that is, $T_e = 10,000$ K, $\mathrm{Log}\, g = 4.0$, and the chemical composition μ equals that of the sun). Explicitly define what *you* mean by the "rate of convergence."

3. As one moves deeper into a stellar atmosphere, the dependence of the source function becomes more linear with optical depth. Is this a result of the opacity becoming more gray (i.e., independent of wavelength), or does the result follow from the directional randomization of the radiation field? Give explicit evidence to support your conclusion.

4. Compute $F_\lambda/F_\lambda(\lambda 5560)$ for a nongray atmosphere where
(*a*) $\sigma_\lambda = 0$, $\kappa_\lambda = a + b\lambda$, and the effective temperature $T_e = 5000$ K and
(*b*) the same as in (*a*) but with $\kappa_\lambda = a$. Assume that the Eddington approximation is sufficiently accurate to solve the equation of radiative transfer.
(*c*) How do $F_\lambda(\lambda 2000)$ and $F_\lambda(\lambda 5560)$ vary with the optical depth?

References and Supplemental Reading

1. Mihalas, D., *Methods in Computational Physics* (Eds.: B. Alder, S. Fernbach, and M. Rotenberg), vol. 7, Academic, New York, 1967, pp. 24–27.

2. Kurucz, R. L., "ATLAS: A Computer Program for Calculating Model Stellar Atmospheres," SAO Special Report 309, 1970.

3. Avrett, E. H., and Krook, M. "The Temperature Distribution in a Stellar Atmosphere," *Ap. J.* 137, 1963, pp. 874–880.

For further insight into temperature correction procedures, see

• Mihalas, D.: *Stellar Atmospheres*, W. H. Freeman, San Francisco, 1970, pp. 169–186.

Additional Reading on the Subject of Stellar Atmospheres for Chapters 9 through 12

Often some of the oldest references remain the most illuminating in any subject. This is particularly true of stellar atmospheres. The computer has made it too easy to avoid thinking about the interaction of the physical processes going on in the atmosphere and to rely instead on the output of the machine. Those references that predate the computer often focus on this physics, for the authors had no other option. Since knowledge and understanding of these interactions are central to the understanding of stellar atmospheres, we strongly recommend that the serious student make some effort to at least peruse some of these references:

• Aller, L.: *The Atmospheres of the Sun and Stars*, 2d ed., Ronald, New York, 1963.

• Ambartsumyan, V. A.: *Theoretical Astrophysics* (Trans.: J. B. Sykes), Pergamon, New York, 1959, pp. 1–106.

• Barbier, D.: "Theorie Generale des Atmospheres Stellaire," *Handbuch der Physik*, vol. 51, Springer-Verlag, Berlin, 1958, pp. 274–397.

• Chandrasekhar, S.: *Radiative Transfer*, Dover, New York, 1949/1960.

• Greenstein, J. L.: *Stars and Stellar Systems*, vol. 6, *Stellar Atmospheres* (Ed.: J. L. Greenstein), University of Chicago Press, Chicago, 1960, particularly the articles by G. Münch and A. D. Code, pp. 1–86.

• Kourganoff, V.: *Basic Methods in Transfer Problems*, Dover, New York, 1952/1963.

• Menzel, D. H. (ed.): *Selected Papers on the Transfer of Radiation*, Dover, New York, 1966.

• Pagel, B. E. J.: "The Surface of a Star," *Quart. J. R. astr. Soc.* 1, 1960, pp. 66–72.

• Sobolev, V. V.: *A Treatise on Radiative Transfer* (Trans.: S. I. Gaposchkin), Van Nostrand, Princeton, N.J., 1963.

• Underhill, A. B.: "Some Methods for Computing Model Stellar Atmospheres," *Quart. J. R. astr. Soc.* 3, 1963, pp. 7–24.

• Unsöld, A.: *Physik der Sternatmosphären*, 2d ed., Springer-Verlag, Berlin, 1955.

• Waldmier, M.: *Einführung in die Astrophysik*, Birkhäuser Verlag, Basel, Switzerland, 1948.

• Woolley, R. v. d. R., and D. W. N. Stibbs: *The Outer Layers of a Star*, Oxford University Press, London, 1953.

13

Formation of Spectral Lines

. . .

Certainly the existence of such striking features as the dark spectral lines that break up the spectra of stars implies the presence of absorption processes that operate in a highly selective manner. The most obvious candidates for this selective absorption are the bound-bound atomic transitions occurring in the abundant species of common elements. Although we saw in Chapter 11 that bound-bound atomic transitions could, when they occur in very large numbers, depress large regions of the spectrum, some transitions will produce lines that dominate the nearby spectrum in a very singular manner. The contrast between these lines and the neighboring spectrum is often so marked that the investigator tends to make a distinction between a specific line and the neighboring spectrum by denoting the spectrum at nearby wavelengths as the "continuum" spectrum.

This choice often causes some grief, for there is rarely a sharp transition between where the line absorption dominates the continuum absorption and vice versa. Indeed, the neighboring absorption of the continuum is often not even dominated by continuum processes, but represents an unresolved blend of discrete and continuous sources. Thus, the assumed location and the subsequent interpretation of the continuum are one of the largest sources of error in quantities resulting from the study of spectral lines. This problem, and the advent of relatively fast computing machines, has led modern analysis away from the discussion of single spectral lines to a synthesis of the entire spectrum by including all the relevant opacity sources. Although this approach undoubtedly yields more accurate results, it is difficult to appreciate the relative contribution of the various constituents of the atmosphere to the resultant spectrum.

Therefore, we follow the traditional development and assume that a clear distinction can be made between the processes that produce a specific atomic spectral line and the absorption processes that control the spectrum at adjacent frequencies.

13.1   Terms and Definitions Relating to Spectral Lines

a   *Residual Intensity, Residual Flux, and Equivalent Width*

Now that the notion of a continuum has been established, we can use it to provide a normalization of the spectrum so that the resulting line strength is measured in units of the continuum (see Figure 13.1). Some authors call this normalized

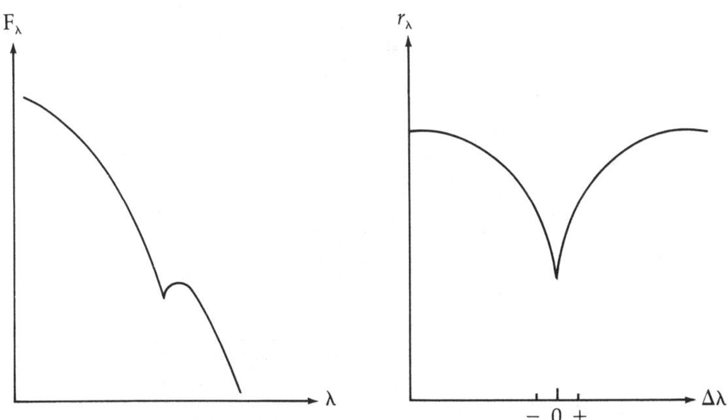

Figure 13.1 (*a*) Shape of a spectral line as it might be observed in units of the absolute flux in the spectrum. (*b*) The same line after normalization by the continuum flux.

flux the *residual intensity*; however, all that can be observed from stars (other than the sun) is the flux of radiation emitted from all points on the stellar surface in the direction of the observer. Only for the sun can the specific intensity of a particular part of the stellar disk be directly observed. For that reason, we reserve the term *residual intensity* for the normalized intensity spectra obtainable from the sun, and we use the term *residual flux* to describe the normalized spectra from stars. Thus, in terms of the emergent intensity and flux of the line and continuum, we have

$$f_\nu(\mu) \equiv \frac{I_\nu(\mu, 0)}{I_c(\mu, 0)} \equiv \text{residual intensity}$$

$$r_\nu \equiv \frac{F_\nu(0)}{F_c(0)} \equiv \text{residual flux}$$

$$(13.1.1)$$

After the wavelength of the center of the line, probably the most common quantity used to describe a spectral line is the *equivalent width*. For absorption lines, this is the width of a rectangularly shaped "line," completely black at the center, that absorbs the same number of photons as the spectral line of interest (see Figure 13.2). We may formally express this definition by

$$W_\lambda \equiv \int_0^\infty (1 - r_\lambda)\, d\lambda \qquad (13.1.2)$$

It is customary to write integrals of this type as ranging from 0 to ∞ largely for convenience. What is meant in reality is that the integral should cover those wave-

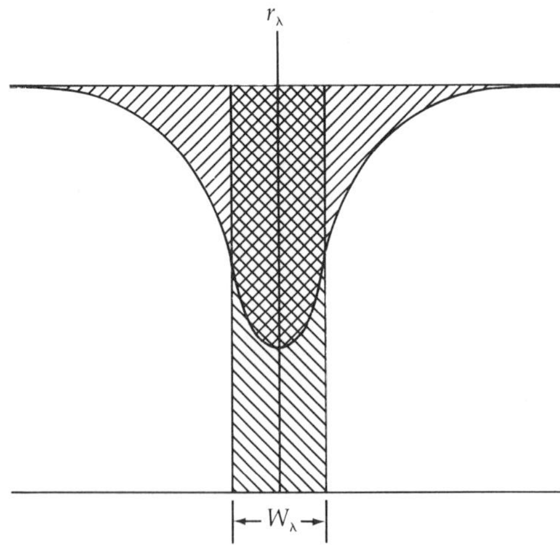

Figure 13.2 Equivalent "black" line profile (rectangular area) appropriate for a specific spectral line (curved area). The areas of the two profiles are equal, and so the rectangular one can be characterized by its width alone. Such equivalent widths are usually measured in angstroms.

lengths for which $1 - r_\lambda$ is significantly different from zero. As long as the line is relatively narrow (that is, $\Delta\lambda \ll \lambda_0$),

$$\frac{W_\lambda}{\lambda_0} \approx \frac{W_\nu}{\nu_0} \tag{13.1.3}$$

b Selective (True) Absorption and Resonance Scattering

It is useful to divide the processes by which a photon interacts with an atom into two idealized cases. The first is called *true*, or *pure, absorption*. In this instance, the emission of photons is completely uncorrelated with the previous absorption of photons. In a sense, any photon that is emitted has "lost all memory of what it was." Such a process is called a true absorption process, and it can occur in stellar atmospheres when an atom suffers numerous collisions between the time that the photon is absorbed and reemitted. These collisions can deexcite or further excite the atom, and with the assumption of LTE, these collisions will be randomly distributed so that any subsequently emitted photon is, indeed, totally uncorrelated with the one that was absorbed. Spectral lines formed in this manner are known as *pure absorption lines*. They exist in stellar spectra since the enhanced opacity provided by the line implies that the line will become optically thick higher in the atmosphere, where it is generally cooler. Since it is cooler, the source function is smaller and the emergent intensity is less than that of the neighboring continuum.

The second type of process is called *resonance scattering*, and it results in the loss of photons in a specialized and indirect manner. Here the photon has a "perfect memory" of its origin. The emitted photon is completely correlated in frequency with the absorbed photon. In Chapter 9, we described such a process as a coherent scattering, and lines for which this is true are known as scattering lines. In contrast to the case of pure absorption, a scattering line is formed when the emitted photon is created so soon after the prior absorption that there is no time for the atom to be perturbed by collisions, and the probability of a transition to the prior state is very great. Such cases occur from those states that have very short lifetimes and only one lower level to which the electron can jump. The resonance line transitions meet all these conditions, and hence any resonance line is likely to be a strong scattering line. However, it is possible for any strong line to behave as a scattering line if the probability of returning to the previous state is very large. The same photon that was absorbed is then reemitted, with no net energy exchange with the atom. This is essentially the condition for an interaction to be termed a scattering. The scattering process does not directly result in any loss of energy from the radiation field, but by changing the direction of the photon the process lengthens the stay of the photon in the atmosphere, making it subject to destruction by continuum absorption processes.

Thus, we can divide spectral lines into two types, the pure absorption lines, where the absorbed energy of the photon is fully shared with the gas, and the resonance scattering lines, where it is not. In Chapter 9 we showed that nature is really more complicated than this and in reality most lines can be viewed as a mixture of these two extreme states. However, the radiative transfer of these two kinds of lines is quite different, and understanding the behavior of these two limiting cases will provide a comprehensive basis for understanding the behavior of spectral lines in general. The different behavior of these two processes is clearly seen by noting that the energy of a pure absorption process is shared immediately with the gas while that of a resonance scattering process is not. Scattering is a fully conservative process and therefore cannot, by itself, result in the destruction of photons. However, scattering does change the direction of a photon, thereby increasing the distance that the photon must travel through the atmosphere before escaping into interstellar space. Any process that lengthens the path of a photon through the atmosphere also enhances the probability that the photon will be absorbed by some other process such as continuum absorption. Thus, the redirection of line photons that results from resonant scattering also produces a net loss of these photons relative to those in the neighboring continuum. This, then, is the origin of the resonance scattering lines in stellar spectra.

c *Equation of Radiative Transfer for Spectral Line Radiation*

It is customary to denote the part of the mass extinction coefficient that results from pure absorption processes by the letter κ, while the part that results from scattering is represented by the Greek letter σ. Photon interactions that occur as a result of absorptions within the line are subscripted by the letter v. Since the continuum processes generally vary quite slowly across a spectral line, we omit the subscript v entirely. Thus,

$$\kappa \equiv \text{the pure absorption coefficient for the continuum}$$

$$\sigma \equiv \text{the scattering coefficient for the continuum}$$

$$\sigma_v \equiv \text{the resonance scattering coefficient for the line} \qquad (13.1.4)$$

$$\kappa_v \equiv \text{the pure absorption coefficient for the line}$$

$$\ell_v \equiv \kappa_v + \sigma_v = \text{the line extinction coefficient}$$

Since the origin of the equation of radiative transfer was dealt with extensively in Section 9.2, we provide only a brief derivation here. Basically we balance the energy passing in and out of a differential volume along a specific path through the atmosphere. If we do this for a plane-parallel atmosphere where we keep the line and continuum processes separate, we get

$$\frac{\mu}{\rho}\frac{dI(x,\mu)}{dx} = -[(\kappa_\nu + \kappa) + (\sigma + \sigma_\nu)]I_\nu + j_\nu$$

$$+ \frac{1}{4\pi}\int_{4\pi}(\sigma + \sigma_\nu)I_\nu(\mu',\phi',x)\,d\Omega \qquad (13.1.5)$$

where the first term on the right-hand side represents the energy lost from the beam. The second and third terms on the right-hand side denote the contributions to the beam within the differential volume. The first of these is just due to thermal emission, while the second results from scattering by both line and continuum processes. By making the usual identification between the Planck function and the processes of thermal emission and absorption, the equation of radiative transfer for line radiation becomes

$$\mu\frac{dI_\nu(\mu,\tau_\nu)}{d\tau_\nu} = +I_\nu(\mu,\tau_\nu) - \frac{(\kappa + \kappa_\nu)B_\nu}{\kappa + \kappa_\nu + \sigma + \sigma_\nu} - \frac{(\sigma + \sigma_\nu)J_\nu}{\kappa + \kappa_\nu + \sigma + \sigma_\nu} \qquad (13.1.6)$$

where the optical depth in the line τ_ν is given by

$$d\tau_\nu = -(\kappa + \kappa_\nu + \sigma + \sigma_\nu)\rho\,dx \qquad (13.1.7)$$

13.2 Transfer of Line Radiation through the Atmosphere

Calculating a stellar spectral line is rather simpler than constructing a model stellar atmosphere since the structure of the atmosphere can be assumed to be known. We need only bring those methods discussed in Chapter 10 for the solution of the equation of transfer to bear on the solution of equation (13.1.6). However, to obtain the line profile, we also have to solve the same equation with $\kappa_\nu = \sigma_\nu = 0$ so that we may determine the continuum flux with which to normalize the line profile. Although this procedure will indeed work and is in fact used for most modern line profile calculations, it is very difficult to obtain any insight into the behavior of scattering and absorption lines from the numerical output. However, their behavior can be seen in some older semianalytic solutions to simple models of line transfer.

a   *Schuster-Schwarzschild Model Atmosphere for Scattering Lines*

The Schuster-Schwarzschild model atmosphere is perhaps the simplest model that one can suggest for line formation. It is to spectral line transfer theory what the gray atmosphere is to atmosphere theory. The model is basically appropriate for strong resonance lines which are formed in a thin layer overlying the photosphere (see Figure 13.3).

If the lines of interest are quite strong, then the opacity in the continuum is negligible compared to the line opacity. Furthermore, since the process of scattering is fully conservative and the photons do not exchange energy with the local constituents of the atmosphere, we need not worry about the physical conditions in the cool gas that overlies the photosphere. Just as with the gray atmosphere, pure scattering decouples the radiation field from the physics of the gas. We further assume that the optical depth in the line corresponding to the location of the photosphere is finite. The plane-parallel equation of radiative transfer appropriate for this model can be obtained from equation (13.1.6) by specifying the values for the absorption and scattering coefficients. Thus,

$$\mu \frac{dI_v}{d\tau_v} = I_v - J_v \tag{13.2.1}$$

where

$$d\tau_v = -\sigma_v \rho \, dx \tag{13.2.2}$$

Since the line extinction coefficient is entirely due to scattering, it is not surprising that equation (13.2.1) looks like the transfer equation for the gray atmosphere [equation (10.2.1)]. This means that radiative equilibrium requires the flux to

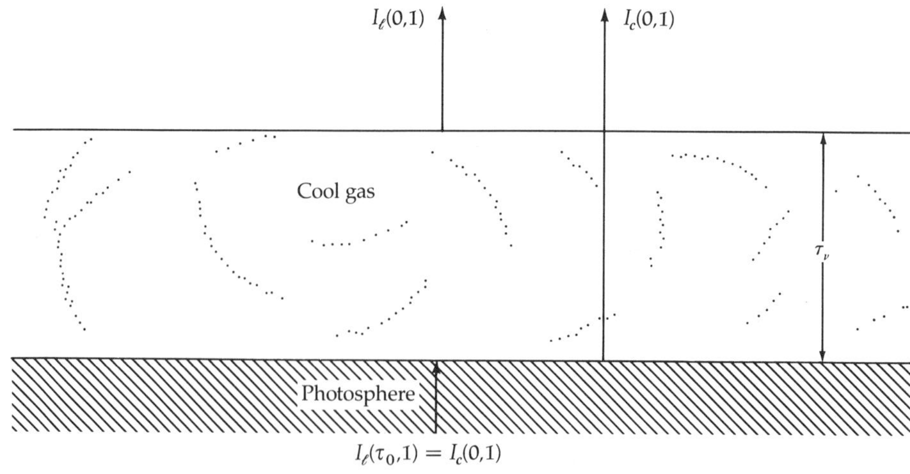

Figure 13.3 Schematic representation of the Schuster-Schwarzschild model atmosphere for the formation of strong scattering lines.

be constant at each frequency throughout the line. To be sure, the constant will be different for each fequency, but the flux at any particular frequency will not vary with the depth. The condition of monochromatic flux constancy can be expressed as

$$F_\nu(\tau_\nu) = \text{const} \tag{13.2.3}$$

Solution of the Radiative Transfer Equation for the Schuster-Schwarzschild Model   Since the equation of transfer for line radiation in this model formally resembles the gray atmosphere equation, we may use the results of Chapter 10 to find the solution. Specifically, equation (10.2.31) gives a general result for the solution of the plane-parallel finite gray atmosphere. However, to keep the discussion simple, we take $n = 2$. Then the appropriate zeros of the Legendre polynomials require that $\mu_i = \pm 1/\sqrt{3}$, and equation (10.2.31) becomes

$$I_+(\tau_\nu) = \frac{3F_\nu(\tau_\nu + 1/\sqrt{3} + Q)}{4} \qquad I_-(\tau_\nu) = \frac{3F_\nu(\tau_\nu - 1/\sqrt{3} + Q)}{4} \tag{13.2.4}$$

where the subscripts $+$ and $-$ refer to the outward- and inward-directed streams of radiation, respectively. Applying the surface boundary condition $I_-(0) = 0$, we find

$$Q = \frac{1}{\sqrt{3}} \tag{13.2.5}$$

and the complete solution becomes

$$I_+(\tau_\nu) = \frac{3F_\nu(\tau_\nu + 2/\sqrt{3})}{4} \qquad I_-(\tau_\nu) = \frac{3F_\nu\tau_\nu}{4} \tag{13.2.6}$$

This solution for the gray atmosphere is sometimes called the *Chandrasekhar two-stream approximation*. It conceptually replaces the entire radiation field by two streams of radiation directed along a line oriented about 54° ($\mu = \pm 1/\sqrt{3}$) to the normal of the atmosphere. The result is nearly identical to that obtained from the Eddington approximation, only the angle is slightly different.

Residual Flux and Intensity for the Schuster-Schwarzschild Model   The ratio of the emergent flux in the line to that of the continuum can be obtained immediately by requiring that the line intensity incident on the base of the cool gas be the same as the emergent intensity in the neighboring continuum, so that

$$I_+(\tau_0) = \frac{3F_c(0 + 2/\sqrt{3})}{4} = \frac{3F_\nu(\tau_0 + 2/\sqrt{3})}{4} \tag{13.2.7}$$

Here τ_0 is the optical depth at any frequency in the line measured at the base of the atmosphere. This value is zero for all frequencies corresponding to the continuum. The quantity F_c is just the continuum flux. Thus, the residual flux is

$$r_\nu = \frac{F_\nu}{F_c} = \left(1 + \frac{\sqrt{3}\tau_0}{2}\right)^{-1} \tag{13.2.8}$$

However, to complete the description of the Schuster-Schwarzschild model, we would like an expression for the residual intensity f_ν. To obtain the angle dependence of the intensity, we have to appeal to the classical solution of the equation of transfer for a finite atmosphere, so that the emergent intensity is

$$I_\nu(\mu, 0) = \int_0^{\tau_0} \frac{J_\nu(t_\nu)e^{-t\nu/\mu}\,dt_\nu}{\mu} + I_c(\mu, 0)e^{-\tau_0/\mu} \tag{13.2.9}$$

The value for the mean intensity $J_\nu(\tau_\nu)$ can be obtained directly from the two-stream approximation as

$$J_\nu(\tau_\nu) = \tfrac{1}{2}[I_+(\tau_\nu) + I_-(\tau_\nu)] = \frac{3F_\nu(\tau_\nu + 1/\sqrt{3})}{4} \tag{13.2.10}$$

Substitution of this into equation (13.2.9) and then into the definition for the residual intensity [equation (13.1.1)] yields

$$f_\nu(\mu) = \frac{3F_c/4}{(1 + \sqrt{3}\tau_0/2)I_c(\mu, 0)}\left[\mu + \frac{1}{\sqrt{3}} - \left(\mu + \tau_0 + \frac{1}{\sqrt{3}}\right)e^{-\tau_0/\mu}\right] + e^{-\tau_0/\mu} \tag{13.2.11}$$

Remember that τ_0 is a function of frequency, and so its behavior with frequency will determine the line shape or profile.

To see how scattering lines will vary in strength across the surface of a stellar disk, consider some limiting cases:

$$f_\nu(\mu) \approx \begin{cases} 1 - \tau_0\dfrac{3F_c}{4I_c(\mu, 0)} & \tau_0 \ll 1 \text{ (weak lines)} \\[3mm] \dfrac{\{\sqrt{3}F_c/[2I_c(\mu, 0)]\}(\mu + 1/\sqrt{3})}{\tau_0} & \tau_0 \gg \text{ (strong lines)} \end{cases} \tag{13.2.12}$$

The first of equations (13.2.12) represents relatively weak spectral lines (or the wings of strong lines). Here there is no angular dependence whatsoever except that dictated by the limb-darkening of the continuum. Thus, we can expect that even weak scattering lines will be visible at all points on the stellar disk with equal strength. While the strong lines described by the second of equations (13.2.12) do show some limb-darkening through the dependence on μ, that dependence is not great. The range in line strength to be expected for a strong scattering line between one formed near the center of the disk and one formed at the limb is about a factor of 2. As we shall see, this contrasts greatly with the behavior of spectral lines formed by pure absorption processes. However, a model that includes absorption processes must include information about the atmospheric structure and so will be somewhat more sophisticated. One such model atmosphere is known as the Milne-Eddington model.

b Milne-Eddington Model Atmosphere for the Formation of Spectral Lines

To appreciate the importance of pure absorption in the formation of a spectral line, we must acknowledge the fact that pure absorption processes imply an interaction between the radiation field and the particles that make up the gas. Thus, we will have to specify something about the behavior of the opacity and source function with optical depth. The trick is to place as few limitations as possible so as to preserve generality and to make those limitations "reasonable" and yet specify the situation sufficiently to guarantee a unique solution.

The Milne-Eddington model is considerably more sophisticated than the Schuster-Schwarzschild model and is correspondingly more complicated. It attempts to simultaneously include the effects of absorption and scattering in the line extinction coefficient (that is, $\kappa_v + \sigma_v$). Let us define the following parameters in terms of the absorption and scattering coefficients of the line and continuum:

$$\epsilon_v \equiv \frac{\kappa_v}{\kappa_v + \sigma_v} \qquad \eta_v \equiv \frac{\kappa_v + \sigma_v}{\kappa}$$

$$\mathscr{L}_v \equiv \frac{\kappa_v + \kappa}{\kappa + \kappa_v + \sigma_v} = \frac{1 + \eta_v \epsilon_v}{1 + \eta_v} \qquad (13.2.13)$$

$$\sigma = 0$$

The parameter ϵ_v measures the relative importance of pure absorption to total extinction for processes involving the spectral line, and η_v is a measure of the line strength, since it is a ratio of the total line extinction coefficient to that of the continuum. The Milne-Eddington model does make the somewhat restrictive assumption that scattering processes are relatively unimportant for continuum photons. Hence, $\sigma = 0$. The parameter $\mathscr{L}$ is clearly not linearly independent of ϵ_v and η_v, but is useful to introduce because it measures the relative importance of total absorption to total extinction for all processes of the line and continuum that operate on the photons passing through the atmosphere.

By substituting these values into the equation for the transfer of line radiation [equation (13.1.6)], we have the appropriate equation of transfer for the Milne-Eddington model atmosphere

$$\mu \frac{dI_v}{d\tau_v} = I_v - \mathscr{L}_v B_v - (1 - \mathscr{L}_v)J_v$$

$$d\tau_v = -(\kappa + \kappa_v + \sigma_v)\rho\, dx$$

(13.2.14)

A simpler equation of transfer for the continuum radiation can be obtained simply by letting κ_v and σ_v go to zero, so that

$$\mu \frac{dI_c}{dt} = I_v - B_v \qquad dt = -\kappa\rho\, dx \qquad (13.2.15)$$

Now if we relate the structure of the atmosphere to the source function in the line through its behavior in the continuum, we can write

$$B(t) = a + bt \qquad (13.2.16)$$

Here, we have made use of the Eddington approximation where the asymptotic behavior of the source function with optical depth is linear [see equation (10.2.32)]. To find the behavior of the residual intensity and flux for the line, we follow basically the same steps as for the Schuster-Schwarzschild model.

Solution of the Equation of Radiative Transfer for the Milne-Eddington Model Atmosphere Before we can solve equations (13.2.14), we must make one additional assumption regarding the behavior of the opacity coefficients with atmospheric depth. We assume that both ϵ_v and η_v (and hence $\mathscr{L}_v$) are not functions of depth in the atmosphere and can therefore be regarded as constants in the equation of transfer [equations (13.2.14)]. While it is certainly unreasonable to expect that any of the absorption or scattering coefficients are independent of depth, since they depend strongly on temperature, it is not unreasonable to think that their ratios may be approximately constant.

If we now solve equation (13.2.14) by taking moments of the equation, we obtain

$$\frac{dF_v}{d\tau_v} = 4\mathscr{L}_v(J_v - B_v) \qquad \frac{dK_v}{d\tau_v} = \frac{F_v}{4} \qquad (13.2.17)$$

Using that part of the Eddington approximation that says $K_\nu \approx J_\nu/3$ and differentiating the second of equations (13.2.17), we get

$$\frac{d^2 J_\nu}{d\tau_\nu^2} = \frac{3}{4}\frac{dF_\nu}{d\tau_\nu} = 3\mathscr{L}_\nu(J_\nu - B_\nu) \qquad (13.2.18)$$

Now we can relate $B_\nu(t)$ to the optical depth in the line by noting that

$$\frac{dt}{\kappa} = -\rho\, dx = \frac{d\tau_\nu}{\kappa + \kappa_\nu + \sigma_\nu} \qquad (13.2.19)$$

so that

$$t = \frac{\tau_\nu}{1 + \eta_\nu} \qquad (13.2.20)$$

This allows us to write

$$B_\nu(\tau_\nu) = a + \frac{b\tau_\nu}{1 + \eta_\nu} \qquad (13.2.21)$$

Substitution of this depth dependence of the Planck function admits a solution of equation (13.2.18) of the form

$$J_\nu(\tau_\nu) - B_\nu(\tau_\nu) = c_1 e^{+\sqrt{3\mathscr{L}_\nu}\tau_\nu} + c_2 e^{-\sqrt{3\mathscr{L}_\nu}\tau_\nu} \qquad (13.2.22)$$

where the constants c_1 and c_2 are to be determined from the boundary conditions.

Unlike the Schuster-Schwarzschild model atmosphere, the Milne-Eddington model is a semi-infinite atmosphere so that at large depths we may be assured that $J_\nu \to B_\nu$ which requires $c_1 = 0$. To determine c_2, we apply the part of the Eddington approximation that we have not used [that is, $J_\nu(0) = \frac{1}{2}F_\nu(0)$]. This can be combined with the assumption about K_ν to determine a value for the derivative of J_ν at the surface, namely,

$$\left.\frac{dJ_\nu(\tau_\nu)}{d\tau_\nu}\right|_{\tau_\nu = 0} = \frac{3}{2}J_\nu(0) = \frac{3(c_2 + a)}{2} \qquad (13.2.23)$$

The right-hand side of this result is obtained from the solution [equation (13.2.22)] itself. If we differentiate that solution and evaluate the result at the surface, we have

$$\left.\frac{dJ_\nu(\tau_\nu)}{d\tau_\nu}\right|_{\tau_\nu = 0} = -\sqrt{3\mathscr{L}_\nu}c_2 + \frac{b}{1 + \eta_\nu} \qquad (13.2.24)$$

which yields the following value of c_2:

$$c_2 = \left[\frac{b}{\sqrt{3}(1 + \eta_v)} - \frac{\sqrt{3}a}{2} \right] \left(\sqrt{\mathcal{L}_v} + \frac{\sqrt{3}}{2} \right)^{-1} \tag{13.2.25}$$

Had we used the Chandrasekhar two-stream approximation to solve the equation of transfer, we would have gotten

$$c_2(\text{2-stream}) = \left[\frac{b}{\sqrt{3}(1 + \eta_v)} - a \right] (\sqrt{\mathcal{L}_v} + 1)^{-1} \tag{13.2.26}$$

This indicates that the solution is not too sensitive to the mode of solution of the equation of transfer.

Residual Flux and Intensity for the Milne-Eddington Model   The residual flux can be obtained from its definition and the Eddington approximation so that

$$r_v \equiv \frac{F_v(0)}{F_c(0)} = \frac{J_v(0)}{J_c(0)} \tag{13.2.27}$$

In the continuum, $\eta_v = 0$ and $\mathcal{L}_v = 1$ so that

$$J_c(0) = \frac{a + b/\sqrt{3}}{1 + \sqrt{3}/2} \tag{13.2.28}$$

This leads to a residual flux given by

$$r_v = \left(1 + \frac{\sqrt{3}}{2} \right) \frac{a\sqrt{3\mathcal{L}_v} + b/(1 + \eta_v)}{(a\sqrt{3} + b)(\sqrt{3}/2 + \sqrt{\mathcal{L}_v})} \tag{13.2.29}$$

To evaluate the residual intensity, we must again use the classical solution to the equation of transfer. However, to do so, we must have expressions for the source function in the line and continuum. It is clear from the equation of transfer [equation (13.2.14)] that the line source function is

$$S_v(\tau_v) = \mathcal{L}_v B_v(\tau_v) + (1 - \mathcal{L}_v) J_v(\tau_v) \tag{13.2.30}$$

Again, remembering that in the continuum $\mathcal{L}_v = 1$, we see that the source function for the continuum is just

$$S_c(t) = B_v(t) \tag{13.2.31}$$

Substitution of these two source functions into the classical solution for the equation of transfer, and some algebra, gives the residual intensity as

$$f_v(\mu) = \frac{a + b\mu/(1 + \eta_v)}{a + b\mu} - \frac{(1 - \mathscr{L}_v)\{\sqrt{3}a/2 - b/[\sqrt{3}(1 + \eta_v)]\}}{(a + b\mu)(1 + \mu\sqrt{3\mathscr{L}_v})(\sqrt{3}/2 + \sqrt{\mathscr{L}_v})} \qquad (13.2.32)$$

Some aspects of this solution should not surprise us. For example, the term $a + b\mu$ is simply $1/\mu$ times the Laplace transform of the continuum source function $a + bt$. Similarly, the numerator of the first term is the Laplace transform of the Planck function in the line, so that the lead term of equation (13.2.32) is just the ratio of the Laplace transforms of the absorption components of the line to the continuum source functions. This is to be expected from the result obtained for limb-darkening in Chapter 10 [equations (10.1.19) and (10.1.20)]. Since the second term vanishes for the case of pure absorption (that is, $\mathscr{L} \to 1$), this term must represent the contribution of scattering in the line to the residual intensity.

Asymptotic Behavior of the Residual Flux and Intensity   Because of the increased generality of the Milne-Eddington model over the Schuster-Schwarzschild model, we can investigate the asymptotic behavior of the line not only with strength but also as the line extinction coefficient ranges from pure absorption to pure scattering. In the case of pure absorption $\epsilon_v = 1$, $\mathscr{L}_v = 1$, and equations (13.2.29) and (13.2.32) become

$$r_v = \frac{a\sqrt{3} + b/(1 + \eta_v)}{a\sqrt{3} + b} \qquad f_v(\mu) = \frac{a + b\mu/(1 + \eta_v)}{a + b\mu} \qquad (13.2.33)$$

Note that in an isothermal atmosphere (that is, $b = 0$) both the residual intensity and the flux are asymptotic to unity and the line disappears. Thus, as one might expect, if there are no temperature gradients, there can be no spectral absorption lines. The radiation field would then be in STE, and the source function as well as the radiation field would be given by the Planck function regardless of the dependence of the absorption coefficient on frequency.

For stellar atmospheres, this has the more immediate implication that the stronger the source function gradient, the stronger the line. This is the simple explanation of why the central depths of the lines in late-type stars are so much darker than those for the early-type stars. For the later-type stars, the visible part of the spectrum tends to lie at wavelengths shorter than that of the stellar energy maximum. For these wavelengths, the source function varies as a large power of the temperature so that the stellar temperature gradient produces a very steep source function gradient. For the early-type stars, the visible part of the spectrum lies well to the red end of the energy maximum on what is generally called the *Rayleigh-Jeans tail*   343 of the energy distribution. Here the source function varies quite slowly so that the temperature gradient produces a more gentle source function gradient and weaker absorption lines.

If we investigate the case for strong absorption lines, we have

$$r_v = \frac{a\sqrt{3}}{a\sqrt{3} + b} \qquad f_v(\mu) = \frac{a}{a + b\mu} \qquad \eta_v \gg 1 \qquad (13.2.34)$$

and as one expects, the above results hold if $b = 0$. However, in a normal stellar atmosphere where $b \neq 0$, even the strongest line must vanish as $\mu \to 0$ at the limb. Thus we expect that the pure absorption component of the line extinction coefficient will play no role in determining the line strength at the limb of the star. This is easier to understand if we consider the physical situation encountered at the limb. The observer's line of sight just grazes the atmosphere passing through a region of reasonably constant conditions including the temperature. Thus, there are no temperature or source function gradients along the line of sight. If no photons are scattered from other locations into the line of sight, then the only contribution to the observed intensity comes from a nearly isothermal sight line through the atmosphere. If there are no gradients, there are no pure absorption lines. In the event that $\eta_v \ll 1$, which prevails for weak absorption lines, the residual flux and intensity are given by

$$r_v = 1 - \frac{\eta_v b}{a\sqrt{3} + b}$$

$$\qquad (13.2.35)$$

$$f_v(\mu) = 1 - \frac{\eta_v b\mu}{a + b\mu}$$

As one might expect, the line strength is simply proportional to η_v, which is proportional to κ_v which in turn is proportional to the number abundance of absorbers. Hence for weak lines we expect that the strength of the line would primarily depend on the abundance of the element giving rise to the line.

In the case of pure scattering, $\epsilon_v = 0$, which requires that

$$\mathcal{L}_v = \frac{1}{1 + \eta_v} \qquad \text{for } \kappa_v = 0 \qquad (13.2.36)$$

For the situation where $\eta_v \gg 1$ and we have very strong scattering lines, the residual flux and intensity become

$$r_v = \frac{(1 + 2/\sqrt{3})a\sqrt{3}\mathcal{L}_v}{a\sqrt{3} + b}$$

$$\qquad (13.2.37)$$

$$f_v(\mu) = \frac{a\sqrt{\mathcal{L}}(\mu\sqrt{3} + 2/\sqrt{3})}{a + b\mu}$$

Even in the event that the atmosphere is isothermal, scattering lines will persist. Indeed, even at the limb of an isothermal atmosphere, the residual intensity is still of the order of one-half of the residual flux which nicely illustrates the ubiquitous property of scattering lines. The physical reason for this is that these lines do not depend on the thermodynamic properties of the gas for their existence. The lines result from the existence of a boundary or a surface to the atmosphere. If there were no surface, then again conditions of STE would prevail and there would be no lines of any kind. However, the presence of a boundary permits the selective escape of photons. Those that are heavily scattered will wander about in the atmosphere for a greater time than those photons that do not experience scattering. Hence, the likelihood that the photons will be absorbed by continuum processes and thereby removed from the beam is increased. This will happen in an isothermal atmosphere as well as in a normal atmosphere. Thus, scattering lines will always be present in a stellar spectrum.

For the case of weak scattering lines, $\eta_v \to 0$ and $\mathscr{L}_v \to 1$, causing the residual flux and intensity to take the form

$$r_v = 1 - \frac{\eta_v b}{a\sqrt{3} + b}$$

$$f_v(\mu) = 1 - \frac{\eta_v b\mu}{a + b\mu} - \frac{\eta_v(a\sqrt{3}/2 - b/\sqrt{3})}{(a + b\mu)(1 + \mu\sqrt{3})(1 + \sqrt{3}/2)} \qquad \eta_v \ll 1 \quad (13.2.38)$$

Under these conditions, the residual flux takes on the form of the result for the case of weak pure absorption. However, even in this case the limb-darkening behavior of the line as given by f_v is different from the corresponding expression [equation (13.2.35)] for weak absorption. Even at the limb of an isothermal atmosphere, a weak scattering line will be visible.

The main purpose of studying these approximate atmospheres is to gain some feeling for the manner in which line strengths vary and to see which parameters are most instrumental in determining the extent of that variation. Lines formed under the conditions of pure absorption behave in a qualitatively different manner from lines where the extinction coefficient is dominated by scattering. You must keep this idea clearly in mind whenever you wish to relate an observed line strength or shape to a theoretically determined model. You must do the radiative transfer correctly. Knowledge of the behavior of spectral lines also will serve you well when you try to understand the results of a complex model atmosphere computer code. If the line does not behave in a manner described by these simple models, then there must be a good and compelling reason that should be understood by the investigator.

Although the 1940s and 1950s generated additional approaches to the problem of radiative transfer of line radiation, most of that work has been superceded by the advent of high-speed computers and large complex atmosphere codes. The contemporary approach basically treats the line absorption coefficient as an additional opacity source that has a strongly variable frequency dependence. Thus, no special distinction is made between opacity due to lines and that of the continuum. The problem of locating the continuum for the purposes of generating a line profile is then about the same for the model maker as for the observer. The many complex physical processes that contribute to the shape of the line, and whose importance varies with position in the atmosphere, are automatically included in the calculations. However, for obtaining insight into the processes of line formation, these simple models remain most useful.

Problems

1. Find equivalent expressions for the asymptotic behavior of the residual intensity and flux as given by equations (13.2.33) through (13.2.37) for the case where the radiative transfer equation is solved by the Chandrasekhar two-stream approximation.

2. Find expressions for the residual intensity and flux of a Schuster-Schwarzschild atmosphere if the equation of transfer is to be solved by using the Chandrasekhar nth-order approximation.

3. Show that the expression for f_ν can be used to obtain the expression for r_ν for the Milne-Eddington atmosphere.

4. Find an expression for r_ν in a Milne-Eddington atmosphere where the source function in the continuum is given by

$$B(\tau) = B^{(0)} + B^{(1)}\tau + B^{(2)}\tau^2$$

5. Derive expressions for a Milne-Eddington type of model atmosphere of finite optical depth τ_0 for $f_\nu(\mu)$ and r_ν. Assume the atmosphere is illuminated from below by a uniform isotropic intensity I_0.

6. Compare the results of Problem 5 to the results obtained for (*a*) the semi-infinite Milne-Eddington atmosphere for strong absorption lines and strong scattering lines and (*b*) the Schuster-Schwarzschild atmosphere for strong scattering lines and weak scattering lines.

7. If the law of limb-darkening for lines formed in a Schuster-Schwarzschild atmosphere can be written as

$$I_{\nu_0}(\mu, 0) = a + b\mu$$

  find a and b in terms of the continuum flux and the optical depth of the line τ_0.

8. Find an expression for the residual flux from a Schuster-Schwarzschild atmosphere if the continuum photospheric intensity has the form

$$I_c(\mu, 0) = a + b\mu$$

9. Use a model atmosphere code to generate r_ν for the spectral line of your choice, and compare the results to those of a Milne-Eddington atmosphere with the same effective temperature. Clearly state all assumptions and approximations that you make.

Supplemental Reading

A reasonable derivation of the solution of the equation of radiative transfer for the Milne-Eddington model atmosphere can be found in

- Aller, L. H.: *The Atmospheres of the Sun and Stars*, 2d ed., Ronald, New York, 1963, pp. 349–352.

An excellent general reference for the formation of spectral lines is

- Jefferies, J. T.: *Spectral Line Formation*, Blaisdell, Waltham, Mass., 1968.

In particular, Jefferies discusses the Schuster-Schwarzschild atmosphere in Section 3.2 (p. 30). A quite complete discussion of classical line transport theory can be found in these two books:

- Mihalas, D.: *Stellar Atmospheres*, 2d ed., W. H. Freeman, San Francisco, 1978, pp. 308–316.
- Mihalas, D.: *Stellar Atmospheres*, W. H. Freeman, San Francisco, 1970, pp. 323–332.

14

Shape of Spectral Lines

. . .

If we took the classical picture of the atom as the definitive view of the formation of spectral lines, we would conclude that these lines should be delta functions of frequency and appear as infinitely sharp black lines on the stellar spectra. However, many processes tend to broaden these lines so that the lines develop a characteristic shape or profile. Some of these effects originate in the quantum mechanical description of the atom itself. Others result from perturbations introduced by the neighboring particles in the gas. Still others are generated by the motions of the atoms giving rise to the line. These motions consist of the random thermal motion of the atoms themselves which are superimposed on whatever large-scale motions may be present. The macroscopic motions may be highly ordered, as in the case of stellar rotation, or show a high degree of randomness, as is characteristic of turbulent flow.

In practice, all these effects are present and give the line its characteristic shape. The correct representation of these effects allows for the calculation of the observed line profile and in the process reveals a great deal about the conditions in the star that give rise to the spectrum. Of course, the photons that give rise to the absorption lines in the stellar spectrum have their origins at different locations in the atmosphere. So the conditons giving rise to a spectral line are really an average of a range of conditions. Thus, when we talk of the excitation temperature or the kinetic temperature appropriate for a specific spectral line, it must be clear that we are referring to some sort of average temperature appropriate for that portion of the atmosphere in which most of the line photons originate. For strong lines with optical depths much greater than the optical depth of the adjacent continuum, the physical depth of the line-forming region is quite small, and the approximation of the physical conditions by their average value is a good one. Unfortunately, for very strong lines, the optical depths can range to such large values that the line-forming region is located in the chromosphere, where most of the assumptions that we have made concerning the structure of the stellar atmosphere break down. A discussion of such lines will have to wait until we are ready to relax the condition of LTE.

In describing the shape or profile of a spectral line, we introduce the notion of the atomic line absorption coefficient. This is a probability density function that describes the probability that a given atom in a particular state of ionization and excitation will absorb a photon of frequency v in the interval between v and $v + dv$. We then assume that an ensemble of atoms will follow the probability distribution function of the single atom and produce the line. To make the connection between the behavior of a single atom and that of a collection of atoms, we use the Einstein coefficients introduced in Section 11.3.

14.1 Relation between the Einstein, Mass Absorption, and Atomic Absorption Coefficients

Since the Einstein coefficient B_{ik} is basically the probability that an atom will make a transition from the ith state to the kth state in a given time interval, the relationship to the mass absorption coefficient can be found by relating all upward transitions to the total absorption of photons that must take place. From the definition of the Einstein coefficient of absorption, the total number of transitions that take place per unit time is

$$N_{i \to k} = n_i B_{ik} I_v \, dt \qquad (14.1.1)$$

where n_i is the number density of atoms in the *i*th state. Since the number of photons available for absorption at a particular frequency is $(I_v/hv)\,dv$, the total number of upward transitions is also

$$N_{i \rightarrow k} = 4\pi n_i \int \kappa_v \rho \, \frac{I_v}{hv} \, dv \, dt \qquad (14.1.2)$$

If we assume that the radiation field seen by the atom is relatively independent of frequency throughout the spectral line, then the integral of the mass absorption coefficient over the line is

$$\int \kappa_v \, dv = \frac{1}{4\pi} \frac{hv_0}{\rho} B_{ik} \qquad (14.1.3)$$

where v_0 is the frequency of the center of the line.

For the remainder of this chapter, we will be concerned with the determination of the frequency dependence of the line absorption coefficient. Thus, we will be calculating the absorption coefficient of a single atom at various frequencies. We call this absorption coefficient the *atomic line absorption coefficient*, which is related to the mass absorption coefficient by

$$S_v = \frac{S_\omega}{2\pi} = \kappa_v \rho \qquad (14.1.4)$$

Note that we will occasionally use the circular frequency ω instead of the frequency v, where $\omega = 2\pi v$.

14.2 Natural or Radiation Broadening

Of all the physical processes that can contribute to the frequency dependence of the atomic line absorption coefficient, some are intrinsic to the atom itself. Since the atom must emit or absorb a photon in a finite time, that photon cannot be represented by an infinite sine wave. If the photon wave train is of finite length, it must be represented by waves of frequencies other than the fundamental frequency of the line center v_0. This means that any photon can be viewed in terms of a "packet" of frequencies ranging around the fundamental frequency. So the photon will consist of energy occupying a range of wavelengths about the line center. The extent of this range will depend on the length of the photon wave train. The longer the wave train, the narrower will be the range of frequencies or wavelengths required to represent it.

Since the length of the wave train will be proportional to the time required to emit or absorb it, the characteristic width of the range will be proportional to the transition probability (i.e., the inverse of the transition time) of the atomic transition. This will be a property of the atom alone and is known as the *natural width* of the transition. It is always present and cannot be removed. Its existence depends only on the finite length of the wave train and so is not just the result of the quantum nature of the physical world. Indeed, there are two effects to estimate: the classical effect relying on the finite nature of the wave train and the quantum mechanical effect that can be obtained for a specific atom's propensity to emit photons. The former will be independent of the type of atom, while the latter will yield a larger broadening that depends specifically on the type of atom and its specific state.

a   Classical Radiation Damping

The classical approach to the problem of absorption relies on a picture of the atom in which the electron is seen to oscillate in response to the electric field of the passing photon. There is a strong analogy here between the behavior of the electrons in the atom and the free electrons in an antenna. The energy of the passing wave is converted to oscillatory motion of the electron(s), which in the antenna produce a current that is subsequently amplified to signal the presence of the photon. It then makes sense to use classical electromagnetic theory to estimate this effect for the single optical electron of an atom. The oscillation of this electron can then be viewed as a classical oscillating dipole.

Since an oscillating electron represents a continuously accelerating charge, the electron will radiate or absorb energy. In the classical picture, the processes of emission and absorption are interchangeable. The emission simply requires the presence of a driving force, which is the ultimate source of the energy that is emitted, while the energy source for the absorption processes is the passing photon itself. If we let $\overline{W}$ represent the energy gained or lost over one cycle of the oscillating dipole, then any good book on classical electromagnetism (i.e., W. Panofsky and M. Phillips[1] or J. Slater and N. Frank[2]) will show that

$$\frac{d\overline{W}}{dt} = -\frac{2e^2}{3c^3}\left(\frac{d^2x}{dt^2}\right)^2 \tag{14.2.1}$$

where d^2x/dt^2 is the acceleration of the oscillating charge. Now if we assume that the oscillator is freely oscillating, then the instantaneous acceleration is simply

$$\frac{d^2x}{dt^2} = -\omega_0^2 x \tag{14.2.2}$$

This is a good assumption as long as the energy is to be absorbed on a time scale that is long compared to the period of oscillation. Since the driving frequency of the oscillator is that of the line center, this is equivalent to saying that the spread or range of absorbed frequencies is small compared to the frequency of the line center.

Equation (14.2.2) can be used to replace the mean square acceleration of equation (14.2.1) to get

$$\frac{d\overline{W}}{dt} = -\frac{2e^2}{3c^3}\,\omega_0^4\overline{x^2} \tag{14.2.3}$$

The mean position of the oscillator $\overline{x^2}$ can, in turn, be replaced with the mean total energy of the oscillator from

$$\overline{W} = \langle T \rangle + \langle \Phi \rangle = m\omega_0^2\overline{x^2} \tag{14.2.4}$$

so that the differential equation for the absorption or emission of radiation from a classical oscillating dipole is

$$\frac{d\overline{W}}{dt} = -\frac{2e^2\omega_0^2\overline{W}}{3m_e c^3} = -\gamma\overline{W} \tag{14.2.5}$$

The quantity γ is known as the *classical damping constant* and is

$$\gamma = \frac{2e^2\omega_0^2}{3m_e c^3} \tag{14.2.6}$$

The solution of equaton (14.2.5) shows that the absorption of the energy of the passing photon will be

$$I = I_0 e^{-\gamma t} \tag{14.2.7}$$

where I_0 is the presumably sinusoidally varying energy field of the passing photon. The result is that energy of the absorbed or emitted photon resembles a damped sine wave (see Figure 14.1).

But we are interested in the behavior of the absorption with wavelength or frequency, for that is what yields the line profile. Since we are interested in the behavior of an uncorrelated collection of atoms, their combined effect will be proportional to the combined effect of the squares of the electric fields of their emitted photons. Thus, we must calculate the Fourier transform of the time-dependent

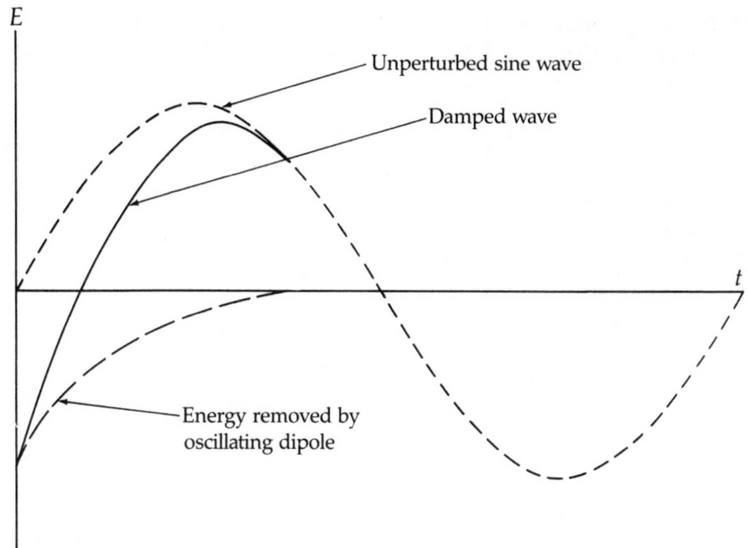

Figure 14.1 Schematic representation of the effect of radiation damping on the wave train of an emitted (absorbed if t is replaced with $-t$) photon. The pure sine wave is assumed to represent the photon without interaction, while the exponential dotted line depicts the effects of radiation damping by the classical oscillator. The solid curve is the combined result in the time domain.

behavior of the electric field of the photon so that

$$E(\omega) = \frac{1}{\sqrt{2\pi}} \int_{-\infty}^{+\infty} E(t)e^{i\omega t}\, dt \tag{14.2.8}$$

If we assume that the photon encounters the atom at $t = 0$ so that $E(t) = 0$ for $t < 0$, and that it has a sinusoidal behavior $E(t) = E_0 e^{-\omega_0 t}$ for $t \geq 0$, then the frequency dependence of the photon's electric field will be

$$E(\omega) = \frac{E_0}{\sqrt{2\pi}} \int_0^\infty e^{-[t\gamma/2 + i(\omega - \omega_0)t]}\, dt \tag{14.2.9}$$

Thus the power spectrum of the energy absorbed or emitted by this classical oscillator will be

$$I_\omega \propto E_\omega^2 = (\text{const})\left[(\omega - \omega_0)^2 + \left(\frac{\gamma}{2}\right)^2\right]^{-1} \tag{14.2.10}$$

It is customary to normalize this power spectrum so that the integral over all frequencies is unity so that

$$I_\omega = \frac{\gamma/(2\pi)}{(\omega - \omega_0)^2 + (\gamma/2)^2} \tag{14.2.11}$$

This normalized power spectrum occurs frequently and is known as a *damping profile* or a *Lorentz profile*. Since the atomic absorption coefficient will be proportional to the energy absorbed,

$$S_\omega = \frac{\pi e^2}{m_e c} \frac{\gamma}{(\omega - \omega_0)^2 + (\gamma/2)^2} \tag{14.2.12}$$

Here the constant of proportionality can be derived from dispersion theory.[3] A plot of S_ω shows a hump-shaped curve with very large "wings" characteristic of a damping profile (see Figure 14.2). At some point in the profile, the absorption coefficient drops to one-half of its peak value. If we denote the full width at this half-power point by $\Delta\lambda_c$, then

$$\Delta\lambda_c = \frac{2\pi c\gamma}{\omega_0^2} = \frac{4\pi e^2}{3m_e c^2} \approx 1.18 \times 10^{-4} \text{ Å} \tag{14.2.13}$$

This is known as the *classical damping width* of a spectral line and is independent of the atom or line. It is also very much smaller than the narrowest lines seen in the laboratory, and to see why, we must turn to a quantum mechanical representation of radiation damping.

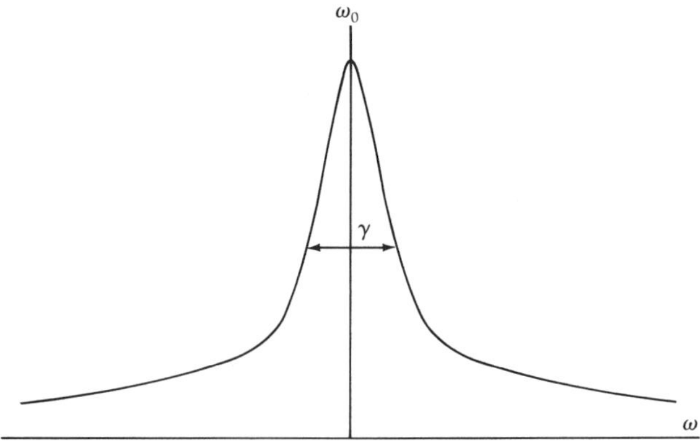

Figure 14.2 Variation of the classical damping coefficient with wavelength. The damping coefficient drops to one-half of its peak value for wavelength shifts equal to $\Delta\lambda_c/2$ on either side of the central wavelength. The overall shape is known as the Lorentz profile.

b   *Quantum Mechanical Description of Radiation Damping*

The quantum mechanical view of the emission or absorption of a photon is rather different from the classical view since it is intimately connected with the nature of the atom in question. The basic approach involves the Heisenberg uncertainty principle as the basis of the broadening. If we consider an atom to be in a certain state, then the length of time that it can remain in that state is related to the uncertainty of the energy of that state by

$$\Delta E \, \Delta t \simeq \hbar \qquad (14.2.14)$$

If there are a large number of states to which the atom can make a transition, then the probability of it doing so is great, Δt is small, and the uncertainty of the energy level is large. A large uncertainty in the energy of a specific state means that a wide range of frequencies can be involved in the transition into or out of that state. Thus any line resulting from such a transition will be unusually broad. Thus any strong line resulting from frequent transitions will also be quite broad.

This view of absorption and emission was quantified by Victor Weisskopf and Eugene Wigner[4,5] in 1930. They noted that the probability of finding an atom with a wave function Ψ_j in an excited state j after a transition from a state i is

$$P_j(t) = \Psi_j^* \Psi_j e^{-\Gamma t} \qquad (14.2.15)$$

where Γ is the Einstein coefficient of spontaneous emission A_{ji}. The exponential behavior of $P_j(t)$ ensures that the power spectrum of emission will have the same form as the classical result, namely,

$$I(\omega) = \frac{\Gamma/(2\pi)}{(\omega - \omega_0)^2 + (\Gamma/2)^2} \qquad (14.2.16)$$

If the transition takes place between two excited levels, which can be labeled u and l, the broadening of which can be characterized by transitions from those levels, then the value of gamma for each level will have the form

$$\Gamma_u = \sum_{i<u} A_{ui} \qquad \Gamma_l = \sum_{i<l} A_{li} \qquad (14.2.17)$$

The power spectrum of the transition between them will then have the form of equation (14.2.16), but with the value of gamma determined by the width of the two levels so that

$$\Gamma = \Gamma_l + \Gamma_u \qquad (14.2.18)$$

c Ladenburg f Value

Since the power spectrum from the quantum mechanical view of absorption has the same form as that of the classical oscillator, it is common to write the form of the atomic absorption coefficient as similar to equation (14.2.12) so that

$$S_\omega = \frac{\pi e^2}{m_e c} f_{ik} \frac{\Gamma_{ik}}{(\omega - \omega_0)^2 + (\Gamma_{ik}/2)^2} \qquad (14.2.19)$$

The quantity f_{ik} is then the equivalent number of classical oscillators that the transition from $i \to k$ can be viewed as representing. If you like, it is the number that brings the quantum mechanical calculation into line with the classical representation of radiation damping. If the energy levels are broad, then the transition is much more likely to occur than one would expect from classical theory, the absorption coefficient will be correspondingly larger, and $f_{ik} > 1$. The quantity f_{ik} is known as the *Ladenburg f value* or the *oscillator strength*. However, the line profile will continue to have the characteristic lorentzian shape that we found for the classical oscillator.

Since the f value characterizes the entire transition, we expect it to be related to other parameters that specify the transition. Thus, the f value and the Einstein coefficient of absorption are not independent quantities. We may quantify this relation by integrating equation (14.2.19) over all frequencies and by using equation (14.1.4), substituting into equation (14.1.3) to get

$$\frac{\pi e^2 f_{ik}}{m_e c} \frac{1}{2\pi} \int_0^\infty \frac{\Gamma_{ik}\, d\omega}{(\omega - \omega_0)^2 + (\Gamma_{ik}/2)^2} = \frac{e^2 f_{ik}}{m_e c} \int_{-\omega_0}^\infty \frac{a\, dx}{a^2 + x^2}$$

$$= B_{ik} \frac{h\nu_0}{4\pi} \qquad (14.2.20)$$

where $a = \Gamma_{ik}/2$ and ν_0 is the frequency of the line center. If we make the assumption that the line frequency width is small compared to the line frequency, then $\Gamma_{ik}/\omega_0 \ll 1$ and equation (14.2.20) becomes

$$f_{ik} = \frac{m_e c}{\pi e^2} \frac{h\nu_0 B_{ik}}{4\pi} \qquad (14.2.21)$$

Thus the classical atom can be viewed as radiating or absorbing a damped sine wave whose Fourier transform contains many frequencies in the neighborhood of the line center. These frequencies are arranged in a symmetric pattern known as a *Lorentz* or *damping profile* characterized by a specific width. The quantum mechanical view changes very little of this except that the transition can be viewed as being made up of a number of classical oscillators determined by the Einstein coefficient of the transition. In addition, the classical damping constant is replaced by a damping constant that depends on all possible transitions in and out of the levels involved in the transition of interest. The term that describes this form of broadening is *radiation damping*, and it is derived from the damped form of the absorbed or emitted photon wave train, as is evident from the classical description.

The broadening of spectral lines by this process is independent of the environment of the atom and is a result primarily of the probabilistic behavior of the atom itself. In cases where external forms of broadening are small or absent, radiation damping may be the dominant form of broadening that effectively determines the shape of the spectral line. When this is the case, little about the nature of the environment can be learned from the line shape. However, for normal stellar atmospheres and most lines, perturbations caused by the surrounding medium cause changes in the energy levels that far outweigh the natural broadening of the uncertainty principle. We now consider these forms.

14.3 Doppler Broadening of Spectral Lines

The atoms that make up the gas of the stellar atmosphere are constantly in motion, and this motion shifts the wavelengths, seen by an observer, at which the atoms can absorb radiation. This motion may be only the thermal motion of the gas, or it may include the larger-scale motions of turbulence or rotation. Whatever the combination, the shifting of the rest wavelengths by varying amounts for different populations of atoms will usually result in the observed line's being broadened by an amount significantly greater than the natural width determined by atomic properties.

The shifting of the rest wavelength caused by the motion of the atoms not only produces a change as seen by the observer, but also may expose the atom to a somewhat different radiation field. This will be true if the motion is locally random so that the motion of each atom is uncorrelated with that of its neighbors. However, should the motions be large-scale, then entire collections of atoms will have their rest wavelengths shifted by the same amount with respect to the observer and the star. If these collections of atoms constitute an optically thick ensemble, then the radiation field of the ensemble will be shifted along with the rest wavelength. To atoms within such a "cloud" there will be no effect of the motion on the atoms themselves. It will be as if a "miniatmosphere" were moving, and no additional photons will be absorbed as a result of the motion. Such motions will not affect the equivalent widths of lines but may change the profiles considerably.

Contrast this with the situation resulting from an atom whose motion is uncorrelated with that of its neighbors. Imagine a line with an arbitrarily sharp atomic absorption coefficient [that is, $S_v = \delta(v - v_0)$]. If there were no motion in the atmosphere, the lowest-lying atoms would absorb all the photons at frequency v_0, leaving none to be absorbed by the overlying atoms. Such a line is said to be saturated because the addition of absorbing material will make no change in the line profile or equivalent width. But allow some motion, and the rest frequency of these atoms is changed slightly from v_0. Now these atoms will be capable of absorbing photons at the neighboring frequencies, and the line will appear wider and stronger. Its equivalent width will be increased simply as a result of the Doppler shifts experienced by some atoms. Thus, if the motion consists of collections of atoms that are optically thin, we can expect changes in the line strengths as well as in the profiles. However, if those collections of atoms are large enough to be optically thick, then no change in the equivalent width will occur in spite of marked changes in the line profile. We refer to the motions of the first case as *microscopic* motions so as to contrast them with the second case of *macroscopic* motion.

a   Microscopic Doppler Broadening

Again, it is useful to make a further subdivision of the classes of microscopic motions based on the nature of those motions. In the case of thermal motions, we may make plausible assumptions regarding the velocity field of the atoms.

Thermal Doppler Broadening   The assumption of LTE from Section 9.1b stated that the particles that make up the gas obeyed Maxwell-Boltzmann statistics appropriate for the local values of temperature and density. For establishing the Saha-Boltzmann ionization and excitation formulas, it was really only necessary that the electrons dominating the collision spectrum exhibit a maxwellian energy spectrum. However, we now insist that the ions also obey Maxwell-Boltzmann statistics so that we may specify the velocity field for the atoms. With this assumption, we may write

$$\frac{dN(v)}{N} = \frac{1}{\sqrt{\pi}} e^{-v^2/v_0^2} \frac{dv}{v_0} \tag{14.3.1}$$

where dN/N is just the fraction of particles having a speed lying between v and $v + dv$, and so it is a probability density frunction of the particle energy distribution. It is properly normalized since the integrals of both sides of equation (14.3.1) are unity. The second moment of this energy distribution gives

$$\langle v^2 \rangle = \frac{1}{\sqrt{\pi}} \int_{-\infty}^{\infty} v^2 e^{-(v/v_0)^2} \frac{dv}{v_0} = \tfrac{1}{2} v_0^2 \tag{14.3.2}$$

which we may relate to the kinetic energy of the gas.

Now we wish to pick the speed used in equations (14.3.1) and (14.3.2) to be the radial or line-of-sight velocity. Since there is no preferred frame of reference for the random velocities of thermal motion, this choice is as good as any other. However, the mean square velocity $\langle v^2 \rangle$ calculated in equation (14.3.2) is then only averaged over line-of-sight or radial motions and thus represents only 1 degree of freedom for the particles of the gas. So the energy associated with that motion is equal to $\frac{1}{2}kT$ for a monatomic gas, and

$$v_0^2 = \frac{2kT}{m} \tag{14.3.3}$$

With the aid of the first-order (classical) Doppler shift, we define the Doppler width of a line in terms of v_0 as

$$\frac{\Delta\lambda_d}{\lambda_0} = \frac{\Delta v_d}{v_0} = \frac{v_0}{c} \tag{14.3.4}$$

Using equation (14.3.4), we may rewrite the particle distribution function for velocity as one for the fraction of atoms capable of absorbing at a frequency shift Δv (or wavelength shift $\Delta\lambda$)

$$\frac{dN(\Delta\lambda)}{N} = \frac{1}{\sqrt{\pi}} e^{-(\Delta\lambda/\Delta\lambda_d)^2} \frac{d\lambda}{\Delta\lambda_d} = \frac{1}{\sqrt{\pi}} e^{-(\Delta v/\Delta v_d)^2} \frac{dv}{\Delta v_d} = \frac{dN(\Delta v)}{N} \tag{14.3.5}$$

Since the atomic line absorption coefficient is basically the probability of an atom's absorbing a photon at a given frequency, that probability should be proportional to the number of atoms capable of absorbing at that frequency. Thus,

$$S_v \, dv = \frac{A}{\sqrt{\pi}} e^{-(\Delta v/\Delta v_d)^2} \frac{dv}{\Delta v_d} \tag{14.3.6}$$

where A is simply a constant of proportionality. This constant can be related to the Einstein coefficient by equations (14.1.3) and (14.1.4), with the result that

$$S_v \, dv = \frac{h v_0 B_{ik}}{4\pi\sqrt{\pi}\,\Delta v_d} e^{-(\Delta v/\Delta v_d)^2} \, dv = \frac{\sqrt{\pi} e^2 f_{ik}}{m_e c\,\Delta v_d} e^{-(\Delta v/\Delta v_d)^2} \, dv \tag{14.3.7}$$

To get the result on the far right-hand side, we used the relationship between the f value for a particular transition and the Einstein coefficient given by equation (14.2.21). This is the expression for the atomic line absorption coefficient for thermal Doppler broadening. It differs significantly from the Lorentz profile of radiation damping by exhibiting a much stronger frequency dependence. A spectral line where both broadening mechanisms are present will posses a line core that is dominated by Doppler broadening while the far wings of the line will be dominated by the damping profile as the gaussian profile of the Doppler core rapidly goes to zero.

Microturbulent Broadening In addition to the thermal velocity field, the atoms in the atmospheres of many stars experience motion due to turbulence. Unfortunately, the theory of turbulent flow is insufficiently developed to enable us to make specific predictions concerning the velocity distribution function of the turbulent elements. So, for simplicity, we assume that they also exhibit a maxwellian velocity distribution, but one having a characteristic velocity different from the thermal velocity. Thus, the form of the probability density distribution function for turbulent elements is the same as equation (14.3.1) except that the velocity is the radial velocity of the turbulent cell:

$$\frac{dN}{N} = \frac{1}{\sqrt{\pi}} e^{-(v/v_0)^2} \frac{dv}{v_0} \tag{14.3.8}$$

If there were no other processes to consider, the atomic absorption coefficient for a turbulently broadened line would have the same form as one that is thermally broadened except for a minor change in the interpretation of the Doppler half-width. However, we are interested in the combined effects of thermal and turbulent broadening, and so we consider how this combination may be carried out.

Since equation (14.3.1) represents the fraction of particles with a thermal velocity within a particular range, we may write the probability that a given atom will have a thermal velocity lying between v and $v + dv$ as

$$P_{\text{th}}(v) = \frac{N}{v_0\sqrt{\pi}} e^{-(v/v_0)^2} \tag{14.3.9}$$

The probability that this same atom will reside in a particular turbulent element having a turbulent velocity lying between v and $v + dv$ can be obtained, in a similar manner, from equation (14.3.8) and is

$$P_{\text{turb}}(v) = \frac{1}{v_0\sqrt{\pi}} e^{-(v/v_0)^2} \tag{14.3.10}$$

However, the observer does not regard these velocities as being independent since she or he is interested only in those combinations of velocities that add to produce a particular radial velocity v which yields a Doppler-shifted line. So we must regard the thermal and turbulent velocities to be constrained by

$$v = v + v \tag{14.3.11}$$

Now the joint probability that an atom will have a velocity v lying between v and v + dv resulting from specific thermal and turbulent velocities v and υ, respectively, is given by the product of equations (14.3.9) and (14.3.10). But we are not interested in just the probability that a thermal velocity v *and* a turbulent velocity υ will yield an observed velocity v; rather we are interested in all combinations of v and υ that will yield v. Thus we must sum the product probability over all combinations of v and υ subject to the constraint given by equation (14.3.11). With this in mind, we can write the combined probability that a given atom will have combined thermal and turbulent velocities that yield a specific observed radial velocity as

$$p(\text{v}) = \frac{dN}{d\upsilon} = \int_{-\infty}^{+\infty} P_{\text{th}}(\upsilon) p_{\text{turb}}(\text{v} - \upsilon)\, d\upsilon$$

$$= \frac{N}{\pi \upsilon_0 \upsilon_0} \int_{-\infty}^{+\infty} e^{-(\upsilon/\upsilon_0)^2 - [(\text{v} - \upsilon)/\upsilon_0]^2}\, d\upsilon \qquad (14.3.12)$$

Since the velocities involved in equation (14.3.12) are radial velocities, they may take on both positive and negative values. Thus the range of integration must run from $-\infty$ to $+\infty$. After some algebra, equation (14.3.13) yields the fraction of atoms with a combined velocity v to be

$$\frac{dN}{N} = \frac{e^{-(\text{v}/\text{v}_0)^2}}{\text{v}_0 \sqrt{\pi}}\, d\text{v} \qquad (14.3.13)$$

where

$$\text{v}_0^2 = v_0^2 + \upsilon_0^2 \qquad (14.3.14)$$

The similarity of the form of equation (14.3.13) to that of equations (14.3.1) and (14.3.8) is no accident. The integral in equation (14.3.12) is known as a *convolution integral*. The combined probability of $p(a)$ *and* $p(b)$ involves taking the product $p(a) \times p(b)$. If, in addition, one has a constraint $q(c) = q(a, b)$, then one must consider all combinations of a and b that yield c and sum over them. That is, one wants the probability of (a_1, b_1) or (a_2, b_2) etc. that yields c. Combining probabilities of *A or B* involves summing those probabilities. So, in general, if one wishes to find the combined probability of two processes subject to an additional constraint, one "convolves" the two probabilities. It is a general property of convolution integrals where the probability distributions have the same form that the resultant probability will also have the same form with a variance that is just the sum of the variances of the two initial probability distribution functions. Thus the convolution of any two gaussian distribution functions will itself be a gaussian distribution function having a variance that is just the sum of the two initial variances. This explains the form of equation (14.3.14). As a result, we may immediately write the atomic absorption coefficient for the combined effects of thermal and turbulent Doppler broadening as

$$S_v = \frac{\sqrt{\pi} e^2 f_{ik}}{m_e c\, \Delta v_d}\, e^{-(\Delta v/\Delta v_d)^2} \tag{14.3.15}$$

where

$$\Delta v_d = \frac{v_0 v_0}{c} \tag{14.3.16}$$

and

$$v_0^2 = \frac{2kT}{m} + v_0^2 \tag{14.3.17}$$

It is now clear why we assumed the turbulent broadening to have a maxwellian velocity distribution. If this were not the case, the convolution integral would be more complicated. If the turbulent velocity distribution function had the form

$$p_{\text{turb}}(v) = \Phi(v) \tag{14.3.18}$$

then the convolution integral with thermal broadening would become

$$p(v) = N \int_{-\infty}^{+\infty} \frac{\Phi(x)}{\sqrt{\pi} v_0}\, e^{-(x-v)^2/v_0^2}\, dx \tag{14.3.19}$$

If the function $\Phi(x)$ is sufficiently simple, the integral may be expressed in terms of analytic functions. If not, then the integral must be evaluated numerically as part of the larger calculation for finding the line profile.

Combination of Doppler Broadening and Radiation Damping   Any spectral line will be subject to the effects of radiation damping or some other intrinsic broadening mechanism as well as the broadening introduced by Doppler motions. So to get a reasonably complete description of the atomic absorption coefficient, we have to convolve the Doppler profile with the classical damping profile given by equation (14.2.19). However, since the atomic absorption coefficient is expressed in terms of frequency, the constraint on the independent variables of velocity and frequency must contain the Doppler effect of that velocity on the observed frequency. Thus the frequency v' at which the atom will absorb in terms of the rest frequency v_0 is

$$v' = v_0 + \frac{v_0 v}{c} \tag{14.3.20}$$

For an atom moving with a line-of-sight velocity v, the atomic absorption for radiation damping is

$$S_v = \frac{\pi e^2 f_{ik} \Gamma_{ik}}{4\pi^2 m_e c} \frac{1}{(v_0 + v_0 v/c - v)^2 + [\Gamma_{ik}/(4\pi)]^2} \tag{14.3.21}$$

This atomic absorption coefficient is essentially the probability that an atom having velocity v will absorb a photon at frequency v. To get the total absorption coefficient, we must multiply by the probability that the atom will have the velocity v [equation (14.3.13)] and sum all possible velocities that can result in an absorption at v. Thus,

$$S_v = \frac{\pi e^2 f_{ik} \Gamma_{ik}}{4\pi m_e c} \int_{-\infty}^{+\infty} \frac{1}{v_0 \sqrt{\pi}} \frac{e^{-(v/v_0)^2}\,dv}{(v_0 + v_0 v/c - v)^2 + [\Gamma_{ik}/(4\pi)]^2} \tag{14.3.22}$$

The convolution integral represented by equation (14.3.22) is clearly not a simple one. When one is faced with a difficult integral, it is advisable to change variables so that the integrand is made up of dimensionless quantities. This fact will remove all the physical parameters to the front of the integral, clarifying their role in the result, and reduce the integral to a dimensionless weighting factor. This also facilitates the numerical evaluation of the integral since the relative values of all the parameters of the integrand are clear. With this in mind, we introduce the following traditional dimensionless variables:

$$u \equiv \frac{c(v - v_0)}{v_0 v_0}$$

$$y \equiv \frac{v}{v_0} \tag{14.3.23}$$

$$a \equiv \frac{c\Gamma_{ik}}{4\pi v_0 v_0} = \frac{\Gamma_{ik}}{4\pi \,\Delta v_d}$$

Substituting these into equation (14.3.22), we get

$$S_v(u) = \frac{e^2 f_{ik} a}{m_e v_0 v_0 \sqrt{\pi}} \int_{-\infty}^{+\infty} \frac{e^{-y^2}\,dy}{a^2 + (u - y)^2} \tag{14.3.24}$$

It is common to absorb all the physical parameters on the right-hand side of equation (14.3.24) into a single constant that has the units of an absorption coefficient so that

$$S_0 \equiv \frac{\sqrt{\pi}e^2 f_{ik}}{m_e v_0 v_0} \tag{14.3.25}$$

The remaining dimensionless function can be written as

$$H(a,\,u) \equiv \frac{a}{\pi} \int_{-\infty}^{+\infty} \frac{e^{-y^2}\,dy}{a^2 + (u-y)^2} \tag{14.3.26}$$

This is known as the *Voigt function*, and it allows us to write the atomic absorption coefficient in the following simple way:

$$S_\omega(u) = S_0 H(a,\,u) \tag{14.3.27}$$

For small values of the damping constant ($a < 0.2$), the Voigt function is near unity at the line center (that is, $u = 0$) and falls off rapidly for increasing values of $|u|$. For values of $|u|$ near zero, the Voigt function is dominated by the exponential that corresponds to the Doppler core of the line. However, at larger values of $|u|$, the denominator dominates the value of the integral. This corresponds to the damping wings of the line profile.

Considerable effort has gone into the evaluation of the Voigt function because it plays a central role in the calculation of the atomic line absorption coefficient. One of the earliest attempts involved expressing the Voigt function as

$$H(a,\,u) = \sum_{i=0}^{\infty} a^i H_i(u) \tag{14.3.28}$$

where the functions $H_i(u)$ are known as the *Harris functions*.[6] More commonly one finds the alternative function

$$V(a,\,u) = \frac{H(a,\,u)}{\sqrt{\pi}} \tag{14.3.29}$$

whose integral over all u is unity. This function is known as the *normalized Voigt function*. Extensive tables of this function were calculated by D. Hummer[7] and a reasonably efficient computing scheme has been given by G. Finn and D. Mugglestone.[8] However, with the advent of fast computers emphasis has been put on finding a fast and accurate computational algorithm for the Voigt function. The best to date is that given by J. Humíček.[9] This has been expanded by S. McKenna[10] to include functions closely related to the Voigt function. All this effort has made it possible to obtain accurate values for the Voigt function with great speed, making the inclusion of this function in computer codes little more difficult than including trigonometric functions.

b Macroscopic Doppler Broadening

The fact that each atom was subject to all the broadening mechanisms described above caused most of the problems in calculating the atomic absorption coefficient through the introduction of a convolution integral. This approach assumed that each atom could "see" other atoms subject to the different velocity sources. However, if the turbulent elements were sufficiently large that they themselves were optically thick, then each element would optically behave independently of the others. The line profiles of each would be similar, but shifted relative to the others by an amount determined by the turbulent velocity of the element. Indeed, this would be the case if any motions involving optically thick sections of the atmosphere were present.

The proper approach to this problem involves finding the locally emitted specific intensity, convolving it with the velocity distribution function, and integrating the result over the visible surface of the star to obtain the integrated flux. This flux can the be normalized to produce the traditional line profile. However, since the macroscopic motions can affect the structure of the atmosphere, the problem can become exceedingly difficult and solvable only with the aid of large computers. In spite of this, much can be learned about the qualitative behavior of these broadening mechanisms from considering some greatly simplified examples. We discuss just two; the first involves motions of large sections of the atmosphere in a presumably uncorrelated fashion, and the second involves the correlated motion of the entire star.

Broadening by Macroturbulence   It would be a mistake to assume that turbulent elements only come in sizes that are either optically thick or thin. However, to gain some insight into the degree to which turbulence can affect a line profile, we divide the phenomena into these two cases. We have already discussed the effects that small turbulent elements have on the resulting atomic line absorption coefficient (i.e., microturbulence), and we have seen that they lead to an increase in value of that parameter for all frequencies. Such is not the case for macroturbulence. The motion of optically thick elements cannot change the value of the atomic line absorption coefficient because the environment of a particular atom concealed within the turbulent element is unaffected by the motion of that element. Thus, each element behaves as a separate "atmosphere," producing its own line profile, which contributes to the stellar profile by an amount proportional to the ratio of the visible area of the element to that of the apparent disk of the star. Thus, the combining (or convolution) of line profiles occurs not on the atomic level, as with microturbulent Doppler broadening, but after the radiative transfer has been locally solved to yield a local line profile. This requires that we make assumptions that apply globally to the entire star in order to relate one turbulent element to another.

To demonstrate the nature of this effect, we consider a particularly simple situation where there is no limb-darkening in or out of the line. In addition, we assume that the local line profile is given by a Dirac delta function of frequency and that the macroturbulent motion is purely radial with a velocity $\pm v_m$. Under these conditions, zones of constant radial velocity will appear as concentric circles on the apparent disk (see Figure 14.3).

Since the intrinsic line profile is a delta function of frequency, the line profile originating at a ring of constant radial velocity located at an angle θ measured from the center of the disk will be Doppler-shifted by an amount

$$\Delta v = v - v_0 = \pm \frac{v_m v_0 \mu}{c} \qquad (14.3.30)$$

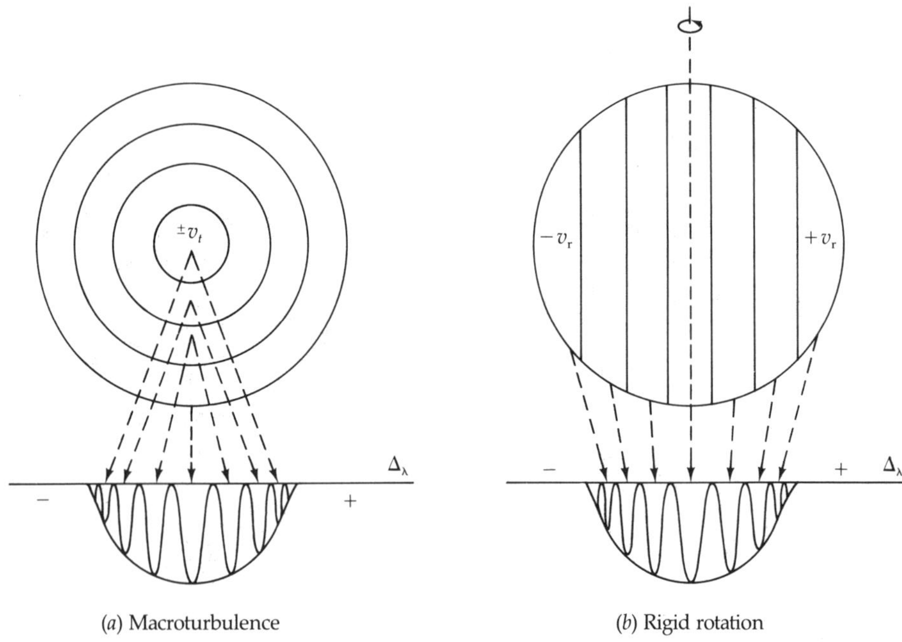

(a) Macroturbulence   (b) Rigid rotation

Figure 14.3 Schematic representation of the apparent disks of two idealized stars. (a) Lines of constant line-of-sight velocity for a macroturbulent stellar atmosphere where the turbulent motion is assumed to be along the stellar radius and of a fixed magnitude v_m. (b) Lines of constant radial velocity for a star that is not distorted by, but exhibits, rigid rotation.

where, as usual,

$$\mu = \cos\theta \qquad (14.3.31)$$

The amount of energy removed from the total continuum flux by the local line absorption will simply be proportional to the area of the differential annulus located at the particular value of μ corresponding to Δv. Thus,

$$\frac{F_c - F_v(\text{line})}{F_c} = 1 - r_v \propto 1 - \mu^2 = 1 - \left(\frac{\Delta v\, c}{v_0 v_m}\right)^2 \qquad (14.3.32)$$

Therefore, the line profile would be given by

$$r_v = 1 - (1 - r_0)\left[1 - \left(\frac{\Delta v\, c}{v_m v_0}\right)^2\right]^{1/2} \qquad \Delta v \le \frac{v_0 v_m}{c} \qquad (14.3.33)$$

This line profile is dish-shaped and is characteristic of this type of mass atmospheric motion. Since the equivalent width remains constant for macroscopic broadening, the central depth of the line will decrease for increasing v_m.

Clearly a real situation replete with limb-darkening, a velocity dispersion of the turbulent elements, an anisotropic velocity field, along with a spectrum of sizes for the turbulent eddies, would make the problem significantly more difficult. A great deal of work has been done to treat the problem of turbulence in a more complete manner, but the results are neither simple to discuss or easy to review. D. Mihalas[11] gives an introduction and excellent references to this problem.

Broadening by Stellar Rotation   As we saw in Chapter 7, rapid rotation of the entire star will lead to significant distortion of the star and a wide variation of the parameters that define a stellar atmosphere over its surface. In such a situation, most of the assumptions we have made for the purpose of modeling the atmosphere no longer apply, and recourse must be made to a more numerical approach (see G. Collins[12] and J. Cassinelli[13]). However, as with macroturbulence, some insight may be gained by considering the effects of rotation on the line profile of a slowly rotating star. Such a model is originally due to G. Shajn and O. Struve[14] and is now commonly referred to as the *Struve model*.

Consider a uniformly bright spherical star which is rotating as a solid body. Except for the rotation, this is essentially the same model as that used for the discussion of macroturbulence (see Figure 14.3). If we define θ and ϕ, respectively, to be the polar and azimuthal angles of a spherical coordinate system with its polar axis aligned with the rotation axis of the star, then the velocity toward the observer's line of sight is

$$v_r = v_{\text{eq}} \sin\theta \sin\phi \sin i \qquad (14.3.34)$$

where v_{eq} is the equatorial velocity of the star and i is the angle between the line of sight and the rotation axis, called the *inclination*. An inspection of Figure 14.3 and some geometry leads one to the conclusion that for spherical stars the product $\sin \theta \sin \phi$ is constant on the stellar surface along any plane parallel to the meridian plane. Thus, any chord on the apparent disk that is parallel to the central meridian is a locus of constant radial velocity (see Figure 14.3). Any profile formed along this cord will be displaced in frequency by an amount

$$\Delta v = \frac{v_{eq} v_0}{c} (1 - \mu^2)^{1/2} \sin i = a(1 - \mu^2)^{1/2} \qquad (14.3.35)$$

For a sphere of unit radius, the length of the chord is 2μ. If we make the same assumptions about the intrinsic line profile as were made for the case of macro-turbulence (i.e., it can be locally represented by a delta function), then the amount of flux removed from the continuum intensity by any profile located on one of these chords will just be proportional to the length of the chord. Therefore,

$$\frac{F_c - F_v(\text{line})}{F_c} = 1 - r_v \propto 2\mu = 2\left(1 - \frac{\Delta v^2}{a^2}\right)^{1/2} \qquad (14.3.36)$$

which leads to a profile of the form

$$r_v = 1 - (1 - r_0)\left(1 - \frac{\Delta v^2 c^2}{v_0^2 v_{eq}^2 \sin^2 i}\right)^{1/2} \qquad \Delta v \le \frac{v_0 v_{eq} \sin i}{c} \qquad (14.3.37)$$

Except for the replacement of the turbulent velocity by the equatorial velocity, the rotational profile has the same form as the profile for macroturbulence [equation (14.3.33)]. This points out a fundamental problem of Doppler broadening by mass motions. In general, it is not possible to infer the velocity field from the line profile alone. To be sure, the presence of limb-darkening would affect these two cases differently, as would the introduction of gravity darkening for the case of rotation. But the nonuniqueness remains for the general case, and any determination of the velocity field from the analysis of line profiles is strongly model-dependent and usually relies on some assumed symmetry.

Many of the simplifying assumptions of these models for macroturbulence and rotation can be removed for a modest increase in complexity. In the case of rotation, if the local line profile were not given by a delta function but had an intrinsic shape $r'(x)$, where

$$x = \frac{\Delta v \, c}{v_0 v_{eq}} \qquad (14.3.38)$$

then the observed line profile would be given by the convolution integral

$$1 - r(x) = \int_{-1}^{+1} [1 - r'(x - y)] Q(y)\, dy \tag{14.3.39}$$

Here $Q(y)$ is known as the *rotational broadening function* which, if limb-darkening is included, is given by A. Unsöld[15] as

$$Q(y) = \frac{3}{3 + 2\beta} \left[\frac{2}{\pi} (1 - y^2)^{1/2} + \frac{\beta}{2} (1 - y^2) \right] \tag{14.3.40}$$

The parameter β is the first-order limb-darkening coefficient. Consider the case for $\beta = 0$ and that the intrinsic line profile is a delta function. It is clear that equations (14.3.39) and (14.3.40) will yield equation (14.3.37) as long as the integral of $Q(y)$ is normalized to unity. Integration of equation (14.3.40) will satisfy the skeptic that this is indeed the case. It is also clear that the general effects of rotation are not qualitatively very different from those implied by equation (14.3.37). While quantitative comparison with observation will clearly be affected by such things as the intrinsic line profile and limb-darkening, a truly useful comparison will have to go even further and include the effects of the variation of the atmospheric structure over the surface on the line profile.

While macroturbulence and rotation constitute the most important forms of macroscopic broadening, there are others. The presence of magnetic fields can split atomic lines through the Zeeman effect. In some instances, this can lead to anomalously broad spectral lines and subsequent errors in the abundances derived from these lines. In some instances, the broadening is sufficiently large to allow the estimation of the magnetic field itself. Fortunately, strong magnetic fields appear to be sufficiently rare among normal stellar atmsopheres to allow us to ignore their effects most of the time. However, we should be ever mindful of the possibility of their existence and of the effects that they can introduce in the shaping of spectral lines.

14.4 Collisional Broadening

To this point, we have described the broadening of spectral lines arising from intrinsic properties of the atom and the collective effects of the motion of these atoms. However, in all but the most extreme cases of macroscopic broadening, the most prominent source of broadening of spectral lines results from the interaction of the absorbing atom with neighboring particles of the gas. Since these particles are often charged (even the neutral atoms possess the potential field of an electric dipole), their potential will interact with that of the atomic nucleus which binds the orbiting electrons. This interaction will perturb the energy levels of the atom in a time-dependent fashion. The collective action of these perturbations on an ensemble of absorbing atoms is to broaden the spectral line. The details of this broadening depend on the nature of the atom and energy level being perturbed and the properties of the dominant perturber. All phenomena that fall into this general class of broadening mechanisms are usually gathered under the generic term *collisional broadening*. However, some authors refer to this concept or a subset of it as *pressure broadening*, on the grounds that there can be no collisions unless the gas has some pressure. The use of the different terms is usually not of fundamentally importance, and the basic notion of what is behind them should always be kept in mind.

There is some confusion in the literature (and much more among students) regarding the terminology for describing these processes. Some of this results from a genuine confusion among the authors, but most derives from an unfortunate choice of terms used to describe some aspects of the problem. You should keep clearly in mind what is being described during any discussion of this topic—the broadening of atomic energy levels resulting from the perturbations of neighboring particles. We adopt a variety of theoretical approaches to this problem, each of which has its own name. Care must be taken lest the name of the theoretical approach be confused with a qualitatively different type of broadening. We discuss perturbations introduced by different types of perturbers, each of which will produce a characteristic line profile for the absorbing gas. Each of these profiles has its own name so as to delineate the type of perturbation. However, they are all just perturbations of the energy levels. Each type will generally be discussed in a "vacuum," in that we assume that it is the only form of perturbation that exists, when in reality virtually all types of perturbation are present at all times and affect the energy levels. Fortunately, one of them usually does dominate the level broadening.

There are two main theoretical approaches to collisional broadening. One deals with the weak, but numerous, perturbations that cause small amounts of broadening. The other is concerned with the large, but infrequent, perturbations that determine the shape of the wings of the line. It is somewhat unfortunate that the former theoretical approach is known as *impact phase-shift theory*, while the latter is called the *statistical* or *static broadening theory*. The word *impact* conjures up visions of violence, yet the theoretical approach labeled by this word is concerned only with the weakest and least violent of the interactions. Similarly, the term *static* implies calm, but this approach deals with the most violent perturbations. So be it. We try to justify this apparent anomaly during the specific discussions of these approaches. In addition, we clearly label the myriad terms as they are introduced so that those which are synonymous are clearly separated from those which have unique meanings.

To estimate the perturbation to the atom that changes the energy of the transition and thereby broadens the line, we must characterize the nature of the collision.

The two theoretical approaches to collisional broadening differ in this description. Both approaches are largely classical in form so that whatever is true for absorption is also true for emission. So we often deal with the effects of a collision on a radiating atom with the full intention of applying the results to absorption.

a Impact Phase-Shift Theory

The approach of impact phase-shift theory assumes that the collision is of a very short duration compared to the span of time during which the atom is actually radiating (or absorbing) the photon. Thus,

$$t_{col} \ll t_{rad} \tag{14.4.1}$$

It is the short duration of the collision that is responsible for the name *impact* for the theoretical approach.

Determination of the Atomic Line Absorption Coefficient Suppose that the atom radiates in an undisturbed manner between collisions with a frequency ω_0. The electric field of the emitted photon will vary as

$$E(t) = E_0 e^{-i\omega_0 t} \qquad -\frac{T}{2} \le t \le \frac{T}{2} \tag{14.4.2}$$

where T is the time between collisions. Further assume that the radiation of the photon does not continue before or after successive collisions, so that

$$E(t) = 0 \qquad t > \left|\frac{T}{2}\right| \tag{14.4.3}$$

It is this interruption in the emission of the photon, or at least a complete and discontinuous change in the phase of the emitted photon, that terminates the wave train and provides the motivation for the second half of the name for this approach. Since a sine wave of finite length must contain wave components of higher frequency introduced by the discontinuity of the wave train, the emitted photon will have more components than the fundamental frequency and thus the line will appear to be broadened. To find this distribution in frequencies, we must take the Fourier transform of the temporal description of the electric field of the photon. So

$$E(\omega) = \frac{1}{\sqrt{2\pi}} \int_{-\infty}^{+\infty} E(t) e^{i\omega t}\, dt = \frac{E_0}{\sqrt{2\pi}} \int_{-T/2}^{+T/2} e^{i(\omega-\omega_0)t}\, dt \tag{14.4.4}$$

or

$$E(\omega) = + \frac{E_0}{\sqrt{2\pi}} \frac{2 \sin\left[(\omega - \omega_0)T/2\right]}{\omega - \omega_0} \tag{14.4.5}$$

Since the power spectrum of the emitted photon will depend on the square of the electric field,

$$I(\omega) = \frac{A^2 E_0^2}{2\pi} \frac{\sin^2\left[T(\omega - \omega_0)/2\right]}{\left[(\omega - \omega_0)/2\right]^2} \tag{14.4.6}$$

Here we have assumed that we will be dealing with emissions and absorptions that are totally uncorrelated, which for random collisions occuring in a sea of un-related atoms is a perfectly reasonable assumption.

Now to determine the effects of multiple collisions (or numerous atoms), we must combine the effects of these collisions, which means that we must have some esti-mate of the time between them T. Let $P(t)$ be the probability that a collision has *not* occurred in a time t measured from the last collision. Now if the collisions are indeed random, the differential probability dp that a collision will occur in a time interval dt is

$$dp = \frac{dt}{T_0} \tag{14.4.7}$$

where T_0 is the average time between collisions. Thus, the differential change in the probability $P(t)$ is

$$d[1 - P(t)] = -dP(t) = P(t)\, dp = \frac{P(t)\, dt}{T_0} \tag{14.4.8}$$

or

$$P(t) = Be^{-t/T_0} \tag{14.4.9}$$

Since a collision must occur at some time, we can determine the constant in equa-tion (14.4.9) by normalizing that expression to unity and integrating over all time. Thus we see that the collision frequency distribution is a Poisson distribution of the form

$$P(t) = \frac{1}{T_0} e^{-t/T_0} \tag{14.4.10}$$

 and the constant of proportionality in equation (14.4.9) is $1/T_0$.

To obtain the total energy distribution or power spectrum resulting from a multitude of collisions, we must sum the power spectra of the individual collisions multiplied by the probability of their occurrence. Thus,

$$I_t(\omega) = \frac{A^2 E_0^2}{2\pi T_0} \int_0^\infty \frac{\sin^2[T(\omega - \omega_0)/2]}{[(\omega - \omega_0)/2]^2} e^{-T/T_0} dT$$

$$= \frac{A^2 E_0^2}{\pi T_0} \frac{1}{(\omega - \omega_0)^2 + (1/T_0)^2} \qquad (14.4.11)$$

We can use the same normalization process implied by equations (14.1.3), (14.1.4), and (14.2.21) to write the atomic absorption coefficient as

$$S_\nu(\text{col}) = \frac{\pi e^2 f_{ik}}{2\pi T_0 m_e c} \frac{1}{(\nu - \nu_0)^2 + (1/T_0)^2} \qquad (14.4.12)$$

In going from equation (14.4.11) to (14.4.12), we have changed from circular frequency ω to frequency ν so that the appropriate factors of 2π must be introduced. The quantity $2/T_0$ is usually called the *collisional damping constant* so that

$$\Gamma_c = \frac{2}{T_0} \qquad (14.4.13)$$

Since the form of equation (14.4.12) is identical to that of equation (14.3.21), we can immediately obtain the convolution of the collisional damping absorption coefficient with that for radiation damping by simply adding the respective damping constants:

$$\Gamma = \Gamma_{ik} + \Gamma_c \qquad (14.4.14)$$

The combined absorption coefficient could then be convolved with that appropriate for microturbulent Doppler broadening, producing a total line profile that is still a Voigt profile but with

$$a = \frac{c\Gamma}{4\pi\nu_0 v_0} \qquad (14.4.15)$$

where Γ is the combined damping constant for radiation and collisional damping. However, before this result can be of any practical use, we must have an estimate of the collisional damping constant in terms of the state variables of the atmosphere.

Determination of Collisional Damping Constant   Determining the collisional damping constant is equivalent to determining the average time between collisions T_0. To do this, it is necessary to be quite specific about exactly what constitutes a collision. We follow a method originally due to Victor Weisskopf[16] and described by many authors.[17–19] Consider that the perturbation of an energy level ΔE caused by a passing perturber has the distance dependence

$$\Delta h\,v = \hbar\,\Delta\omega \simeq r^{-n} \qquad (14.4.16)$$

which will produce a change in the frequency of the emitted photon of

$$\Delta\omega = \frac{2\pi C_n}{r^n} \qquad (14.4.17)$$

The constant C_n is known as the *interaction constant*, and it must be determined empirically from laboratory experiments involving the kinds of particles found in the collisions. Since all these collisions are mediated by the electromagnetic force, the typical interaction can be viewed as a "long-range" one so that the short collisions [see equation (14.4.1)] refer to distant collisions where the colliding particle is located near its point of closest approach. This distance is commonly referred to as the *impact distance*, or *impact parameter*. Since the collision is short and the interaction weak, we can assume that the perturbing particle is largely unaffected by the encounter, and its path can be viewed as a straight line (see Figure 14.4). This assumption is usually referred to as the *classical path approximation*, and it appears in one form or another in all theories of collisional broadening.

We wish to calculate the entire frequency shift caused by the collision because when the accumulated phase shift becomes large enough, it is reasonable to say that the wave train has been interrupted and a collision has occurred. To estimate

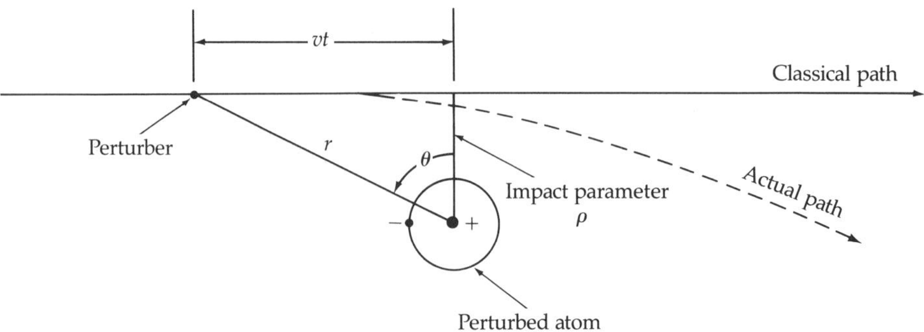

Figure 14.4 Classical path taken by a perturbing particle under the Weisskopf approximation. The point of closest approach ρ is the impact parameter.

this total phase shift, it is necessary to describe the path taken by the particle, so we use the classical path approximation. It is clear from Figure 14.4 that

$$r^2 = \rho^2 + (vt)^2 \tag{14.4.18}$$

This enables the total phase shift η caused by the encounter to be calculated from

$$\eta = \int_{-\infty}^{+\infty} \Delta\omega \, dt = 2\pi C_n \int_{-\infty}^{+\infty} \frac{dt}{[\rho^2 + (vt)^2]^{n/2}} = \frac{2\pi C_n a_n}{v\rho^{n-1}} \tag{14.4.19}$$

where

$$a_n = \frac{\sqrt{\pi}\,\Gamma[(n-1)/2]}{\Gamma(n/2)} \tag{14.4.20}$$

Here $\Gamma(x)$ is the gamma function, and it should not be confused with the symbol for the damping constant. Before using this for the determination of the average time between collisions, we must decide what constitutes an interruption in the wave train. Weisskopf took this value of η to be 1 radian. The smaller the value for the phase shift, the larger the value of the impact parameter may be that will produce that phase shift. The value of the impact parameter ρ_0 that produces the minimum phase shift η which constitutes an interruption in the wave train is known as the *Weisskopf radius*:

$$\rho_0 = \left(\frac{2\pi C_n a_n}{\eta_0 v}\right)^{1/(n-1)} \tag{14.4.21}$$

Since the Weisskopf radius defines the distance inside of which any encounters will produce a large enough phase shift to be considered a collision, it may be used to calculate a collision cross section $\sigma = \pi\rho_0^2$ and an average time between collisions T_0. The collision frequency is

$$\frac{1}{T_0} = \frac{\langle v\rangle_{\text{rel}}}{l} = \pi\rho_0^2 N\langle v\rangle_{\text{rel}} = \frac{\Gamma_c}{2} \tag{14.4.22}$$

where N is the number density of the perturbing particles, l is the mean free path between collisions, and $\langle v\rangle_{\text{rel}}$ is the relative velocity between the perturber and the perturbed atom. That relative velocity is

$$\langle v\rangle_{\text{rel}} = \left[\frac{8kT}{\pi m_h}\left(\frac{1}{A_1} + \frac{1}{A_2}\right)\right]^{1/2} \tag{14.4.23}$$

375

where A_i is the atomic weight of the constituents of the collision in units of the mass of the hydrogen atom. Thus, we can write the collisional damping constant as

$$\Gamma_c = (\text{const})\langle v \rangle_{\text{rel}}^{1 - 2/(n-1)} \eta_0^{-2/(n-1)} N \tag{14.4.24}$$

where N is the number density of perturbers and

$$\text{const} = (2\pi)^{(n+1)/(n-1)}(a_n C_n)^{2/(n-1)} \tag{14.4.25}$$

All that remains is to specify the power law that describes the perturbing force and the interaction constant C_n. Since the force that mediates the collision is electromagnetic, the exponent of the perturbing field is determined by the electric field of the perturber. A simple way of understanding this is to view the passage of the perturber as interposing a "screening" potential energy between an optical electron and the nucleus. The screening potential energy will depend on the locally interposed energy density of the perturber's electric field, which is proportional to E^2, so that for a perturbing ion or electron $n = 4$. This is called the *quadratic Stark effect* because it depends quadratically on the perturber's electric field. If the perturber is a neutral atom, it still possesses a dipole moment that produces a measurable field near the particle. However, this field varies as r^{-3}, so that the perturbing energy density varies as r^{-6} and $n = 6$. Broadening of this type is called *van der Waal's broadening*, and it will play a role in relatively cool gases where there are few ions.

If the atomic energy level of interest is degenerate, the interposition of an external electric field will result in the removal of the degeneracy and a splitting of the energy level into a set of discrete energy levels. The amount of the splitting is proportional to the electric field. Since a time-dependent splitting is equivalent to a broadening brought about by a shift of the energy level itself, the broadening of degenerate levels will occur, but their broadening will be directly proportional to the electric field of the perturber rather than to its square. Thus the broadening of a degenerate level by ions or electrons will produce $n = 2$. This form of broadening is known as the *linear Stark effect*. This form of broadening creates an interesting problem for the impact phase-shift theory since the integral for the minimum phase shift [equation (14.4.19)] will not converge for $n = 2$ and the theory is not applicable. Since the energy levels of hydrogen are degenerate, and the hydrogen lines are among the most prominent in stellar spectra, we are left with the somewhat embarrassing result that these lines can not be dealt with by the impact phase-shift theory, and we have to resort to some other description of collisional broadening to obtain line profiles for hydrogen. The problem basically arises from the $1/r^2$ nature of the perturbing field and is not restricted to the theory of line broadening. Since the number of perturbers increases as r^2 while the perturbation from any one of them declines as r^2, the contribution to the total perturbation from particles at a given distance is independent of distance. Thus some cutoff of the distances to be considered must be invoked. This problem arises frequently in gravitation theory where there can be no screening of the potential field and the fundamental force is also long-range. In our case, the Heisenberg uncertainty principle sets a limit on the smallest perturbation that can matter and hence an upper limit on the volume of space to be considered.

The case of the broadening of degenerate levels by neutral atoms does present a situation that can be dealt with by the impact phase-shift theory. Here the electric field near the perturber varies as r^{-3}, so that the proper value of n is $n = 3$. In the special case where the broadening is by atoms of the same species as the atom being perturbed, a significant enhancement of the broadening occurs. Indeed, for astrophysical cases, the broadening of spectral lines arising from degenerate levels by neutral particles is of interest only when the broadening occurs from collisions with atoms of the same species. For that reason, this kind of broadening is known as *self-broadening*. These considerations are summarized in Table 14.1.

An important improvement was made by Lindholm[20] and Foley[21] which included the effects of multiple collisions on the line. Although the multiple collisions are weak, they are frequent. The result of their work is that the secondary collisions introduce a slight shift in the line center of the atomic absorption coefficient so that

$$S_\omega(\text{col}) = \frac{2\pi e^2 f_{ik}}{mc} \frac{\Gamma_c/2}{(\omega - \omega_0 - \beta)^2 + (\Gamma_c/2)^2} \qquad (14.4.26)$$

where

$$\beta = 2\pi N \langle v \rangle_{\text{rel}} \int_0^\infty \sin\left[\eta(\rho)\right] \rho \, d\rho \qquad (14.4.27)$$

Limits of Validity for Impact Phase-Shift Theory In developing the impact phase-shift theory, we tacitly assumed that the collisions were adiabatic. By that we mean that all the perturbing energy was contained in the perturbation and none was lost to other processes. There were no collisional transitions within the energy level or between the split levels of the degenerate cases. This will be a reasonable approximation as long as the splitting of the degenerate levels or the width

Table 14.1 Types of Collisional Broadening

	Type of Energy Level	
Type of Perturber	*Degenerate*	*Nondegenerate*
Ion or electron	Linear Stark effect $n = 2$	Quadratic Stark effect $n = 4$
Neutral atom	Self-broadening $n = 3$	van der Waal's broadening $n = 6$

of the perturbed level is greater than the uncertainty of energy of the perturber due to the Heisenberg uncertainty principle. Since the duration of the collision is of the order of ρ/v, the uncertainty of the colliding particle's energy is of the order

$$\Delta E \frac{\rho}{v} \simeq \hbar \tag{14.4.28}$$

In equation (14.4.16) we estimated the energy of the perturbation itself so that

$$\frac{v\hbar}{\rho} < \Delta\omega\,\hbar = \frac{2\pi C_n\hbar}{r^n} \simeq \frac{2\pi C_n\hbar}{\rho^n} \tag{14.4.29}$$

which requires that

$$\rho < \left(\frac{2\pi C_n}{v}\right)^{1/(n-1)} \simeq \rho_0 \tag{14.4.30}$$

So it appears that only collisions that occur inside the Weisskopf radius will be adiabatic, and all the energy of the collision goes into perturbing the energy level.

However, if the impact parameter is too small, the classical path approximation will be violated and the duration of the collision will exceed the radiation time [equation (14.4.1)]. The extent of this constraint can be estimated by noting that the duration of the collision is of the order of ρ/v. The radiation time is of the order of $1/\Delta\omega$, so that

$$\frac{\rho\,\Delta\omega}{v} \ll 1 \tag{14.4.31}$$

if the classical path approximation is to be valid. However, equation (14.4.29) places a constraint of the impact parameter ρ that must be met if the collisions are to be adiabatic. Using equation (14.4.29) to eliminate ρ/v from equation (14.4.31), we get

$$\Delta\omega \ll \frac{v^{n/(n-1)}}{(2\pi C_n)^{1/(n-1)}} \tag{14.4.32}$$

Obviously the impact phase-shift theory will be valid only for the inner part of the line. For the outer part we must turn to another description of collisional broadening.

b Static (Statistical) Broadening Theory

In some real sense, the impact phase-shift theory follows the life history of a single radiating (or absorbing) atom which is subject to numerous weak collisions of short duration. The atomic absorption coefficient is then represented by the average of many atoms in various phases of that temporal history. In static broadening theory, the atomic absorption coefficient is constructed from the average of many atoms that are subject to the electric field of perturbers scattered randomly about. The opposite assumption is made concerning the duration of the collision compared to the radiation time. That is, the collision time is much longer than the radiation time, so that

$$t_{col} \gg t_{rad} \tag{14.4.33}$$

It is as if we took a picture of the perturbed atom with a shutter duration of the radiation time for the photon. In the impact phase-shift theory, we would see a blur of colliding tracks of the perturbers, while in the case of statistical broadening the picture would show individual perturbers fixed in space and some might be quite close to the atom in question. We are most interested in these near perturbers for they are responsible for the largest perturbations to the atomic energy levels which in turn generate the broadest part of the line. This is precisely the part of the line for which the impact phase-shift theory fails.

Remember that the perturbation arises from the presence of an external electric field. In the static theory of line broadening, all particles are fixed in space, and the perturbing electric field is the vector sum of the electric fields of all the perturbers (see Figure 14.5). However, since we are concerned mostly about the strongest perturbations that form the wings of the line, we address only the perturbers closest to the atom in question.

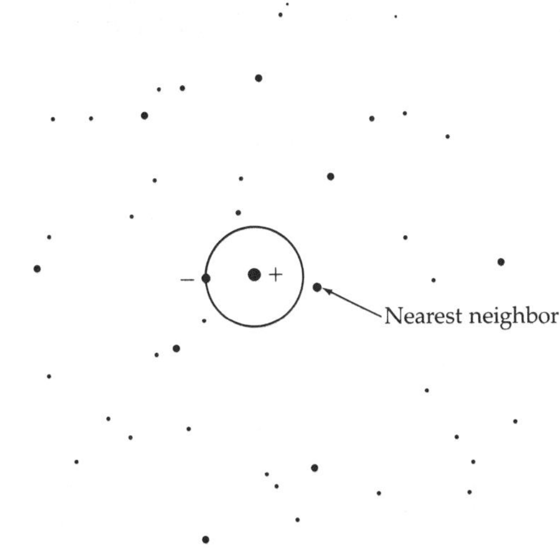

Figure 14.5 Schematic view of the universe of perturbers under the assumptions of the static theory of broadening. The perturbers are randomly distributed in space, but only the nearest neighbor will be used for calculating the perturbing electric field.

Nearest-Neighbor Approximation and Distribution of Electric Fields The assumptions required for the development of the static theory of broadening are similar in form and content to those required of the impact phase-shift theory. The collision time is assumed to be much larger than the radiation time [equation (14.4.33)] so that from the point of view of the radiating atom, the universe is frozen in time. Implicit in this assumption is the notion that every perturber has a well-defined position and momentum (zero) with regard to the perturbed atom. This is equivalent to the classical path approximation of the impact phase-shift theory in that the position and momentum of the perturber are specified throughout the interaction time and are such that they are unaffected by the interaction.

To these complementary assumptions we add one more. Let us assume that the perturbative electric field can be represented by the electric field of the perturber closest to the atom and by that perturber alone. This is known as the *nearest-neighbor approximation*. Our task, then, is to find the probability distribution function for the perturber lying within a specified distance and thereby producing a perturbing electric field of a particular strength. Consider a spherical shell of thickness dr located a distance r from the perturbed atom (see Figure 14.5), and let the probability that the *nearest neighbor* is located within that shell be $P(r)\,dr$. Then the probability that the nearest neighbor lies within a sphere of radius r is $\int_0^r P(r)\,dr$. Since the universe is not empty, there must be a nearest neighbor somewhere, so that the probability that the nearest neighbor does not lie within that sphere is $1 - \int_0^r P(r)\,dr$. Now if the region around the perturbed atom is of uniform density, the probability of finding any perturber within the spherical shell of thickness dr located at r is $4\pi r^2 n\,dr$, where n is the perturber density. Thus, the probability that the particle in that shell is the nearest neighbor is just the probability that there is a particle there multiplied by the probability that there is no particle nearer to the perturbed atom. So

$$P(r)\,dr = (4\pi r^2 n\,dr)\left[1 - \int_0^r P(r)\,dr\right] \qquad (14.4.34)$$

This is an integral equation for the distribution function of nearest neighbors $P(r)$. We can solve it most easily by differentiating with respect to r and forming a differential equation for $P(r)/(4\pi r^2 n)$. The solution to this equation is

$$P(r)\,dr = 4\pi r^2 n e^{-4\pi r^3 n/3}\,dr \qquad (14.4.35)$$

However, we need the probability distribution of perturbing electric fields, so we assume that the perturber has a field that behaves as

$$E = \frac{a}{r^m} \qquad (14.4.36)$$

Then by substituting this dependence of the electric field on r into equation (14.4.35), the probability that an atom will see a perturbing electric field of strength E is

$$W_m(E)\,dE = \frac{3}{m}\,\frac{E_0^{3/m}}{E^{(m+3)/m}}\,e^{-(E_0/E)^{3/m}}\,dE \qquad (14.4.37)$$

where

$$E_0 = a\left(\frac{4\pi n}{3}\right)^{m/3} \qquad (14.4.38)$$

and it is sometimes called the *normalizing field strength*. If we further define the dimensionless quantity

$$\beta = \frac{E}{E_0} \qquad (14.4.39)$$

we can write the probability distribution for this dimensionless field strength as

$$W_m(\beta)\,d\beta = \frac{3}{m}\,\beta^{-(m+3)/m}e^{-\beta^{-3/m}}\,d\beta \qquad (14.4.40)$$

Finally, if we consider the case for broadening by ions or electrons, then $m = 2$ and we have

$$W_2(\beta)\,d\beta = \frac{3}{2}\,\beta^{-5/2}e^{-\beta^{-3/2}}\,d\beta \qquad (14.4.41)$$

which is usually called the *Holtsmark distribution function*. As can be seen from Figure 14.6, the probability of finding a weak field due to the nearest neighbor is very small simply because it is unlikely that the nearest neighbor can be so far away and still be the nearest neighbor. As the field strength rises, so does the probability of its being the perturbing field, peaking between 1 and 2 times the normalized field strength. Stronger fields become less likely because the volume of space surrounding the atom within which the perturber would have to exist becomes just too small.

Behavior of the Atomic Line Absorption Coefficient   If we assume that the perturbative change in the atomic energy level is proportional to the electric field to

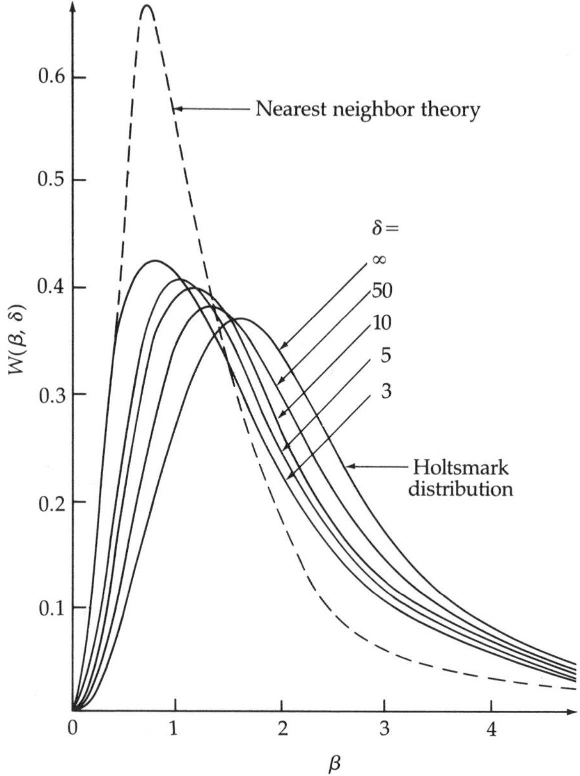

Figure 14.6 Nearest-neighbor distribution function for the perturbing electric field of the nearest neighbor, assuming that it is an ion or electron as the solid line. The dashed line is for the Holtsmark distribution that includes the contribution from the rest of the gas. The parameter δ is a measure of the screening potential of the nearest neighbor [see Mihalas[11] (pp. 292–295)].

some power, then we can write

$$\Delta E = \Delta(h\nu) = h\,\Delta\nu \propto E^q \propto \beta^q \tag{14.4.42}$$

We can then use the nearest-neighbor distribution function to generate a probability density distribution function for the absorption of photons at a particular frequency shift $\Delta\nu$ as

$$P(\Delta\nu)\,d\nu \propto \Delta\nu^{-[1 + 3/(mq)]} e^{-\Delta\nu^{-3/(mq)}}\,d\nu \tag{14.4.43}$$

For large frequency or wavelength shifts, the argument of the exponential approaches zero, so the wavelength-dependent probability of absorption becomes

$$P(\Delta\lambda)\,d\lambda \propto \Delta\lambda^{-[1 + 3/(mq)]} \tag{14.4.44}$$

This is precisely the range for which the static theory through the nearest-neighbor approximation was expected to be accurate. Since the atomic line absorption coefficient is indeed proportional to the probability of photon absorption, its behavior in the far wings of a line is

$$S_{\Delta\lambda} \propto \Delta\lambda^{-[1+3/(mq)]} \tag{14.4.45}$$

Table 14.2 provides a brief summary of the asymptotic dependence of the absorption coefficient in the wings of the line for the various types of interactions discussed.

Finally, we may find the constant of proportionality for the atomic line absorption coefficient in terms of the interaction constant for the force law C_l. This is analogous to the constant C_n that appears in equation (14.4.17) and is usually determined empirically. In terms of this constant, the atomic line absorption coefficient becomes

$$S_\nu = \frac{3C_l}{(4\pi N/3)^{1/3} a_m} \Delta\nu^{-(1+3/l)} e^{-\Delta\nu^{-3/l}} \tag{14.4.46}$$

where a_m is given by equation (14.4.20) and

$$l \equiv mq \tag{14.4.47}$$

In the broadening of degenerate levels, the splitting of the energy levels is so large that the line should be considered to consist of individual linear Stark components, each of which is quadratically Stark-broadened. Under these conditions, the atomic line absorption coefficient for the combined Stark components becomes

$$S_\nu = \sum_k \frac{3C_{kl}}{(4\pi N/3)^{1/3} a_m} \Delta\nu_k^{-(1+3/l)} e^{-\Delta\nu_k^{3/l}} \tag{14.4.48}$$

If one were to improve on the static theory, the most obvious place would be to relax the nearest-neighbor approximation. The problem of including an ensemble

Table 14.2 Asymptotic Behavior of the Atomic Line Absorption Coefficient

	Type of Energy Level	
Type of Perturber	*Degenerate $q = 1$*	*Nondegenerate $q = 2$*
Ion or electron $m = 2$	Linear Stark effect $S_\lambda \sim \Delta\lambda^{-5/2}$	Quadratic Stark effect $S_\lambda \sim \Delta\lambda^{-7/4}$
Neutral atom $m = 3$	Self-broadening $S_\lambda \sim \Delta\lambda^{-2}$	van der Waal's broadening $S_\lambda \sim \Delta\lambda^{-3/2}$

of perturbers, all with their electric fields adding vectorially, was considered by J. Holtsmark[22] and solved by S. Chandrasekhar,[23] who also provides tables of the results. As one might expect, the resultant form is similar to that of the nearest-neighbor distribution in shape but somewhat more spread out (see Figure 14.6). Unfortunately the result takes the form of an integral, so a complete description must be obtained numerically. However, for β in the vicinity of 1 we get the following asymptotic formula for $W(\beta)$:

$$W(\beta) = 1.496\beta^{-5/2}(1 + 5.107\beta^{-3/2} + 14.43\beta^{-3} + \cdots +) \quad (14.4.49)$$

As we might expect, the lead term of this series is just that of the nearest-neighbor approximation [see equation (14.4.41)].

Limits of Validity and Further Improvements for the Static Theory   Since the assumption relating the collision time to the radiation time led to a limit on the range of validity for the impact phase-shift theory [equation (14.4.32)], we should not be surprised if the same were true for the static theory. This is indeed the case, and the result, known as the *Holstein relation*, can be deduced from equation (14.4.32) almost by inspection:

$$\Delta\omega \gg \frac{v^{n/(n-1)}}{(2\pi C_n)^{1/(n-1)}} \quad (14.4.50)$$

So, as we hoped at the outset of the development of the static theory, it will be valid for precisely those regions of the line profile for which the impact phase-shift theory fails.

Of course, any microscopic inspection of a problem usually finds phenomena that provide additional complications for the solution. For example, we have assumed that the perturbers interact with the atom in question but do not interact among themselves. In reality, an ion will attract electrons so as to create a neutral plasma on as small a scale as possible. In effect, then, the plasma will try to shield the ions from even more distant perturbers. This phenomenon is known as *Debye shielding*, and it is discussed in some detail by Mihalas[11] (pp. 292–295). The basic effect is density-dependent and tends to flatten the Holtsmark distribution still further, thereby broadening the line even more (see Figure 14.6). Fortunately, for normal stellar atmospheres the densities are not large enough to make Debye shielding a major effect until one reaches optical depths in the line that are quite remote from the boundary.

The treatment of collisional line broadening described so far has been based on purely classical considerations and has now been largely replaced by quantum mechanical calculations of the atomic line absorption coefficient for the more important stellar spectral lines. However, the quantum mechanical treatment is considerably less transparent than the classical one, so we give only the basic form. The power spectrum for the line is given by

$$I(\omega) = \frac{2\omega^4}{3\pi c^2} \, Tr\left[\mathbf{P} \int_{-\infty}^{+\infty} \int_{-\infty}^{+\infty} \mathbf{Q}(t)e^{-i\omega t}\mathbf{Q}(t')e^{i\omega t'} \, dt \, dt' \right] \qquad (14.4.51)$$

where $\mathbf{P}$ is the probability density matrix for the atomic states involved in the formation of the spectral line and $\mathbf{Q}$ is the matrix of dipole moments with elements.

$$Q_{ij} = \int \Psi_i^* q \Psi_j \, dV \qquad (14.4.52)$$

The wave functions Ψ_i must include the effects of the perturbers as well as the atomic states of interest. A very complete discussion of the quantum theory of spectral line formation is given by Hans Griem.[24]

For simple lines the classical theories of collisional broadening produce line profiles that agree well with observation. However, for the stronger lines of hydrogen and helium, any serious model should involve an atomic absorption coefficient based on the quantum mechanical description. While tables of these coefficients exist for many important lines (see Griem[24] and references there), much remains to be done to produce accurate values for many lines of astrophysical interest.

14.5 Curve of Growth of the Equivalent Width

While we have discussed the most important aspects of the formation of spectral lines, we have said little about the most important contributor to the appearance of the line in the spectrum—the abundance of the atomic species giving rise to the line. Obviously the more absorbers present in the atmosphere, the stronger the associated spectral line will appear. However, the quantitative relationship between the abundance and the equivalent width is not simple and is worthy of some discussion. Although most contemporary determinations of elemental abundances rely on detailed atmospheric modeling with the abundance as a parameter to be determined from comparison with observation, the classical picture of the relation between the equivalent width and the abundance is quite revealing about what to expect from such models. That classical quantitative relationship is known as the *curve of growth*. Some students have wondered what is growing in the curve of growth. The answer is that the equivalent width increases or "grows" with increasing abundance.

a Schuster-Schwarzschild Curve of Growth

To create a curve of growth, we must relate the equivalent width to the atomic abundance. This requires some model of the atmosphere in which the atoms reside.

For purposes of illustration, we take the simplest model possible. In Chapter 13 we set up the equation of radiative transfer for line radiation [equation (13.1.6)], and we solved it for some special cases. For the Schuster-Schwarzschild atmosphere, this led to a line profile given by equation (13.2.8):

$$r_v = \frac{F_v}{F_c} = \left(1 + \frac{\sqrt{3}\tau_0}{2}\right)^{-1} \tag{14.5.1}$$

The definition of τ_0 allows us to write

$$\tau_0 = \int_0^{\tau_0} dt_v = \int_0^{x_0} \kappa_v \rho \, dx = \int_0^{x_0} n_i S_v \, dx = \langle S_v \rangle \int_0^{x_0} n_i \, dx = N_i \langle S_v \rangle \tag{14.5.2}$$

where N_i is the column density of the atom giving rise to the line and $\langle S_v \rangle$ is the line absorption coefficient averaged over depth. Since for this simple model the atmospheric conditions are considered constant throughout the cool gas, we drop the average-value symbols for the remainder of this section. We have already seen [equation (14.3.27)] that for many atomic lines the atomic line absorption coefficient, including the effects of radiation damping, collisional damping, and Doppler broadening, can be written as

$$S_v = S_0 H(a, u) \tag{14.5.3}$$

where $H(a, u)$ can be either the Voigt or the normalized Voigt function depending on what constants have been absorbed into S_0. Thus the line profile for the Schuster-Schwarzschild atmosphere is

$$r_v = \left(\frac{1 + \sqrt{3}\tau_0}{2}\right)^{-1} = \left[1 + \frac{\sqrt{3}S_0 H(a, u)N_i}{2}\right]^{-1} \tag{14.5.4}$$

To relate this to the equivalent width, equation (14.5.4) must be integrated over the frequencies contained in the line so that

$$W_\lambda = 2 \int_0^\infty \frac{\sqrt{3}\tau_0/2}{1 + \sqrt{3}\tau_0/2} \, d\lambda = 2 \int_0^\infty \frac{[\sqrt{3}S_0 H(a, u)N_i/2] \, d\lambda}{1 + \sqrt{3}S_0 H(a, u)N_i/2} \tag{14.5.5}$$

It is convenient to express the frequency-dependent optical depth in the line in terms of the optical depth at the line center χ_0 so that

$$\tau_0(v) = \frac{\chi_0 H(a, u)}{H(a, 0)} \tag{14.5.6}$$

From equations (14.5.2) and (14.5.3),

$$\chi_0 = S_0 H(a, 0) N_i \approx \frac{\sqrt{\pi} e^2 f_{jk} N_i \lambda_0}{m_e v_0 c} \qquad a < 0.2 \qquad (14.5.7)$$

Consider the case where the damping constant is small compared to Doppler broadening so that $a < 0.2$. Then the Doppler core will dominate the line profile, and we can write the optical depth in the line as

$$\tau_0(\Delta\lambda) = \chi_0 e^{-(\Delta\lambda/\Delta\lambda_d)^2} = \chi_0 e^{-\xi^2} \qquad (14.5.8)$$

where $\xi \equiv \Delta\lambda/\Delta\lambda_d$. Substitution into equation (14.5.5) yields

$$W_\lambda = \sqrt{3} \, \Delta\lambda_d \, \chi_0 \int_0^\infty \left(e^{\xi^2} + \frac{\sqrt{3}\chi_0}{2} \right)^{-1} d\xi \qquad (14.5.9)$$

The integral can be expanded in a series so that

$$\int_0^\infty \left(e^{\xi^2} + \frac{\sqrt{3}\chi_0}{2} \right)^{-1} d\xi \approx \int_0^\infty \left[e^{-\xi^2} - \frac{\sqrt{3}\chi_0}{2} e^{-2\xi^2} + \frac{3\chi_0^2}{4} e^{-3\xi^2} \right.$$

$$\left. + \cdots + (-1)^k \left(\frac{\sqrt{3}}{2} \right)^k \chi_0^k e^{-(k+1)\xi^2} \right] d\xi \quad (14.5.10)$$

But

$$\int_0^\infty e^{-k\xi^2} d\xi = \frac{1}{2} \sqrt{\frac{\pi}{k}} \qquad (14.5.11)$$

Thus, we can write the equivalent width in the line as

$$W_\lambda = \frac{\sqrt{3\pi}}{2} \chi_0 \, \Delta\lambda_d \left[1 - \chi_0 \sqrt{\tfrac{3}{8}} + \frac{\chi_0^2 \sqrt{3}}{4} - \frac{9\chi_0^3}{8} + \cdots \right] \qquad (14.5.12)$$

This, then, represents the first part of the curve of growth, and the equivalent width is indeed directly proportional to χ_0 and hence the abundance N_i. This is a commonsense result that simply says that the number of photons removed from the beam is proportional to the number of atoms doing the absorbing, so that section of the curve of growth is known as the *linear* section.

However, the seeds of difficulties are apparent in the higher-order terms in equation (14.5.12). As the number of absorbers increases, we would expect that some 387 atoms high in the atmosphere to be "shadowed" by atoms lower in the atmosphere. When all the photons at a given frequency have been absorbed, then the further addition of atoms that can absorb at those frequencies will make no change in the equivalent width. When this happens, the line is said to be saturated. As the optical depth in the line center χ_0 increases, the term in brackets will fall below unity and the curve of growth will increase more slowly than the linear growth. For $0 \leq \chi_0 \leq 0.5$, the series may be terminated after the first term. However, for larger values of χ_0, a somewhat different expression of the integral on the left-hand side of equation (14.5.10) is in order. If we make the transformation

$$\frac{\sqrt{3}\chi_0}{2} = e^b \qquad \xi = \zeta^{1/2} \tag{14.5.13}$$

the equivalent width becomes

$$W_\lambda = \Delta\lambda_d \int_0^\infty \frac{\zeta^{-1/2}\, d\zeta}{1 + e^{\zeta - b}} = 2\,\Delta\lambda_d \sqrt{b}\left(1 - \frac{\pi^2}{24b^2} - \frac{\pi^4}{384b^4} - \cdots\right) \tag{14.5.14}$$

If $\chi_0 > 55$, all but the lead term of the approximation may be ignored. However, in the region where $0.5 < \chi_0 < 55$, the series given by either equation (14.5.12) or (14.5.14) must be used. From the lead term of equation (14.5.14) it is clear that as the Doppler core saturates, the equivalent width grows very slowly as

$$\frac{W_\lambda}{\lambda} \propto \sqrt{\ln N_i} \tag{14.5.15}$$

This is known as the "flat" part of the curve of growth.

As the abundance increases still further, a significant number of atoms will exist that can absorb in the damping wings of the line and the equivalent width will again begin to increase, but at a rate that will depend on the damping constant appropriate for the line (see Figure 14.7).

Once more we will need a different representation of the optical depth that is appropriate for the damping wings of the line. From the definition of the dimensionless variables of the Voigt function [see equation (14.3.23)]

$$dv = \frac{\Delta v_d\, du}{c} \tag{14.5.16}$$

so that we can rewrite equation (14.5.5) with the aid of (14.5.6) to obtain

$$W_\lambda = \frac{2\,\Delta v_d}{c} \int_0^\infty \frac{du}{1 + 2H(a, 0)/[\sqrt{3}\chi_0 H(a, u)]} \tag{14.5.17}$$

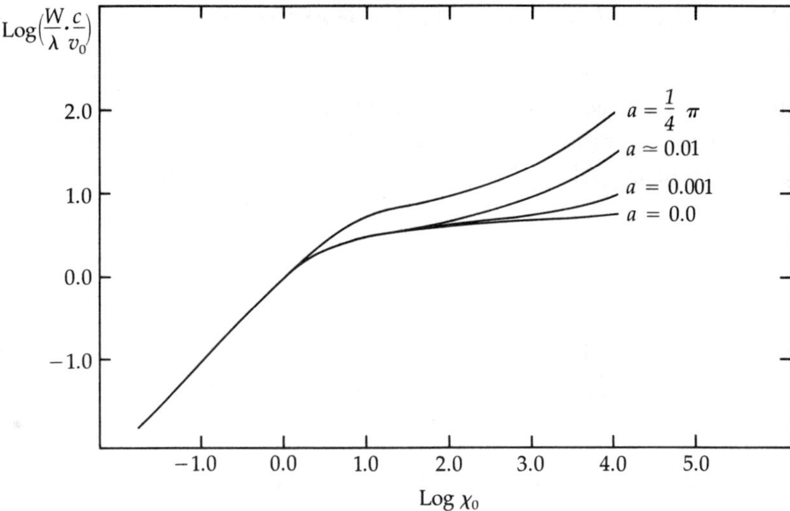

Figure 14.7 Curve of growth for the classical Schuster-Schwarzschild model atmosphere.

The Voigt function as given in equation (14.3.26) can be approximated for large u as

$$H(a, u) \approx \frac{a}{\pi} \int_{-\infty}^{+\infty} \frac{e^{-y^2}}{u^2}\, dy = \frac{a}{\sqrt{\pi}u^2} \tag{14.5.18}$$

For modest values of the damping parameter a, $H(a, 0)$ is near unity so that

$$W_v \approx \left(\frac{2\,\Delta v_d}{c}\right)(3^{1/4}\pi^{3/4}\sqrt{\chi_0 a}) \tag{14.5.19}$$

So for large abundances the curve of growth will again increase in a manner that depends on the square root of the damping constant as well as the square root of the abundance. Except for the separation brought about by the growth of the damping wings of the line, the curve of growth is a single-valued function of $W_\lambda/\Delta\lambda_d$ versus the optical depth at the line center χ_0. Both these parameters are dimensionless, so for this model a single curve satisfies all problems. However, it is worth remembering that the Schuster-Schwarzschild model is correct for scattering lines only, and very few spectral lines that go into abundance calculations are scattering lines. Thus, the classical curve of growth can give only very approximate results even if it is calculated exactly.

b   More Advanced Models for the Curve of Growth

There are several ways to improve the accuracy of the curve of growth. First, we could use a more accurate solution to the equation of radiative transfer such as the Chandrasekhar discrete ordinate method. The use of the equations of condition on the boundary values [equation (10.2.31)] enables us to obtain a profile of the form

$$r_v = \frac{\sum\limits_{\alpha=1}^{n-1} L_{c,+\alpha} + Q_c}{\sum\limits_{\alpha=1}^{n-1} L_{+\alpha} e^{-k_\alpha \tau_0(v)} + L_{-\alpha} e^{+k_\alpha \tau_0(v)} + \tau_0(v) + Q} \tag{14.5.20}$$

The behavior of the optical depth could then be substituted into equation (14.5.20) and from there into equation (14.5.5), thereby relating the equivalent width to the optical depth at the line center. However, this would only improve the details of the radiative transfer without improving the model itself. Since we know that the errors of the two-stream (Eddington) approximation are of the order of 12 percent, this is a small improvement indeed for the additional work involved.

A significant improvement could be made by using the Milne-Eddington model atmosphere. Here the line profile is given by equation (13.2.29), where the frequency dependence is entirely contained in the behavior of $\mathscr{L}_v$, ϵ_v, and η_v with frequency. In addition, the parameters a and b which describe the surface temperature and temperature gradient need to be specified. Laborious as the task of constructing these more sophisticated curves of growth is, it was done by Marshall Wrubel[25-27] in a series of papers. Although the additional parameters required by the model are annoying, the improvement in the representation of the star by these models is usually worth the effort. It is probably not worth the trouble to generate more sophisticated classical models than these. Direct modeling by a model atmosphere code is the appropriate approach, for one can remove virtually all the assumptions required for the classical models so that the accuracy is largely determined by the accuracy of the atomic constants characterizing the line.

c   Uses of the Curve of Growth

Determination of the Doppler Velocity and Abundance   We already indicated that the curve of growth can be used to estimate stellar abundances. However, it is possible (in principle) to learn a great deal more about the conditions in the atmosphere of the star from the curve of growth. Imagine that we have measured equivalent widths for a collection of lines that all arise from the same lower level for which the atomic parameters and damping constants are accurately known.

Further suppose that the values for the lines span a reasonable range of the curve of growth. Thus we have empirical values for $W(\lambda_i)/\lambda_i$ and $\sqrt{\pi e^2 f_i \lambda_i/(m_e c)}$. The second of these two quantities, which we call X_i, is given by

$$\text{Log } X_i = \text{Log } \chi_0 - \text{Log } \frac{v_0}{N} \tag{14.5.21}$$

Thus, a plot of $\text{Log } X_i$ versus $\text{Log }[W(\lambda_i)/\lambda_i]$ will yield an empirical curve of growth that differs from the theoretical curve given in Figure 14.7 by a shift in both the ordinate and the abscissa. Since the lines all arise from the same lower level, N is the same for all points. The horizontal shift then specifies $\text{Log }(v_0/N)$, while the vertical shift specifies $\text{Log }(c/v_0)$. Thus both the abundance and the Doppler velocity are determined independently. To the extent that the kinetic temperature is known, we know the microturbulent velocity. If the span of the curve of growth is large enough to determine a, an average value of Γ_c may also be found.

Determination of the Excitation Temperature Consider the situation where, in addition to the information given above, we know the equivalent widths for a number of lines arising from different states of excitation. Further assume that LTE holds so that the populations of those excited states are given by the Boltzmann formula. Then

$$\text{Log } X_i = \text{Log } \chi_0 - \text{Log } \frac{v_0}{N} - \text{Log } \frac{g_i e^{-\epsilon_i/(kT)}}{U(T)} \tag{14.5.22}$$

We have already determined v_0, so we may correct the observed equivalent widths so that the observed values are brought into correspondence with the theoretical ordinate of the curve of growth $W_\lambda/\Delta\lambda_d$. The horizontal points will now miss the theoretical curve of growth by an amount

$$\aleph_i = \text{Log } X_i - \left(\text{Log } \chi_0 - \text{Log } \frac{v_0}{N} \right) + \text{Log } \frac{g_i e^{-\epsilon_i/(kT)}}{U(T)} \tag{14.5.23}$$

or

$$\text{Log } \frac{g_i e^{-\epsilon_i/(kT)}}{U(T)} + \text{const} = \aleph_i \tag{14.5.24}$$

Since everything about the lines in equation (14.5.24) is known, only the constant and the temperature are unknowns, and they can be determined by least squares.

Important parameters concerning the structure of a stellar atmosphere can be estimated from the classical curve of growth. Not only can the abundance of the elements that make up the atmosphere be measured, but also the turbulent velocity and excitation temperature can be roughly determined. However, to use the classical curve of growth is to make some very restrictive assumptions. The assumption that the parameters determining the lines are independent of optical depth is a poor assumption and is usually the reason that the excitation temperature does not agree with the effective temperature. In addition, the thickness of the atmosphere is probably not the same for all the lines used. Finally, the lines are usually not scattering lines. Nevertheless, the method should be used prior to undertaking any detailed analysis in order to set the ranges for the expected solution. Any sophisticated analysis that produces answers wildly different from those of the curve of growth should be regarded with suspicion.

Finally, sooner or later, we must be wary of the assumption of LTE. In the upper layers of the atmosphere, the density will become low enough that collisions will no longer occur frequently enough to overcome the nonequilibrium effects of the radiation field, and the level populations of the various atomic states will depart from that given by the Saha-Boltzmann ionization-excitation formulas. This will particularly affect the strong spectral lines that are formed very high up in the atmosphere. In the next chapter, we survey what is to be done when LTE fails.

Problems

1. Imagine a line whose intensity profile is

$$f_\nu = 1 + \delta(\nu - \nu_0)$$

Calculate the observed line profile for a radially expanding atmosphere which exhibits a velocity gradient

$$\frac{d\nu}{d\tau_c} = \text{const} = c\,\frac{\Delta\lambda_0}{\lambda_0}$$

State any assumptions that you make in solving the problem.

2. Consider a line generated by atoms constrained to move perpendicular to a radius vector from the center of the star. Find an expression for the atomic absorption coefficient due to Doppler broadening alone.

3. Find the natural width for (*a*) Hβ, (*b*) Mg II(4481), and (*c*) Fe I(3720).

4. Estimate the transition times from the natural widths of the lines in Problem 3, and compare them with a crude estimate of the collision rates for atoms in these states. State clearly any assumptions you make. In what kind of star would you expect to find these spectral lines?

5. If both the atoms of a radiating gas and the particles perturbing them are in statistical equilibrium, show that the average relative velocity between them is given by

$$\langle v \rangle = \left[\left(\frac{8kT}{\pi \mu_h} \right) \left(\frac{1}{A_1} + \frac{1}{A_2} \right) \right]^{1/2}$$

where μ_h = the mass of a unit atomic weight and A_1 and A_2 are the atomic weight of the atom and perturber, respectively.

6. Find the far-wing dependence of the line absorption coefficient of an atom having nondegenerate energy levels which are broadened by perturbers having only octopole moments of their charge configurations.

7. Compute a line profile for S II($\lambda\lambda 6347.10$) for an A0V star. Use a model atmosphere code if possible.

8. Show that $W_\lambda/\lambda = W_v/v$.

9. Use a model atmosphere code such as ATLAS to generate "curves of growth" for Fe I($\lambda\lambda 4476$), Mg II($\lambda\lambda 4481$), and Si II($\lambda\lambda 4130$). Include a microturbulent velocity of 2 km/s. Consider the reference atmosphere to be one with $T_e = 10^4$ K, Log $g = 4.0$, and solar abundance (except for Fe, Mg, and Si). Compare your results with the classical curve of growth for a Schuster-Schwarzschild model atmosphere and obtain values for $\Delta \lambda_d$, the kinetic temperature, microturbulent velocity, and Γ for each line. Compare your results with the values used to generate the line profiles and discuss any differences.

10. Consider the following situation. A 1-mm beam of neutral hydrogen gas with an internal kinetic temperature of 10^4 K is accelerated to an energy of 10^{-3} eV per atom. The beam enters a 10-m vacuum chamber and is directed toward a 1-cm bar located in the center of the chamber and oriented at right angles to the beam. The bar has been charged to 10^7 V. The beam passes through a 1-mm hole in the bar and proceeds out the opposite side of the chamber. A spectrograph is placed so that it "looks" along the beam and sees the beam against a 2×10^4 K continuum blackbody source located near where the beam enters the chamber. Assuming that the beam density is sufficiently low to ensure that it is optically thin, but high enough to establish LTE, find the line profile for Hβ. Further assume that the central depth of the line is 0.6. Find the equivalent width of Hβ and the density of hydrogen. On the basis of your results, discuss the validity of the assumptions you used.

11. Consider a Schuster-Schwarzschild model atmosphere populated with several types of atoms having different atomic absorption coefficients. Find the theoretical curves of growth for each of these atoms.

(a) $S_1(v) = a + \dfrac{b}{|\Delta v|}$

(b) $S_2(v) = \delta(\Delta v)$

(c) $S_3(v) = \begin{pmatrix} a & \text{for } |\Delta v| \le \Delta v_0 \\ 0 & \text{for } |\Delta v| > \Delta v_0 \end{pmatrix}$

(d) $S_4(v) = \begin{pmatrix} a + b(\Delta v)^2 & \text{for } |\Delta v| \le \left(\dfrac{-a}{b} \right)^{1/2} \\ 0 & \text{for } |\Delta v| > \left(\dfrac{-a}{b} \right)^{1/2} \end{pmatrix} \quad b < 0$

(e) $S_5(v) = \begin{pmatrix} a + b|\Delta v| & \text{for } |\Delta v| \le \dfrac{-a}{b} \\ 0 & \text{for } |\Delta v| > \dfrac{-a}{b} \end{pmatrix} \quad b < 0$

Compare with the classical solutions for the curve of growth.

12. Let the probability of finding a value of the turbulent velocity projected along the line of sight v be uniform in the range $-v_0 \le v \le v_0$. The probability of finding a value of v outside this range is zero. In addition to turbulence, there are thermal Doppler motions present which correspond to a temperature T. Assuming that f and Γ are known, derive an expression for the atomic line absorption coefficient. Leave your answer in the form of a definite integral containing an error function.

13. Consider a certain atom in the solar atmosphere at a point where the hydrogen abundance $N_h = 10^{17}$ cm^{-3} and $T = 5500$ K. The atom has a strong resonance line at $\lambda = 5000$ Å with an Einstein A coefficient of 9.7×10^7 s^{-1}. The atom has interacted with a neutral hydrogen atom so that a frequency shift of $\Delta \omega = 2 \times 10^6/r^2$ s^{-1} of the line frequency has resulted. Here, r is in angstroms.

(a) Make a reasonable estimate of how long the collision lasts.

(b) Qualitatively justify the type of broadening theory you would use to describe the atomic absorption coefficient.

(c) What is the approximate cross section for this event?

(d) What is the value of Γ you would obtain from the impact phase-shift theory of line broadening?

14. Suppose the data below are observed in a certain star. They all pertain to the lines of the neutral state of the same element which has a partition function of 2.0. The parameter ϵ_i refers to the lower level of the transition.

j	λ, Å	ϵ_i, eV	g_i	W_λ, Å	f_{ij}
1	4,930	2.70	3	0.330	0.50
2	3,470	2.70	3	0.0155	0.050
3	5,676	2.70	3	0.0562	0.0667
4	3,360	1.80	6	0.0190	0.00316
5	10,050	1.80	6	1.349	1.060
6	3,862	1.80	6	0.0617	----

Using the Schuster-Schwarzschild model atmosphere, find (*a*) the number of atoms per square centimeter above the photosphere, (*b*) the missing f value, and (*c*) the value for the Doppler velocity v_0.

References and Supplemental Reading

1. Panofsky, W., and M. Phillips, *Classical Electricity and Magnetism*, Addison-Wesley, Reading, Mass., 1955, p. 222.

2. Slater, J. C., and N. H. Frank, *Electromagnetism*, McGraw-Hill, New York, 1947, p. 159.

3. Mihalas, D., *Stellar Atmospheres*, W. H. Freeman, San Francisco, 1970, pp. 86–92.

4. Weisskopf, V., and E. Wigner, "Berechnung der natürlichen Linienbreite auf Grund der Diracschen Lichttheorie," *Zs. f. Physik* 63, 1930, pp. 54–73.

5. Weisskopf, V., and E. Wigner, "Über die natürliche Linienbreite in der Strahlung des harmonischen Ozillators," *Zs. f. Physik* 65, 1930, pp. 18–29.

6. Harris, D., "On the Line-Absorption Coefficient due to Doppler Effect and Damping," *Ap. J.* 108, 1948, pp. 112–115.

7. Hummer, D., "The Voigt Function—An Eight Significant Figure Table and Generating Procedure," *Mem. R. astr. Soc.* 70, 1965, pp. 1–32.

8. Finn, G., and D. Mugglestone, "Tables of the Line-Broadening Function $H(a, v)$," *Mon. Not. R. astr. Soc.* 129, 1965, pp. 221–235.

9. Humíček, J., "An Efficient Method for Evaluation of the Complex Probability Function: The Voigt Function and Its Derivatives," *J. Quant. Spectrosc. & Rad. Trans.* 21, 1979, pp. 309–313.

10. McKenna, S., "A Method of Computing the Complex Probability Function and Other Related Functions over the Whole Complex Plane," *Astrophys. & Sp. Sci.* 107, 1984, pp. 71–83.

11. Mihalas, D., *Stellar Atmospheres*, 2d ed., W. H. Freeman, San Francisco, 1978, pp. 463–471.

12. Collins II, G. W., "The Effects of Rotation on the Atmospheres of Early-Type Main Sequence Stars," *Stellar Rotation* (Ed.: A. Slettebak), Reidel, Dordrecht, Holland, 1970, pp. 85–109.

13. Cassinelli, J. P. "Rotating Stellar Atmospheres," *The Physics of Be Stars* (eds.: A. Slettebak and T. Snow), Cambridge University Press, Cambridge, 1987, pp. 106–122.

14. Shajn, G., and O. Struve, "On the Rotation of Stars," *Mon. Not. R. astr. Soc.* 89, 1929, pp. 222–239.

15. Unsöld, A., *Physik der Sternatmosphären*, 2d ed., Springer-Verlag, Berlin, 1955, pp. 508–518.

16. Weisskopf, V., "Zur Theorie der Kopplungsbreite und der Stossdampfung," *Z. F. Physik* 75, 1932, pp. 287–301.

17. Aller, L. H., *The Atmospheres of the Sun and Stars*, 2d ed., Ronald, New York, 1963, p. 317.

18. Unsöld, A., *Physik der Sternatmosphären*, 2d ed., Springer-Verlag, Berlin, 1955, p. 302.

19. Mihalas, D., *Stellar Atmospheres*, 2d ed., W. H. Freeman, San Francisco, 1978, pp. 281–284.

20. Lindholm, E., "Zur Theorie der Verbreiterung von Spektrallinien," *Arkiv. F. Math. Astron. och Fysik* 28B (no. 3), 1942, pp. 1–11.

21. Foley, H., "The Pressure Broadening of Spectral Lines," *Phys. Rev.* 69, 1946, pp. 616–628.

22. Holtsmark, J., "Über die Verbreiterung von Spektrallinien," *Ann. der Physik* 58, 1919, pp. 577–630.

23. Chandrasekhar, S., "Stochastic Problems in Physics and Astronomy," *Rev. Mod. Phy.* 15, 1943, pp. 1–89, 70–74.

24. Griem, H., *Spectral Line Broadening by Plasmas*, Academic, New York, 1974.

25. Wrubel, M., "Exact Curves of Growth for the Formation of Absorption Lines according to the Milne-Eddington Model, I: Total Flux," *Ap. J.* 109, 1949, pp. 66–75.

26. Wrubel, M., "Exact Curves of Growth for the Formation of Absorption Lines according to the Milne-Eddington Model, II: Center of the Disk," *Ap. J.* 111, 1950, pp. 157–164.

27. Wrubel, M., "Exact Curves of Growth, III: The Schuster-Schwarzschild Model," *Ap. J.* 119, 1954, pp. 51–57.

In addition to the references listed above, an excellent overall reference to line broadening theory can be found in

• Böhm, K.-H.: "Basic Theory of Line Formation" *Stellar Atmospheres* (Ed.: J. Greenstein, *Stars and Stellar Systems: Compendium of Astronomy and Astrophysics Vol VI*, University of Chicago Press, Chicago, 1960, pp. 88–155.

For a somewhat different approach to the problem of line broadening, the interested student should consult

• Jefferies, J. T.: *Spectral Line Formation*, Blaisdell, New York, 1968, pp. 46–91.

A more complete treatment than we have given here can be found in

• Mihalas, D.: *Stellar Atmospheres*, 2d ed., W. H. Freeman, San Francisco, 1978, pp. 273–331.

• Griem, H. R.: *Plasma Spectroscopy*, McGraw-Hill, New York, 1964.

15

Breakdown of Local Thermodynamic Equilibrium

. . .

Thus far we have made considerable use of the concepts of equilibrium. In the stellar interior, the departures from a steady equilibrium distribution for the photons and gas particles were so small that it was safe to assume that all the constituents of the gas behaved as if they were in STE. However, near the surface of the star, photons escape in such a manner that their energy distribution departs from that expected for thermodynamic equilibrium, producing all the complexities seen in stellar spectra. However, the mean free path for collisions between the particles that make up the gas remained short compared to that of the photons, and so the collisions could be regarded as random. More importantly, the majority of the collisions between photons and the gas particles could be viewed as occurring between particles in thermodynamic equilibrium. Therefore, while the radiation field departs from that of a blackbody, the interactions determining the state of the gas continue to lead to the establishment of an energy distribution for the gas particles characteristic of thermodynamic equilibrium. This happy state allowed the complex properties of the gas to be determined by the local temperature alone and is known as local thermodynamic equilibrium.

However, in the upper reaches of the atmosphere, the density declines to such a point that collisions between gas particles and the remaining "equilibrium" photons will be insufficient for the establishment of LTE. When this occurs, the energy-level populations of the excited atoms are no longer governed by the Saha-Boltzmann ionization-excitation formula, but are specified by the specific properties of the atoms and their interactions.

Although the state of the gas is still given by a time-independent distribution function and can be said to be in steady or statistical equilibrium, that equilibrium distribution is no longer the maximal one determined by random collisions. We have seen that the duration of an atom in any given state of excitation is determined by the properties of that atomic state. Thus, any collection of similar atoms will attempt to rearrange their states of excitation in accordance with the atomic properties of their species. Only when the interactions with randomly moving particles are sufficient to overwhelm this tendency will the conditions of LTE prevail. When these interactions fail to dominate, a new equilibrium condition will be established that is different from LTE. Unfortunately, to find this distribution, we have to calculate the rates at which excitation and deexcitation occur for each atomic level in each species and to determine the population levels that are stationary in time. We must include collisions that take place with other constituents of the gas as well as with the radiation field while including the propensity of atoms to spontaneously change their state of excitation. To do this completely and correctly for all atoms is a task of monumental proportions and currently is beyond the capability of even the fastest computers. Thus we will have to make some approximations. In order for the approximations to be appropriate, we first consider the state of the gas that prevails when LTE first begins to fail.

A vast volume of literature exists relating to the failure of LTE, and it would be impossible to cover it all. Although the absorption of some photons produced by bound-bound transitions occurs in that part of the spectrum through which the majority of the stellar flux flows, only occasionally is the absorption by specific lines large enough to actually influence the structure of the atmosphere itself. However, in these instances, departures from LTE can affect changes in the atmosphere's structure as well as in the line itself. In the case of hydrogen, departures in the population of the excited levels will also change the "continuous" opacity coefficient and produce further changes in the upper atmosphere structure. To a lesser extent, this may also be true of helium. Therefore, any careful modeling of a stellar atmosphere must include these effects at a very basic level. However, the understanding of the physics of non-LTE is most easily obtained through its effects on specific atomic transitions. In addition, since departures from LTE primarily occur in the upper layers of the atmosphere and therefore affect the formation of the stellar spectra, we concentrate on this aspect of the subject.

15.1 Phenomena Which Produce Departures from Local Thermodynamic Equilibrium

a Principle of Detailed Balancing

Under the assumption of LTE, the material particles of the gas are assumed to be in a state that can be characterized by a single parameter known as the temperature. Under these conditions, the populations of the various energy levels of the atoms of the gas will be given by Maxwell-Boltzmann statistics regardless of the atomic parameters that dictate the likelihood that an electron will make a specific transition. Clearly the level populations are constant in time. Thus the flow into any energy level must be balanced by the flow out of that level. This condition must hold in any time-independent state. However, in thermodynamic equilibrium, not only must the net flow be zero, so must the net flows that arise from individual levels. That is, every absorption must be balanced by an emission. Every process must be matched by its inverse. This concept is known as the *principle of detailed balancing.*

Consider what would transpire if this were not so. Assume that the values of the atomic parameters governing a specific set of transitions are such that absorptions from level 1 to level 3 of a hypothetical atom having only three levels are vastly more likely than absorptions to level 2 (see Figure 15.1). Then a time-independent equilibrium could only be established by transitions from level 1 to level 3 followed by transitions from level 3 to level 2 and then to level 1. There would basically be a cyclical flow of electrons from levels $1 \rightarrow 3 \rightarrow 2 \rightarrow 1$. The energy to supply the absorptions would come from either the radiation field or collisions with other particles. To understand the relation of this example to LTE, consider a radiationless gas where all excitations and deexcitations result from collisions. Then such a cyclical flow would result in energy corresponding to the $1 \rightarrow 3$ transition being systematically transferred to the energy ranges corresponding to the transitions $3 \rightarrow 2$ and $2 \rightarrow 1$. This would lead to a departure of the energy momentum distribution from that required by Maxwell-Boltzmann statistics and hence a departure from LTE. But since we have assumed LTE, this process cannot happen and the upward transitions must balance the downward transitions. Any process that tends to drive the populations away from the values they would have under the principle of detailed balancing will generate a departure from LTE.

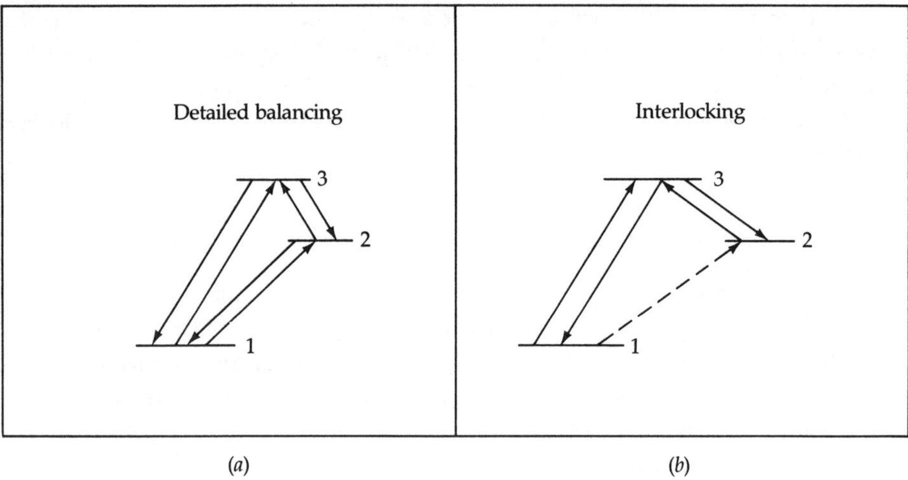

Figure 15.1 Conditions that must prevail in the case of (*a*) detailed balancing and (*b*) interlocking. The transition from level 1 to level 3 might be a resonance line and hence quite strong. The conditions that prevail in the atmosphere can then affect the line strengths of the other lines which otherwise might be accurately described by LTE.

In the example, we considered the case of a radiationless gas so that the departures had to arise in the velocity distributions of the colliding particles. In the upper reaches of the atmosphere, a larger and larger fraction of the atomic "collisions" are occurring with photons that are departing further and further from the Planck function representing their thermodynamic equilibrium distribution. These interactions will force the level populations to depart from the values they would have under LTE.

b Interlocking

Consider a set of lines that have the same upper level (see Figure 15.1). Any set of lines that arise from the same upper level is said to be *interlocked* (see R. Woolley and D. Stibbs[1]). Lines that are interlocked are subject to the cyclical processes such as we used in the discussion of detailed balancing and are therefore candidates to generate departures from LTE. Consider a set of lines formed from transitions such as those shown in Figure 15.1. If we assume that the transition $1 \rightarrow 3$ is a resonance line, then it is likely to be formed quite high up in the atmosphere where the departures from LTE are the largest. However, since this line is interlocked with the lines resulting from transitions $3 \rightarrow 2$ and $3 \rightarrow 1$, we can expect the departures affecting the resonance line to be reflected in the line strengths of the other lines. In general, the effect of a strong line formed high in the atmosphere under conditions of non-LTE that is interlocked with weaker lines formed deeper in the atmosphere is to fill in those lines, so that they appear even weaker than would otherwise be expected. A specific example involves the red lines of CaII($\lambda\lambda 8498$, $\lambda\lambda 8662$, $\lambda\lambda 8542$), which are interlocked with the strong Fraunhofer H and K resonance lines. The red lines tend to appear abnormally weak because of the photons fed into them in the upper atmosphere from the interlocked Fraunhofer H and K lines.

c Collisional Ionization versus Photoionization

We have suggested that it is the relative dominance of the interaction of photons over particles that leads to departures from LTE that are manifest in the lines. Consider how this notion can be quantified. The number of photoionizations from a particular state of excitation that takes place in a given volume per second will depend on the number of available atoms and the number of ionizing photons. We can express this condition as

$$N_i R_{ik} = 4\pi \int_{v_0}^{\infty} \frac{\kappa_v \rho J_v}{hv} \, dv = 4\pi \int_{v_0}^{\infty} \frac{\alpha_v J_v}{hv} \, dv \qquad (15.1.1)$$

The frequency v_0 corresponds to the energy required to ionize the atomic state under consideration. The integral on the far right-hand side is essentially the number of ionizing photons (modulo 4π), so that this expression really serves as a definition of R_{ik} as the rate coefficient for photoionizations from the ith state to the continuum. In a similar manner, we may describe the number of collisional ionizations by

$$N_i C_{ik} \equiv N_i N_e \int_{v_0}^{\infty} \sigma(v) f(v) \, dv = N_i N_e \Omega_{ik} \qquad (15.1.2)$$

Here, C_{ik} is the rate at which atoms in the ith state are ionized by collisions with particles in the gas. The quantity $\sigma(v)$ is the collision cross section of the particular atomic state, and it must be determined either empirically or by means of a lengthy quantum mechanical calculation; and $f(v)$ is the velocity distribution function of the particles.

In the upper reaches of the atmosphere, the energy distribution functions of the constituents of the gas depart from their thermodynamic equilibrium values. The electrons are among the last particles to undergo this departure because their mean free path is always less than that for photons and because electrons suffer many more collisions per unit time than the ions. Under conditions of thermodynamic equilibrium, the speeds of the electrons will be higher than those of the ions by $(m_h A/m_e)^{1/2}$ as a result of the equipartition of energy. Thus we may generally ignore collisions of ions of atomic weight A with anything other than electrons. Since the electrons are among the last particles to depart from thermodynamic equilibrium, we can assume that the velocity distribution $f(v)$ is given by Maxwell-

Boltzmann statistics. Under this assumption Ω_{ik} will depend on atomic properties and the temperature alone. If we replace J_v with $B_v(T)$, then we can estimate the ratio of photoionizations to collisional ionizations R_{ik}/C_{ik} under conditions that prevail in the atmospheres of normal stars. Karl Heintz Böhm[2] has used this procedure along with the semiclassical Thomson cross section for the ion to estimate this ratio. Böhm finds that only for the upper-lying energy levels and at high temperatures and densities, will collisional ionizations dominate over photoionizations. Thus, for most lines in most stars we cannot expect electronic collisions to maintain the atomic-level populations that would be expected from LTE. So we are left with little choice but to develop expressions for the energy-level populations based on the notion that the sum of *all* transitions into and out of a level must be zero. This is the weakest condition that will yield an atmosphere that is time-independent.

15.2 Rate Equations for Statistical Equilibrium

The condition that the sum of all transitions into and out of any specific level must be zero implies that there is no net change of any level populations. This means that we can write an expression that describes the flow of electrons into and out of each level, incorporating the detailed physics that governs the flow from one level to another. These expressions are known as the *rate equations for statistical equilibrium.* The unknowns are the level populations for each energy level, which will appear in every expression for which a transition between the respective states is allowed. Thus we have a system of n simultaneous equations for the level populations of n states. Unfortunately, as we saw in estimating the rates of collisional ionization and photoionization, it is necessary to know the radiation field to determine the coefficients in the rate equations. Thus any solution will require self-consistency between the radiative transfer solution and the statistical equilibrium solution. Fortunately, a method for the solution of the radiative transfer and statistical equilibrium equations can be integrated easily in the iterative algorithm used to model the atmosphere (see Chapter 12). All that is required is to determine the source function in the line appropriate for the non-LTE state.

Since an atom has an infinite number of allowed states as well as an infinite number of continuum states that must be considered, some practical limit will have to be found. For the purpose of showing how the rate equations can be developed, we consider two simple cases.

a Two-Level Atom

It is possible to describe the transitions between two bound states just as we did for photo- and collisional ionization. Indeed, for the radiative processes, basically we have already done so in Section 11.3 through the use of the Einstein coefficients.

However, since we are dealing with only two levels, we must be careful to describe exactly what happens to a photon that is absorbed by the transition from level 1 to level 2. Since the level is not arbitrarily sharp, there may be some redistribution of energy within the level. Now since the effects of non-LTE will affect the level populations at various depths within the atmosphere, we expect these effects to affect the line profile as well as the line strength. Thus, we must be clear as to what other effects might change the line profile. For that reason, we assume complete redistribution of the line radiation. This is not an essential assumption, but rather a convenient one.

If we define the probability of the absorption of a photon at frequency v' by

$$\phi(v') \, dv' = \left[\int_{-\infty}^{+\infty} R(v', v) \, dv \right] dv' \qquad (15.2.1)$$

and the probability of reemission of a photon at frequency v as

$$\psi(v) \, dv = \left[\int_{-\infty}^{+\infty} R(v', v) \, dv' \right] dv \qquad (15.2.2)$$

then the redistribution function describes to what extent these photons are correlated in frequency. In Chapter 9, we introduced a fairly general notion of complete redistribution by stating that v' and v would not be correlated. Thus,

$$\phi(v') = \psi(v) \qquad (15.2.3)$$

Under the assumption of complete redistribution, we need only count radiative transitions by assuming that specific emissions are unrelated to particular absorptions. However, since the upward radiative transitions in the atom will depend on the availability of photons, we will have to develop an equation of radiative transfer for the two-level atom.

Equation of Radiative Transfer for the Two-Level Atom   In Chapter 11 [equations (11.3.6) and (11.3.7)], we described the emission and absorption coefficients, j_v and κ_v, respectively, in terms of the Einstein coefficients. Using these expressions, or alternatively just balancing the radiative absorptions and emissions, we can write an equation of radiative transfer as

$$\mu \frac{dI_v}{dx} = -N_1 B_{12} \left(\frac{hv}{4\pi} \right) \phi_v I_v + N_2 (A_{21} + B_{21} I_v) \left(\frac{hv}{4\pi} \right) \phi_v \qquad (15.2.4)$$

This process of balancing the transitions into and out of levels is common to any order of approximation in dealing with statistical equilibrium. As long as all the processes are taken into account, we will obtain an expression like equation (15.2.4) for the transfer equation for multilevel atoms [see equation (15.2.25)]. Equation (15.2.4) can take on a somewhat more familiar form if we define

$$d\tau_\nu = \frac{-(N_1 B_{12} - N_2 B_{21})h\nu\,dx}{4\pi} \tag{15.2.5}$$

Then the equation of transfer becomes

$$\mu\frac{dI_\nu}{d\tau_\nu} = \phi_\nu(I_\nu - S_\ell) \tag{15.2.6}$$

where

$$S_\ell \equiv \frac{N_2 A_{21}}{N_1 B_{12} - N_2 B_{21}} \tag{15.2.7}$$

Making use of the relationships between the Einstein coefficients determined in Chapter 11 [equation (11.3.5)], we can further write

$$S_\ell = \frac{2h\nu^3}{c^2}\left(\frac{N_1 g_2}{N_2 g_1} - 1\right)^{-1} \tag{15.2.8}$$

Under conditions of LTE

$$\frac{N_1 g_2}{N_2 g_1} = e^{\epsilon_i/(kT)} = e^{h\nu_{12}/(kT)} \tag{15.2.9}$$

so that we recover the expected result for the source function, namely,

$$S_\ell = B_\nu(T) \tag{15.2.10}$$

Two-Level-Atom Statistical Equilibrium Equations   The solution to equation (15.2.6) will provide us with a value of the radiation field required to determine the number of radiative transitions. Thus the total number of upward transitions in the two-level atom is

$$N_{1\to2} = N_1 B_{12}\int \phi_\nu J_\nu\,d\nu + N_1 N_e \Omega_{12} \tag{15.2.11}$$

Similarly, the number of downward transitions is

$$N_{2\to1} = N_2 A_{21} + N_2 B_{21}\int \phi_\nu J_\nu\,d\nu + N_2 N_e \Omega_{21} \tag{15.2.12}$$

The requirement that the level populations be stationary means that

$$N_{1 \to 2} = N_{2 \to 1} \tag{15.2.13}$$

so that the ratio of level populations is

$$\frac{N_1}{N_2} = \frac{A_{21} + B_{21} \int \phi_v J_v \, dv + N_e \Omega_{21}}{B_{12} \int \phi_v J_v \, dv + N_e \Omega_{12}} \tag{15.2.14}$$

Now consider a situation where there is no radiation field and the collisions are driven by particles characterized by a maxwellian energy distribution. Under these conditions, the principle of detailed balancing requires that

$$N_1 \Omega_{12} = N_2 \Omega_{21} \tag{15.2.15}$$

or

$$\Omega_{12} = \frac{g_2}{g_1} \Omega_{21} e^{-hv/(kT)} \tag{15.2.16}$$

This argument is similar to that used to obtain the relationships between the Einstein coefficients, and since the collision coefficients depend basically on atomic constants, equation (15.2.16) must hold under fairly arbitrary conditions. Specifically, the result will be unaffected by the presence of a radiation field. Thus we may use it and the relations between the Einstein coefficients [equations (11.3.5)] to write the line source function as

$$S_\ell = \frac{\int \phi_v J_v \, dv + \dfrac{N_e \Omega_{21}}{A_{21}} \dfrac{2hv^3}{c^2} e^{-hv/(kT)}}{1 + \dfrac{N_e \Omega_{21}}{A_{21}} (1 - e^{-hv/(kT)})} \tag{15.2.17}$$

If we let

$$\epsilon = \frac{N_e \Omega_{21}(1 - e^{-hv/(kT)})/A_{21}}{1 + N_e \Omega_{21}(1 - e^{-hv/(kT)})/A_{21}} \tag{15.2.18}$$

then the source function takes on the more familiar form

$$S_\ell = \epsilon B_v + (1 - \epsilon) \int \phi_{v'} J_{v'} \, dv' \tag{15.2.19}$$

The quantity ϵ is, in some sense, a measure of the departure from LTE and is sometimes called the *departure coefficient*. A similar method for describing the departures from LTE suffered by an atom is to define

$$b_j = \frac{N_j}{\tilde{N}_j} \qquad (15.2.20)$$

where $\tilde{N}_j$ is the level population expected in LTE so that b_j is just the ratio of the actual population to that given by the Saha-Boltzmann formula. From that definition, equation (15.2.14), and the relations among the Einstein coefficients we get

$$1 - \frac{b_1}{b_2} = \frac{\int \phi_v J_v \, dv - B_v}{(1 - e^{-h v/(kT)})^{-1} \int \phi_v J_v \, dv + (N_e \Omega_{21}/A_{21}) B_v(T)} \qquad (15.2.21)$$

b Two-Level Atom plus Continuum

The addition of a continuum increases the algebraic difficulties of the above analysis. However, the concepts of generating the statistical equilibrium equations are virtually the same. Now three levels must be considered. We must keep track of transitions to the continuum as well as the two discrete energy levels. Again, we assume complete redistribution within the line so that the line source function is given by equation (15.2.8), and the problem is to find the ratio of the populations of the two levels.

We begin by writing the rate equations for each level which balance all transitions into the level with those to the other level and the continuum. For level 1,

$$N_1 \left(B_{12} \int \phi_v J_v \, dv + N_e \Omega_{12} + R_{1k} + N_e \Omega_{1k} \right)$$

$$= N_2 \left(A_{21} + B_{21} \int \phi_v J_v \, dv + N_e \Omega_{21} \right) + N_1^*(R_{k1} + N_e \Omega_{1k}) \qquad (15.2.22)$$

The parameter R_{ik} is the photoionization rate defined in equation (15.1.1), while R_{ki} is the analogous rate of photorecombination. When the parameter Ω contains the subscript k, it refers to collisional transitions to or from the continuum. The term on the left-hand side describes all the types of transitions from level 1, which are photo- and collisional excitations followed by the two terms representing photo- and collisional ionizations, respectively. The two large terms on the right-hand side contain all the transitions into level 1. The first involves spontaneous and stimulated radiative emissions followed by collisionly stimulated emissions. The second term describes the recombinations from the continuum. The parameter N_i^* will in general represent those ions that have been ionized from the ith state.

We may write a similar equation

$$N_2\left(A_{21} + B_{21} \int \phi_v J_v \, dv + N_e\Omega_{21} + R_{2k} + N_e\Omega_{2k}\right)$$
$$= N_1\left(B_{12} \int \phi_v J_v \, dv + N_e\Omega_{12}\right) + N_2^*(R_{k2} + N_e\Omega_{2k}) \quad (15.2.23)$$

for level 2 by following the same prescription for the meaning of the various terms. Again letting the terms on the left-hand side represent transitions out of the two levels while terms on the right-hand side denote inbound transitions, we find the rate equation for the continuum is

$$N_1^*(R_{k1} + N_e\Omega_{1k}) + N_2^*(R_{k2} + N_e\Omega_{2k})$$
$$= N_1(R_{1k} + N_e\Omega_{1k}) + N_2(R_{2k} + N_e\Omega_{2k}) \quad (15.2.24)$$

However, this equation is not linearly independent from the other two and can be generated simply by adding equations (15.2.22) and (15.2.23). This is an expression of continuity and will always be the case regardless of how many levels are considered. There will always be one less independent rate equation than there are levels. An electron that leaves one state must enter another, so its departure is not independent from its arrival. If all allowed levels are counted, as they must be if the equations are to be complete, this interdependence of arrivals and departures of specific transitions will make the rate equation for one level redundant. Noting that the same kind of symmetry described by equation (15.2.15) also holds for the collisional ionization and recombination coefficients, we may solve equations (15.2.22) and (15.2.23) for the population ratio required for the source function given by equation (15.2.8). The algebra is considerably more involved than for the two levels alone and yields[3] a source function of the form

$$S_\ell = \frac{\int \phi_v J_v \, dv + \tilde{\epsilon} B_v(T) + \eta B^*}{1 + \tilde{\epsilon} + \eta}$$

$$\tilde{\epsilon} = \frac{N_e\Omega_{21}}{A_{21}} (1 - e^{-hv/(kT)})$$

$$\eta = \frac{1}{A_{21}} \frac{(R_{2k} + N_e\Omega_{2k})N_1^*(R_{k1} + N_e\Omega_{1k}) - (g_1/g_2)(R_{1k} + N_e\Omega_{1k})N_2^*(R_{2k} + N_e\Omega_{2k})}{N_1^*(R_{k1} + N_e\Omega_{1k}) + N_2^*(R_{k2} + N_e\Omega_{2k})}$$

$$B^* = \frac{2hv^3}{c^2} \left[\frac{N_1^* g_2(R_{2k} + N_e\Omega_{2k})(R_{1k} + N_e\Omega_{1k})}{N_2^* g_1(R_{1k} + N_e\Omega_{1k})(R_{k2} + N_e\Omega_{2k})} - 1\right]^{-1} \quad (15.2.25)$$

Table 15.1 Types of Solar Spectral Lines

Collisionly Dominated	Dominated by Photoionization
Resonance lines of singly ionized metals	Resonance lines of neutral metals
Resonance lines of hydrogen	Balmer lines of hydrogen
Resonance lines of nonmetals	

If the terms involving $\tilde{\epsilon}$ dominate the source function, the line is said to be collisionly dominated, while if the terms involving η are the largest, the line is said to be dominated by photoionization. If $\tilde{\epsilon} B_\nu(T) > \eta B_\nu^*$ but $\eta > \tilde{\epsilon}$ (or vice versa), the line is said to be mixed. Some examples of lines in the solar spectrum that fall into these categories are given in Table 15.1.

c Multilevel Atom

A great deal of effort has gone into approximating the actual case of many levels of excitation by setting up and solving the rate equations for three and four levels or approximating any particular transition of interest by an "equivalent two-level atom" (see D. Mihalas,[3] pp. 391–394). However, the advent of modern, swift computers has made most of these approximations absolete. Instead, one considers an n-level atom (with continuum) and solves the rate equations directly. We have already indicated that this procedure can be integrated into the standard algorithm for generating a model atmosphere quite easily. Consider the generalization of equations (15.2.22) through (15.2.24). Simply writing equations for each level, by balancing the transitions into the level with those out of the level, will yield a set of equations which are linear in the level populations. However, as we have already indicated, these equations are redundant by 1. So far we have only needed population ratios for the source function, but if we are to find the population levels themselves, we will need an additional constraint. The most obvious constraint is that the total number of atoms and ions must add up to the abundance specified for the atmosphere. Mihalas[4] suggests using charge conservation, which is a logically equivalent constraint. Whatever additional constraint is chosen, it should be linear in the level populations so that the linear nature of the equations is not lost.

It is clear that the equations are irrevocably coupled to the radiation field through the photoexcitation and ionization terms. It is this coupling that led to the rather messy expressions for the source functions of the two-level atom. However, if one takes the radiation field and electron density as known, then the rate equations have the form

$$\mathbf{A}\vec{N} = \vec{B}$$

(15.2.26)   409

where $\mathbf{A}$ is a matrix whose elements are the coefficients multiplying the population levels and $\vec{N}$ is a vector whose elements are the populations of the energy levels for all species considered in the calculation. The only nonzero element of the constant vector $\vec{B}$ arises from the additional continuity constraint that replaced the redundant level equation. These equations are fairly sparse and can be solved quickly and accurately by well-known techniques.

Since the standard procedure for the construction of a model atmosphere is an iterative one wherein an initial guess for the temperature distribution gives rise to the atmospheric structure, which in turn allows for the solution of the equation of radiative transfer, the solution of the rate equations can readily be included in this process. The usual procedure is to construct a model atmosphere in LTE that nearly satisfies radiative equilibrium. At some predetermined level of accuracy, the rate equations are substituted for the Saha-Boltzmann excitation and ionization equations by using the existing structure (electron density and temperature distribution) and radiation field. The resulting population levels are then used to calculate opacities and the atmospheric structure for the next iteration. One may even chose to use an iterative algorithm for the solution of the linear equations, for an initial guess of the LTE populations will probably be quite close to the correct populations for many of the levels that are included. The number of levels of excitation that should be included is somewhat dictated by the problem of interest. Depending on the state of ionization, four levels are usually enough to provide sufficient accuracy. However, some codes routinely employ as many as eight. One criterion of use is to include as many levels as is necessary to reach those whose level populations are adequately given by the Saha-Boltzmann ionization-excitation formula.

Many authors consider the model to be a non-LTE model if hydrogen alone has been treated by means of rate equations while everything else is obtained from the Saha-Boltzmann formula. For the structure of normal stellar atmospheres, this is usually sufficient. However, should specific spectral lines be of interest, one should consider whether the level populations of the element in question should also be determined from a non-LTE calculation. This decision will largely be determined by the conditions under which the line is formed. As a rule of thumb, if the line occurs in the red or infrared spectral region, consideration should be given to a non-LTE calculation. The hotter the star, the more this consideration becomes imperative.

d Thermalization Length

Before we turn to the solution of the equation of radiative transfer for lines affected by non-LTE effects, we should consider an additional concept which helps characterize the physical processes that lead to departures from LTE. It is known as the *thermalization length*. In LTE all the properties of the gas are determined by the local values of the state variables. However, as soon as radiative processes become important in establishing the populations of the energy levels of the gas, the problem becomes global. Let ℓ be the mean free path of a photon between absorptions or scatterings and $\mathscr{L}$ be the mean free path between collisional destructions. If scatterings dominate over collisions, then $\mathscr{L}$ will not be a "straight-line" distance through the atmosphere. Indeed, $\mathscr{L} \gg \ell$ if $A_{ij} \gg C_{ij}$. That is, if the probability of radiative deexcitation is very much greater than the probability of collisional deexcitation, then an average photon will have to travel much farther to be destroyed by a collision than by a radiative interaction. However, $\mathscr{L} \approx \ell$ if $C_{ij} \gg A_{ij}$. In this instance, all photons that interact radiatively will be destroyed by collisions.

If the flow of photons is dominated by scatterings, then the character of the radiation field will be determined by photons that originate within a sphere of radius $\mathscr{L}$ rather than ℓ. However, $\mathscr{L}$ should be regarded as an upper limit because many radiative interactions are pure absorptions that result in the thermalization of the photon as surely as any collisional interaction. In the case when $\mathscr{L} \gg \ell$, some photons will travel a straight-line distance equal to $\mathscr{L}$, but not many. A better estimate for an average length traveled before the photon is thermalized would include other interactions through the notion of a "random walk." If n is the ratio of radiative to collisional interactions, then a better estimate of the thermalization length would be

$$l_{\text{th}} = \ell \sqrt{n} = l\sqrt{\mathscr{L}/\ell} = \sqrt{\mathscr{L}\ell} \tag{15.2.27}$$

If the range of temperature is large over a distance corresponding to the thermalization length l_{th}, then the local radiation field will be characterized by a temperature quite different from the local kinetic gas temperature. These departures of the radiation field from the local equilibrium temperature will ultimately force the gas out of thermodynamic equilibrium. Clearly, the greatest variation in temperature within the thermalization sphere will occur as one approaches the boundary of the atmosphere. Thus it is no surprise that these departures increase near the boundary. Let us now turn to the effects of non-LTE the transfer of radiation.

15.3 Non-LTE Transfer of Radiation and the Redistribution Function

While we did indicate how departures of the population of the energy levels from their LTE values could be included in the construction of a model atmosphere so that any structural effects are included, the major emphasis of the effects of non-LTE has been on the strengths and shapes of spectral lines. During the discussion of the two-level atom, we saw that the form of the source function was somewhat different from what we discussed in Chapter 10. Indeed, the equation of transfer [equation (15.2.6)] for complete redistribution appears in a form somewhat different from the customary plane-parallel equation of transfer. Therefore, it should not be surprising to find that the effects of non-LTE can modify the profile of a spectral line. The extent and nature of this modification will depend on the nature of the redistribution function as well as on the magnitude of the departures from LTE. Since we already introduced the case of complete redistribution [equations (15.2.1) and (15.2.2)], we begin by looking for a radiative transfer solution for the case where the emitted and absorbed photons within a spectral line are completely uncorrelated.

a Complete Redistribution

In Chapter 14, we devoted a great deal of effort to developing expressions for the atomic absorption coefficient for spectral lines that were broadened by a number of phenomena. However, we dealt tacitly with absorption and emission processes as if no energy were exchanged with the gas between the absorption and reemission of the photon. Actually this connection was not necessary for the calculation of the atomic line absorption coefficient, but this connection is required for calculating the radiative transfer of the line radiation. Again, for the case of pure absorption there is no relationship between absorbed and emitted photons. However, in the case of scattering, as with the Schuster-Schwarzschild atmosphere, the relationship between the absorbed and reemitted photons was assumed to be perfect. That is, the scattering was assumed to be completely coherent. In a stellar atmosphere, this is rarely the case because microperturbations occurring between the atom and surrounding particles will result in small exchanges of energy, so that the electron can be viewed as undergoing transitions *within* the broadened energy level. If those transitions are numerous during the lifetime of the excited state, then the energy of the photon that is emitted will be uncorrelated with that of the absorbed photon. In some sense the electron will "lose all memory" of the details of the transition that brought it to the excited state. The absorbed radiation will then be completely redistributed throughout the line. This is the situation that was described by equations (15.2.1) through (15.2.3) and led to the equation of transfer (15.2.6) for complete redistribution of line radiation.

Although this equation has a slightly different form from what we are used to, it can be put into a familiar form by letting

$$d\tau_x = \phi_x \, d\tau \tag{15.3.1}$$

It now takes on the form of equation (10.1.1), and by using the classical solution discussed in Chapter 10, we can obtain an integral equation for the mean intensity in the line in terms of the source function:

$$J(\tau_x) = \tfrac{1}{2} \int_0^\infty S_\ell(t) E_1 \left| \int_{\tau_x}^t \phi_x(t')\,dt' \right| \phi_x(t)\,dt \tag{15.3.2}$$

This can then be substituted into equation (15.2.19) to obtain an integral equation for the source function in the line:

$$S_\ell(\tau_x) = \epsilon B_\nu(T) + \tfrac{1}{2}(1 - \epsilon)$$

$$\times \int_0^\infty S_\ell(t) \left[\int_{-\infty}^{+\infty} \phi_x(\tau_x)\phi_x(t) E_1 \left| \int_{\tau_x}^t \phi_x(t')\,dt' \right| dx \right] dt \tag{15.3.3}$$

The integral over x results from the integral of the mean intensity over all frequencies in the line [see equation (15.2.19)]. Note the similarity between this result and the integral equation for the source function in the case of coherent scattering [equation (9.1.14)]. Only the kernel of the integral has been modified by what is essentially a moment in frequency space weighted by the line profile function $\phi_x(t)$. This is clearly seen if we write the kernel as

$$K(\tau_x, t) = \int_{-\infty}^{+\infty} \phi_x(\tau_x)\phi_x(t) E_1 \left| \int_{\tau_x}^t \phi_x(t')\,dt' \right| dx \tag{15.3.4}$$

so that the source function equation becomes a Schwarzschild-Milne equation of the form

$$S_\ell(\tau_x) = \epsilon B_\nu(T) + \tfrac{1}{2}(1 - \epsilon) \int_{-\infty}^{+\infty} S_\ell(t) K(\tau_x, t)\,dt \tag{15.3.5}$$

Since

$$\left| \int_{\tau_x}^t \phi_x(t')\,dt' \right| = \left| \Phi(t) - \Phi(\tau_x) \right| \tag{15.3.6}$$

the kernel is symmetric in τ_x and t, so that $K(\tau_x, t) = K(t, \tau_x)$. This is the same symmetry property as the exponential integral $E_1|\tau - t|$ in equation (10.1.14). Unfortunately, for an arbitrary depth dependence of $\phi_x(t)$, equations (15.3.4) and (15.3.5) must be solved numerically. Fortunately, all the methods for the solution of Schwarzschild-Milne equations discussed in Chapter 10 are applicable to the solution of this integral equation.

While it is possible to obtain some insight into the behavior of the solution for the case where $\phi_x(t) \neq f(t)$ (see Mihalas,[3] pp. 366–369), the insight is of dubious value because it is the solution for a special case of a special case. However, a property of such solutions, and of noncoherent scattering in general, is that the core

413

of the line profile is somewhat filled in at the expense of the wings. As we saw for the two-level atom with continuum, the source function takes on a more complicated form. Thus we turn to the more general situation involving partial redistribution.

b Hummer Redistribution Functions

The advent of swift computers has made it practical to model the more complete description of the redistribution of photons in spectral lines. However, the attempts to describe this phenomenon quantitatively go back to L. Henyey[5] who carried out detailed balancing *within an energy level* to describe the way in which photons are actually redistributed across a spectral line. Unfortunately, the computing power of the time was not up to the task, and this approach to the problem has gone virtually unnoticed. More recently, D. Hummer[6] has classified the problem of redistribution into four main categories which are widely used today. For these cases, the energy levels are characterized by Lorentz profiles which are appropriate for a wide range of lines. Regrettably, for the strong lines of hydrogen, many helium lines as well as most strong resonance lines, this characterization is inappropriate (see Chapter 14), and an entirely different analysis must be undertaken. This remains one of the current nagging problems of stellar astrophysics. However, the Hummer classification and analysis provides considerable insight into the problems of partial redistribution and enables rather complete analyses of many lines with Lorentz profiles produced by the impact phase-shift theory of collisional broadening.

Let us begin the discussion of the Hummer redistribution functions with a few definitions. Let $p(\xi', \xi)\,d\xi$ be the probability that an absorbed photon having frequency ξ' is scattered into the frequency interval $\xi \to \xi + d\xi$. Furthermore, let the probability density function $p(\xi', \xi)$ be normalized so that $\int p(\xi', \xi)\,d\xi' = 1$. That is, the absorbed photon must go somewhere. If this is not an appropriate result for the description of some lines, the probability of scattering can be absorbed in the scattering coefficient (see Section 9.2). In addition, let $g(\hat{n}', \hat{n})$ be the probability density function describing scattering from a direction $\hat{n}'$ into $\hat{n}$, also normalized so that the integral over all solid angles $[\int g(\hat{n}', \hat{n})\,d\Omega]/(4\pi) = 1$. For isotropic scattering, $g(\hat{n}', \hat{n}) = 1$, while in the case of Rayleigh scattering $g(\hat{n}', \hat{n}) = 3[1 - \cos(\hat{n}' \cdot \hat{n})]/4$. We further define $f(\xi')\,d\xi'$ as the relative [that is, $\int f(\xi')\,d\xi' = 1$] probability that a photon with frequency ξ' is absorbed. These probability density functions can be used to describe the redistribution function introduced in Chapter 9.

In choosing to represent the redistribution function in this manner, it is tacitly assumed that the redistribution of photons in frequency is independent of the direction of scattering. This is clearly not the case for atoms in motion, but for an observer located in the rest frame of the atom it is *usually* a reasonable assumption. The problem of Doppler shifts is largely geometry and can be handled separately. Thus, the probability that a photon will be absorbed at frequency ξ' and reemitted at a frequency ξ is

$$R(\xi', \xi, \hat{n}', \hat{n})\, d\xi'\, d\xi\, \frac{d\omega'}{4\pi}\frac{d\omega}{4\pi} = f(\xi')p(\xi', \xi)g(\hat{n}', \hat{n})\, d\xi'\, d\xi\, \frac{d\omega'}{4\pi}\frac{d\omega}{4\pi} \qquad (15.3.7)$$

David Hummer[6] has considered a number of cases where f, p, and g take on special values which characterize the energy levels and represent common conditions that are satisfied by many atomic lines.

Emission and Absorption Probability Density Functions for the Four Cases Considered by Hummer   Consider first the case of coherent scattering where both energy levels are infinitely sharp. Then the absorption and reemission probability density functions will be given by

$$\begin{aligned} f(\xi') &= \delta(\xi' - v_0) \\ p(\xi', \xi) &= \delta(\xi' - \xi) \end{aligned} \qquad \text{Hummer's case I} \qquad (15.3.8)$$

If the lower level is broadened by collisional radiation damping but the upper level remains sharp, then the absorption probability density function is a Lorentz profile while the reemission probability density function remains a delta function, so that

$$\begin{aligned} f(\xi') &= \frac{\gamma/\pi}{(\xi' - v_0)^2 + \gamma^2} \\ p(\xi', \xi) &= \delta(\xi' - \xi) \end{aligned} \qquad \text{Hummer's case II} \qquad (15.3.9)$$

If the lower level is sharp but the upper level is broadened by collisional radiation damping, then both probability density functions are given Lorentz profiles since the transitions into and out of the upper level are from a broadened state. Thus,

$$\begin{aligned} f(\xi') &= \frac{\gamma_u/\pi}{(\xi' - v_0)^2 + \gamma_u^2} \\ p(\xi', \xi) &= \frac{\gamma_u/\pi}{(\xi' - v_0)^2 + \gamma_u^2} \end{aligned} \qquad \text{Hummer's case III} \qquad (15.3.10)$$

Since ξ' and ξ are uncorrelated, this case represents a case of complete redistribution of noncoherent scattering. Hummer gives the joint probability of transitions from a broadened lower level to a broadened upper level and back again as

$$f(\xi')p(\xi', \xi) = \frac{\gamma_u\gamma_l/\pi^2}{[(\xi' - \xi)^2 + \gamma_l^2][(\xi - v_0)^2 + \gamma_u^2]} \qquad \text{Hummer's case IV} \qquad (15.3.11)$$

This probability must be calculated as a unit since ξ' is the same for both f and p. Unfortunately a careful analysis of this function shows that the lower level is considered to be sharp for the reemitted photon and therefore is inconsistent with the assumption made about the absorption. Therefore, it will not satisfy detailed balancing in an environment that presupposes LTE. A correct quantum mechanical analysis[7] gives

$$f(\xi')p(\xi', \xi) =$$

$$\frac{\gamma_u(2\gamma_l + \gamma_u)\gamma_l/\pi^2}{\{(\xi' - \nu_0)^2 + [(\gamma_l + \gamma_u)/2]^2\}\{(\xi - \nu_0)^2 + [(\gamma_l + \gamma_u)/2]^2\}[(\xi' - \xi)^2 + \eta_i^2]}$$

$$+ \frac{\gamma_l\gamma_u}{\{(\xi' - \nu_0)^2 + [(\gamma_u + \gamma_l)/2]^2\}[(\xi' - \xi)^2 + \gamma_l^2]}$$

$$+ \frac{\gamma_l\gamma_u}{\{(\xi - \nu_0)^2 + [(\gamma_l + \gamma_u)/2]^2\}[(\xi' - \xi)^2 + \gamma_l^2]}$$

$$+ \frac{\gamma_1^2}{\{(\xi' - \nu_0)^2 + [(\gamma_u + \gamma_l)/2]^2\}\{(\xi' - \nu_0)^2 + [(\gamma_u + \gamma_l)/2]^2\}} \qquad (15.3.12)$$

A little inspection of this rather messy result shows that it is symmetric in ξ' and ξ, which it must be if it is to obey detailed balancing. In addition, the function has two relative maxima at $\xi' = \xi$ and $\xi' = \nu_0$. Since the center of the two energy levels represents a very likely transition, transitions from the middle of the lower level and back again will be quite common. Under these conditions $\xi' = \xi$ and the scattering is fully coherent. On the other hand, transitions from the exact center of the lower level ($\xi' = \nu_0$) will also be very common. However, the return transition can be to any place in the lower level with frequency ξ. Since the function is symmetric in ξ' and ξ, the reverse process can also happen. Both these processes are fully noncoherent so that the relative maxima occur for the cases of fully coherent and noncoherent scattering with the partially coherent photons being represented by the remainder of the joint probability distribution function.

Effects of Doppler Motion on the Redistribution Functions Consider an atom in motion relative to some fixed reference frame with a velocity $\vec{v}$. If a photon has a frequency ξ' as seen by the atom, the corresponding frequency in the rest frame is

$$v' = \xi' + \frac{\nu_0}{c}\vec{v} \cdot \hat{n}' \qquad (15.3.13)$$

Similarly, the photon emitted by the atom will be seen in the rest frame, Doppler-shifted from its atomic value by

$$v = \xi + \frac{v_0}{c}\,\vec{v} \cdot \hat{n} \tag{15.3.14}$$

Thus the redistribution function that is seen by an observer in the rest frame is

$$R(v', v, \hat{n}', \hat{n}) = f\left(v' - \frac{v_0\vec{v}\cdot\hat{n}'}{c}\right) p\left(v' - \frac{v_0\vec{v}\cdot\hat{n}'}{c}, v - \frac{v_0\vec{v}\cdot\hat{n}}{c}\right) g(\hat{n}', \hat{n}) \tag{15.3.15}$$

We need now to relate the scattering angle determined from $\hat{n}' \cdot \hat{n}$ to the angle between the atomic velocity and the directions of the incoming and outgoing scattered photon. Consider a coordinate frame chosen so that the xy plane is the scattering plane and the x axis lies in the scattering plane midway between the incoming and outgoing photon (see Figure 15.2). In this coordinate frame, the directional unit vectors $\hat{n}$ and $\hat{n}'$ have cartesian components given by

$$\hat{n}' = \cos\left(\frac{\psi}{2}\right)\hat{i} - \sin\left(\frac{\psi}{2}\right)\hat{j} \qquad \hat{n} = \cos\left(\frac{\psi}{2}\right)\hat{i} + \sin\left(\frac{\psi}{2}\right)\hat{j} \tag{15.3.16}$$

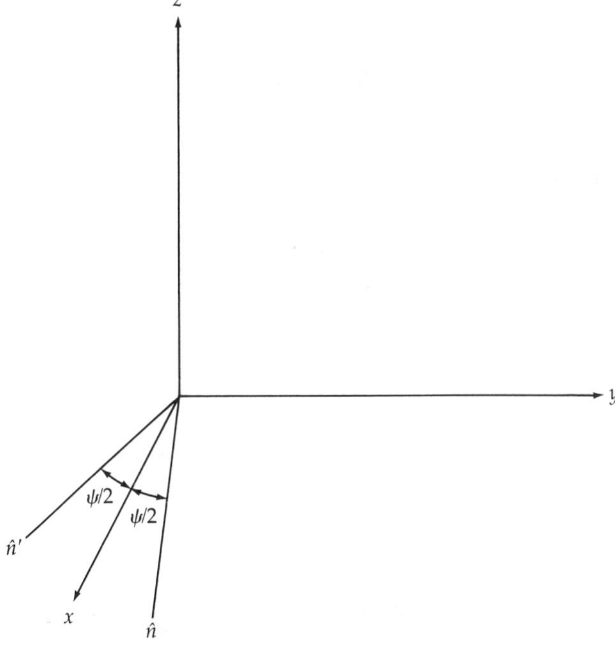

Figure 15.2 Cartesian coordinate frame where the x axis bisects the angle between the incoming and outgoing photons, and the xy plane is the scattering plane.

Now if we assume that the atoms have a maxwellian velocity distribution

$$P(\vec{v}) = \left(\frac{m}{2\pi kT}\right)^{3/2} e^{-[m/(2kT)](v_x^2 + v_y^2 + v_z^2)} \tag{15.3.17}$$

we can obtain the behavior of an ensemble of atoms by averaging equation (15.3.15) over all velocity. First it is convenient to make the variable transformations

$$\vec{u} \equiv \sqrt{\frac{m}{2kT}}\,\vec{v} = \frac{\vec{v}}{v_{\text{th}}}$$

$$\alpha \equiv \cos\frac{\psi}{2} \qquad \tilde{\alpha} \equiv \cos\psi$$

$$\beta \equiv \sin\frac{\psi}{2} \qquad \tilde{\beta} \equiv \sin\psi$$

$$w \equiv \frac{v_0 v_d}{c} = \frac{v_0}{c}\sqrt{\frac{2kT}{m}} \tag{15.3.18}$$

so that the components of the particle's velocity projected along the directions of the photon's path become

$$\vec{v} \cdot \hat{n}' = \sqrt{\frac{2kT}{m}}(\alpha u_x - \beta u_y) \qquad \vec{v} \cdot \hat{n} = \sqrt{\frac{2kT}{m}}(\alpha u_x + \beta u_y) \tag{15.3.19}$$

and the velocity distribution is

$$P(\vec{u})\,d\vec{u} = \pi^{-3/2} e^{-\vec{u} \cdot \vec{u}}\,d\vec{u} \tag{15.3.20}$$

The symbol $d\vec{u}$ means $du_x\,du_y\,du_z$.

We define the ensemble average over the velocity of the redistribution function as

$$\langle R(v', v, \hat{n}', \hat{n}) \rangle \equiv \int_{\vec{v}} R(v', v, \hat{n}', \hat{n}, \vec{v}) P(\vec{v})\,d\vec{v} \tag{15.3.21}$$

or

$$\langle R(v', v, \hat{n}', \hat{n}) \rangle = \frac{g(\hat{n}', \hat{n})}{\pi} \int_{-\infty}^{+\infty} \int_{-\infty}^{+\infty} e^{-(u_x^2 + u_y^2)} f[v' - w(\alpha u_x - \beta u_y)]$$

$$\times\, p[v' - w(\alpha u_x - \beta u_y), v - w(\alpha u_x + \beta u_y)]\,du_x\,du_y \tag{15.3.22}$$

A coordinate rotation by $\psi/2$ about the y axis so that $\hat{n}'$ is aligned with $\hat{x}$ (see Figure 15.2) leads to the equivalent, but useful, form

$$\langle R(v, v', \hat{n}', \hat{n}) \rangle = \frac{g(\hat{n}', \hat{n})}{\pi} \left[\int_{-\infty}^{+\infty} e^{-u_x^2} f(v' - wu_x)\, du_x \right.$$

$$\left. \times \left\{ \int_{-\infty}^{+\infty} e^{-u_y^2} p[(v' - wu_x),\, v - w(\tilde{\alpha}u_x + \tilde{\beta}u_y)]\, du_y \right\} \right. \quad (15.3.23)$$

We are now in a position to evaluate the effects of thermal Doppler motion on the four cases given by Hummer,[6] represented by equations (15.3.8) through (15.3.12). The substitution of these forms of $f(\xi')$ and $p(\xi', \xi)$ into equation (15.3.22) or (15.3.23) will yield the desired result. The frequencies ξ' and ξ must be Doppler-shifted according to equations (15.3.13) and (15.3.14), and some difficulty may be encountered for the case of direct forward or back scattering (that is, $\beta = 0$) and when one of the distribution functions is a delta function (i.e., for a sharp energy level). The fact that $\beta = 0$ for these cases should be invoked before any variable transformations are made for the purposes of evaluating the integrals.

Making a final transformation to a set of dimensionless frequencies

$$x \equiv \frac{v - v_0}{w} \qquad x' \equiv \frac{v' - v_0}{w} \qquad\qquad (15.3.24)$$

we can obtain the following result for Hummer's case I:

$$\langle R_{\mathrm{I}}(x', x, \hat{n}', \hat{n}) \rangle = \frac{g(\hat{n}', \hat{n})}{\pi \sin \psi} e^{-[x^2 + (x' - x \cos \psi)^2/\sin^2 \psi]} \qquad (15.3.25)$$

Consider what the emitted radiation would look like for an ensemble of atoms illuminated by an isotropic uniform radiation field I_0. Substitution of such a radiation field into equation (9.2.29) would yield

$$S_x = \frac{1}{4\pi} \int_{4\pi} \int_{-\infty}^{+\infty} I_0 \langle R_{\mathrm{I}}(x', x, \hat{n}', \hat{n}) \rangle\, dx'\, d\omega' \qquad (15.3.26)$$

which after some algebra gives

$$S_x = \frac{I_0 e^{-x^2}}{\sqrt{\pi}} = I_0 \phi(x) \qquad (15.3.27)$$

This implies that the emission of the radiation would have exactly the same form as the absorption profile. But this was our definition of complete redistribution [see equation (15.2.3)]. Thus, although a single atom behaves coherently, an ensemble of thermally moving atoms will produce a line profile that is equivalent to one suffering complete redistribution of the radiation over the Doppler core. Perhaps this is not too surprising since the motion of the atoms is totally uncorrelated, so that the Doppler shifts produced by the various motions will mimic complete redistribution.

As one proceeds with the progressively more complicated cases, the results become correspondingly more complicated to derive and express. Hummer's cases II and III yield

$$\langle R_{\mathrm{II}}(x', x, \hat{n}', \hat{n})\rangle = \frac{g(\hat{n}', \hat{n})}{\pi \sin \psi} H\left(a \sec \frac{\psi}{2}, \frac{x + x'}{2} \sec \frac{\psi}{2}\right)$$

$$\times \exp - \left(\frac{x - x'}{\sqrt{2}} \csc \frac{\psi}{2}\right)^2 \tag{15.3.28}$$

and

$$\langle R_{\mathrm{III}}(x', x, \hat{n}', \hat{n})\rangle = \frac{g(\hat{n}', \hat{n})}{\pi} \left(\frac{a \csc \psi}{\pi}\right)$$

$$\times \int_{-\infty}^{+\infty} \frac{e^{-u^2} H[a \csc \psi, (x \csc \psi - u \cot \psi)]\, du}{(x' - u)^2 + a^2} \tag{15.3.29}$$

respectively. There is little point in giving the result for case IV, as given by equation (15.3.11). But the result for the correct case IV (sometimes called case V) that is obtained from equation (15.3.12) is of some interest and is given by McKenna[8] as

$$\langle R_{\mathrm{IV}}(x', x, \hat{n}', \hat{n})\rangle = \frac{g(\hat{n}', \hat{n})\gamma_u}{w^2 \pi^3} \int_{-\infty}^{+\infty} \left(\frac{e^{-u^2}}{(x' - x - 2\beta u)^2 + (\gamma_u/\pi)^2}\right.$$

$$\times \left\{\frac{\gamma_u}{\alpha^4}\left[\frac{(\gamma_u + \gamma_1)^2}{\pi^2} + (x' - x - 2\beta u)^2\right]\right.$$

$$\times Y\left(\frac{\gamma_u + \gamma_1}{4\pi}, \frac{x' - \beta u}{\alpha}, \frac{\gamma_u + \gamma_1}{\alpha}, \frac{a + \beta u}{\alpha}\right) + \frac{\pi \gamma_l}{\alpha(\gamma_u + \gamma_l)}$$

$$\times \left[H\left(\frac{\gamma_u + \gamma_1}{4\pi\alpha}, \frac{x' - \beta u}{\alpha}\right) + H\left(\frac{\gamma_u + \gamma_1}{4\pi\alpha}, \frac{x - \beta u}{\alpha}\right)\right]\right\}\right) du \tag{15.3.30}$$

where

$$Y(a, x, b, y) \equiv \int_{-\infty}^{+\infty} \frac{e^{-u^2}\, du}{[(x - u)^2 + a^2][(y - u)^2 + b^2]}$$

$$= \frac{\pi}{2a^3}\left[(1 - 2a^2)H(a, x) - 2axK(a, x) + \frac{2a}{\sqrt{\pi}}\right] \tag{15.3.31}$$

and the function $K(a, x)$ which is known as the shifted Voigt function, is defined by

$$K(a, x) \equiv \frac{1}{\pi} \int_{-\infty}^{+\infty} \frac{(x - u)e^{-u^2}\, du}{(x - u)^2 + a^2} \tag{15.3.32}$$

Unfortunately, all these redistribution functions contain the scattering angle ψ explicitly and so by themselves are difficult to use for the calculation of line profiles. Not only does the scattering angle appear in the part of the redistribution function resulting from the effects of the Doppler motion, but also the scattering angle is contained in the phase function $g(\hat{n}', \hat{n})$. Thus, the Doppler motion can be viewed as merely complicating the phase function. While there are methods for dealing with the angle dependence of the redistribution function (see McKenna[9]), they are difficult and beyond the present scope of this discussion. They are, however, of considerable importance to those interested in the state of polarization of the line radiation.

For most cases, the phase function is assumed to be isotropic, and we may remove the angle dependence introduced by the Doppler motion by averaging the redistribution function over all angles, as we did with velocity. These averaged forms for the redistribution functions can then be inserted directly into the equation of radiative transfer. As long as the radiation field is nearly isotropic and the angular scattering dependence (phase function) is also isotropic, this approximation is quite accurate. However, always remember that it is indeed an approximation.

Angle-Averaged Redistribution Functions We should remember from Chapter 13 [equation (13.2.14)], and the meaning of the redistribution function [see equation (9.2.29)], that the equation of transfer for line radiation can be written as

$$\mu \frac{dI_\nu}{d\tau_\nu} = I_\nu - \mathscr{L}_\nu B_\nu - (1 - \mathscr{L}_\nu) \int_0^\infty \int_{4\pi} I_{\nu'}(\mu')R(\nu', \nu, \mu', \mu)\, d\omega'\, d\nu' \tag{15.3.33}$$

Here the parameter $\mathscr{L}_\nu$ is not to be considered constant with depth as it was for the Milne-Eddington atmosphere. If we assume that the radiation field is nearly isotropic, then we can integrate the equation of radiative transfer over μ and write

$$\frac{dF_\nu}{d\tau_\nu} = 4(J_\nu - \mathscr{L}_\nu B_\nu) - 2(1 - \mathscr{L}_\nu) \int_0^\infty J_{\nu'} \oint_{4\pi} \oint_{4\pi} \langle R(\nu', \nu, \mu', \mu) \rangle\, d\omega'\, d\omega\, d\nu' \tag{15.3.34}$$

If we define the angle-averaged redistribution function as

$$R_A(v', v) \equiv \frac{1}{(4\pi)^2} \oint_{4\pi} \oint_{4\pi} R(v', v, \hat{n}', \hat{n}) \, d\Omega' \, d\Omega \qquad (15.3.35)$$

then in terms of the absorption and reemission probabilities $f(\xi')$ and $p(\xi', \xi)$ it becomes

$$R_A(v', v) \equiv \frac{1}{(4\pi)^2} \int_0^{2\pi} \int_0^{2\pi} \int_{-1}^{+1} \int_{-1}^{+1} f(v' - wu\mu')p(v' - wu\mu', v - wu\mu)$$

$$\times \, g(\hat{n}', \hat{n}) \, d\mu' \, d\mu \, d\phi' \, d\phi \qquad (15.3.36)$$

The phase function $g(\hat{n}', \hat{n})$ must be expressed in the coordinate frame of the observer, that is, in terms of the incoming and outgoing angles that the photon makes with the line of sight (see Figure 15.3).

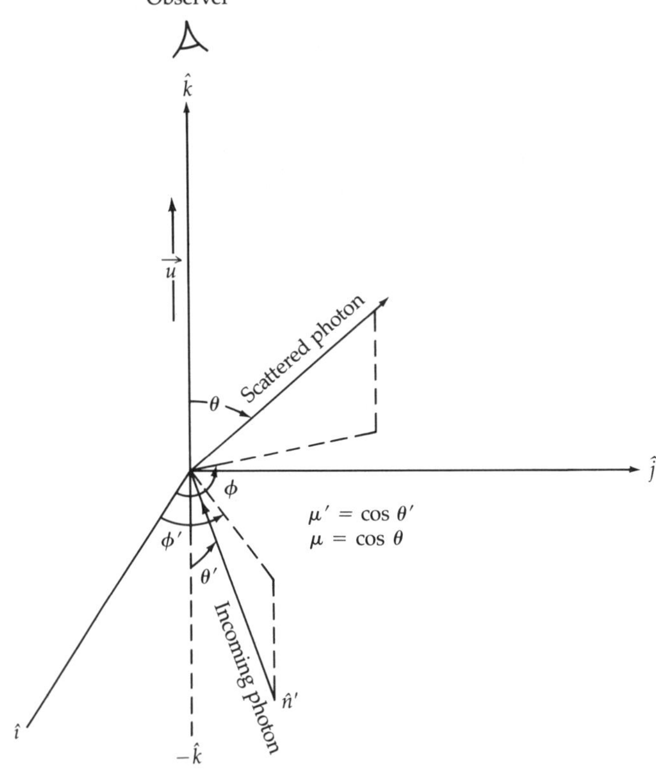

Figure 15.3 Scattering event as seen from the coordinate frame of the observer. The $\hat{k}$ axis points along the normal to the atmosphere or the observer's line of sight. Angle θ is the angle between the scattered photon and the observer's line of sight, while angle θ' is the corresponding angle of the incoming photon. The quantities μ and μ' are just the cosines of these respective angles.

We may write the phase function $g(\hat{n}', \hat{n})$ as

$$g(\mu', \mu) = \frac{1}{4\pi} \int_0^{2\pi} g(\mu', \mu, \phi')\,d\phi' \tag{15.3.37}$$

so that the angle-averaged redistribution function becomes

$$R_A(v', v) = \tfrac{1}{2} \int_{-1}^{+1} f(v' - wu\mu') \int_{-1}^{+1} p(v' - wu\mu', v - wu\mu) g(\mu', \mu)\,d\mu\,d\mu' \tag{15.3.38}$$

The two most common types of phase functions are isotropic scattering and Rayleigh scattering. Although the latter occurs more frequently in nature, the former is used more often because of its simplicity. Evaluating these phase functions in terms of the observer's coordinate frame yields

$$g_{\text{iso}}(\mu', \mu) = \tfrac{1}{2} \qquad g_{\text{Ray}}(\mu', \mu) = \frac{3(3 - \mu^2 - \mu'^2 + 3\mu^2\mu'^2)}{16} \tag{15.3.39}$$

In general, the appropriate procedure for calculating the angle-averaged redistribution functions involves carrying out the integrals in equation (15.3.38) and then applying the effects of Doppler broadening so as to obtain a redistribution function for the four cases described by Hummer. For the first two cases, the delta function representing the upper and lower levels requires that some care be used in the evaluation of the integrals (see Mihalas,[4] pp. 422–433). In terms of the normalized frequency x, the results of all that algebra are, for case I,

$$\langle R_{\text{I},A}(x', x)\rangle = \frac{1}{\sqrt{\pi}} \int_{x_m}^{\infty} e^{-x^2}\,dx = \tfrac{1}{2}\text{erfc}(x_m) \tag{15.3.40}$$

$$x_m \equiv \text{Max}\,(|x|, |x'|)$$

For case II the result is somewhat more complicated where

$$\langle R_{\text{II},A}(x', x)\rangle = \frac{1}{\pi^{3/2}} \int_{|x-x'|/2}^{\infty} e^{-u^2}\left(\text{Tan}^{-1}\frac{\bar{x} + u}{a} - \text{Tan}^{-1}\frac{\bar{x} - u}{a}\right)du \tag{15.3.41}$$

$$\bar{x} \equiv \text{Max}\,(x, x') \qquad \underline{x} \equiv \text{Min}\,(x, x')$$

while for case III it is more complex still;

$$\langle R_{\text{III},A}(x', x)\rangle = \frac{1}{\pi^{5/2}} \int_0^{\infty} e^{-u^2}\left(\text{Tan}^{-1}\frac{x' + u}{a} - \text{Tan}^{-1}\frac{x' - u}{a}\right)$$

$$\times \left(\text{Tan}^{-1}\frac{x + u}{a} - \text{Tan}^{-1}\frac{x - u}{a}\right)du \tag{15.3.42}$$

Note that for all these cases the redistribution function is symmetric in x and x'. From equations (15.2.1) through (15.2.3), it is clear that the angle-averaged redistribution functions will yield a complete redistribution profile in spite of the fact that case I is completely coherent.

To demonstrate the effect introduced by an anisotropic phase function, we give the results for redistribution by electrons. Although we have always considered electron scattering to be fully coherent in the atom's coordinate frame, the effect of Doppler motion can introduce frequency shifts that will broaden a spectral line. This is a negligible effect when we are calculating the flow of radiation in the continuum, but it can introduce significant broadening of spectral lines. If we assume that the scattering function for electrons is isotropic, then the appropriate angle-averaged redistribution function has the form

$$\langle R_{e,A}(x', x)\rangle = \mathrm{ierfc}\left|\frac{x' - x}{2}\right| \qquad \mathrm{ierfc}\,(z) \equiv \pi^{-1/2}e^{-z^2} - z\,\mathrm{erfc}\,(z) \qquad (15.3.43)$$

However, the correct phase function for electron scattering is the Rayleigh phase function given in the observer's coordinate frame by the second of equations (15.3.39). The angle-averaged redistribution function for this case has been computed by Hummer and Mihalas[10] and is

$$\langle R_{e,B}(x', x)\rangle = \frac{(11 + 4\beta^2 + \tfrac{1}{2}\beta^4)e^{-\beta^2/4} - \tfrac{1}{2}\beta\sqrt{\pi}(15 + 5\beta^2 + \tfrac{1}{2}\beta^4)\,\mathrm{erfc}\,(\beta/2)}{10\sqrt{\pi}}$$

$$\beta \equiv \left|(x' - x)/2\right| \qquad\qquad\qquad (15.3.44)$$

Clearly the use of the correct phase function causes a significant increase in the complexity of the angle-averaged redistribution function. Since the angle-averaged redistribution function itself represents an approximation requiring an isotropic radiation field, one cannot help but wonder if the effort is justified.

We must also remember that the entire discussion of the four Hummer cases relied on the absorption and reemission profiles being given by Lorentz profiles in the more complicated cases. While considerable effort has been put into calculating the Voigt functions and functions related to them that arise in the generation of the redistribution functions,[11] some of the most interesting lines in stellar astrophysics are poorly described by Lorentz profiles. Perhaps the most notable example is the lines of hydrogen. At present, there is no quantitative representation of the redistribution function for any of the hydrogen lines. While noncoherent scattering is probably appropriate for the cores of these lines, it most certainly is not for the wings. Since a great deal of astrophysical information rests on matching theoretical line profiles of the Balmer lines to those of stars, greater effort should be made on the correct modeling of these lines, including the appropriate redistribution functions.

The situation is even worse when one tries to estimate the polarization to be expected within a spectral line. It is a common myth in astrophysics that the radiation in a spectral line should be locally unpolarized. Hence, the global observation of spectral lines should show no net polarization. While this is true for simple lines that result only from pure absorption, it is not true for lines that result from resonance scattering. The phase function for a line undergoing resonance scattering is essentially the same as that for electron scattering—the Rayleigh phase function. While noncoherent scattering processes will tend to destroy the polarization information, those parts of the line not subject to complete redistribution will produce strong local polarization. If the source of the radiation does not exhibit symmetry about the line of sight, then the sum of the local net polarization will not average to zero as seen by the observer. Thus there should be a very strong wavelength dependent polarization through such a line which, while difficult to model, has the potential of placing very tight constraints on the nature of the source. Recently McKenna[12] has shown that this polarization, known to exist in the specific intensity profiles of the sun, can be successfully modeled by proper treatment of the redistribution function and a careful analysis of the transfer of polarized radiation. So it is clear that the opportunity is there, remaining to be exploited. The existence of modern computers now makes this feasible.

15.4 Line Blanketing and Its Inclusion in the Construction of Model Stellar Atmospheres

In Chapter 10, we indicated that the presence of myriads of weak spectral lines could add significantly to the total opacity in certain parts of the spectrum and virtually blanket the emerging flux, forcing it to appear in other less opaque regions of the spectrum. This is particularly true for the early-type stars for which the major contribution from these lines occurs in the ultraviolet part of the spectrum, where most of the radiative flux flows from the atmosphere. Although it is not strictly a non-LTE effect, the existence of these lines generally formed high in the atmosphere can result in structural changes to the atmosphere not unlike those of non-LTE. The addition of opacity high up in the atmosphere tends to heat the layers immediately below and is sometimes called *back warming*.

Because of their sheer number, the inclusion of these lines in the calculation of the opacity coefficient poses some significant problems. The simple approach of including sufficient frequency points to represent the presence of all these lines would simply make the computational problem unmanageable with even the largest of computing machines that exist or can be imagined. Since the early attempts of Chandrasekhar,[13] many efforts have been made to include these effects in the modeling of stellar atmospheres. These early efforts incorporated approximating the lines by a series of frequency "pickets." That is, the frequency dependence would be represented by a discontinuous series of opaque regions that alternate with transparent regions. One could then average over larger sections of the spectrum to obtain a mean line opacity for the entire region. However, this did not represent the effect on the photon flow through the alternatingly opaque and relatively transparent regions with any great accuracy. Others tried using harmonic mean line opacities to reduce this problem. Of these attempts, two have survived and are worthy of consideration.

a Opacity Sampling

This conceptually simple method of including line blanketing takes advantage of the extremely large number of spectral lines. The basic approach is to represent the frequency-dependent opacity of all the lines as completely as possible. This requires tabulating a list of all the likely lines and their relative strengths. For an element like iron, this could mean the systematic listing of several million lines. In addition, the line shape for each line must be known. This is usually taken to be a Voigt function for it represents an excellent approximation for the vast majority of weak lines. However, its use requires that some estimate of the appropriate damping constant be obtained for each line. In many cases, the Voigt function has been approximated by the Doppler broadening function on the assumption that the damping wings of the line are relatively unimportant. At any frequency the total line absorption coefficient is simply the sum of the significant contributions of lines to the opacity at that frequency, weighted by the relative abundance of the absorbing species. These abundances are usually obtained by assuming that LTE prevails and so the Saha-Boltzmann ionization-excitation equation can be used.

If one were to pick a very large number of frequencies, this procedure would yield an accurate representation of the effects of metallic line blanketing. However, it would also require prodigious quantities of computing time for modeling the atmosphere. Sneden et al.[14] have shown that sufficient accuracy can be obtained by choosing far fewer frequency points than would be required to represent each line accurately. Although the choice of randomly distributed frequency points which represent large chunks of the frequency domain means that the opacity will be seriously overestimated in some regions and underestimated in others, it is possible to obtain accurate structural results for the atmosphere if a large enough sample of frequency points is chosen. This sample need not be anywhere near as large as that required to represent the individual lines, for what is important for the structure is only the net flow of photons. Thus, if the frequency sampling is sufficiently large to describe the photon flow over reasonably large parts of the spectrum, the resulting structure and the contribution of millions of lines will be accurately represented. This procedure will begin to fail in the higher regions of the atmosphere where the lines become very sharp and non-LTE effects become increasingly important. In practice, this procedure may require the use of several thousand frequency points whereas the correct representation of several million spectral lines would require tens of millions of frequency points. For this reason (and others), this approach has been extremely successful as applied to the structure of late-type model atmospheres where the opacity is dominated by the literally millions of bound-bound transitions occurring in molecules. The larger the number of weak lines and the more uniform their distribution, the more accurate this procedure becomes. However, the longer the lists of spectral lines, the more computer time will be required to carry out the calculation. This entire procedure is generally known as *opacity sampling*, and it possesses a great degree of flexibility in that all aspects of the stellar model that may affect the line broadening can be included *ab initio* for each model. This is not the case with the competing approach to line blanketing.

b Opacity Distribution Functions

This approach to describing the absorption by large numbers of lines also involves a form of statistical sampling. However, here the statistical representation is carried out over even larger regions of the spectra than was the case for the opacity sampling scheme. This approach has its origins in the mean opacity concept alluded to earlier. However, instead of replacing the complicated variation of the line opacity over some region of the spectrum with its mean, consider the fraction of the spectral range that has a line opacity less than or equal to some given value. For small intervals of the range, this may be a fairly large number since small intervals correspond to the presence of line cores. If one considers larger fractions of the interval, the total opacity per unit frequency interval of this larger region will decrease, because the spaces between the lines will be included. Thus, an opacity distribution function represents the probability that a randomly chosen point in the interval will have an opacity less than or equal to the given value (see Figure 15.4). The proper name for this function should be the *inverse cumulative opacity probability distribution function* but in astronomy it is usually referred to as just the *opacity distribution function* (ODF). Carpenter[15] gives a very complete description of the details of computing these functions, while a somewhat less complete picture is given by Kurucz and Petrymann[16] and by Mihalas[4] (pp. 167–169).

The ODF gives the probability that the opacity is a particular fraction of a known value for any range of the frequency interval, and the ODF may be obtained from a graph that is fairly simple to characterize by simple functions. This approach allows the contribution to the total opacity due to spectral lines appropriate for that range of the interval to be calculated. Unfortunately, the magnitude of that given value will depend on the chemical composition and the details of the individual line-broadening mechanisms. Thus, any change in the chemical composition, turbulent broadening, etc., will require a recalculation of the ODF. In addition,

427

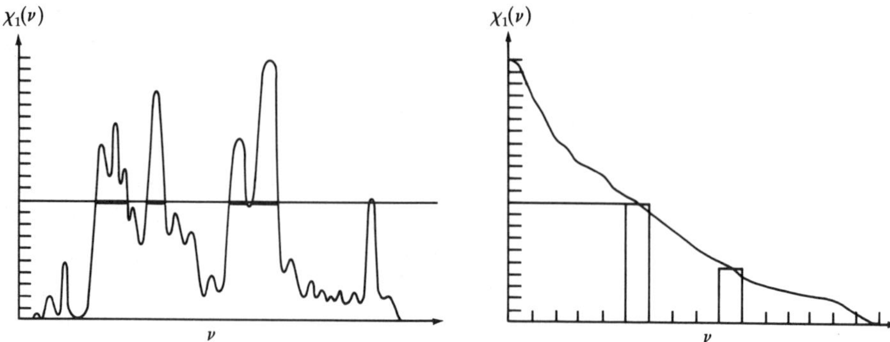

Figure 15.4 Opacity of a region of the spectrum represented in terms of (*a*) the actual line opacity and (*b*) the opacity distribution function.

ODFs must be calculated as a function of temperature and pressure (or alternatively, electron density), and so their tabular representation can be extremely large. Their calculation also represents a significant computational effort. However, once ODFs exist, their inclusion in a stellar atmosphere code is fairly simple and the additional computational load for the construction of a model atmosphere is not great, particularly compared to the opacity sampling technique. This constitutes the primary advantage of this approach for the generation of model stellar atmospheres. For stars where the abundances and kinematics of the atmospheres are well known, ODFs provide by far the most efficient means of including the effects of line blanketing. This will become increasingly true as the number of spectral lines for which atomic parameters are known grows—although the task of calculating the opacity distribution functions will also increase.

Considerations such as these will enable the investigator to include the effects of line blanketing and thereby to create reasonably accurate models of the stellar atmosphere which will represent the structure correctly through the line-forming region of a normal star. These, when combined with the model interiors discussed in the first six chapters of the book, will allow for the description of normal stars from the center to the surface. While this was the goal of the book, we cannot resist the temptation to demonstrate to the conservative student that the concepts developed so far will allow the models to be extended into the region above stars and to determine some properties of the stellar radiation field that go beyond what is usually considered to be part of the normal stellar model. So in the last chapter we consider a few extensions of the ideas that have already been developed.

Problems

1. Estimate the ratio of collisional ionization to photoionization for hydrogen from the ground state, and compare it to the ratio from the second level. Assume the pressure is 300 bars. Obtain the physical constants you may need from the literature, but give the appropriate references

2. Calculate the Doppler-broadened angle-averaged redistribution function for Hummer's case I, but assuming a Rayleigh phase function [i.e., find $\langle R(x, x')\rangle_{I,B}$], and compare it to $\langle R(x, x')\rangle_{I,A}$ and the result for electron scattering.

3. Show that

$$J(\tau_x) = \tfrac{1}{2} \int_0^\infty S_\ell(t) E_1 \left| \int_{\tau_x}^t \phi_x(t') \, dt' \right| \phi_x \, dt$$

is indeed a solution to

$$\mu \frac{dI_x}{d\tau_x} = \phi_x(I_x - S_\ell)$$

and obtain an integral equation for S_ℓ.

4. Describe the mechanisms which determine the $Ly\alpha$ profile in the sun. Be specific about the relative importance of these mechanisms and the parts of the profile that they affect.

5. Given a line profile of the form

$$\phi_x(\tau_x) = \begin{cases} 0 & \text{for } |x| > x_0 \\ 1 & \text{for } |x| \le x_0 \end{cases}$$

find S_ℓ. Assume complete redistribution of the line radiation. State what further assumptions you may need; indicate your method of solution and your reasons for choosing it.

6. Show explicitly how equation (15.2.21) is obtained.

7. Show how equation (15.3.24) is implied by equation (15.3.15).

8. How does equation (15.3.27) follow from equation (15.3.26).

9. Derive equations (15.3.28) and (15.3.29).

10. Use equation (15.3.30) to obtain the angle-averaged form of $\langle R_{IV,A}(x', x)\rangle$.

11. Show explicitly how equations (15.3.40) and (15.3.41) are obtained.

References and Supplemental Reading

1. Woolley, R. v. d. R., and D. W. N. Stibbs, *The Outer Layers of a Star*, Oxford University Press, London, 1953, p. 152.

2. Böhm, K. H., "Basic Theory of Line Formation," *Stellar Atmospheres* (Ed.: J. Greenstein), *Stars and Stellar Systems: Compendium of Astronomy and Astrophysics*, vol. 6, University of Chicago Press, Chicago, 1960, pp. 88–155.

3. Mihalas, D., *Stellar Atmospheres*, W. H. Freeman, San Francisco, 1970, pp. 337–378.

4. Mihalas, D., *Stellar Atmospheres*, 2d ed., W. H. Freeman, San Francisco, 1978, p. 138.

5. Henyey, L., " Near Thermodynamic Radiative Equilibrium," *Ap. J.* 103, 1946, pp. 332–350.

6. Hummer, D. G., " Non-Coherent Scattering," *Mon. Not. R. astr. Soc.* 125, 1962, pp. 21–37.

7. Omont, A., E. R. Smith, and J. Cooper, " Redistribution of Resonance Radiation, I: The Effect of Collisions," *Ap. J.* 175, 1972, pp. 185–199.

8. McKenna, S., "A Reinvestigation of Redistribution Functions R_{III} and R_{IV} ," *Ap. J.* 175, 1980, pp. 283–293.

9. McKenna, S., " The Transfer of Polarized Radiation in Spectral Lines: Formalism and Solutions in Simple Cases," *Astrophys. & Sp. Sci.* 108, 1985, pp. 31–66.

10. Hummer, D. G., and D. Mihalas, "Line Formation with Non-Coherent Electron Scattering in O and B Stars," *Ap. J. Lett.* 150, 1967, pp. 57–59.

11. McKenna, S., "A Method of Computing the Complex Probability Function and Other Related Functions over the Whole Complex Plane," *Astrophys. & Sp. Sci.* 107, 1984, pp. 71–83.

12. McKenna, S., "The Transfer of Polarized Radiation in Spectral Lines: Solar-Type Stellar Atmospheres," *Astrophys. & Sp. Sci.* 106, 1984, pp. 283–297.

13. Chandrasekhar, S., "The Radiative Equilibrium of the Outer Layers of a Star with Special Reference to the Blanketing Effect of the Reversing Layer," *Mon. Not. R. astr. Soc.* 96, 1936, pp. 21–42.

14. Sneden, C., H. R. Johnson, and B. M. Krupp, "A Statistical Method for Treating Molecular Line Opacities," *Ap. J.* 204, 1976, pp. 281–289.

15. Carpenter, K. G., "A Study of Magnetic, Line-Blanketed Model Atmospheres," a doctoral dissertation, Ohio State University, Columbus, 1983.

16. Kurucz, R., and E. Petrymann, "A Table of Semiempirical *gf* Values, Part 3," Smithsonian Astrophysical Observatory Special Report 362, 1975.

Although they have been cited frequently, the serious student of departures from LTE should read both of these:

• Mihalas, D.: *Stellar Atmospheres*, W. H. Freeman, San Francisco, 1970, chaps. 7–10, 12, 13.

• Mihalas, D.: *Stellar Atmospheres*, 2d ed., W. H. Freeman, San Francisco, 1978, chaps. 11–13.

A somewhat different perspective on the two-level and multilevel atom can be found in

• Jefferies, J. T.: *Spectral Line Formation*, Blaisdell, New York, 1968, chaps. 7, 8.

Although the reference is somewhat old, the physical content is such that I would still recommend reading the entire chapter:

• Böhm, K.-H.: "Basic Theory of Line Formation," *Stellar Atmospheres* (Ed.: J. Greenstein), *Stars and Stellar Systems: Compendium of Astronomy and Astrophysics*, vol. 6, University of Chicago Press, Chicago, 1960, chap. 3.

16

Beyond the Normal Stellar Atmosphere

. . .

With the exception of a few subjects in Chapter 7, we have dealt exclusively with spherical stars throughout this book. In addition, these stars have been "normal" stars; that is, they are stars without strange peculiarities or large time-dependent effects in their spectra. These are stars, for the most part, with interiors evolving on a nuclear time scale and atmospheres which can largely be said to be in LTE. The prototypical star that fits this description is a main sequence star, and for such stars the theoretical framework laid down in this book will serve quite well as a basis for understanding their structure and spectra. However, the notion of a spectral peculiarity is somewhat "in the eye of the beholder". Someone who looks closely at any star will find grounds for calling it peculiar or atypical. This is true of even the main sequence stars, and we would delude ourselves if we believed that all aspects of any star are completely understood.

Phenomena are known to exist in the atmospheres of stars that we have not included in the basic theory. For example, convection does exist in the upper layers of late-type stars and contributes to the transport of energy. However, the efficiency of this transport is vastly inferior to that which occurs in the stellar interior, because the mean free path for a convective element is of the order of that for a photon. Nevertheless, in the lower sections of the atmosphere, convection may carry a significant fraction of the total energy. Unfortunately, the crude mixing-length theory of convective transport that was effective for the stellar interior is insufficient for an accurate description of atmospheric convection.

The turbulent motion exhibited by the atmospheres of many giant stars also represents the transport of energy and momentum in a manner similar to convection. Just the kinetic energy of the turbulent elements themselves should contribute significantly to the pressure equilibrium in the upper atmosphere. However, if one naively assigned a turbulent pressure of ρv^2 to the elements, the measured turbulent velocities would imply a pressure greater than the total pressure of the gas and the atmosphere would be unstable. Clearly a better theory of turbulence is required before any significant progress can be made in this area.

In the last 20 years, it has become clear that the atmospheres of many stars are indeed unstable. Ultraviolet observations of the spectra of early-type stars suggest that they all possess a stellar wind of outflowing matter of significant proportions. As one moves upward in the atmosphere of a star, departures from LTE increase to such a point that they dominate the atmospheric structure. In addition, energy sources such as magnetic fields, atmospheric mass motions, and possibly acoustic waves associated with convection or turbulence may contribute significantly to the higher-level structure. While these energy contributions are tiny compared to the energy densities in the photosphere, the contributions may supply the majority of the interaction energy between the star and its outermost layers which are basically transparent to photons.

In the hottest stars, the pressure of radiation, transmitted to the rarefied upper layers of the atmosphere through absorption by the resonance lines of metals, will cause this low-density region to become unstable, expand, and eventually be driven away. This is undoubtedly the source of the stellar winds for stars on the upper main sequence and in the early-type giants and supergiants. The extended atmospheres of these stars pose an entirely new set of problems for the theory of stellar atmospheres.

Many stars exist in close binary systems so that the atmosphere of each is exposed to the radiation from the other. In some instances, the illuminating radiant flux from the companion may exceed that of the star. Under these conditions we can expect the local atmospheric structure to be dominated by the external source of radiation. In addition, such stars are liable to be severely distorted from the spherical shape of a normal star. This implies that the defining parameters of the stellar atmosphere (μ, g, and T_e) will vary across the surface. Even in the absence of a companion, rapid rotation of the entire star can provide a similar distortion with the accompanying variation of atmospheric parameters over the surface.

These are some of the areas of contemporary research, and to even recount the present state of progress in them would require another book. Instead of skimming them all, in this last chapter, we consider one or two in the hopes of pricking the reader's interest to find out more and perhaps contribute to the solution of some of these problems.

16.1 Illuminated Stellar Atmospheres

Although the majority of stars exist in binary systems, and many of them are close binary systems, little attention has been paid to the effects on a stellar atmosphere by the illuminating radiation of the companion. The general impact of this illumination on the light curve of an eclipsing binary system has usually been lumped under the generic term *"reflection effect."* This is a complete misnomer, for most of the incident energy is absorbed by the atmosphere and then reradiated. Only that fraction which is scattered could, in any sense, be considered to be reflected, and then only that fraction that is scattered in the direction of the observer.

a   Effects of Incident Radiation on the Atmospheric Structure

The presence of incident radiation introduces an entirely new set of parameters into the problem of modeling a stellar atmosphere. Basically the incident radiation will be absorbed in the atmosphere, causing local heating. The amount of this heating will depend on the intensity, direction, and frequency of the incident radiation as well as the fraction of the visible sky covered by the source. This local heating can totally alter the atmospheric structure, causing the appearance of the illuminated star to change greatly. However, if radiative equilibrium is to apply, all the radiation that falls on the star must eventually emerge. The interaction of the incident radiation with the atmosphere will cause its spectral energy distribution to be significantly altered. In any event, the total emergent flux must simply be the sum of the incident flux and that which would be present in the unilluminated star. In addition to this being the commonsense result, it is a consequence of the linearity of the equation of radiative transfer.

Before we can even formulate an equation of radiative transfer appropriate for an illuminated atmosphere, we must know the angular distribution of the incident energy. For simplicity, let us assume that the incident radiation comes from some known direction (θ_0, ϕ_0) in the form of a plane wave (see Figure 16.1).

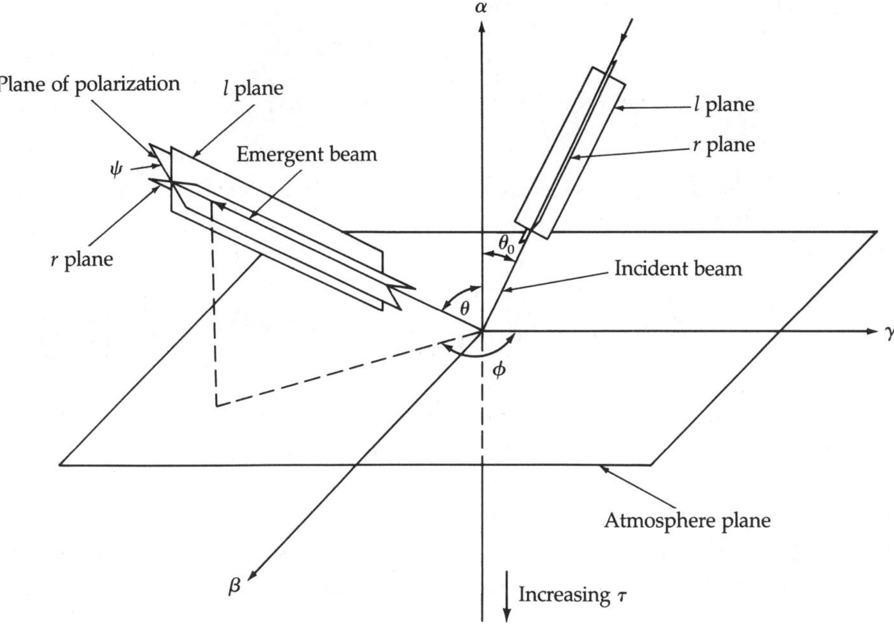

Figure 16.1 Radiation incident on a plane-parallel atmosphere. The radiation can be specified as coming from a specific direction, indicated by the coordinates (θ_0, ϕ_0).

The specific intensity incident on the atmosphere can then be characterized by

$$I_{i,0}(\mu, \phi) = 4\pi F_i\, \delta(\phi - \phi_0)\, \delta(\mu - \mu_0) \qquad (16.1.1)$$

where F_i is the incident flux and the $\delta(x)$ is the Dirac delta function indicating the direction of the beam. Along an optical path τ_v/μ_0, this radiation will be attenuated by $e^{-\tau_v/\mu_0}$. Chandrasekhar[1] makes a distinction between the direct transmitted intensity and that part of the radiation field (the diffuse field) that has been scattered at least once. The contribution to the diffuse radiation field from the attenuated incident radiation field is

$$S_i = (1 - \epsilon_v)\left(\frac{F_i}{4}\right) e^{-\tau_v/\mu_0} \qquad (16.1.2)$$

where ϵ_v is the fraction of the intensity absorbed in a differential length of the optical path and has the same meaning as in equation (10.1.8). This can be regarded as the scattering contribution to the diffuse source function from the attenuated incident radiation field. Thus the equation of radiative transfer for the illuminated

435

atmosphere is

$$\mu \frac{dI_v}{d\tau_v} = I_v - \epsilon_v B_v - (1 - \epsilon_v)J_v - (1 - \epsilon_v)\left(\frac{F_i}{4}\right)e^{-\tau_v/\mu_0} \qquad (16.1.3)$$

By applying the classical solution to the equation of transfer, we can generate an integral equation for the diffuse field source function, as we did in Section 10.1, so that

$$S_v(\tau_v) = \epsilon_v B_v[T(\tau_v)] + \frac{1}{2}(1 - \epsilon_v)\int_0^\infty S_v(t)E_1|t - t_v|\,dt - (1 - \epsilon_v)\left(\frac{F_i}{4}\right)e^{-\tau_v/\mu_0}$$

$$(16.1.4)$$

This is a Schwarzschild-Milne integral equation of the same type as we considered in Chapter 10, and it can be solved in the same manner. The only difference is that the term that makes the equation inhomogeneous has been augmented by the last term on the right in equation (16.1.4). Thus the presence of incident radiation will even make a pure scattering atmosphere inhomogeneous and subject to a unique solution.

Modification of the Avrett-Krook Iteration Scheme for Incident Radiation
Buerger[2] has solved this equation, using the ATLAS atmosphere program for a variety of cases. He found that the standard Avrett-Krook iteration scheme described in Section 11.4 would no longer lead to a converged atmosphere. This is not surprising since the Eddington approximation was used to obtain the specific perturbation formulas [see equations (12.4.17), (12.4.25), and (12.4.29)], and the approximation $J(0) = \frac{1}{2}F(0)$ will no longer be correct. Buerger therefore adopted a modification of the Avrett-Krook scheme due to Karp[3] which replaces the ad hoc assumption of equations (12.4.14) and (12.4.20) with

$$J_v^{(1)'}(t) = aF_v^{(1)'}(t) \qquad (16.1.5)$$

which implies that

$$J_v^{(1)}(t) = aF_v^{(1)}(t) \qquad (16.1.6)$$

He continues by making the plausible assumption that perturbed flux variations have the same frequency dependence as the zeroth-order flux variations, so that

$$F_v^{(1)} = \frac{F^{(1)}F_v^{(0)}}{F^{(0)}} \qquad (16.1.7)$$

Following the same procedure as Karp, Buerger[2] finds the perturbation formulas for $\tau^{(1)}$ and $T^{(1)}$ analogous to equations (12.4.21) and (12.4.29), respectively, to be

$$\tau^{(1)'} + \tau^{(1)} \int_0^\infty \frac{k_\nu^{(0)'}}{k_\nu^{(0)}} \frac{F_\nu^{(0)}}{F^{(0)}} \, d\nu = 1 - \frac{\mathbf{F}}{F^{(0)}} - \frac{a F^{(0)'}}{\int_0^\infty k_\nu^{(0)} F_\nu^{(0)} \, d\nu}$$

$$
\begin{aligned}
T^{(1)}(t) \int_0^\infty k_\nu^{(0)}(\tau)\epsilon_\nu^{(0)}(t)\dot{B}_\nu^{(0)} \, d\nu &= -a\left[1 - \frac{\mathbf{F}}{F^{(0)}}\right] \int_0^\infty k_\nu^{(0)}(t)\epsilon_\nu^{(0)}(t)F_\nu^{(0)} \, d\nu \\
&+ \int_0^\infty k_\nu^{(0)}(t)\epsilon_\nu^{(0)}(t)[J_\nu^{(0)}(t) - B_\nu^{(0)}(t) + \tfrac{1}{4}F_\nu^{(i)}e^{-\tau\nu/\mu_0}] \, d\nu \\
&+ \tau^{(1)} \int_0^\infty [k_\nu^{(0)}(t)\epsilon_\nu^{(0)}(t)]'[J_\nu^{(0)}(t) - B_\nu^{(0)}(t)] + \{k_\nu^{(0)}(t)[1 - \epsilon_\nu^{(0)}(t)]\}' \\
&\times \tfrac{1}{4}F_\nu^{(i)}e^{-\tau\nu/\mu_0} \, d\nu + \tau^{(1)'} \int_0^\infty k_\nu^{(0)}(t)\epsilon_\nu^{(0)}(t)[J_\nu^{(0)}(t) - B_\nu^{(0)}(t) + \tfrac{1}{4}F_\nu^{(i)}e^{-\tau\nu/\mu_0}] \, d\nu
\end{aligned}
$$

$$\tag{16.1.8}$$

where

$$\mathbf{F} = \frac{\sigma T_e^4}{\pi} + \mu_0 \int_0^\infty F_\nu^{(i)}e^{-\tau\nu/\mu_0} \, d\nu \tag{16.1.9}$$

While these equations do provide a convergent iteration algorithm, the rate of convergence can be quite slow if the amount of incident radiation is large and differs greatly from that which the illuminated star would have in the absence of incident radiation.

If the incident flux has an energy distribution significantly different from that of the illuminated star and large compared to the stellar flux, the Eddington approximation will fail rather badly at some frequencies and in that part of the atmosphere where the majority of the incident flux is absorbed. Under these conditions, the Avrett-Krook iteration scheme will fail badly. Steps must be taken to directly estimate those parameters obtained from the Eddington approximation. In some instances it is possible to use results from a previous iteration.

Initial Temperature Distribution   As with any iteration process, the closer the initial guess, the faster a correct answer will be found. In Chapter 12, we suggested that the model-making process could be initiated with the unilluminated gray atmosphere temperature distribution in the absence of anything better. In the case of incident radiation, such a choice will yield an initial temperature distribution that departs rather badly from the correct one. It is tempting to suggest that a suitable first guess could be obtained by merely scaling up the gray atmosphere temperature distribution so as to include the incident flux in the definition of the effective temperature. Thus,

$$\frac{\sigma T_e^4}{\pi} = \frac{\sigma T_e^4(0)}{\pi} + F_i \qquad (16.1.10)$$

where $T_e(0)$ is the effective temperature and the temperature distribution is

$$T^4(\tau) = \tfrac{3}{4}T_e^4[\tau + q(\tau)] \qquad (16.1.11)$$

Here $q(\tau)$ is the Hopf function defined in Chapter 10 [equation (10.2.33)]. For the Eddington approximation,

$$q(\tau) = Q_0 = \tfrac{2}{3} \qquad (16.1.12)$$

However, there are two effects that make this a poor choice. First, if the sky of the illuminated star is partially filled with the illuminating star, then much of the incident radiation will enter at a grazing angle and preferentially heat the upper atmosphere. This will raise the surface temperature over what would normally be expected from a gray atmosphere and will flatten the temperature gradient. The more radiation that enters from grazing angles, the greater this effect will be. This effect may also be understood by considering how radiation can ultimately escape from the atmosphere. If the source of radiation is a close companion, filling much of the sky, then a smaller and smaller fraction of the sky is "black," representing directions of possible escape. As the possible escape routes for photons diminish, the atmosphere will heat up since the photons have nowhere to go. Anderson and Shu[4] suggest that this may be compensated for by

$$Q_0 \approx \frac{2}{3}\left[\frac{1 + \Omega_0/(2\pi)}{1 - \Omega_0/(2\pi)}\right] \qquad (16.1.13)$$

where Ω_0 is the solid angle subtended by the source of radiation in the sky. As $\Omega_0 \to 0$, we recover the Eddington approximation's value for Q_0. As Ω_0 increases, so does the value of Q_0, implying that the surface temperature will rise relative to the effective temperature. In the limit, the atmosphere will approach being isothermal.

The second aspect of incident radiation that can markedly affect the temperature distribution is the frequency distribution. If the source of the incident radiation contains a large quantity of ionizing radiation and is incident on a relatively cool star, then almost all the radiation will be absorbed by the neutral hydrogen of the upper reaches of the atmosphere, raising the local temperature significantly. Similarly, if the incident radiation is highly penetrating radiation, such as x-rays, the heating may occur much lower in the atmosphere. Since this energy must be reradiated, this will result in a general elevation of the entire atmospheric temperature distribution. This dependence of the atmosphere's temperature gradient on the frequency distribution makes predicting the outcome extremely difficult since the process is highly nonlinear. That is, the local heating may change the opacity, which will in turn change the heating.

In general, incident radiation will preferentially heat the surface layers, and a modification like that suggested by Anderson and Shu[4] should be used for the initial guess. An additional approach that some investigators have found useful is to slowly increase the amount of incident radiation during the iteration processes from an initially low value until the desired level is reached. The rate at which this can be done depends on the spectral energy distribution of the incident radiation and always tends to lengthen the convergence process.

b Effects of Incident Radiation on the Stellar Spectra

If the amount of this illuminating radiation is large, then the heating of the upper atmosphere will cause a decrease in the temperature gradient. As we have seen in Chapter 13, the strength of a pure absorption line will depend strongly on the temperature gradient in the atmosphere. Thus, aside from changes in the environment of the radiation field caused by the external heating, we should expect spectral lines to appear weaker simply because of the change in the source function gradient. The change in line strength may cause an attendant change in the spectral type. Although such changes should be phase-dependent in a binary system, as different aspects of the illuminated surface are presented to the observer, such changes can be confused with the contribution to the light by the companion.

Although Milne[5] first laid the foundations for estimating these effects in 1930, very little has been done in the intervening half-century to quantify them. The atmospheres investigated by Buerger[2] did indeed show the general decrease in the equivalent widths of the lines investigated due to heating of the higher levels of the atmosphere. However, his investigation was restricted to point-source illumination which tends to minimize the upper-level heating of the atmosphere. A more realistic modeling of the atmosphere of an early-type binary system by Kuzma[6] showed that the effects can be quite pronounced. Local changes of the atmospheric structure can produce spectral changes of several spectral subtypes with the integrated effect being as large as several tenths of a subtype with phase.

The effects of the incident radiation on the spectra become deeply intertwined with the structure. The inclusion of metallic line blanketing brought about a reversal of the effects of non-LTE on the hydrogen lines in that, in the upper levels of the atmosphere, the departure coefficients for the first two levels of hydrogen (b_1 and b_2) went from about 0.6 to 1.5 when the effects of line blanketing were included. Similar anomalies occurred for the departure coefficients for the higher levels as well. This fact largely serves to show the interconnectedness of the phenomena which earlier we were able to consider separately. This results primarily from the dominance of the interactions of the gas with the radiation field over interactions with itself. That is, the radiation field and its departures from statistical equilibrium begin to determine completely the statistical equilibrium of the gas. If the illuminating radiation has a spectral energy distribution significantly different from that of the illuminating star, the departures from LTE may be very large indeed.

16.2   Transfer of Polarized Radiation

Even before the advent of classical electromagnetism in 1864 by James Clerk Maxwell, it was generally accepted that light propagated by means of a transverse wave. Any such wave would have a preferred plane of vibration as seen by an observer, and any collection of such beams that exhibited such a preferred plane was said to be polarized. In 1852, George Stokes[7] showed that such a collection of light beams could be characterized by four parameters that now bear his name. Normally a heterogeneous collection of light beams will have planes of oscillation that are randomly oriented as seen by an observer and will be called unpolarized. However, should a light beam exhibit some degree of polarization, a measure of that polarization potentially contains a significant amount of information about the nature of the source.

This property of light has been largely ignored in stellar astrophysics until relatively recently. The probable reason can be found in the assumption that all stars are spherical. Any such object would be completely symmetric about the line of sight from the observer to the center of the apparent disk so if even local regions on the surface emitted 100 percent polarized light, the symmetric orientation of their planes of polarization would yield no net polarization to an observer viewing the integrated light of the disk. However, many astrophysical processes produce polarized light, and numerous stars are not spherical. In particular, eclipsing binary systems present a source of integrated light which is not symmetric about the line of sight. Early work by Chandrasekhar[8,9] and Code[10] implied that there might be measurable polarization that would place significant constraints on the nature of the source. However, only after the development of very sensitive detectors and the discovery of intrinsic circular polarization in white dwarfs did intrinsic stellar polarization attract any great interest. In the hope that this interest will continue to grow, we discuss some of the theoretical approaches to the problems posed by the transfer of polarized light.

a   Representation of a Beam of Polarized Light and the Stokes Parameters

Imagine a collection of largely uncorrelated electromagnetic waves propagating in the same direction through space. The combined effect of these beams can be obtained by adding the squares of the amplitudes of their oscillating electric and magnetic vectors. If the waves were completely correlated, as are those from a laser, the amplitudes themselves would vectorially add directly. Indeed, the representation of the addition of the squares of the amplitudes can be taken as a definition of a completely uncorrelated set of waves. Now consider a situation where the collection of waves interacts with some object that produces a systematic effect on the waves such as a phase shift that depends on the orientation of the wave to the object. A reflection from a mirror produces such an effect. If we introduce a coordinate system to represent this interaction (see Figure 16.2), the net effect of summing components of the amplitudes of the combined waves could produce an ellipse whose orientation changes as it propagates through space. Such a beam is said to be *elliptically polarized*. If the tip of the combined electric vectors appears to rotate counterclockwise as viewed by the observer, the beam is said to have

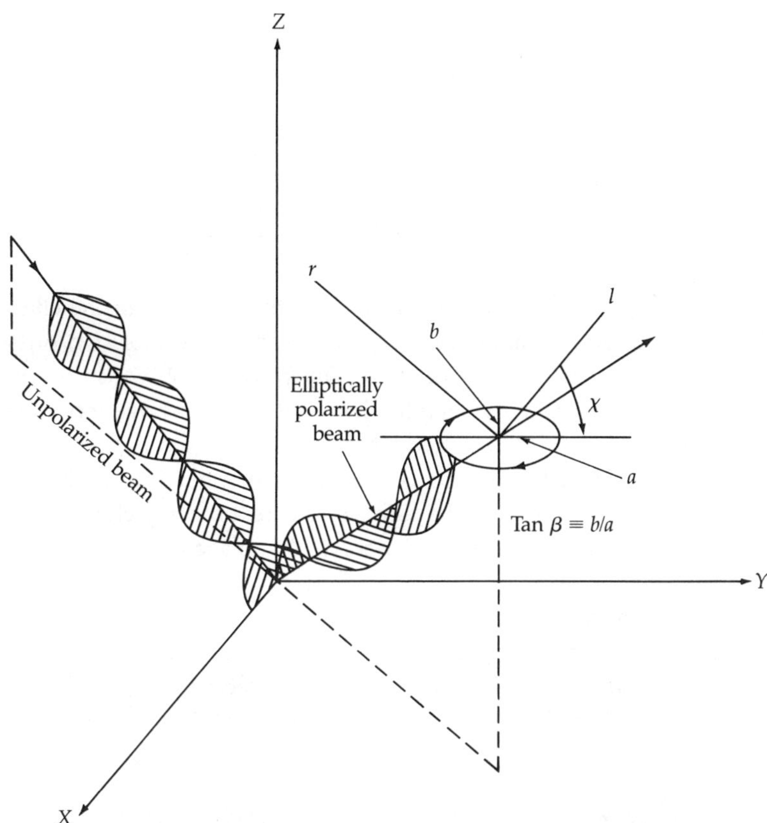

Figure 16.2 Two beams of light incident on a reflecting surface where a phase shift is introduced between those with electric vectors vibrating parallel to the surface and those vibrating normal to the surface. The combined effect on the two beams is to have initially unpolarized light converted to an elliptically polarized beam.

positive helicity or to be *left circularly polarized*. However, some authors define the state of positive polarization to be the reverse of this convention, and students should be careful, when reading polarization literature, which definition is being employed.

In 1852, George Stokes[7] showed that such a beam of light could be represented by four quantities. Although we now use a somewhat different notation than that of Stokes, the meaning of those parameters is the same. Basically, the four parameters required to describe a polarized beam are the total intensity; the difference between two orthogonal components of the intensity, which is related to the degree of polarization; the orientation of the ellipse, which specifies the plane of polarization; and the degree of ellipticity, which is related to the magnitude of the systematic phase shifts introduced by the interaction that gives rise to the polarized beam. Since these quantities presuppose the existence of the coordinate system for their definition, they can be defined most easily in terms of those coordinates (see Figure 16.2) as

$$I \equiv I_l + I_r = E_l^2 + E_r^2$$

$$Q \equiv I_l - I_r = E_l^2 - E_r^2$$

$$U \equiv (I_l - I_r)\operatorname{Tan} 2\chi = E_l E_r \cos \epsilon$$

$$V \equiv (I_l - I_r)\operatorname{Sec} 2\chi \operatorname{Tan} 2\beta = E_l E_r \sin \epsilon$$

$$(16.2.1)$$

where $\tan \beta$ is the ratio of the semiminor to the semimajor axis of the ellipse. The quantities E_l and E_r are the amplitudes of the electric vectors of two orthogonal waves that are phase-shifted by ϵ. The vector superposition of two such waves will produce a single wave with an amplitude that rotates as the wave moves through space. We will take $0 \leq |\beta| \leq \pi/2$, and whether the sign of β is positive or negative will determine if the beam is to be called left or right elliptically polarized, respectively. If $Q = U = V = 0$, the light is said to be completely unpolarized, while if $V = 0$, the beam is linearly polarized with a plane of polarization making an angle χ with the l axis.

Since the Stokes parameters contain some information which is purely geometric, we should not be surprised if, for a completely polarized beam, the parameters are not linearly independent and that relationships exist between them involving their geometric representation. Specifically, from equations (16.2.1) we get

$$I^2 = Q^2 + U^2 + V^2 \qquad \tan 2\chi = \frac{U}{Q} \qquad \sin 2\beta = \frac{V}{(Q^2 + U^2 + V^2)^{1/2}} \qquad (16.2.2)$$

Since the Stokes parameters already involve combining the squares of the amplitudes of an arbitrary beam, we should be able to add a natural or unpolarized beam directly to the completely polarized beam. For such a beam $I_l = I_r$ in all coordinate systems and $Q = U = V = 0$, so that

$$I^2 \geq Q^2 + U^2 + V^2 \tag{16.2.3}$$

Thus for a collection of light beams, some of which may be polarized, all four Stokes parameters are indeed linearly independent and must be specified to completely characterize the beam.

It is clear from the definitions of the Stokes parameters that I and V are invariant to a coordinate rotation about the axis of propagation while Q and U are not. If we rotate the lr coordinate frame through an angle ϕ in a *counterclockwise* direction (see Figure 16.2), then

$$Q' = Q\cos 2\phi - U\sin 2\phi = I\cos 2\beta \cos 2(\chi + \phi)$$
$$U' = Q\sin 2\phi + U\cos 2\phi = I\cos 2\beta \sin 2(\chi + \phi) \tag{16.2.4}$$

Since I and V are invariant to rotational transformations, we can write the general rotation for the Stokes parameters in matrix-vector notation. If we define an intensity vector to have components $\vec{I}[I, Q, U, V]$ so that

$$\begin{pmatrix} I \\ Q \\ U \\ V \end{pmatrix} = \begin{pmatrix} 1 & 0 & 0 & 0 \\ 0 & \cos 2\phi & -\sin 2\phi & 0 \\ 0 & \sin 2\phi & \cos 2\phi & 0 \\ 0 & 0 & 0 & 1 \end{pmatrix} \begin{pmatrix} I' \\ Q' \\ U' \\ V' \end{pmatrix} \tag{16.2.5}$$

then we can call the matrix the *rotation matrix* for the Stokes parameters. If we choose a coordinate system so that I_l is a maximum, then the degree of linear polarization P is a useful parameter defined by

$$P \equiv \left(\frac{Q}{I} \right)_{\text{max}} \tag{16.2.6}$$

where, since the lr coordinate frame has been chosen so that the l axis lies along this direction of maximum intensity, $U = 0$.

It is also useful to linearly decompose the Stokes parameters I and Q into I_l and I_r so that

$$I_l = \tfrac{1}{2}(I + Q) \qquad I_r = \tfrac{1}{2}(I - Q) \tag{16.2.7}$$

However, neither of these parameters is invariant to a coordinate rotation, so that if we denote a vector $\vec{I}(I_l, I_r, U, V)$ whose components are this different set of Stokes

443

parameters, then

$$\vec{I} = \mathbf{L}(\phi)\vec{I}' \tag{16.2.8}$$

where the rotation matrix for these Stokes parameters is

$$\mathbf{L}(\phi) = \begin{pmatrix} \cos^2\phi & \sin^2\phi & -\frac{1}{2}\sin 2\phi & 0 \\ \sin^2\phi & \cos^2\phi & \frac{1}{2}\sin 2\phi & 0 \\ \sin 2\phi & -\sin 2\phi & \cos 2\phi & 0 \\ 0 & 0 & 0 & 1 \end{pmatrix} \tag{16.2.9}$$

The rotation matrix $\mathbf{L}(\phi)$ can be obtained from equation (16.2.5) by employing the definitions of I and Q in terms of I_l and I_r in equation (16.2.7). Here the angle ϕ is taken as positively increasing for *counterclockwise* rotations. While this definition is the opposite of that used by Chandrasekhar[1] (p. 26), it is the more common notation for a rotational angle. The form of this rotation matrix is a little different from that used to transform vectors because the quantities being transformed are the squares of vector components rather than the components and their composition law is different from that for vectors alone. However, the transformation matrix does share the common group identities of the normal rotation matrices so that

$$\mathbf{L}^{-1}(\phi) = \mathbf{L}(-\phi) \qquad \mathbf{L}(\phi_1)\mathbf{L}(\phi_2) = \mathbf{L}(\phi_1 + \phi_2) \tag{16.2.10}$$

Finally, there is an additional way to conceptualize the Stokes parameters that many have found useful. In general, a polarized beam of light can be imagined as being composed of two orthogonally oscillating electric vectors bearing a relative phase shift ϵ with respect to one another. The vector sum of these two amplitudes will be a vector whose tip traces out an elliptically helical spiral as the vector propagates through space. We have defined the Stokes parameters in terms of the projection of that spiral onto a plane normal to the direction of propagation. However, we can also view them in terms of the motion of the vector through space. The fact that the vector rotates results from the phase difference between the two waves. If the phase difference were an odd multiple of $\pi/2$, then the wave would spiral through space and would be said to contain a significant amount of circular polarization. However, if the phase difference ϵ were zero or a multiple of π, the tip of the vector would merely oscillate in a plane whose orientation is determined by the ratio of the amplitudes of the electric vectors in the original orthogonal coordinate frame defining the beam, and the beam would be said to be linearly polarized.

Since the Stokes parameters are intensity-like (i.e., they have the units of the square of the electric field), one way to identify them is to take the outer (tensor) product of the electric vector with itself. Such an electric vector will have components E_l and E_r so that the outer product will be

$$\vec{E}\vec{E} = \begin{pmatrix} E_l E_l & E_r E_l \\ E_l E_r & E_r E_r \end{pmatrix} \tag{16.2.11}$$

Only three of these components are linearly independent, but the component E_r itself is made up of two linearly independent components $E_r \cos \epsilon$ and $E_r \sin \epsilon$. Thus the expansion of the term $E_l E_r$ leaves us with four intensity-like components which we can identify with the Stokes parameters so that formally

$$\begin{pmatrix} I_l \\ I_r \\ U \\ V \end{pmatrix} = \begin{pmatrix} E_l^2 \\ E_r^2 \\ 2E_l E_r \cos \epsilon \\ 2E_l E_r \sin \epsilon \end{pmatrix} \tag{16.2.12}$$

Thus taking the outer product of a general electric vector provides an excellent formalism for generating the Stokes parameters for that beam.

b Equations of Transfer for Stokes Parameters

The intensity-like nature of the Stokes parameters that describe an arbitrarily polarized beam makes them additive. This and the linear nature of the equation of radiative transfer ensure that we can write equations of transfer for each of them. Thus, if the components of $\vec{I}$ are the Stokes parameters, we can express this result as a vector equation of radiative transfer that has the familiar form

$$\mu \frac{d\vec{I}}{d\tau} = \vec{I} - \vec{S} \tag{16.2.13}$$

As usual, all the problems in the equations of transfer are contained in the source function. If we choose the intensity vector to have components $\vec{I}[I_l, I_r, U, V]$, then the source function can be written as a vector whose components are S_l, S_r, S_U, and S_V. Again, as we did with incident radiation, it is convenient to break up the radiation field into the "diffuse" field consisting of photons that have interacted at least once and the direct transmitted field that could originate from any incident radiation. We concentrate initially on the diffuse field since, in the absence of incident radiation, it exhibits axial symmetry about the normal to the atmosphere.

Nature of the Source Function for Polarized Radiation   The source function that appears in equation (16.2.13) is a vector quantity that must account for all contributions to the four Stokes parameters from thermal and scattering processes. Since thermal emission is by nature a random equilibrium process, we expect it to make no contribution to the Stokes parameters U and V, for these measure the departure of the radiation field from isotropy but carry no energy. In short, thermal radiation is completely unpolarized. For the same reasons, the thermal radiation will make an equal contribution to both I_l and I_r. Therefore the thermal contribution to the vector source function in equation (16.2.13) will have components

$$\epsilon_\nu \vec{B}_\nu(T) = \epsilon_\nu [\tfrac{1}{2} B_\nu(T), \tfrac{1}{2} B_\nu(T), 0, 0] \qquad (16.2.14)$$

where ϵ_ν is the familiar ratio of pure absorption to total extinction defined in equation (10.1.8). In some instances it is possible for ϵ_ν to depend on orientation. This is particularly true if an external magnetic field is present. The role played by scattering is considerably more complex.

Redistribution Phase Function for Polarized Light For the sake of simplicity, let us assume that the scattering is completely coherent so that there are no shifts in photon frequency introduced by the scattering process. However, in some instances scattering can result in photons being scattered from the l plane into the r plane, so that in principle all four Stokes parameters may be affected. In addition, the scattering by the atom or electron is most simply described in the reference frame of the scatterer and most simply involves the angle between the incoming and outgoing photons. But the coordinate frame of the equation of transfer is affixed to the observer. Thus we must express not only the details by which the photons are scattered from one Stokes parameter to another, but also that result in the coordinate frame of the equation of radiative transfer. It is this latter transformation that took the relatively simple form of the Rayleigh phase function and turned it into the more complicated form given by equation (15.3.39).

Since the scattering of a photon may carry it from one Stokes parameter to another, we can expect the redistribution function to have the form of a matrix where each element describes the probability of the scattering carrying the photon from some given Stokes parameter to another. We have already seen that the tensor outer product provides a vehicle for identifying the Stokes parameters in a beam. Let us use this formalism to investigate the Stokes parameters for a polarized beam of light scattered by an electron. While we generally depict a light beam as being composed of a single electric vector oscillating in a plane, this is not the most general form that a light wave can have.

Consider a beam of light composed of orthogonal electric vectors $\vec{E}'_l$ and $\vec{E}'_r$ which are systematically phase-shifted with respect to each other. Let this beam be incident on an electron so that the wave is scattered in the l plane (see Figure 16.3). We are free to pick the l plane to be any plane we wish, and the scattering plane is a convenient one. We can also view the scattering event to result from

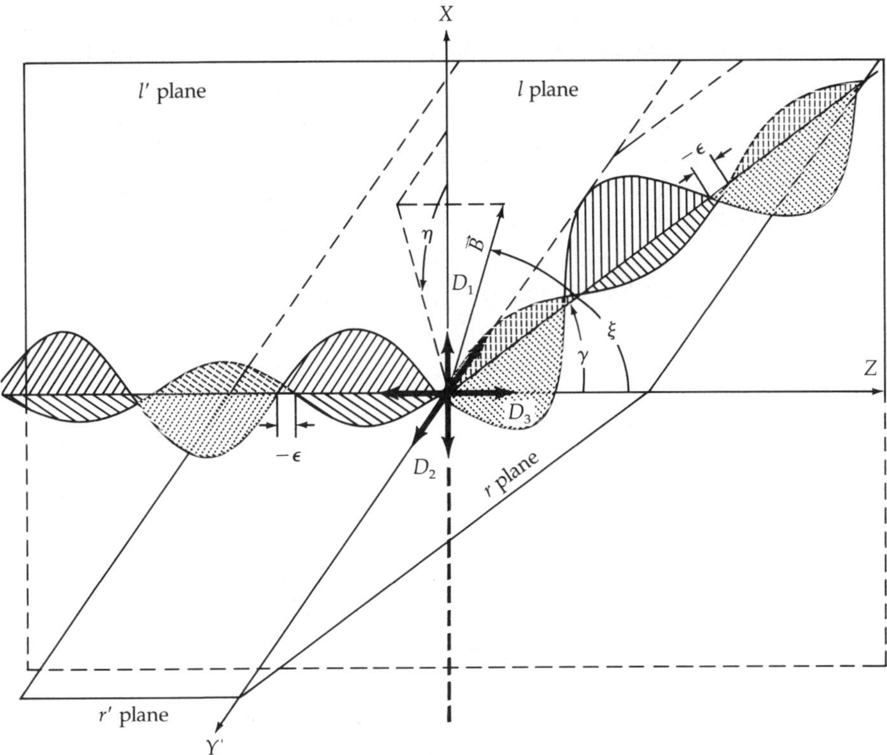

Figure 16.3 Electric vectors of two orthogonal waves differing in phase by ϵ incident on an electron. The scattering of the waves takes place in the same plane as the oscillating $\vec{E}'_l$ wave. The angle through which the waves are scattered is γ. Each wave can be viewed as exciting a dipole which radiates in the scattering plane in accordance with dipole radiation for the orientation of the respective dipoles.

the acceleration of the electron by the incident electric fields $\vec{E}'_l$ and $\vec{E}'_r$, producing oscillating dipoles which reradiate the incident energy. Since the photon is reradiated in the l plane, we may consider it to be made up of electric vectors $\vec{E}_l$ and $\vec{E}_r$ which bear the same relative phase shift to each other as $\vec{E}'_l$ and $\vec{E}'_r$. Now consider how each of the dipole excited by $\vec{E}'_l$ and $\vec{E}'_r$ will reradiate in the scattering plane. In general, electric dipoles will radiate in a plane containing the dipole axis, but the amplitude of that wave will decrease as one approaches the pole, specifically as the sine of the polar angle. Thus the intensity of $\vec{E}_l$ will vary as $\cos \gamma$ while $\vec{E}_r$ will be independent of the scattering (see Figure 16.3). Thus the scattered vector $\vec{E}$ will have components $(\vec{E}'_l \cos \gamma, \vec{E}'_r)$. Taking the outer product of this vector with itself, we can identify the linearly independent components as

$$\vec{E}\vec{E} = [E_l'^2 \cos^2 \gamma, \; E_r'^2, \; (2E_l'E_r' \cos \epsilon) \cos \gamma, \; (2E_l'E_r' \sin \epsilon) \cos \gamma] \qquad (16.2.15)$$

Since the dipoles excited by the incident radiation do not reradiate in a plane orthogonal to the exciting radiation, there can be no mixing of energy from the r plane to the l plane or vice versa. Thus, in any scattering matrix for Stokes parameters, the four independent components of the beam given by equation (16.2.15) must appear on the diagonal. In addition, one would expect a constant of proportionality determined by the conservation of energy. Specifically, the total energy of the incident beam must equal the total energy of radiation scattered through all allowed scattering angles. Thus, factoring out the definitions of the Stokes parameters from the components given in equation (16.2.15), we can write

$$\begin{pmatrix} I_l \\ I_r \\ U \\ V \end{pmatrix} = A_0 \begin{pmatrix} \cos^2 \gamma & 0 & 0 & 0 \\ 0 & 1 & 0 & 0 \\ 0 & 0 & \cos \gamma & 0 \\ 0 & 0 & 0 & \cos \gamma \end{pmatrix} \begin{pmatrix} (E_l')^2 \\ (E_r')^2 \\ 2E_l'E_r' \cos \epsilon \\ 2E_l'E_r' \sin \epsilon \end{pmatrix} \qquad (16.2.16)$$

where A_0 is the constant of proportionality described above. By applying the conservation of energy, A_0 can be found to be $\frac{3}{2}$. Thus the matrix in equation (16.2.16) represents the transformation of $\vec{I}'[I_l', I_r', U', V']$ to $\vec{I}[I_l, I_r, U, V]$. This matrix is generally called the *Rayleigh phase matrix* and is

$$\mathbf{R} = \frac{3}{2} \begin{pmatrix} \cos^2 \gamma & 0 & 0 & 0 \\ 0 & 1 & 0 & 0 \\ 0 & 0 & \cos \gamma & 0 \\ 0 & 0 & 0 & \cos \gamma \end{pmatrix} \qquad (16.2.17)$$

Here γ is the scattering angle between the incoming and outgoing photons. The Stokes parameters in the scattering coordinate system will be given by the vector components of the intensity $\vec{I}[I_l, I_r, U, V]$ where I_l and I_r lie in the scattering plane and perpendicular to it in a plane containing the outgoing photon (see Figure 16.3 and Figure 16.4). The equivalent matrix for scattering of the more common Stokes vector $\vec{I}'[I', Q', U', V']$ to $\vec{I}[I, Q, U, V]$ can be obtained by transforming equation (16.2.17) in accordance with the definitions given in equation (16.2.7) so that

$$R[I, Q, U, V] = \frac{3}{4} \begin{pmatrix} 1 + \cos^2 \gamma & -\sin^2 \gamma & 0 & 0 \\ -\sin^2 \gamma & 1 + \cos^2 \gamma & 0 & 0 \\ 0 & 0 & 2\cos \gamma & 0 \\ 0 & 0 & 0 & 2\cos \gamma \end{pmatrix} \qquad (16.2.18)$$

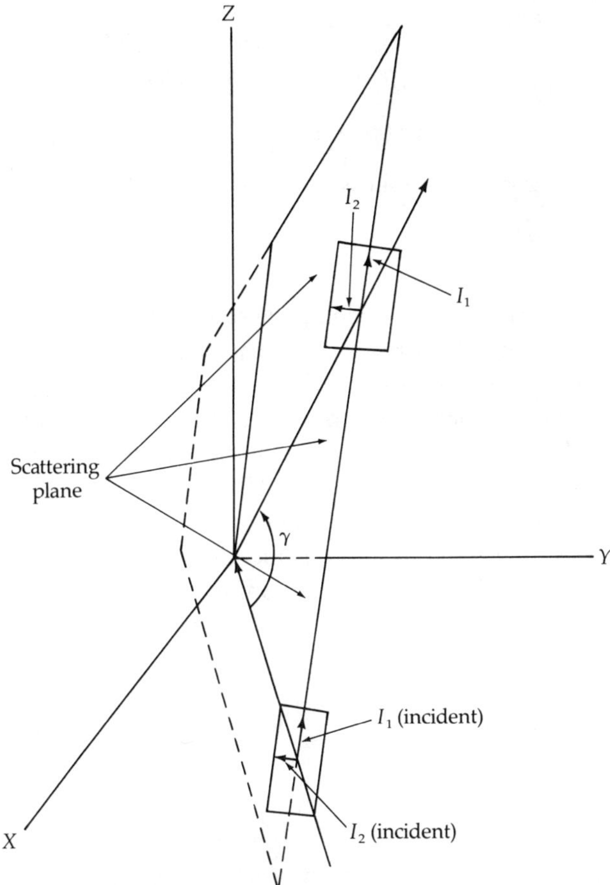

Figure 16.4 Coordinate frame for a single scattering event and the orientation of the polarized components.

To be useful for radiative transfer, these matrices must now be expressed in the observer's coordinate frame by multiplying them by the appropriate rotation matrix as given in equation (16.2.9) or (16.2.5). The final transformation to the coordinate system of the equation of transfer is moderately complicated (see Chandrasekhar,[1] pp. 37−41), and we simply quote the result for equation (16.2.17), using his notation:

$$\mathbf{R} = \frac{3}{2}\begin{pmatrix} (ll)^2 & (rl)^2 & (ll)(rl) & 0 \\ (lr)^2 & (rr)^2 & (lr)(rr) & 0 \\ 2(ll)(lr) & 2(rr)(rl) & (ll)(rr) + (rl)(lr) & 0 \\ 0 & 0 & 0 & (ll)(rr) - (rl)(lr) \end{pmatrix} \quad (16.2.19)$$

where

$$ll = \sin\theta\sin\theta' + \cos\theta\cos\theta'\cos(\phi' - \phi)$$

$$rl = \cos\theta\sin(\phi' - \phi)$$

$$lr = -\cos\theta\sin(\phi' - \phi)$$

$$rr = \cos(\phi' - \phi)$$

$$(16.2.20)$$

The equivalent matrix for the describing the scattering of $\vec{I}'[I', Q', U', V']$ can be obtained by transforming the matrix given in equation (16.2.19) in accordance with the definitions given in equations (16.2.7) (see equation (16.2.37)). Thus the scattering matrix can be obtained in the observer's frame for either representation of the Stokes parameters. In cases where the scattering matrix may have a more complicated representation in the frame of the scatterer, it may be easier to accomplish the transformation to the observer's coordinate frame for the orthogonal electric vectors themselves and then reconstruct the appropriate Stokes parameters. Substitution of these admittedly messy functions into equation (16.2.13) means that we can write the vector source function for Rayleigh scattering as

$$\vec{S}_v = \epsilon_v \vec{B}(T) - (1 - \epsilon_v)\frac{1}{4\pi}\int_0^{2\pi}\int_{-\pi/2}^{+\pi/2} \mathbf{R}(\theta, \theta', \phi, \phi')\vec{I}(\theta', \phi')\sin\theta'\,d\theta'\,d\phi'$$

$$(16.2.21)$$

where the only angles that appear in the source function are those that define the coordinate frame of the equation of transfer (see Figure 16.5).

Although the angle ϕ appears in the equation of transfer, the solution for the diffuse radiation field will not depend on it as long as there is no incident radiation. All the solution methods described in Chapter 10 for the equation of transfer may be used to solve the problem. However, equation (16.2.13) represents four simultaneous equations coupled through the scattering matrix.

More Complicated Phase Functions   The scattering of photons by electrons is one of the simplest yet widespread scattering phenomena in astronomy. However, in a number of cases the scattering source results in more complicated phase functions and produces interesting observational tests of stellar structure. Some examples are the scattering of radiation by resonance lines, the scattering from dust grains, and the effects of global magnetic fields on scattering. Even for these more difficult cases, one may proceed in the manner described for classical Thomson scattering to formulate the Stokes phase matrix. In most cases one can characterize the scattering process as involving the excitation of an electron which then

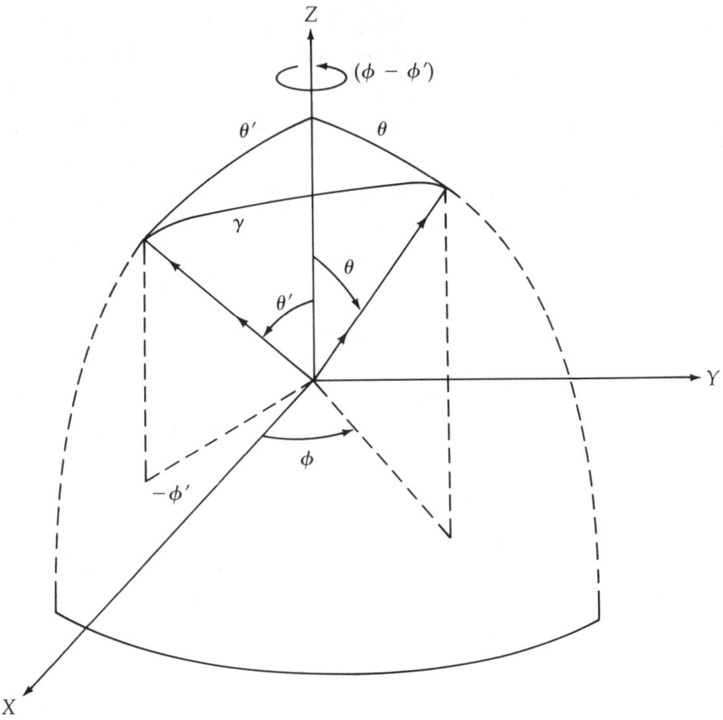

Figure 16.5 Scattering of a photon in the observer's coordinate frame. The Z axis points in the direction of the normal to the atmosphere, and $\cos\theta = \mu$. Angle ϕ is the azimuthal angle and now is a parameter of the problem.

reradiates the photon in a radiation pattern of an electric dipole. In some instances, the electron may be bound so that its oscillatory amplitude and orientation are restricted by external forces. As long as the physics of this restriction is understood, the reradiation can be characterized and the nature of the scattered wave described. The tensor outer product of that wave will then describe the scattering behavior of the Stokes parameters of the incident wave. As long as one is careful to keep track of the plane of origin of the various scattered electric vector amplitudes as well as the planes in which they are radiated, the contribution of the elements of the tensor outer product of the scattered beam with itself to the Stokes parameters of the scattered beam is unique.

For example, consider the case of Thomson scattering by an electron in a macroscopic magnetic field. The motion of the electron produced by the incident electric fields will subject the electron to Lorentz forces orthogonal to both the magnetic field and the induced velocity. This will cause the electron to oscillate and generate a second "electric dipole" whose radiation is added to that of the first. This allows energy to be transferred from one plane of polarization to another by an amount that depends on the magnetic field strength and orientation. This will have a profound effect on all the Stokes parameters. Contributions will be made to the scattered Stokes parameters from the interaction of the radiation by the initial electric dipole with itself, the interaction of that radiation with that of the magnetically induced dipole, and finally the self-interaction of the magnetically induced dipole. If we represent the scattered electric fields as being composed of a component from the initially driven dipole $\vec{E}$ and the magnetically induced dipole $\vec{M}$, then we can write the outer product of the scattered electric field with itself as

$$\vec{E}\vec{E} = \vec{E}\vec{E} + 2\vec{E}\vec{M} + \vec{M}\vec{M} \tag{16.2.22}$$

Each of these tensor outer products will produce a scattering matrix, and the sum will represent the scattering of the Stokes parameters by electrons in the magnetic field. The first will be identical to the Rayleigh phase matrix, where each element has been reduced by an amount equal to the energy that has been fed into the magnetically induced dipole. The second matrix represents the mixing between the electric field of the initial dipole with that of the magnetically induced dipole. Since these dipoles radiate out of phase with one another by $\pi/2$, the information contained here will largely involve the U and V Stokes parameters. The last matrix represents the contribution from the magnetically induced dipole with itself and will depend strongly on the relative orientation of the magnetic field. Keeping track of the relative phases of the various components, we calculate the three scattering matrices to be

$$R(\vec{E}\vec{E}) = \frac{3}{2}\zeta^2 \begin{pmatrix} \cos^2\gamma & 0 & 0 & 0 \\ 0 & 1 & 0 & 0 \\ 0 & 0 & \cos\gamma & 0 \\ 0 & 0 & 0 & \cos\gamma \end{pmatrix}$$

$$R(\vec{E}\vec{M}) = \frac{3\zeta^2 x}{2} \begin{pmatrix} 0 & 0 & 0 & +b_2\cos\gamma \\ 0 & 0 & 0 & a_2 \\ 0 & 0 & 0 & a_1 \\ 2a_2\cos\gamma & +2b_1 & a_1 & 0 \end{pmatrix} \tag{16.2.23}$$

$$R(\vec{M}\vec{M}) = \frac{3}{2}x^2\zeta^2 \begin{pmatrix} a_1^2 & b_1^2 & a_1 b_1 & 0 \\ a_2^2 & 0 & 0 & 0 \\ 2a_1 a_2 & 0 & +a_2 b_1 & 0 \\ 0 & 0 & 0 & -a_2 b_1 \end{pmatrix}$$

where

$$a_1 = \sin\gamma\sin\xi\cos\eta \qquad a_2 = -\cos\xi \qquad b_1 = -\sin\gamma\sin\xi\sin\eta \tag{16.2.24}$$

The coordinates of the magnetic field with respect to the scattering plane of the incident photon are given by η and the polar angle ξ, respectively. The strength of the magnetic field is contained in the parameter x so that

$$x \equiv \frac{\omega_c}{\omega} = \frac{eB/m_e}{\omega} \tag{16.2.25}$$

If we require that the scattering be conservative, then, since energy is required to generate those oscillations, the amplitude of the incident electric vector will be reduced in order to supply the energy to excite them. However, the energy that excites the dipoles must be equal to the energy of the incident beam so that

$$\vec{E}' \cdot \vec{E}' = [\vec{E} + x(\vec{E} \times \hat{b})] \cdot [\vec{E} + x(\vec{E} \times \hat{b})] = \vec{E}^2[1 + x^2 \sin^2(\xi)] \tag{16.2.26}$$

where ξ is the angle between the electric vector and the magnetic field. If we let the magnitude of the electric vector be reduced by ζ, then

$$\zeta^2 = [1 + x^2 \sin^2 \xi]^{-1} = \zeta_0^2[1 - x^2\zeta_0^2 \cos^2 \xi]^{-1} \tag{16.2.27}$$

where

$$\zeta_0^2 = [1 + x^2]^{-1} \tag{16.2.28}$$

If we assume that $x \ll 1$, then $\vec{E}$ will remain nearly parallel to $\vec{E}'$ and we may expand equations (16.2.27), and (16.2.28) to give

$$\left.\begin{array}{l} \zeta^2 \approx \zeta_0^2[1 + x^2\zeta_0^2 \cos^2 \xi] \approx 1 \\[6pt] \zeta_0^2 \approx 1 - x^2 \approx 1 \end{array}\right\} \tag{16.2.29}$$

and we may write

$$\vec{E} = \mathbf{R}[\vec{E}' + x(\vec{E}' \times \hat{b})] \tag{16.2.30}$$

For zero magnetic field, one recovers the Rayleigh phase matrix. As the strength of the field increases, the contribution of the second two matrices increases and normal electron scattering decreases. As the gyrofrequency ω_c is reached, the motion of the electron becomes circular and significant circular polarization is introduced to the beam along the direction of the magnetic field. Since we have ignored the resonance that occurs at the gyrofrequency (that is, $x = 1$), these results are only approximate. For the scattering of natural light, $\vec{I}[\frac{1}{2}I, \frac{1}{2}I, 0, 0]$, the strength of the components from the Rayleigh matrix are reduced while the $R(\vec{EM})$ contributes solely to the V Stokes parameter, indicating that the beam will be completely circularly polarized when scattered along the magnetic field. This contribution diminishes as the scattering angle moves toward $\pi/2$. The contribution from the $R(\vec{MM})$ matrix only has nonzero components that transfer energy equally from I_l to I_r and vice versa (for $\gamma = \pi$), but the contribution to I_l vanishes as $\gamma \to \pi/2$. The fact that U remains zero implies that the scattered radiation for $\gamma = \pi/2$ is linearly polarized with the plane of polarization perpendicular to the magnetic field.

As one approaches the gyrofrequency, the transport of radiation becomes increasingly complex. Since the component of the electric field parallel to the magnetic field produces no Lorentz force on the electron, the interaction of the two orthogonal components of the generalized electromagnetic wave with the magnetic field are unequal. This anisotropic interaction has its optical counterparts in the phenomena known as Faraday rotation and a complex index of refraction. The phenomena enter the scattering problem as x^2 so that our results are valid only for $x \ll 1$.

At frequencies less than the gyrofrequency, scattering becomes more complex as the energy of the photon may be absorbed by the electron in the form of circular motion. This energy may be lost through collisions before it can be reradiated.

Again to utilize these matrices for problems involving radiative transfer requires that these matrices be transferred to the observer's frame. However, as they were developed primarily to indicate the method for generating the Stokes phase matrices, we do not develop them further. Obviously such techniques can be used to develop phase matrices for quite complex problems.

c Solution of the Equations of Radiative Transfer for Polarized Light

While solution of the integrodifferential equations of radiative transfer may be accomplished by direct means, all the various rays characterizing the four Stokes parameters must be solved together so that the integrals required by the scattering part of the source function may be evaluated simultaneously. To obtain the necessary accuracy for low levels of polarization would require a computing power that would tax even the fastest computers. Even if the axial symmetry of the diffuse field can be used to advantage for some problems, cases involving incident radiation will require involving discrete streams in ϕ as well as θ.

There is an approach involving the integral equations for the source function that greatly simplifies the problem. We have seen throughout the book that the development of Schwarzschild-Milne equations for the source function always produced equations that involved moments of the radiation field, usually the mean intensity J. As long as the redistribution function can be expressed as a finite series involving tesseral harmonics of θ and ϕ, the source function can be expressed in terms of simple functions of θ and ϕ which multiply various moments of the radiation field that depend on depth only. As in Chapter 9, it will then be possible to generate Schwarzschild-Milne-like equations for these moments that depend on depth only. Indeed, since a moment of a radiation field cannot depend on angle, by definition, this effectively separates the angular dependence of the radiation field from the depth dependence. This is true even for the case of incident radiation. The solution of these equations will then yield the source function at all depths which, together with the classical solution for the equation of transfer, will provide a complete description of the radiation field in all four Stokes parameters. This conceptually simple approach becomes rather algebraically complicated in practice, even for the relatively simple case of Rayleigh scattering.

The general case of this problem has been worked out by Collins[11] and more succinctly by Collins and Buerger,[12] and we only quote the results. The source functions for nongray Rayleigh scattering can be written as

$$S_l(\tau_\nu, \theta, \phi) = \tfrac{1}{2}\epsilon_\nu(\tau_\nu)B_\nu(\tau_\nu) + \tfrac{3}{4}\{2X(\tau_\nu) - \cos^2\theta\,[T(\tau_\nu) + Z(\tau_\nu)$$

$$\times\,(\cos 2\phi - 2\xi\sin 2\phi)] + M(\tau_\nu)\cos\theta\sin\theta\,(\cos\phi - \zeta\sin 2\phi\}$$

$$S_r(\tau_\nu, \theta, \phi) = \tfrac{1}{2}\epsilon_\nu(\tau_\nu)B_\nu(\tau_\nu) + \tfrac{3}{4}[2X(\tau_\nu) - Y(\tau_\nu) + Z(\tau_\nu)(\cos 2\phi - 2\xi\sin 2\phi)]$$

$$S_U(\tau_\nu, \theta, \phi) = -\tfrac{3}{4}[2Z(\tau_\nu)\cos\theta\,(2\xi\cos 2\phi + \sin 2\phi)$$

$$-\,M(\tau_\nu)\sin\theta\,(\sin\phi - \zeta\cos\phi)]$$

$$S_V(\tau_\nu, \theta, \phi) = \tfrac{3}{8}[P(\tau_\nu)\cos\theta + W(\tau_\nu)\sin\theta\cos\phi] \tag{16.2.31}$$

where the moments of the radiation field $X(\tau)$, $Y(\tau)$, $Z(\tau)$, $M(\tau)$, $P(\tau)$, and $W(\tau)$ depend on depth alone and satisfy these integral equations:

$$X(\tau_\nu) = C_X(\tau_\nu) + \tfrac{1}{4}[\Lambda_1(\tau_\nu) - \Lambda_3(\tau_\nu)]\big|\epsilon_\nu(t)B_\nu(t) + 3X(t)\big|$$

$$-\,\tfrac{3}{8}[\Lambda_3(\tau_\nu) - \Lambda_5(\tau_\nu)]\big|Y(t)\big|$$

$$Y(\tau_\nu) = C_Y(\tau_\nu) + \tfrac{1}{4}[\Lambda_1(\tau_\nu) - 3\Lambda_3(\tau_\nu)]\big|\epsilon_\nu(t)B(t) - 3X(t)\big|$$

$$-\,\tfrac{3}{8}[\Lambda_1(\tau_\nu) - 2\Lambda_3(\tau_\nu) + 3\Lambda_5(\tau_\nu)]\big|Y(t)\big|$$

$$Z(\tau_\nu) = C_Z(\tau_\nu) + \tfrac{3}{16}[\Lambda_1(\tau_\nu) + 2\Lambda_3(\tau_\nu) + \Lambda_5(\tau_\nu)]\big|Z(t)\big| \tag{16.2.32}$$

$$M(\tau_\nu) = C_M(\tau_\nu) + \tfrac{3}{8}[\Lambda_1(\tau_\nu) + \Lambda_3(\tau_\nu) - 2\Lambda_5(\tau_\nu)]\big|M(t)\big|$$

$$W(\tau_\nu) = C_W(\tau_\nu) + \tfrac{3}{8}[\Lambda_1(\tau_\nu) - \Lambda_3(\tau_\nu)]\big|W(t)\big|$$

$$P(\tau_\nu) = C_P(\tau_\nu) + \tfrac{3}{4}\Lambda_3(\tau_\nu)\big|P(t)\big|$$

The Λ_n operator is the same as that defined in Chapter 10 [see equation (10.1.16)] and produces exponential integrals of order n. The constants $C_N(\tau)$ depend on the incident radiation field only and are attenuated by the appropriate optical path. It is assumed that the incident radiation encounters the atmosphere along a path defined by $\phi = 0$. The values of these constants for plane-parallel polarized radiation are

$$
\begin{pmatrix} C_X(\tau_\nu) \\ C_Y(\tau_\nu) \\ C_Z(\tau_\nu) \\ C_M(\tau_\nu) \\ C_W(\tau_\nu) \\ C_P(\tau_\nu) \end{pmatrix} = \frac{[1 - \epsilon_\nu(\tau_\nu)]e^{-\tau_\nu/\mu_0}}{4\pi} \begin{pmatrix} (1 - \mu_0^2)I_l^0 \\ (2 - 3\mu_0^2)I_l^0 - I_r^0 \\ I_r^0 - \mu_0^2 I_l^0 \\ -\mu_0(1 - \mu_0^2)^{1/2}I_l^0 \\ (1 - \mu_0^2)^{1/2}V^0 \\ -\mu_0 V^0 \end{pmatrix}
\tag{16.2.33}
$$

The parameters ξ and ζ also depend on the incident radiation alone and are given by

$$
\xi \equiv \frac{\mu_0 U^0}{2(I_r^0 - \mu_0^2 I_l^0)} \qquad \zeta \equiv \frac{U^0}{2\mu_0 I_l^0}
\tag{16.2.34}
$$

These parameters basically represent the direct contribution to the source function from the incident radiation field. If the incident radiation is unpolarized, they are both zero.

Note that only the first two integral equations for the moments $X(\tau)$ and $Y(\tau)$ are coupled and therefore must be solved simultaneously. In addition, the net radiative flux flowing through the atmosphere involves only these moments, so that only these two equations must be solved to satisfy radiative equilibrium. This can be shown by substituting the equations for the source functions into the definition for the radiative flux and obtaining

$$
F_\nu(\tau_\nu) = 4\Lambda_2(\tau_\nu)\left|\epsilon_\nu(t)B(t) + 3X(t) - \tfrac{3}{4}Y(t)\right| - 3\Lambda_4(\tau_\nu)\left|Y(t)\right|
\tag{16.2.35}
$$

or indifferential form

$$
\frac{dF_\nu(\tau_\nu)}{d\tau_\nu} = \epsilon_\nu(\tau_\nu)B_\nu[T(\tau_\nu)] - 3X(\tau_\nu) + Y(\tau_\nu)
\tag{16.2.36}
$$

These two expressions are all that is needed to apply the Avrett-Krook perturbation scheme for constructing a model atmosphere. The remaining equations need to be solved only when complete convergence of the atmosphere has been achieved, and then only if the state of polarization for the emergent radiation field is required.

d Approximate Formulas for the Degree of Emergent Polarization

The complexity of equations (16.2.31) and (16.2.32) shows that the complete solution of the problem of polarized radiation transfer is quite formidable. Clearly approximations that yielded the degree of polarization and its wavelength dependence would be useful. To get such approximation formulas, we plan to create them from the Stokes parameters $I[I, Q, U, V]$ rather than $I[I_l, I_r, U, V]$ since the degree of polarization is just Q/I. We may generate the appropriate scattering matrix from equation (16.2.19) from the relation for the Stokes parameters $I[I, Q, U, V]$ and $I[I_l, I_r, U, V]$ given in equation (16.2.7). The resulting matrix for the Stokes scattering matrix is then

$$
\mathbf{R}_{IQUV} = \left[
\begin{array}{cc}
\frac{1}{2}[g_{11}^2 + g_{12}^2 + g_{21}^2 + g_{22}^2] & \frac{1}{2}[g_{11}^2 - g_{12}^2 + g_{21}^2 - g_{22}^2] \\
\frac{1}{2}[g_{11}^2 + g_{12}^2 - g_{21}^2 - g_{22}^2] & \frac{1}{2}[g_{11}^2 - g_{12}^2 - g_{21}^2 + g_{22}^2] \\
[g_{11}g_{21} + g_{12}g_{22}] & [g_{11}g_{21} - g_{12}g_{22}] \\
0 & 0
\end{array}
\right.
$$

$$
\left.
\begin{array}{cc}
[g_{11}g_{12} + g_{21}g_{22}] & 0 \\
[g_{11}g_{12} - g_{21}g_{22}] & 0 \\
[g_{11}g_{22} + g_{21}g_{12}] & 0 \\
0 & [g_{11}g_{22} - g_{21}g_{12}]
\end{array}
\right] \tag{16.2.37}
$$

Here we have replaced the notation of Chandrasekhar with

$$
g_{11} = ll \qquad g_{12} = lr \qquad g_{21} = rl \qquad g_{22} = rr \tag{16.2.38}
$$

To obtain the source function for Thomson-Rayleigh scattering, we must carry out the operations implied by equation (16.2.21) for the source function. However, now we are to do it for all four of the Stokes parameters. Thus we will integrate all elements of the scattering matrix over ϕ' and θ'. However, the geometry of the plane-parallel atmosphere means that the solution can have no ϕ dependence so that all odd functions of ϕ' must vanish as well as those involving $\sin 2\phi'$. Similarly, the resulting source function can have no explicit ϕ dependence. Also the elements of $\mathbf{R}[I, Q, U, V]$ contain a limited number of squares and products of the g_{ij}'s. If we simply average these quantities over ϕ', we get

$$
\overline{g_{11}^2} = \sin^2 \theta - \sin^2 \theta \cos^2 \theta' + \tfrac{1}{2} \cos^2 \theta \cos^2 \theta'
$$

$$
\overline{g_{12}^2} = \tfrac{1}{2} \cos \theta \qquad \overline{g_{21}^2} = \tfrac{1}{2} \cos \theta' \qquad \overline{g_{22}^2} = \tfrac{1}{2}
$$

$$
\overline{g_{12}g_{21}} = -\tfrac{1}{2} \cos \theta \cos \theta' \tag{16.2.39}
$$

$$
\overline{g_{11}g_{22}} = +\tfrac{1}{2} \cos \theta \cos \theta'
$$

$$
\overline{g_{12}g_{22}} = \overline{g_{21}g_{22}} = \overline{g_{11}g_{12}} = \overline{g_{11}g_{21}} = 0
$$

Integration of these terms times the appropriate Stokes parameter over θ' will then yield familiar moments of the Stokes parameters such as J_P and K_P (Section 9.3).

If we let the subscript P stand for the different Stokes parameters, then expressing the product of the g_{ij}'s in equation (16.2.39) and the Stokes vector in terms of the moments of the Stokes parameters yields

$$\langle I_P g_{11}^2 \rangle = J_P - K_P - (\cos^2 \theta)\left(J_P - \frac{3K_P}{2}\right)$$

$$\langle I_P g_{12}^2 \rangle = \tfrac{1}{2} J_P \cos^2 \theta$$

$$\langle I_P g_{21}^2 \rangle = \tfrac{1}{2} K_P$$

$$\langle I_P g_{22}^2 \rangle = \tfrac{1}{2} J_P \tag{16.2.40}$$

$$\langle I_P g_{12} g_{21} \rangle = -\frac{(\cos \theta)F_P}{4}$$

$$\langle I_P g_{11} g_{22} \rangle = +\frac{(\cos \theta)F_P}{4}$$

Substituting these average values into the respective elements of equation (16.2.37), we calculate the scattering fraction of the source functions for the Stokes parameters. Adding the contribution to the I source function from thermal emission, we can then write the source functions for all the Stokes parameters as

$$S_I(\tau_\nu) = \epsilon_\nu B_\nu(t) + \frac{3(1 - \epsilon_\nu)}{8}$$

$$\times \left\{ [(3J_I - K_I) + (J_Q - K_Q)] - (\cos^2 \theta)\,[(J_I - 3K_I) + 3(J_Q - K_Q)] \right\}$$

$$S_Q(\tau_\nu) = \frac{3(1 - \epsilon_\nu)}{8}(\sin^2 \theta)[(J_I - 3K_I) + 3(J_Q - K_Q)] \tag{16.2.41}$$

$$S_U(\tau_\nu) = 0$$

$$S_V(\tau_\nu) = \frac{3(1 - \epsilon_\nu)}{4} F_V \cos \theta$$

where, as before,

$$\epsilon_\nu(\tau_\nu) \equiv \frac{\kappa_\nu(\tau_\nu)}{\kappa_\nu(\tau_\nu) + \sigma_\nu(\tau_\nu)} \tag{16.2.42}$$

For an isotropic radiation field such as one could expect deep in the star note that $J_I = 3K_I$ and $J_Q = K_Q = 0$ since there is no preferred plane in which to measure Q. Under these conditions all fluxes vanish, and the source functions become

$$S_I(\tau_v) = \epsilon_v(\tau_v)B_v(T) + (1 - \epsilon_v)J_v(\tau_v)$$

$$S_Q(\tau_v) = S_U(\tau_v) = S_V(\tau_v) = 0$$

(16.2.43)

which is the expected result for isotropic scattering and is consistent with an isotropic radiation field.

For anisotropic scattering, the source function is no longer independent of θ. However, the θ-dependent terms depend on a collection of moments that measure the departure of the radiation field from isotropy. For a one-dimensional beam $J_Q = K_Q$, while for an isotropic radiation field $J_I = 3K_I$. This is true for both S_I and S_Q. Indeed, the coefficients of the angular terms are the same for both source functions. The fact that S_U is zero means that U is zero throughout the atmosphere. This result is not surprising since for U to be nonzero would imply that polarization had to have a maximum in some plane other than that containing the normal to the atmosphere and the observer (or one perpendicular to it). From symmetry, there can be no such plane, and hence U must be zero everywhere. We are now in a position to generate integral equations for quantities that will explicitly determine the variation of the source functions with depth.

Approximation of S_Q in the Upper Atmosphere To begin our analysis of the polarization expected from a stellar atmosphere, we consider the complete source functions for such an atmosphere. These can be written as

$$S_I(\tau_v) = \epsilon_v(\tau_v)B_v(T) + X(\tau_v) - Y(\tau_v)\cos^2\theta$$

$$S_Q(\tau_v) = Y(\tau_v)\sin^2\theta$$

(16.2.44)

where

$$X(\tau_v) \equiv \tfrac{3}{8}[1 - \epsilon_v(\tau_v)]\{[3J_I(\tau_v) - K_I(\tau_v)] + [J_Q(\tau_v) - K_Q(\tau_v)]\}$$

$$Y(\tau_v) \equiv \tfrac{3}{8}[1 - \epsilon_v(\tau_v)]\{[J_I(\tau_v) - 3K_I(\tau_v)] + 3[J_Q(\tau_v) - K_Q(\tau_v)]\}$$

(16.2.45)

Let us first consider the pure scattering gray case so that $\epsilon_v(\tau_v) = 0$. Our problem then basically reduces to finding an approximation for $Y(\tau_v)$. For this part of the discussion, we drop the subscript v because radiative equilibrium guarantees constant flux at all frequencies. The quantity $J_I - 3K_I$ occurs in $Y(\tau)$ and normally is the dominant term. However, in the Eddington approximation, this term would be zero. This clearly demonstrates that polarization is a second-order effect and will rely on the departure of the radiation field from isotropy for its existence.

The entire problem of estimating the polarization then boils down to estimating $J - 3K$ or the Eddington factor. The Eddington factor was defined in equation

459

(10.4.9); it is a measure of the extent to which the radiation field departs from isotropy. If the radiation field is isotropic, then $f(\tau) = \frac{1}{3}$. If the radiation field is strongly forward-directed, then $f(\tau) = 1$. If the radiation field is somewhat flattened with respect to the forward direction, then $f(\tau)$ can fall below $\frac{1}{3}$. Thus we may write the source function for Q as

$$S_Q(\tau) = 3\sin^2\theta \,\frac{J_I(\tau)[1 - 3f_I(\tau)] + 3J_Q(\tau)[1 - f_Q(\tau)]}{8} \tag{16.2.46}$$

It is fairly easy to show that if $J_I(\tau)$ is given by the gray atmosphere solution

$$J_I(\tau) = \tfrac{3}{4}F_I[\tau + q(\tau)] \tag{16.2.47}$$

where $q(\tau)$ is the Hopf function, then

$$J_Q(\tau) = C_1 e^{-\alpha\tau} + C_2 e^{-k_1\tau} \tag{16.2.48}$$

To obtain this result, we have simply taken moments of the Q equation of transfer, assumed $f_Q(\tau) = \frac{1}{3}$, and used the Chandrasekhar $n = 1$ approximation for $q(\tau)$. Thus k_1 is the $n = 1$ eigenvalue for the gray equation of transfer (see Table 10.1) and $\alpha^2 = \frac{3}{2}$. And C_1 and C_2 have opposite signs. Thus $J_Q(\tau)$ will indeed vanish rapidly with optical depth. This rapid decline will also extend to $S_Q(\tau)$. Thus we will assume that

$$Q(0, \mu) \approx S_Q(0, \mu) \tag{16.2.49}$$

This means that

$$J_Q(0) = \frac{1}{2}\int_0^1 Q(0, \mu)\,d\mu = \frac{1}{5}J_I(0)\,\frac{1 - 3f_I(0)}{1 + 3f_Q(0)/5} \tag{16.2.50}$$

This may be substituted back into equation (16.2.46) to yield

$$S_Q(0) = \frac{3[1 - 3f_I(0)]}{5[1 + 3f_Q(0)/5]}\,J_I(0)\sin^2\theta \tag{16.2.51}$$

Approximate Formulas for the Degree of Polarization in a Gray Atmosphere   Since we can expect the maximum value of the polarization to occur at the limb (that is, $\theta = \pi/2$), both $Q(0, 0)$ and $I(0, 0)$ will be given by the value of their source functions at the surface. If for $I(0, 0)$ we take that to be $J_I(0)$, we get for the degree of polarization

$$P_1(\mu) = \frac{3[1 - 3f_I(0)]}{5[1 + 3f_Q(0)/5]} (1 - \mu^2) \tag{16.2.52}$$

Just as we calculated $J_Q(0)$ in equation (16.2.50), so we can calculate $K_Q(0)$ and determine $f_Q(0) = \frac{1}{5}$. Using the best value for $f_I(0)$ of 0.4102 (i.e., the exact approximation of Chandrasekhar), we get that the limb polarization for a gray atmosphere should be 12.35 percent. This is to be compared with Chandrasekhar's (1949) value of 11.713 percent. Although the μ dependence given by equation (16.2.52) will be approximately correct, we can hardly expect it to be accurate. The solution for $\mu > 0$ requires integrating the source function over a range of $\mu\tau$, which means that we will have to know the source function as a function of τ. Physically this amounts to including the limb-darkening in the expressions for $I(0, \mu)$ and $Q(0, \mu)$. This can be done by using expressions for the source functions obtained from the same kind of analysis that was used to get $J_Q(\tau)$. This approach can make a significant difference since the limb-darkening effects act in opposite ways for I and for Q. It is basically the increasing source function for I that produces a decrease of the intensity at the limb. The limb is simply a less bright region than the center of the disk. However, the source function for Q declines rapidly as one enters the star, so that the region of the star that is the brightest in Q is the limb. Hence Q will suffer "limb-brightening." The ratio of these two effects will cause the polarization to drop much faster than $1 - \mu^2$ as one approaches the center of the disk. Determining the source function's τ dependence in the same level of approximation that was used to get $J_Q(\tau)$ yields

$$P_2(\mu) = \frac{3\{(1 - \mu^2)[1 - 3f_I(0)]q(0)\}(1 + k_1\mu)}{\{5[q(0) + \mu(1 + k_1 q(\infty)) + k_1\mu^2][1 + 3f_q(0)/5]\}} \tag{16.2.53}$$

or substituting in values of the constants for the same order of approximation we get

$$P_2(\mu) = -0.107(1 - \mu^2)(1 + k_1\mu)/(1 + 4.026\mu + 3.416\mu^2) \tag{16.2.54}$$

The value of the polarization for $\mu = \frac{1}{2}$ from equation (16.2.52) is $P_1(\mu = 0.5) = 9.26$ percent. In contrast, the value given by equation (16.2.54) is $P_2(\mu = 0.5) = 4.1$ percent. This is to be compared with Chandrasekhar's value of 2.252 percent. Thus the limb-darkening is important in determining the center-limb variation of the polarization, and it drops much faster than would be anticipated from the simple surface approximation.

Wavelength Dependence of Polarization   The gray atmosphere, being frequency-independent, tells us little about the way in which the polarization can be expected to vary with wavelength. However, the approximation formulas for the center-limb variation of the gray polarization clearly show that the polarization is largely determined by the quantity $1 - 3f_I(0)$. This quantity basically measures the departure from isotropy of the radiation field. In any stellar atmosphere, this departure will be determined by the gradient of the source function. Since the polarization is largely determined by the atmosphere structure above an optical depth of unity, we may approximate the source function by a surface term and a gradient so that

$$S_I(\tau) \approx S_0(1 + b\tau) \tag{16.2.55}$$

Thus,

$$J_I(0) = \frac{1}{2} \int_0^1 \int_0^\infty \frac{S_I(t)e^{-t/\mu}\,dt}{\mu} = \frac{1}{2} S_I(0)\left(1 + \frac{b}{2}\right)$$

$$K_I(0) = \frac{1}{2} \int_0^1 \int_0^\infty \frac{\mu^2 S_I(t)e^{-t/\mu}\,dt}{\mu} = \frac{1}{2} S_I(0)\left(\frac{1}{3} + \frac{b}{4}\right) \tag{16.2.56}$$

This enables us to write the term that measures the isotropy of the radiation field as

$$J_I(0)[1 - 3f_I(0)] = \frac{-\frac{1}{2}S_0 b}{4} \tag{16.2.57}$$

Thus replacing the term that measures the anisotropy of the radiation field with b yields

$$P_1(\mu) = -(1 - \epsilon_\nu)\left(\frac{21b}{250}\right)\frac{\sin^2\theta}{1 + \mu b} \tag{16.2.58}$$

If, for purposes of determining the wavelength dependence of the polarization, we take the source function to be the Planck function, then the normalized gradient b_ν becomes

$$b_\nu = \frac{1}{B_\nu(0)}\left(\frac{dB_\nu}{d\tau_\nu}\right)\bigg|_{\tau_\nu = 0} = \frac{1}{B_\nu(0)}\left(\frac{dB_\nu}{dT}\frac{dT}{d\tau_\nu}\right)\bigg|_{\tau_\nu = 0} = \frac{1}{B_\nu(0)}\left(\frac{dB_\nu}{dT}\frac{dT}{d\tau_0}\frac{d\tau_0}{d\tau_\nu}\right)\bigg|_{\tau_\nu = 0} \tag{16.2.59}$$

The frequency ν_0 is the reference frequency at which the temperature gradient is to be evaluated. In this manner, the temperature gradient is the same for all frequencies, and the frequency dependence is then determined by the derivative of the

Planck function and the term $d\tau_0/d\tau_v$. This term is simply the ratio of the extinction coefficients at the reference frequency and the frequency of interest. This term is largely responsible for the frequency dependence of the polarization in hot stars.

If for purposes of approximation we continue to use the Planck function as the source function, we can evaluate much of the expression for b_v explicitly so that

$$b_v = \beta_v \frac{\sigma_0(1 - \epsilon_v)}{\sigma_v(1 - \epsilon_0)} \tag{16.2.60}$$

where

$$\beta_v = \frac{hv}{kT}(1 - e^{-hv/(kT)})^{-1}\frac{1}{T}\frac{dT}{d\tau_0} \tag{16.2.61}$$

We can write the polarization as

$$P_v(\mu) \approx \frac{-21(1 - \mu^2)(1 - \epsilon_v)^2\beta_v}{250(1 - \epsilon_0)(1 + b_v\mu)} \tag{16.2.62}$$

For the gray atmosphere $\epsilon_v = \epsilon_0 = 0$, and the maximum polarization will be achieved where β_v is a maximum. Letting $\alpha = hv/(kT)$, setting $P_v(0)$ equal to the gray atmosphere result, and using the gray atmosphere temperature distribution, we find that the gray result will occur near $\alpha = 4$. This value lies between the maximum of B_v ($\alpha = 2.82$) and B_λ ($\alpha = 4.97$) and is virtually identical to the frequency at which vB_v ($\alpha = 3.92$) is a maximum, which is practically the same as the frequency for which dB_v/dT ($\alpha = 3.83$) is a maximum. Thus, to the accuracy of the approximation, the gray polarization will occur at the frequency for which the source function gradient is a maximum. This results from the fact that Q will attain its largest value when $J_I - 3K_I$ is a maximum. This will occur when the source function gradient is a maximum which, since the radiative flux is driven by that gradient, is also the energy maximum. For larger values of the frequency, $f_I(0)$ will become larger, meaning that the degree of polarization can exceed the gray value, but the magnitude of Q_v will decrease.

In the nongray case, the situation is somewhat more complicated. Here, in addition to the frequency dependence of the source function gradient, the variation of the opacity with frequency strongly influences the value of the polarization. If we assume that the gray result is actually realized for hot stars in the vicinity of the energy maximum, then the polarization in the visible range will be roughly given by

$$P_v(0) = P_{v_0}\left[\frac{(1 - \epsilon_v)^2}{1 - \epsilon_0}\right] \tag{16.2.63}$$

That is, it will drop quadratically with $1 - \epsilon_v$ as one moves into the visible range. The change in ϵ_v with frequency will be largely due to the change in κ_v. To the extent that this is largely due to hydrogen, we can expect the opacity to vary as v^{-3}. Thus, in stars where the dominant extinction is from absorption by hydrogen and the energy maximum is in the far ultraviolet radiation, we can expect the polarization to drop by several powers of 10 from its maximum value. In stars where electron scattering is the dominant opacity source, a significant decrease in the degree of polarization will still occur in the visible as a result of the increase in the pure absorption component of the extinction.

In the limit of $\alpha = hv/(kT) \ll 1$, the source function asymtotically approaches a constant given by the temperature gradient, and the wavelength dependence of the polarization depends on only the opacity effects. In the case where $\alpha \gg 1$, β is proportional to α times the normalized temperature gradient, and the polarization will continue to rise into the ultraviolet end until there is a change in the opacity. Thus we should expect the early-type stars to show significant polarization when approaching or exceeding the gray value in the vicinity of the Lyman jump. Little polarization will be in evidence in the Lyman continuum as a result of the marked increase in the absorptive opacity. The polarization will also decline as one moves into the optical part of the spectrum to values typically of the order of 0.01 percent at the limb and will slowly decline throughout the Paschen continuum.

Everything done so far has presupposed that the dependence of the source function near the surface can be expressed in terms of a first-order Taylor series. In many stars, this expression is not adequate because there is a very steep plunge in the temperature near the surface. This plunge will require a substantial second derivative of opposite sign from the first derivative to adequately describe the surface behavior of the source function. It is clear from equations (16.2.55) and (16.2.56) that the presence of such a term will increase the value of $J_I - 3K_I$ and could, in principle, reverse its sign, resulting in a positive degree of polarization (i.e., polarization with the maximum electric vector aligned with the atmospheric normal). Detailed models of nongray atmospheres indeed show this positive polarization, and it can be expected any time the second-order limb-darkening coefficient is positive and greater than one-half the magnitude of the first-order limb-darkening coefficient.

e *Implications of Transfer of Polarization for Stellar Atmospheres*

It is clear from equations (16.2.33) that, in the absence of illumination on the atmosphere, all the C_N's in the integral equations for the moments [equations (16.2.32)] vanish and all but the first two integral equations become homogeneous. The last two of equations (16.2.18) deal with the propagation of the ellipticity of the polarization, and they make it clear that, in the absence of incident elliptically polarized light and sources of such light in the atmosphere, Rayleigh scattering will make no contribution to the Stokes parameter V. The trivial solution $V = 0$ will prevail throughout the atmosphere, and no amount of circularly polarized light is to be expected. In the absence of incident radiation, the resulting homogeneity of the integral equations for $Z(\tau)$ and $M(\tau)$ simply means that it is always possible to choose a coordinate system aligned with the plane of the resulting linear polarization, so that $U = 0$ is also true throughout the atmosphere. The only parameter that prevents the integral equations for $X(\tau)$ and $Y(\tau)$ from becoming homogeneous is the presence of the Planck function tying the radiation field of the gas to the thermal field of the particles.

Interpretation of Polarization Moments of the Radiation Field   From the behavior of the integral equations (16.2.32), it is possible to understand the properties of the Stokes parameters that they represent. For numerical reasons, the scattering fraction $1 - \epsilon(\tau)$ has been absorbed into the definition of the moments. Clearly $W(\tau)$ and $P(\tau)$ describe the propagation of the Stokes parameter V throughout the atmosphere. Since both U and V are basically geometric, two equations are required to determine the magnitude and quadrant of the angular parameters which they contain. Since the scattering matrix [equation (16.2.19)] is reducible with respect to V (i.e., only the diagonal element describing scatterings from V' into V is nonzero), it is not surprising that these two equations stand alone and can generally be ignored. The two equations for $Z(\tau)$ and $M(\tau)$ describe the orientation of the plane of linear polarization throughout the atmosphere. Since the orientation of the plane of polarization is also a geometric quantity, no energy is involved in the propagation of $Z(\tau)$ and $M(\tau)$ so their absence in the equations for radiative equilibrium is explained. The moments $X(\tau)$ and $Y(\tau)$ describe the actual flow of the energy field associated with the photons. An inspection of the $X(\tau)$ equation of equations (16.2.32) shows that $X(\tau)$ is very like the mean intensity J. The moment $Y(\tau)$ measures the difference between I_l and I_r and therefore describes the propagation of the degree of linear polarization in the atmosphere. Since both these moments appear in the flux equations, the amount of linear polarization does affect the condition of radiative equilibrium and hence the atmospheric structure. However, in the stellar atmosphere, the effect is not large.

The reason that there should be any effect at all can be found in the explanation for the existence of scattering lines in stars. Anisotropic scattering redirects the flow of photons from that which would be expected for isotropic scattering. This redirection increases the distance that a photon must travel to escape the atmosphere, thereby increasing the probability that the photon will be absorbed. This effectively increases the extinction coefficient and the opacity of the gas. While the effect is small for Rayleigh scattering, it has been found to play a significant role in the stellar interior when electron scattering is a major source of the opacity. In the stellar atmosphere, the effect on the stellar structure is small because the photon mean free path is comparable to the dimensions of the atmosphere (by definition), so that there is simply not enough space for anisotropic scattering to significantly affect the photon flow.

However, Rayleigh scattering will always produce linear polarization of the emergent radiation field. Since the Rayleigh phase function correctly describes electron scattering as resonance scattering, we can expect that polarization will be ubiquitous in stellar spectra, even in the absence of incident radiation.

Effects on the Continuum   The early work by Chandrasekhar[8,9] on the gray atmosphere showed that one might expect up to 11 percent linear polarization at the limb of a star with a pure scattering atmosphere. As was pointed out earlier, even this rather large polarization would average to zero for observers viewing spherical stars. Hence, Chandrasekhar suggested that the effect be searched for in eclipsing binary systems. Such a search led Hiltner[13] and Hall[14] to independently discover the interstellar polarization. It was not until 1984 that Kemp and colleagues[15] verified the existence of the effect in binary systems. However, the measured result was significantly smaller than that anticipated by the gray atmosphere study.

The reason can be found by considering the effects that tend to destroy intrinsic stellar polarization. The most significant is the near-axial symmetry about the line of sight of most stars. This tends to average any locally generated polarization to much lower values. The second reason results from the sources of the polarization itself. It is obvious that to generate polarization locally on the surface of a star, it is necessary to have opacity sources that produce polarization. In the hot early-type stars (earlier than B3), electron scattering is the dominant source of opacity, while in the late-type stars Rayleigh scattering from molecules is a major source of continuous opacity. Unfortunately, the spectra of these stars are so blanketed by atomic and molecular lines that it is difficult to find a region of the spectrum dominated by the continuous opacity alone. For stars of spectral type between late B and early K, there is a general lack of anisotropic scatterers making a major contribution to the total opacity, so that polarization is generally absent even locally.

In the early-type stars where anisotropic scattering electrons are abundant, the third reason comes into play and is basically a radiative transfer effect. To understand this rather more subtle effect, consider the following extreme and admittedly contrived examples. Remember that an electron acting as an oscillating dipole will scatter photons which are polarized in a plane perpendicular to the scattering plane. Consider a region of the atmosphere so that the observer sees the last scattering and is therefore in the scattering plane. The distribution of polarized photons that the observer sees will then reflect the distribution of the photons incident on the electrons. Further consider a point near the limb so that photons emerging from deep in the atmosphere will be scattered through 90 degrees and be 100 percent linearly polarized. If the *source function gradient* is steep, most of the photons scattered toward the observer will indeed come from deep in the atmosphere and exhibit such polarization and the observer will see a beam of light, strongly polarized in a plane tangent to the limb. However, if the source function gradient is shallow or nearly nonexistent, the majority of last scattered photons will originate not from deep within the star, but from positions near the limb. There will then be a surplus of scattered photons, with their scattering plane parallel to the limb and therefore polarized at right angles to the limb. Somewhere between these two extremes there exists a source function gradient which will produce zero net polarization locally at the limb. This argument can be generalized for any point on the surface of the star. Unfortunately, in the early-type stars that abound with anisotropically scattering electrons, the energy maximum lies far in the ultraviolet part of the spectrum. Thus the source function in the optical part of the spectrum is weakly dependent on the temperature and hence has a rather flat gradient. So these stars exhibit less than 1 percent or so of linear polarization in the optical part of the spectrum, even at the limb. The integrated effect, even from a highly distorted star, is minuscule. Such would not be the case for the late-type stars where the energy maximum lies to longer wavelengths than the visible light. These stars have very strong source function gradients in the visible part of the spectrum and should produce large amounts of polarization of the local continuum flux.

There remains a class of stars for which significant amounts of polarization may be found in the integrated light. These are the *close binaries*. Here the source of the polarization is not the star itself, but the scattered light of the companion. Although little has been done in a rigorous fashion to investigate these cases, Kuzma[16] has found that measurable polarization should be detected in at least some of these stars. Study of these stars offers the significant advantage that the polarization will be phase-dependent and so the effects of interstellar polarization may be removed in an unambiguous manner. Much work remains to be done in this area, but it must be done very carefully. All the effects that determine the degree of polarization of the emergent light depend critically on the temperature distribution of the upper atmosphere. Thus, such studies must include nongray and non-LTE effects. Kuzma[16] also included the effect that the incident radiation comes not from a point source, but from a finite solid angle and found that this significantly affected the result. While the task of modeling these stars is formidable, the return for determining the structure of the upper layers of the atmosphere will be great.

Effects of Polarization on Spectral Lines   While a line formed in pure absorption will be completely unpolarized except for contributions from electron scattering, this is not the case for scattering lines. For resonance lines where an upward transition must be followed quickly by a downward transition, the reemission of the photon is not isotropic, but follows a phase function not unlike the Rayleigh phase function. The scattering process can change the state of polarization in much the same fashion as electron scattering does. Thus, these lines can behave in much the same way as electrons in introducing or modifying the polarization of a beam of light. Since any absorption has a finite probability of being followed immediately by the reemission of a similar photon, any spectral line has the possibility of introducing some polarization into the beam. For example, under normal atmospheric conditions, the formation of Hβ results in a scattered photon about 25 percent of the time. For Hα the fraction can approach 50 percent. In other strong lines such as the Fraunhofer H and K lines of calcium, the fraction is much larger. Thus one should not be surprised if large amounts of local polarization are present in such lines. Unfortunately, the effect will be wavelength-dependent.

The majority of the wavelength dependence arises from the redistribution function by linking one frequency within the line to another. The introduction of a frequency-dependent redistribution function greatly complicates the problem. To handle this problem, it is necessary to express the redistribution function in some analytic form such as those of Hummer (see Section 15.3). Furthermore, one cannot use the angle-averaged forms because the angle dependence is crucial to the transfer of polarized radiation. McKenna[17] has developed the formalism from the standpoint of integral equations for the moments of the radiation field for a fairly general class of spectral lines. In the case of coherent scattering, these equations reduce to those presented here. For a general redistribution function, the equations are much more complicated.

Using the corrected Hummer $R_{\rm IV}$ for the case of the sun, McKenna[18] found that weak resonance lines showed a slowly decreasing degree of polarization from the core to the wings. The structure of the line as well as the state of polarization agreed well with observation and reproduced the center-limb variation observed for such lines. For strong resonance lines, a sharp spike in the polarization at the line core was found that rapidly diminished as one moved away from the line core before rising again and then diminishing into the wings, as in the case of the weak lines. Again, this effect is seen in the sun and provides a useful diagnostic tool for the upper atmosphere structure.

In most stars, except those distorted by rotation, binary systems, and nonradially pulsating stars, symmetry will tend to destroy these effects. But for these other classes of stars, line polarization may provide a useful constraint on their shape and structure. Unfortunately, the correct redistribution function for hydrogen has yet to be worked out. Since the hydrogen lines are readily observed and are often found in situations where the geometry would indicate little symmetry, an analysis of this type extended to these lines would provide an interesting constraint on the physical nature of the system.

16.3 Extended Atmospheres and the Formation of Stellar Winds

So far we have limited the subject of this book to the interiors and atmospheres of normal stars. However, we cannot leave the subject of stellar astrophysics without offering some comments about the transition between the outer layers of the stars and the interstellar medium that surrounds the stars. This is the domain where all the simplifying assumptions, which made the description of the inner regions of the star possible, fail. As one moves outward from the photosphere, through the chromosphere and beyond, she or he will encounter regions that involve some of the most difficult physics in astrophysics. Here the regions become sufficiently large that the plane-parallel assumption which so simplified radiative transfer is no longer valid, for the transition region may extend for many stellar radii. The density of matter is so low that LTE can no longer be applied to any elements of the gas. Because of the low density, the radiative interactions, while remaining crucial to the understanding of the structure of the medium, receive competition from previously neglected modes of energy transport. From observation, we know that turbulence is present in most stars. In some instances, the mechanical energy of the turbulent motion can be systematically transmitted to the outer layers of the star in amounts that compete with the energy from the radiation field. For some stars, the coupling of the magnetic field of the underlying star may provide a mechanism for the transmission of the rotational energy and momentum of the star to the surrounding low-density plasma.

The influx of energy from all sources seems to be sufficient to cause much of the surrounding plasma to be driven into the interstellar medium. Thus, we must deal with a medium in which hydrostatic equilibrium no longer applies. Continuous absorption is no longer of great significance so that the primary coupling of the radiation field to the gas is through the bound-bound transitions that produce the line spectra of stars. However, as we have seen, the frequency dependence of the line absorption coefficient is quite large, so Doppler shifts supplied by the mass motions of the gas will yield an absorption coefficient that is strongly dependent on the location of the gas. Thus the coupling of the radiation field is further complicated.

The region surrounding the star where these processes take place is generally known as the *extended atmosphere* of the star, while the material that is driven away is referred to as the *stellar wind*. It is likely that all stars possess such winds, but the magnitude of the mass loss produced by them may vary by 8 powers of 10 or more. The primary processes responsible for their origin differ greatly with the type of star. For example, the wind from the sun originates from the inability of the hot corona to come into equilibrium with the interstellar medium, so that the outer layers "boil" away. The details of the solar wind are complicated by coupling of the rotational energy of the sun to the corona through the solar magnetic field. Thus the outer regions of the solar atmosphere are accelerated outward at an average mass loss rate of the order of $10^{-14} M_\odot$ per year. The heating of the corona is likely to result from coupling with the convective envelope of the sun. Thus corona-like winds are unlikely to originate in the early-type stars that have radiative envelopes. For the hotter stars on the left-hand side of the *H-R* diagram, the acceleration mechanism is most likely radiative in origin. The coupling of the turbulent motions of the photosphere to the outer regions of the atmosphere is so poorly understood that its role is not at all clear. However, in the case of some giants and supergiants, this coupling may well be the dominant source of energy driving the stellar wind.

In the face of these added difficulties, it would be both beyond the scope of this book and presumptuous to attempt a thorough discussion of this region. For that the reader is directed to the supplemental reading at the end of the chapter. We merely outline some approaches to some aspects of these problems that have yielded a small measure of understanding of this transition region between the star and the surrounding interstellar space.

a Interaction of the Radiation Field with the Stellar Wind

For those stars where the stellar wind results from the star's own radiation, not only is the wind driven by the radiation, but also the motion of the material influences the radiative transport. Thus one has a highly nonlinear problem involving the formidable physics of hydrodynamics and radiative transfer. This interplay between the two is the source of the major problems in the theory of radiatively driven stellar winds.

Since it is clear that we must abandon most of the assumptions in describing this region, we will have to replace them with others. Thus we assume that these outer regions are spherically symmetric. Although this is clearly not the case for many stars, an understanding of the problems posed by a spherically symmetric star will serve as a foundation for dealing with the more difficult problems of distorted stars. Since we have already dealt with the problems of spherical radiative transport (see Section 10.4) and departures from LTE (see Chapter 15), we concentrate on the difficulties introduced by the velocity field of the outer atmosphere and wind.

One of the earliest indications that some stars were undergoing mass loss came from the observation of the line profiles of the star P Cygni. The Balmer lines (and others) in the spectrum of this star showed strong emission shifted slightly to the red of the stellar rest wavelength and modest absorption centered slightly to the blue side of the rest wavelength. This was recognized as the composite line profile arising from an expanding envelope, with the emission being formed in a relatively transparent outflowing spherical envelope. The "blue" absorption arose from that material that was moving, more or less, toward the observer and was seen silhouetted against the hotter underlying star (see Figure 16.6). Such line profiles are generally known as P-Cygni profiles and are seen during novae outbursts and in many other types of stars.

To understand the formation of such profiles, one must deal with the radiative transport of spectral line radiation in a moving medium. If the envelope surrounding the star cannot be assumed to be optically thin to line radiation, then one is faced with a formidable radiative transfer problem. In 1958, V. V. Sobolev[19] realized that the velocity gradient in such an expanding envelope could actually simplify the

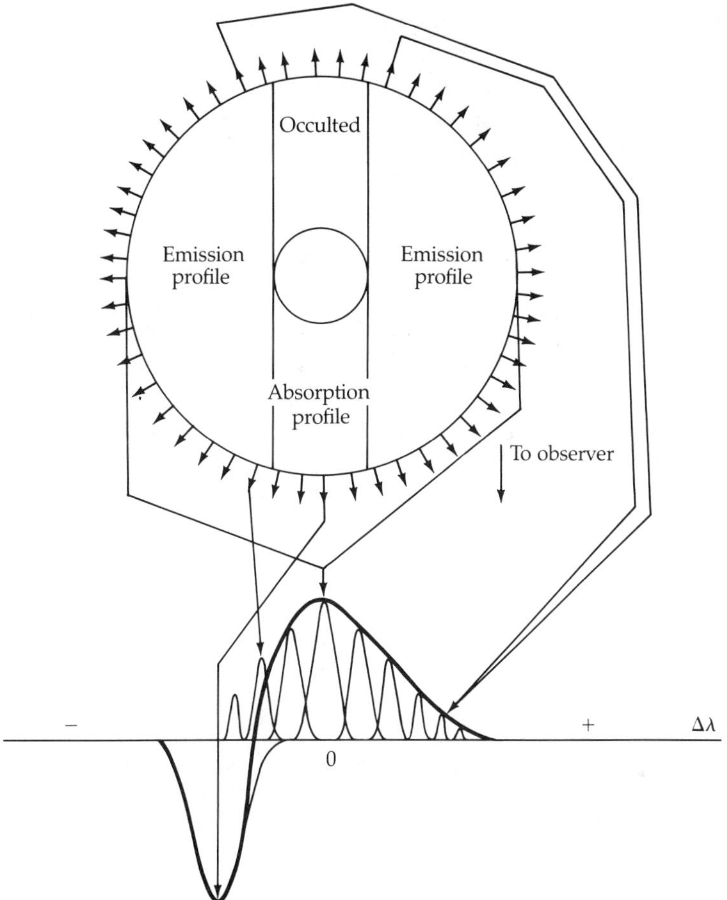

Figure 16.6 Expanding envelope of a star and the regions that give rise to the corresponding P-Cygni profile.

problem. If the gradient is sufficiently large, then the relative Doppler shifts implied by that gradient could effectively decouple line interaction in one part of the envelope from that in another part. Since the outflow velocities often reach several thousand kilometers per second, the region of the envelope where photons are not Doppler-shifted by more than the line width can actually be quite thin. The contribution to the observed line profile at any given frequency therefore arises from emission (or absorption) occurring on surfaces of constant radial (i.e., line-of-sight) velocity. The extent of that contribution will then depend on the probability that a photon emitted on one of these surfaces can escape to the observer.

Consider an envelope which is optically thick to line radiation surrounding a stellar core of radius r_c whose continuum radiation passes freely through the envelope. Let us further assume that the line is such that complete redistribution is an appropriate choice for the redistribution function. Finally, we define β to be the probability that an emitted photon will escape from the envelope. Clearly such a probability will depend on the line width and the velocity gradient. We can then write the local value for the average mean intensity in the line as

$$\bar{J} = (1 - \beta)S(r) + \beta_c I_c \qquad (16.3.1)$$

where $(1 - \beta)S(r)$ is the contribution to the local radiation field from those photons created in the line that did not escape and $\beta_c I_c$ represents those continuum photons that have penetrated to that part of the envelope. An estimate of β and β_c will then provide us with a relationship between $\bar{J}$ and S which we can substitute into equations for the non-LTE source function developed in Chapter 15 [see equations (15.2.19) and (15.2.25)]. The mean escape probabilities β and β_c will depend on local conditions only if the velocity gradient is sufficiently large that the conditions determining the probabilities can be represented by the local values of temperature and density.

Castor[20] has developed this formalism for a two-level atom, and it is nicely described in Mihalas.[21] Basically, the problem becomes one of geometry. Contributions to the local mean intensity will come largely from surfaces within the envelope which exhibit a relative Doppler shift that is less than the Doppler width of the line. Castor finds this to be

$$\beta = \frac{1}{2} \int_{-1}^{+1} \frac{1 - e^{-\tau_0(r)/[1 + \mu^2\Phi(v)]}}{\tau_0(r)/[1 + \mu^2\Phi(v)]} \, d\mu \qquad (16.3.2)$$

where

$$v = \frac{v(r)}{v_{\text{th}}} \qquad (16.3.3)$$

is simply the velocity measured in units of the thermal velocity characterizing the line width and

$$\Phi(v) = \frac{d \ln v}{d \ln r} - 1 \qquad (16.3.4)$$

which occurs from expanding the effective optical path in a power series of the velocity field and retaining only the first-order term. The parameter τ_0 is effectively the optical depth in the line along a radius from the star and is defined by

$$\tau_0 \equiv \frac{\kappa_{\text{line}}(r)\rho(r)}{v(r)/r} \qquad (16.3.5)$$

This quantity is essentially the line extinction coefficient multiplied by some effective length that is obtained by dividing by the dimensionless velocity gradient.

We may approximate β_c by the following argument. Since β_c is essentially an escape probability of a continuum photon leaving the envelope by traveling in any direction, it is essentially the same as β except for the probability of the photon's hitting the stellar core. That probability is given simply by the fraction of 4π steradians occupied by the core itself and is commonly called the *dilution factor*, since it is the amount by which the local radiation density is diluted as one recedes from the star. The dilution factor W is therefore

$$W \equiv \frac{1}{2}\left[1 - \left(1 - \frac{r_c^2}{r^2}\right)^{1/2}\right] \approx \frac{1}{4}\left(\frac{r_c}{r}\right)^2 \qquad r \gg r_c \qquad (16.3.6)$$

so that

$$\beta_c \approx W\beta \qquad (16.3.7)$$

We now have β and β_c defined in terms of the velocity field, position, and parameters of the gaseous envelope that determine the local effective optical depth in the line τ_0. Remembering that

$$\bar{J} \equiv \int_0^\infty \phi(v)J_v \, dv \qquad (16.3.8)$$

we may substitute into either of the non-LTE source functions given by equation (15.2.19) or (15.2.25) and obtain the source function for the line. Once the line source function is known, the solution for the equation of transfer (see Section 10.4 for spherical geometry) can be solved and the line profile obtained.

Solution of this problem also yields the radiant energy deposited in the envelope as a result of the line interactions. By doing this for a number of lines, the local effects of radiation on the envelope can be determined. This is one of the primary factors in determining the dynamics of the envelope.

b *Flow of Radiation and the Stellar Wind*

Most progress in understanding stellar winds has been made with the radiation-driven winds of the early-type stars. While coronal winds of the later-type stars are important, particularly for slowing down the rotational velocity of these stars by magnetic breaking, the uncertainties concerning the formation of their coronal source force us to leave their discussion to another to elucidate. What is presently known about them is nicely summarized by Mihalas[21] (pp. 521–540) and

Cassinelli.[22] Instead, we concentrate on some aspects of the large winds encountered in the early-type stars.

In Chapter 6 [equation (6.5.2)] we showed that if the radiation pressure became sufficiently high, a star would become unstable. While many stars on the upper main sequence approach this limit, they still fall short. Yet these same stars exhibit significant mass loss in the form of a substantial stellar wind. In 1970, Leon Lucy and Philip Solomon[23] showed that, under certain circumstances, the coupling of the radiation field to the matter in the upper atmosphere through the strong resonance lines of certain elements could provide sufficient momentum transport to the envelope to drive it away from the star. This idea was developed more fully by Castor, Abbot, and Klein[24] to provide the foundations for the theory of radiatively driven stellar winds. Castor, Abbot, and Klein found that by including the effects of the large number of weaker lines present in material of the high atmosphere, accelerations 100 times greater than those found by Lucy and Solomon would result. This implied that mass loss from radiatively driven winds could be expected in cooler stars with higher surface gravities than the supergiants considered by Lucy and Solomon. Thus, stellar winds should be a ubiquitous phenomenon throughout the early-type stars.

The basic approach to describing any phenomenon of this type is to write down the conservation laws that must apply. To make the description as simple as possible, we assume that the flow is steady. Again, we assume spherical symmetry and remind the reader that the relevant conservation laws all have their origin in the Boltzmann transport equation developed in Chapter 1. Mass will be continuously lost by the wind, but for a steady flow, the mass flux through a spherical shell of radius r will be constant and given by

$$\frac{dM}{dt} = 4\pi r^2 \rho v = \text{const} \qquad (16.3.9)$$

The conservation of momentum requires

$$\rho v \frac{dv}{dr} + \frac{dP}{dr} + \rho \frac{GM}{r^2} - \rho g_r = 0 \qquad (16.3.10)$$

where g_r is the radiative acceleration. The conservation of energy and thermodynamics can be expressed by

$$v \frac{dU}{dr} - \frac{Pv}{\rho^2} \frac{d\rho}{dr} = \frac{\nabla \cdot \vec{F}_t}{\rho} = \frac{Q_A + Q_R + Q_c}{\rho} \qquad (16.3.11)$$

where $\mathbf{U}$ is the internal energy of the gas and F_t is the total energy flux whose divergence can be described in terms of energies Q_A, Q_R, and Q_C deposited locally in the gas by acoustical, radiative, and conductive processes, respectively. Since both the radiative and gravitational accelerations vary as r^{-2}, it is common to introduce the parameter

$$\Gamma = \frac{g_r}{g} \tag{16.3.12}$$

If the wind is assumed to be radiatively driven, then $Q_A = 0$. Similarly, the material density is sufficiently low that $Q_C = 0$. If the total energy in the flow is significantly less than the stellar luminosity, we may assume that radiative equilibrium may be locally applied, so that

$$Q_R = \int_0^\infty 4\pi\kappa_\nu(J_\nu - B_\nu)\,d\nu \approx 0 \tag{16.3.13}$$

which greatly simplifies the energy equation. The energy and momentum equations can now be combined to yield

$$\frac{d}{dr}\left(\frac{v^2}{2} + \mathbf{U} + \frac{P}{\rho} - \frac{GM}{r}\right) = g_r \tag{16.3.14}$$

which, after the total energy of the gas is defined to be

$$E \equiv \frac{v^2}{2} + \mathbf{U} + \frac{P}{\rho} - \frac{GM}{r} \tag{16.3.15}$$

can be integrated to give

$$E = E_0 + (\dot{M})^{-1} \int_0^\infty 4\pi r^2 v\rho g_r\,dr \tag{16.3.16}$$

Thus even if the total energy of the gas near the star is negative, continual radiative acceleration can introduce sufficient energy to drive the energy positive and thus allow it to escape.

As an example, consider the case where the gas in the wind is isothermal so that we can introduce an equation of state of the form

$$P = c_s^2 \rho \tag{16.3.17}$$

where the speed of sound c_s is constant. We can then differentiate equation (16.3.9) and eliminate dP/dr from equation (16.3.10) to get

$$\frac{r}{v}\frac{dv}{dr} = \frac{2c_s^2 - GM(1 - \Gamma)/r}{v^2 - c_s^2} \qquad (16.3.18)$$

For supersonic flow to occur, the numerator of the right-hand side of equation (16.3.18) must vanish at the point where the speed of sound is reached (that is, $v = c_s$). This places significant constraints on the value that Γ must have at the transonic point. Cassinelli[22] shows that this requires additional radiative heating of the gas as it approaches the transonic point. He suggests that scattering by electrons and resonance lines can do the work required to meet the conditions at the transonic point. Beyond the transonic point, further acceleration is most likely driven by the line opacity.

Since it has been shown that it is possible to drive stellar winds radiatively for stars whose luminosity is well below the Eddington luminosity, it remains for a complete and fully consistent model to be made. The problem here is largely numerical for the basic physical constraints are understood. To be sure, these problems are formidable. Correct representation of the radiative acceleration requires calculation of the effects due to the myriads of weak spectral lines that populate the ultraviolet region of the spectrum where these stars radiate most of their energy. The energy balance equations must be solved to properly incorporate the effects of heating near the transonic point and to determine the temperature structure of the flow. Then the hydrodynamic flow equations will have to be solved numerically, incorporating an appropriate equation of state. Of course, all this must be done in non-LTE so that the proper ionization and excitation structure can be obtained. Undoubtedly the Castor, Abbot, and Klein models have pointed the correct way to understanding radiatively driven winds, but much remains to be done before the models can be used as a probe of the physics of the outer layers of stars.

This brief look at the problems of extended atmospheres and radiatively driven winds points out that the transition region between the normal stellar atmosphere and the interstellar medium is a difficult region to understand. A closer look at the chromospheres and coronas of other stars would have shown the same, or possibly more difficult, problems. Certainly the description of the transition zone for non-spherical stars will include more difficulties and require greater cleverness on the part of astrophysicists to understand them. This will be necessary if we are to understand the evolution of close binary stars in sufficient detail to understand their ultimate fate and the manner in which they influence the interstellar medium. Without that understanding, the evolution of the galaxy will be difficult, if not impossible, to understand. And without a reliable knowledge of galactic evolution, can we ever hope to delineate the evolution of the universe as a whole?

Problems

1. Show that equation (16.1.4) is indeed an equation for the diffuse field source function for an illuminated atmosphere.

2. Show that equations (16.1.8) and (16.1.9) do give the correct form of the Avrett-Krook perturbation scheme for an illuminated stellar atmosphere.

3. Find an integral expression for the radiative flux transmitted horizontally through a plane-parallel stellar atmosphere that is illuminated by a point source located at $[\mu_0 = \cos\theta_0, \phi_0 = 0]$.

4. Show how the degree of polarization P defined by equation (16.2.6) changes with a rotation of the observer's plane of observation.

5. Show that the group properties assigned to the rotation matrix $\mathbf{L}(\phi)$ [defined in equation (16.2.9)] are indeed those given in equation (16.2.10).

6. Show that equation (16.2.33) follows from the definition of the radiative flux and the moments of the polarized radiation field. Derive an analogous expression for the horizontal flux transported through a plane-parallel atmosphere under the conditions specified in Problem 3, assuming that the incident radiation is unpolarized.

7. Show that equation (16.3.18) is a solution to the more general equations of motion given by equation (16.3.10) in the case of an isothermal atmosphere.

References and Supplemental Reading

1. Chandrasekhar, S., *Radiative Transfer*, Dover, New York, 1949/1960, p. 22.

2. Buerger, P. F., "Illuminated Stellar Atmospheres," *Ap. J.* 177, 1972, pp. 657–664.

3. Karp, A. H., "A Modification of the Avrett-Krook Temperature-Correction Procedure," *Ap. J.* 173, 1972, pp. 649–652.

4. Anderson, L., and F. H. Shu, "On the Light Curves of W Ursae Majoris Stars," *Ap. J.* 214, 1977, pp. 798–808.

5. Milne, E. A., "Thermodynamics of Stars," *Handbuch der Astrophysik*, vol. 3 (first half), Springer-Verlag, Berlin, 1930, pp. 134–141.

6. Kuzma, T. J., "On Some Aspects of the Radiative Interaction in a Close Binary System," doctoral thesis, The Ohio State University, Columbus, 1981.

7. Stokes, G. G., "On the Composition and Resolution of Streams of Polarized Light from Different Sources," *Trans. Cam. Phil. Soc.* 9, 1852, pp. 399–416.

8. Chandrasekhar, S., "On the Radiative Equilibrium of a Stellar Atmosphere X," *Ap. J.* 103, 1946, pp. 351–370.

9. Chandrasekhar, S., "On the Radiative Equilibrium of a Stellar Atmosphere XI," *Ap. J.* 104, 1946, pp. 110–132.

10. Code, A., "Radiative Equilibrium in an Atmosphere in which Pure Scattering and Pure Absorption Play a Role," *Ap. J.* 112, 1950, pp. 22–47.

11. Collins, G. W., II, "Transfer of Polarized Radiation," *Ap. J.* 175, 1972, pp. 147–156.

12. Collins, G. W., II, and P. F. Buerger, "Polarization from Illuminated Non-Gray Stellar Atmospheres," *Planets, Stars, and Nebulae Studied with Photopolarimetry* (Ed.: T. Geherels), University of Arizona Press, Tucson, 1974, pp. 663–675.

13. Hiltner, W. A., "Polarization of the Light from Distant Stars by Interstellar Medium," *Science* 109, 1949, p. 165.

14. Hall, J. S., "Observations of the Polarized Light from Stars," *Science* 109, 1949, pp. 166–167.

15. Kemp, J. C., G. D. Henson, M. S. Barbour, D. J. Krause and G. W. Collins II, "Discovery of Eclipse Polarization in Algol," *Ap. J.* 273, 1982, pp. L85–L88.

16. Kuzma, T. J., "On Some Aspects of the Radiative Interaction in a Close Binary System," doctoral thesis, The Ohio State University, Columbus, 1981, p. 135.

17. McKenna, S., "The Transfer of Polarized Radiation in Spectral Lines: Formalism and Solutions in Simple Cases," *Astrophy & Sp. Sci.* 108, 1985, pp. 31–66.

18. McKenna, S., "The Transfer of Polarized Radiation in Spectral Lines: Solar-Type Stellar Atmospheres," *Astrophy & Sp. Sci.* 106, 1984, pp. 283–297.

19. Sobolev, V. V., "The Formation of Emission Lines," *Theoretical Astrophysics* (Ed.: V. A. Ambartsumyan, trans.: J. B. Sykes), Pergammon, New York, 1958, pp. 478–497.

20. Castor, J. I., "Spectral Line Formation in Wolf-Rayet Envelopes," *Mon. Not. R. astr. Soc.* 149, 1970, pp. 111–127.

21. Mihalas, D., *Stellar Atmospheres*, 2d ed., W. H. Freeman, San Francisco, 1978, pp. 478–485.

22. Cassinelli, J. P., "Stellar Winds," *Annual Reviews of Astronomy and Astrophysics* (Ed.: G. Burbidge), vol. 17, Annual Reviews, Palo Alto, Calif., 1979, pp. 300–306.

23. Lucy, L. B., and P. M. Solomon, "Mass Loss by Hot Stars," *Ap. J.* 159, 1970, pp. 879–893.

24. Castor, J. I., D. C. Abbot, and R. I. Klein, "Radiation Driven Winds in Of Stars," *Ap. J.* 195, 1975, pp. 157–174.

For general review on the subject of stellar winds, one should read Cassinelli,[22] pp. 275–308.

Epilogue

. . .

Those who have labored through this entire book will
have noticed that the first half appears to form a more cohesive and understandable unit than the second half. The pedagogy seems more complete for the theory
of stellar interiors than for stellar atmospheres. I believe this to be an intrinsic
property of the material rather than a reflection of my ability. More effort and
minds have been involved in the formulation of the theory of stellar interiors and
evolution than in the development of the theory of stellar atmospheres. More
importantly, more time has elapsed during which the central ideas, important con-

cepts, and their interrelationships can become apparent. This distillation process requires time as well as good minds to produce the final product.

The reason is not simply a temporal accident but has its origin in the relative difficulty of the formulation of the subjects themselves. I have emphasized throughout this book the steady removal of simplifying assumptions as the description of stellar structure moves progressively from the center of the star to the surface and beyond. This loss of assumptions that make the physical description of the star easier to effect also sets the time scale for the development of the subject itself, for people generally try to solve the simple problems first. So the effort to understand the stellar interior began in earnest as soon as nuclear physics delineated the source of the energy in stars. This fact is not to belittle the elegant work with polytropes that began at the end of the nineteenth century. But it is clear that fully credible models of stellar interiors, suitable for the description of stellar evolution, could not be made without certain knowledge of the origin of a star's source of energy. After the elucidation by Hans Bethe of the probable energy sources for the sun, the theory of stellar structure and evolution developed with remarkable speed. In a mere two decades, one of which was dominated by World War II, the basic framework of the theory of stellar evolution was in place.

Although the origin of stellar atmospheres can be traced back nearly as far as that of stellar interiors, its development has been slower. It is tempting to suggest that the theory required the existence of good stellar interior models before it could be developed, but that would be false. The defining parameters for the stellar atmosphere are known sufficiently well from direct observation to permit the development of a full-blown theory of atmospheric structure. The retarded development of stellar atmospheres as compared to stellar interiors can be attributed entirely to the loss of the assumption of STE. All the complications introduced by the myriad complexities of atomic structure now become entwined with the description of the atmospheric structure through the radiative opacity. While this fact was realized early in the century, little could be done about it until the numerical and computational capabilities required to include these complexities developed after World War II. The computational requirements of the theory of stellar atmospheres exceed those of stellar interiors, so that it was logical that the theory of stellar structure and evolution should develop earlier. This, plus the "sifting and winnowing" of ideas of central importance from those of necessary detail by people like Martin Schwarzschild and others, has provided that more lucid pedagogy of stellar interiors and evolution. It is the task of our generation and the next to bring the same level of conceptual understanding to the theory of stellar atmospheres.

While the business of model making, in both stellar interiors and stellar atmospheres, has become almost routine, one must remain wary lest the work of large computers beguile the investigator into believing that the process is simple. It is tempting to think that in the future computer codes will yield all aspects of the structure of a star as easily as one obtains values for trigonometric functions at present. To believe this is to miss a central difference between physical science and mathematics. Mathematics is a logical construct of the human mind, resting on a set of axioms within the framework of which the majority of statements can be adjudged true or false. (The exceptions implied by Gödel's incompleteness theorem are duely noted.) Thus, one can have complete confidence that a competently written computer code designed to calculate trigonometric functions will always deliver the correct answer within some known tolerance. While physical science aspires to the status of an axiomatic discipline, it is far from reaching that goal. Even if such a goal is reached, the complexities of the universe require us to continue to make simplifying assumptions to describe the physical world. These assumptions cannot be deemed either true or false, for they will only yield a correct description of the phenomena of interest within some generally unknown tolerance. The interplay of the tolerances associated with the assumptions and those of numerical and computational approximation is so complex as to always raise some doubt about the model's level of accuracy. Only the naive would ascribe the same level of confidence to the results of such a model description as to the calculation of a trigonometric function.

Nevertheless, we may expect to see a rapid growth in the art of modeling the physical world by computer. We must always be alert, however, to the simplicity engendered by the speed and accuracy of digital computers lest we credit greater accuracy to their results than is warranted. We must also eschew false understanding of results by the explanation that the universe is the way it is "because the computer tells me so." Such is no understanding at all. Instead, we must continually struggle to understand the fundamental laws at work in framing the universe and the manner by which these laws relate to one another. It is the interrelationships among those laws describing the physical world that provide the insight necessary to advance our understanding, and they must be seen as concepts, not merely numbers.

The remaining problems to be solved in stellar astrophysics are legion. But the tools to solve them remain the same: the conservation laws of physcis, the fundamental properties of matter, and the mathematical language to relate the two. To contribute to our understanding of the universe, the student must master them all.

Index

• • •